# Frontiers in Probability and the Statistical Sciences



More information about this series at http://www.springer.com/series/11957

Riten Mitra • Peter Müller
Editors

# Nonparametric Bayesian Inference in Biostatistics

Springer

*Editors*

Riten Mitra
Department of Bioinformatics
and Biostatistics
University of Louisville
Louisville, KY, USA

Peter Müller
Department of Mathematics
University of Texas
Austin, TX, USA

Frontiers in Probability and the Statistical Sciences
ISBN 978-3-319-19517-9 ISBN 978-3-319-19518-6 (eBook)
DOI 10.1007/978-3-319-19518-6

Library of Congress Control Number: 2015945621

Springer Cham Heidelberg New York Dordrecht London


Printed on acid-free paper

Springer International Publishing AG Switzerland is part of Springer Science+Business Media (www.springer.com)

# Preface

Nonparametric Bayesian (BNP) approaches are becoming increasingly more common in biostatistical inference. Many problems involve an abundance of data that allows the use of more flexible and complex probability models beyond traditional parametric families. One of the most traditional application areas for BNP is in survival analysis, including in particular survival regression. The nature of the recorded outcomes makes it natural to target inference on an entire unknown distribution, rather than focus on just a mean function. Many more recent applications of BNP in biostatistics and bioinformatics involve inference on unknown partitions. For example, this could be an arrangement of patients into clinically meaningful subpopulations. This volume covers these and some more applications of BNP in biomedical inference problems. The intention of this book is to provide a good review of and introduction to related application areas.

Part I starts with two introductory chapters. Chapter 1 provides a brief review of the most commonly used basic BNP models, including the Dirichlet process, Dirichlet process mixtures, the dependent Dirichlet process, Polya trees and Gaussian processes. These models and related variations act as the workhorses of BNP methods in biostatistics applications. Chapter 2 discusses some related examples, spanning a wide range of applications.

Part II includes several chapters that consider BNP methods for inference with genomic data, including gene expression, mutations, copy number aberrations and more. Many chapters in this part involve a notion of clustering biomolecular entities, e.g. genes, cells, genomic locations or proteins. The inferred clusters provide insights into underlying biological structure and functionality of these entities. Another major feature for most chapters in this part is that the data are obtained from state-of-the-art experimental platforms, e.g. next generation sequencing counts in Chapter 4. Although the exact form of raw data vary across chapters, all chapters are perfect illustrations of the ubiquitous applicability of the common BNP priors in bioinformatics. For example, Chapter 3 discusses curve clustering methods, focusing on Chinese restaurant process priors to cluster proteins based on their shapes.

The inference is centered around an inner product matrix that is built using a special metric on the shape space. In this way, it efficiently reduces the computational scale of the problem from infinite-dimensional curves to clustering patterns in a finite matrix. Chapter 4 uses a variation of the Indian buffet process prior to address the challenging issue of estimating tumor heterogeneity. Inferring subclones of tumor cells based on genomic profile has important implications in genomics and cancer research. Chapters 9 and 6 deal with clustering not as goals in themselves, but interestingly, as efficient tools for dimension reduction. Chapter 9 clusters a large number of genes based on their expression and eventually uses these clusters to build regression models for predicting severity of multiple myeloma. Chapter 6 addresses the classical problem of SNP disease association through a normalized generalized gamma (NGG) process. This is one of the first applications of this prior to large scale biostatistical inference. A thorough review of Bayesian network models can be found in Chapter 8. In addition, the chapter introduces novel and flexible semi-parametric extensions of current approaches to inference for high dimensional gene networks. Chapter 7 reviews the recent foray of BNP priors into the classical field of population genetics. The chapter discusses the problem of detecting population clusters based on allele frequencies using hierarchical DP priors. Population admixture models are not only relevant for understanding the pattern of human migration and evolution, but are also critical to account for confounding in gene association studies. Finally, we would want the readers to take note that all clustering priors discussed in this part are related to exchangeable distributions. That is, they assume that there is no natural ordering of the variables, which restrains their straightforward application to segmentation problems where time or say, genomic locations can be of great importance. Chapter 5 discusses an elegant way to circumvent this issue by generalizing an exchangeable prior through latent variables. The method is applied to detect regions of copy number variations in the genome and is compared with standard segmentation methods like hidden Markov models.

Chapters in Part III discuss applications of BNP to survival analysis. Inference for survival data were some of the first problems that motivated the early BNP literature. This is the case because for event time data it is natural to focus on the entire distribution, including all detailed features, rather than just location and scale. Chapter 10 motivates Markov processes as a flexible class of priors on hazard rates. It discusses how such priors can be easily adapted to cure rate models and complex multivariate settings, like competing risks and recurrent events. Chapter 11 links the relevant priors for baseline hazard rates with the standard parametric models to provide a comprehensive review of some common semi-parametric approaches in the literature. The chapter concludes with a discussion on spatially correlated survival data. Chapter 12 introduces a fully nonparametric prior to address the challenging and complex issues of interval censoring and misclassification.

Part IV groups together chapters that include a notion of modeling some random function (or response surface). Chapter 13 uses Gaussian process (GP) priors to model the temporal evolution of the firing patterns of a group of neurons. The two main themes of this chapters are adapting GP priors to signaling data and multivariate extensions to capture complex spatio-temporal effects. Next, in Chapter 14 we find an extensive review of general curve registration techniques that can account for phase variability in functional data. These approaches are critical to many public health applications, as illustrated in their application to growth data and pharmacokinetics. Chapter 15 delivers yet another flavor of BNP priors in the context of new age adaptive clinical trials. The chapter focuses on a flexible modeling approach for a clinical outcome over a large covariate space, where the covariates are biomarkers. The complexity of the related response surface is particularly relevant for clinical trial settings where disease status and progression are related to higher order interactions between biomarkers. Part IV finally concludes with a review of non-parametric modeling of ROC curves in Chapter 16. ROC curves are ubiquitous in classification problems, especially in medical diagnostic tests. While providing another illustration of the utility of the DPM prior, this chapter also provides a lucid description of Bayesian bootstrap methods.

Inference for spatio-temporal data gives rise to a specific set of challenges and modeling needs. Part V discusses such models in considerable detail. Chapter 17 starts by providing an in-depth theoretical introduction to Gaussian processes and other BNP priors for spatial data. This review starts from semi-parametric models and gradually builds up to fully non-parametric methods. Chapter 18 discusses two specific priors for spatial data. One of the highlights of this chapter is a clever method of adapting conventional product partition priors to the spatial context. The final chapter in this part, Chapter 19, deals with the specialized problem of detecting boundaries in spatial data. The chapter discusses some specific BNP priors for this problem. In addition, it formulates this problem in the framework of Bayesian multiple hypothesis testing advocating ways to extract multiplicity-adjusted posterior probabilities.

Increasingly stricter ethical standards, increasing cost and complexity of clinical studies make it ever more difficult to carry out randomized studies. This leads to an increased use of non-randomized data. Deriving meaningful conclusion from such data requires adjustment for the lack of randomization by techniques known as "causal inference." Many of these methods deal with missing data as well. Chapters in Part VI discuss this class of methods, including a BNP prior for random discontinuation designs in Chapter 20 and BNP priors for inference with missing data in Chapter 21.

Finally, besides popularizing the use of BNP methods in biostatistics and bioinformatics this volume supports BNP research by donating any royalties to the International Society for Bayesian Analysis (ISBA) in support of travel awards for young researchers for upcoming meetings of the biennial workshop in nonparametric Bayesian inference.

Louisville, KY, USA Riten Mitra
Austin, TX, USA Peter Müller

# Contents

# List of Contributors

**Edoardo Airoldi**
Department of Statistics, Harvard University, Cambridge, MA, USA

**Raffaele Argiento**
CNR-IMATI, Milano, Italy

**Veerabhadran Baladandayuthapani**
Department of Biostatistics, The University of Texas MD Anderson Cancer Center, Houston, TX, USA

**Sayantan Banerjee**
Department of Biostatistics, The University of Texas MD Anderson Cancer Center, Houston, TX, USA

**Sudipto Banerjee**
Department of Biostatistics, U.C.L.A. School of Public Health, Los Angeles, CA, USA

**Federico Bassetti**
Dipartimento di Matematica, Universitá di Pavia, Pavia, Italy

**Sam Behseta**
CSUF, Fullerton, CA, USA

**Miguel de Carvalho**
Departamento de Estadística, Pontifica Universidad Católica de Chile, Santiago, Chile

**Vanda Inácio de Carvalho**
Pontifica Universidad Católica de Chile, Santiago, Chile

**Michael J. Daniels**
Department of Statistics & Data Sciences, University of Texas at Austin, Austin, TX, USA

Department of Integrative Biology, University of Texas at Austin, Austin, TX, USA

**Stefano Favaro**
Department of Economics and Statistics, University of Torino and Collegio Carlo Alberto, Moncalieri, TO, Italy

**Montserrat Fuentes**
Department of Statistics, North Carolina State University, Raleigh, NC, USA

**María José García-Zattera**
Departamento de Estadística, Pontifica Universidad Católica de Chile, Santiago, Chile

**Chiyu Gu**
Department of Statistics, University of Missouri, Columbia, MO, USA

**Alessandra Guglielmi**
Dipartimento di Matematica, Politecnico di Milano, Milano, Italy

**Subharup Guha**
Department of Statistics, University of Missouri, Columbia, MO, USA

**Michele Guindani**
Department of Biostatistics, MD Anderson Cancer Center, University of Texas, Houston, TX, USA

**Kamalakar Gulutoka**
NorthShore University HealthSystem, Evanston, IL, USA

**Timothy Hanson**
Department of Statistics, University of South Carolina, Columbia, SC, USA

**Chuhsing K. Hsiao**
Institute of Epidemiology and Preventive Medicine, National Taiwan University, Taipei, Taiwan

**Maria De Iorio**
Department of Statistical Science, University College, London, UK

**Alejandro Jara**
Departamento de Estadística, Pontifica Universidad Católica de Chile, Santiago, Chile

**Yuan Ji**
NorthShore University HealthSystem/The University of Chicago, Evanston, IL, USA

**Seongil Jo**
Department of Statistics, Korea University, Seoul, South Korea

**Wesley O. Johnson**
University of California, Irvine, CA, USA

**George Karabatsos**
University of Illinois-Chicago, Chicago, IL, USA

**Arnošt Komárek**
Charles University in Prague, Praha 8, Karlín, Czech Republic

**Jaeyong Lee**
Department of Statistics, Seoul National University, Seoul, South Korea

**Juhee Lee**
Department of Statistics, University of California Santa Cruz, Santa Cruz, CA, USA

**Fabrizio Leisen**
School of Mathematics, Statistics and Actuarial Sciences, University of Kent, Canterbury, Kent, UK

**Pei Li**
Medtronic Incorporated, Minneapolis, MN, USA

**Antonio R. Linero**
Department of Statistics, Florida State University, Tallahassee, FL, USA

**Giovanni M. Marchetti**
Dipartimento di Statistica, Informatica, Applicazioni "G. Parenti", University of Florence, Firenze, Italy

**Alexander McBean**
Division of Biostatistics, University of Minnesota, Minneapolis, MN, USA

**Riten Mitra**
University of Louisville, Louisville, KY, USA

**Peter Müller**
University of Texas at Austin, 1 University Station, C1200, Austin, TX, 78712, USA

**Yang Ni**
Department of Statistics, Rice University, Houston, TX, USA

**Luis E. Nieto-Barajas**
ITAM, Progreso Tizapan, D.F., Mexico

**Garritt L. Page**
Departamento de Estadísica, Pontificia Universidad Católica de Chile, Santiago, Chile

**Debdeep Pati**
Florida State University, Tallahassee, FL, USA

**Fernando A. Quintana**
Departamento de Estadísica, Pontificia Universidad Católica de Chile, Santiago, Chile

**Brian J. Reich**
Department of Statistics, North Carolina State University, Raleigh, NC, USA

**Fabrizio Ruggeri**
CNR-IMATI, Milano, Italy

**Subhajit Sengupta**
NorthShore University HealthSystem/The University of Chicago, Evanston, IL, USA

**Babak Shahbaba**
UC Irvine, CA, USA

**Anuj Srivastava**
Florida State University, Tallahassee, FL, USA

**Francesco C. Stingo**
Department of Biostatistics, The University of Texas MD Anderson Cancer Center, Houston, TX, USA

**Yee Whye Teh**
Department of Statistics, University of Oxford, Oxford, UK

**Donatello Telesca**
Department of Biostatistics, University of California Los Angeles, Los Angeles, CA, USA

**Lorenzo Trippa**
Department of Biostatistics, Harvard University, Cambridge, MA, USA

**Alexander Vandenberg-Rodes**
UC Irvine, CA, USA

**Stephen G. Walker**
The University of Texas at Austin, Austin, TX, USA

**Charlotte Wang**
Institute of Epidemiology and Preventive Medicine, National Taiwan University, Taipei, Taiwan

**Yanxun Xu**
Department of Statistics & Data Sciences, The University of Texas at Austin, Austin, TX, USA

**Zhengwu Zhang**
Florida State University, Tallahassee, FL, USA

**Haiming Zhou**
Division of Statistics, Northern Illinois University, DeKalb, IL, USA

# Editors' Biography

**Riten Mitra** is Assistant Professor in the Department of Bioinformatics and Biostatistics at University of Louisville. His research interests include Bayesian graphical models and nonparametric Bayesian methods with a special emphasis on applications in genomics and bioinformatics.

**Peter Müller** is Professor in the Department of Mathematics and the Department of Statistics & Data Science at the University of Texas at Austin. He has published widely on nonparametric Bayesian statistics, with an emphasis on applications in biostatistics and bioinformatics.

# Part I
# Introduction

# Chapter 1
# Bayesian Nonparametric Models

Peter Müller and Riten Mitra

**Abstract** We briefly review some of the nonparametric Bayesians models that are most widely used in biostatistics and bioinformatics. We define the Dirichlet process, Dirichlet process mixtures, the Polya tree, the dependent Dirichlet process and the Gaussian process prior. These few models and variations cover a major part of the models that are used in the literature. The discussion includes references to variations of the basic models that are defined in the chapters of this volume.

## 1.1 Nonparametric Bayesian Inference in Biostatistics and Bioinformatics

The increased complexity of biomedical inference problems requires ever more sophisticated and flexible approaches to statistical inference. The challenges include in particular massive data, high-dimensional sets of potential covariates, highly structured stochastic systems, and complicated decision problems. Some of these challenges can be naturally addressed with a class of inference approaches known as nonparametric Bayesian (BNP) methods. A technical definition of BNP models is that they are probability models on infinite dimensional probability spaces. This includes priors on random probability measures, random mean functions, and more.

BNP methods relax the sometimes restrictive assumptions of traditional parametric methods. A parametric model is indexed by an unknown finite dimensional

P. Müller (✉)
The University of Texas at Austin, 1, University Station, C1200, Austin, TX 78712, USA
e-mail: pmueller@math.utexas.edu

R. Mitra
University of Louisville, Louisville, KY, USA
e-mail: riten82@gmail.com

 
R. Mitra, P. Müller (eds.), *Nonparametric Bayesian Inference in Biostatistics*, Frontiers in Probability and the Statistical Sciences,
DOI 10.1007/978-3-319-19518-6_1

parameter vector $\theta$. Bayesian inference proceeds by assuming a prior probability model $p(\theta)$ which is updated with the relevant sampling model $p(y \mid \theta)$ for the observed data $y$.

For example, consider a density estimation problem, with observed data $y_i \sim G$, $i = 1, \ldots, n$. Inference under the Bayesian paradigm requires a completion of the model with a prior for the unknown distribution $G$. If $G$ is restricted to be in a family $\{G_\theta,\ \theta \in \Re^d\}$, then the prior is specified as a prior probability model $p(\theta)$ for the $d$-dimensional parameter vector $\theta$. In contrast, if $G$ is not restricted to a finite dimensional parametric family, then the prior model $p(G)$ becomes a probability model for the infinite dimensional $G$.

A very common related use of BNP priors on random probability measures is for random effects distributions in mixed effects models. Such generalizations of parametric models are important when the default choice of multivariate normal random effects might understate uncertainties and miss some important structures. Another important class of BNP priors are priors on unknown functions, for example as prior $p(f)$ for the unknown mean function $f(x)$ in a regression model $y_i = f(x_i) + \varepsilon_i$.

The chapters in this volume discuss important research problems in biostatistics and bioinformatics that are naturally addressed by BNP methods. Each chapter introduces and defines the BNP methods and models that are used to address the specific problem. In this introductory chapter we briefly introduce and review some of the most commonly used BNP priors. Posterior inference in many of these models gives rise to challenging computational problems. We review some of most commonly used computational methods and include some references. The brief review in this introduction includes the ubiquitous Dirichlet process (DP) model, the DP mixture model (DPM), the dependent DP (DDP) model, the Polya tree (PT) prior, and the Gaussian process (GP) prior. These models and their variations are the workhorses of BNP inference in biostatistics. The next chapter in this volume discusses some typical examples by reviewing BNP methods in some important applications.

For a more exhaustive discussion of BNP models, see, for example, recent discussions in Hjort et al. (2010), Müller and Rodríguez (2013), Walker et al. (1999), Müller and Quintana (2004), Walker (2013) and Müller et al. (2015).

## 1.2 Dirichlet Process

Let $\delta_x(\cdot)$ denote a point mass at $x$. The DP prior (Ferguson 1973) is a probability model for a random distribution $G$,

$$G = \sum_{h=1}^{\infty} w_h \delta_{m_h}, \tag{1.1}$$

with independent locations $m_h \sim G_0$, i.i.d., and weights that are constructed as $w_h = v_h \prod_{\ell<h}(1 - v_\ell)$ with independent beta fractions $v_h \sim \text{Be}(1, M)$, i.i.d. (Sethurman 1994). The prior on $w_h$ is known as the stick-breaking process. It can be described as breaking off fractions $v_h$ of a stick of initially unit length. The DP prior is characterized by the base measure $G_0$ that generates the locations of the atoms $m_h$ and the total mass parameter $M$ that determines the distribution of the beta fractions $v_h$. We write $G \sim \text{DP}(M, G_0)$. Implied in the constructive definition of the stick breaking construction is an important property of DP random measures. A DP random measure $G \sim \text{DP}(M, G_0)$ is discrete with probability one.

The DP is a conjugate prior under i.i.d. sampling. That is, assume $x_i \mid G \sim G$, i.i.d., $i = 1, \ldots, n$ and $G \sim \text{DP}(M, G_0)$. Let $F_n = \frac{1}{n}\sum_{i=1}^n \delta_{x_i}$ denote the empirical distribution. Then $p(G \mid \boldsymbol{x}) = \text{DP}(M + n, G_1)$ with $G_1 \propto MG_0 + nF_n$. An interesting limiting case occurs for $M \to 0$, when the posterior on $G$ is entirely determined by the empirical distribution. This leads to a construction known as the Bayesian bootstrap, which is discussed in Chap. 16 (Inácio de Carvalho et al. 2015).

One of the reasons for the wide use of the DP prior is ease of computation for posterior inference in models based on the DP. In particular, the DP prior implies a particularly simple predictive probability function $p(x_n \mid x_1, \ldots, x_{n-1})$. Under i.i.d. sampling from a DP random measure the marginal distribution $p(x_1, \ldots, x_n) = \int \prod_{i=1}^n G(x_i)\, dp(G)$ reduces to a simple expression which is easiest characterized as $p(x_1, \ldots, x_n) = \prod_{i=1}^n p(x_i \mid x_1, \ldots, x_{i-1})$ with increasing conditionals

$$p(x_i \mid x_1, \ldots, x_{i-1}) \propto MG_0(x_i) + \sum_{\ell=1}^{i-1} \delta_{x_\ell}. \tag{1.2}$$

With probability $\pi_0 = M/(i - 1 + M)$ the sample $x_i$ is a new draw from $G_0$, and with probability $1/(i - 1 + M)$ the new sample is tied with a previous sample $x_\ell$. The conditional distribution (1.2) is also known as the Polya urn. We will return to it below. Let $\boldsymbol{x}_{-i} = \boldsymbol{x} \setminus \{x_i\}$. For later reference we note that by symmetry the conditional distribution $p(x_i \mid \boldsymbol{x}_{-i})$ takes the same form

### 1.2.1 DP Mixture

The discrete nature of a DP random measure is awkward in many applications and is therefore often avoided by using an additional convolution with a continuous kernel. Let $k(x_i \mid \theta)$ denote a continuous kernel, for example a Gaussian kernel. Without loss of generality we assume in the remaining discussion $k(x_i \mid \theta) = N(x_i \mid \theta, s)$ (with fixed $s$). The DP mixture (DPM) model assumes $G = \int N(x_i \mid \theta, s) dF(\theta)$, with $F \sim \text{DP}(M, F_0)$. We write $G \sim \text{DPM}(M, G_0, k)$. It is often convenient to rewrite the mixture as an equivalent hierarchical model. Instead of $y_i \sim G$ and $G \sim \text{DPM}(M, G_0, k)$ we write

$$y_i \mid \theta_i \sim N(\theta_i, s) \text{ and } \theta_i \sim F \tag{1.3}$$

with $F \sim \text{DP}(M, G_0)$. The DPM model is one of the most widely used BNP priors for random distributions. In this volume we find it, for example, in Chap. 11 (Zhou and Hanson 2015) to construct a semiparametric version of an accelerated failure time model; in Chap. 16 (Inácio de Carvalho et al. 2015) as a prior for the distribution of test outcomes to develop inference on ROC curves; in Chap. 21 (Daniels and Linero 2015) for longitudinal outcomes under different missingness patterns; and many more.

Consider again the $\theta_i$ in (1.3). As a sample from the discrete random measure $F$, the newly introduced latent variables $\theta_i$ include many ties. Let $\boldsymbol{\theta}^\star = \{\theta_1^\star, \ldots, \theta_k^\star\}$ denote the $k \le n$ unique values and let $S_j = \{i : \theta_i = \theta_j^\star\}$ denote the indices $[n] \equiv \{1, \ldots, n\}$ arranged by the configuration of ties. Then $\rho_n \equiv \{S_1, \ldots, S_k\}$ defines a partition of $[n]$. Since the $\theta_i$ were random, as a consequence the partition is random. That is, the DP mixture model (1.3) induces a random partition $p(\rho_n)$. At first glance this seems like a coincidental detail of the model. However, many applications of the DPM model exploit exactly this feature. It features in many chapters in this volume. The implied prior $p(\rho_n)$ on the random partition is also known as Chinese restaurant process (CRP). It is used, for example, in Chap. 3 (Zhang et al. 2015).

Sometimes it is convenient to index the partition $\rho_n$ alternatively by an equivalent set of cluster membership indicators. Let $s_i$ denote indicators with $s_i = j$ if $i \in S_j$, that is when $\theta_i = \theta_j^\star$. Let $n_j = |S_j|$ denote the size of the $j$th cluster, $n_j^- = |S_j \setminus \{i\}|$ and let $k^-$ denote the number of unique values $\theta_\ell$ in $\boldsymbol{\theta}_{-i}$. Then we can rewrite (1.2) as

$$s_i \mid \boldsymbol{s}_{-i} = \begin{cases} j & \text{with prob } \frac{n_j^-}{n-1+M}, \quad j = 1, \ldots, k^- \\ k^- + 1 & \text{with prob } \frac{M}{n-1+M} \end{cases} \tag{1.4}$$

The attraction of model (1.3) is the ease of posterior simulation. Consider a generic model $y_i \sim G$ with DPM prior (1.3) and similar to $k^-$ and $n_j^-$ let $\theta_j^{\star -}$ denote the $j$th unique value among $\boldsymbol{\theta}_{-i}$. Then (1.4) implies

$$\theta_i \mid \boldsymbol{y}, \boldsymbol{\theta}_{-i} = \begin{cases} \theta_j^{\star -} & \text{with prob.} \propto n_j^- \, p(y_i \mid \theta_j^{\star -}) \\ \sim H_1 & \text{with prob.} \propto M \int p(y_i \mid \theta) \, dG_0(\theta) \end{cases} \tag{1.5}$$

with $H_1(\theta) \propto p(y_i \mid \theta) \, G_0(\theta)$. If $p(y_i \mid \theta)$ and $G_0(\theta)$ are chosen as a conjugate pair of sampling model and prior, then generating from (1.5) is straightforward. In the general case, the evaluation of $h_0 \equiv \int p(y_i \mid \theta) \, dG_0(\theta)$ can be computationally challenging. Several MCMC algorithms have been proposed to circumvent the evaluation of an analytically intractable integral $h_0$ (Neal 2000). For a recent review of the DP and related models, see, for example, Ghosal (2010).

### *1.2.2 Generalizations of the DP*

Many generalizations of the DP prior have been proposed in the literature. One example is the Poisson-Dirichlet (PD) process that is used in Chap. 9 (Guha et al. 2015). The PD arises by replacing the $\text{Be}(1, M)$ prior on the fractions $v_h$ in the stick breaking construction by $\text{Be}(1-a, b+ha)$ priors. Other generalizations are specifically focused on the implied random partition model, like the generalized Ottawa sequence introduced in Chap. 5 (Bassetti et al. 2015) or the hierarchical DP (HDP) model in Chap. 7 (Iorio et al. 2015). The latter defines a prior on a family of random probability measures $\{G_j; j = 1, \ldots, j\}$.

## 1.3 Dependent Dirichlet Process

Many problems involve a family of unknown random probability measures $\mathscr{F} = \{F_x; x \in X\}$. For example, in a mixed effects model that includes data from several related studies, $F_j$ might be the random effects distribution for patients in study $j$. More generally, a formalization of non-parametric regression could assume

$$y_i \mid \boldsymbol{x}_i = x, \mathscr{F} \sim F_x \tag{1.6}$$

$i = 1, \ldots, n$. That is, we denote by $F_x$ the sampling model for the response of a subject with covariates $\boldsymbol{x}_i = x$. If we are willing to assume $F_x = N(\boldsymbol{x}_i' \boldsymbol{\beta}, \sigma^2)$, then the problem reduces to parametric inference on the finite dimensional parameter vector $\boldsymbol{\theta} = (\boldsymbol{\beta}, \sigma^2)$. In other words, we restrict $\mathscr{F}$ to the family of probability measures indexed by $\boldsymbol{\theta}$. In the absence of such restrictions Bayesian inference in (1.6) requires a prior probability model $p(\mathscr{F})$ that allows for dependence and borrowing of strength across $x$, short of the strict parametric assumption, but still more than in a model with independent, separate priors on each $F_x$.

One of the most popular models in the recent literature for a family of random probability measures $\mathscr{F}$ is the dependent DP (DDP) and variations of it. The model was first introduced in MacEachern (1999). The idea is simple. We continue to use

$$F_x = \sum_{h=1}^{\infty} w_h \delta_{m_{xh}}, \tag{1.7}$$

with independent locations $m_{xh}$, i.i.d. across $h$ and weights that are constructed with independent beta fractions as before, in (1.7). The only addition is that we now introduce dependence on the point masses $m_{xh}$ *across* $x$. For example, we could assume that $(m_{xh},\ x \in X)$ is a realization of a Gaussian process indexed by $x$. In the simplest implementation the weights $w_h$ are shared across all $x$, as implied in the notation $w_h$ without a subindex for $x$.

Similar to the DP mixture model, the DDP model (1.7) is often combined with a continuous kernel, for example a normal kernel to define

$$G_x(y) = \int N(y \mid \theta, \sigma^2)\, dF_x(\theta) = \sum_{h=1}^{\infty} w_h N(y \mid m_{xh}, \sigma^2). \tag{1.8}$$

with a DDP prior on $\{F_x,\ x \in X\}$. Here $N(y \mid m, s^2)$ denotes a normal kernel in $y$. We refer to (1.8) as a DDP mixture of normals. For categorical covariates $x \in X$ the dependent probability model for $(m_{xh},\ x \in X)$ could be defined, for example, as an ANOVA model. This defines the ANOVA DDP proposed in DeIorio et al. (2002). A version of the same, with a general linear model in place of the ANOVA model is the linear dependent DP (LDDP) (Jara et al. 2010).

### *1.3.1 Variations of the DDP*

The DDP prior and variations of it are used in several chapters in this volume. Chapter 12 (Jara et al. 2015) uses an LDDP to implement survival regression. Chapter 20 (Karabatsos and Walker 2015) constructs a variation of a DDP by introducing the dependence on covariates in (1.7) by a probit regression in the weights $w_h$, rather than the atoms $m_h$.

## 1.4 Polya Tree

The Polya tree (PT) prior (Lavine 1992, 1994) is an attractive alternative BNP prior for a random probability measure. The PT prior is essentially a random histogram. Without loss of generality, assume that we wish to define a random probability measure $G$ on the unit interval $[0,1]$. We could start with a random histogram with two bins $\{B_0, B_1\}$, say over $B_0 = [0, 0.5)$ and $B_1 = [0.5, 1]$. Let $Y_0 = G(B_0)$ and $Y_1 = 1 - Y_0$ denote the (random) probabilities of $B_0$ and $B_1$. Next we refine the histogram by splitting the bins into $B_0 = B_{00} \cup B_{01}$ with $B_{00} = [0, 0.25)$, etc. Let $Y_{00} = G(B_{00} \mid B_0)$, $Y_{10} = G(B_{10} \mid B_1)$, $Y_{01} = 1 - Y_{00}$, and $Y_{11} = 1 - Y_{10}$. We continue refining the histogram to $2^m$ bins, $m = 1, 2, \ldots$ by repeating similar binary splits. The process creates a sequence $\Pi = \{\Pi_m,\ m = 1, 2, \ldots\}$ of nested binary partitions $\Pi_m = \{B_{e_1 \cdots e_m}\}$ with $e_j \in \{0, 1\}$. The PT defines a prior on $G$ by assuming

$$Y_{\varepsilon 0} \sim \mathrm{Be}(\alpha_{\varepsilon 0}, \alpha_{\varepsilon 1}),$$

independently across $\varepsilon$ and $Y_{\varepsilon 1} = 1 - Y_{\varepsilon 0}$. The nested partitions $\Pi$ together with the beta parameters $\mathscr{A} = \{\alpha_\varepsilon\}$ characterize the PT prior. We write $G \sim \mathrm{PT}(\Pi, \mathscr{A})$.

One of the attractions of the PT prior is the ease of centering the model. Let $0.e_1 \cdots e_m = \sum_j e_j 2^{-j}$ denote the number with binary digits $\varepsilon = e_1, \ldots, e_m$ and let $q_\varepsilon$ denote the corresponding quantile of a fixed probability measure $G_0$. That is, for example, $q_1, q_{01}, q_{10}$ are the median and the first and third quartile of $G_0$. Next define $B_\varepsilon$ to denote the corresponding partitioning subsets and let $\prod$ denote the nested partition sequence with partitioning subsets $B_\varepsilon$. If $G \sim \mathrm{PT}(\Pi, \mathscr{A})$ with $\alpha_{\varepsilon 0} = \alpha_{\varepsilon 1}$, then $E(G) = G_0$. We write

$$G \sim \mathrm{PT}(G_0, \mathscr{A}).$$

A particularly attractive choice is $\alpha_{e_1\cdots e_m} = c2^m$ which can be shown to imply a continuous random probability measure $G$. We write $G \sim \text{PT}(G_0, c)$. Alternatively, for an arbitrary nested partitioning sequence $\Pi$, define $\mathscr{A}$ by $\alpha_{\varepsilon e_m} = c\,G_0(B_{\varepsilon e_m} \mid B_\varepsilon)$ and assume $G \sim \text{PT}(\Pi, \mathscr{A})$. Then again $E(G) = G_0$. We write

$$G \sim \text{PT}(\Pi, G_0).$$

For a recent review of the PT prior, see, for example, Müller et al. (2015 Chapter 3). PT priors are used, for example, in Chap. 11 (Zhou and Hanson 2015) to construct a semi-parametric accelerated failure time model.

## 1.5 Gaussian Process

Gaussian Process (GP) priors are widely used in machine learning, medical imaging, ecology, and various disease risk models. A GP is a stochastic process $\{Y(s); s \in S\}$ that extends (finite dimensional) multivariate Gaussians to infinite dimensions. Here Y(.) is a function-valued random variable while $S$ denotes the domain (typically $\Re^e$) of the function. The domain $S$ and thus $Y(.)$ can have very different interpretation and meaning depending upon specific applications. For example, in Chap. 17 (Reich and Fuentes 2015), and typically in the context of spatial models, $S$ refers to all location points in a given region. For machine learning applications, it can be the set of all possible input stimuli. It could even represent the time domain for recording neuronal activity as in the case study provided in Chap. 13 (Shahbaba et al. 2015). $S$ is usually endowed with its own specific metric, e.g. the Euclidean distance in spatial applications. The problem of analyzing the random function $Y(.)$ or predicting its value Y(**s**) at a specific point **s** can be formulated within the framework of non-parametric regression, where the values in $S$ play the role of covariates and $Y(.)$ is the regression function to be estimated. A prior on the random function $Y()$ would simply refer to the probability law of the stochastic process.

We formally characterize a GP as a stochastic process with mean function $m(.)$ and covariance function $k(\cdot,\cdot)$ if every finite sub-collection of this process, $[Y(s_1), Y(s_2) \ldots Y(s_n)]$ is multivariate Gaussian

$$[Y(s_1), \ldots, Y(s_n)] \sim N(\mu, \Sigma) \text{ with } \mu = [m(s_1), \ldots, m(s_n)] \text{ and } \Sigma_{ij} = k(s_i, s_j).$$

We write $Y \sim \text{GP}(m,k)$. The covariance function is sometimes also referred to as the kernel of the GP. The prior on the random $Y(.)$, thus defined, is called a GP prior. Simply put, a GP extends finite multivariate Gaussian models to infinite dimensions. It can be shown that such an extension is possible using Kolmogorov's consistency theorem. Naturally, the infinite process inherits many attractive properties of its finite version. For example, no restrictions are required for the mean function $m$. However, since all finite dimensional subsets are required to be Gaussian, a condition of positive semi-definiteness is implied on $V$ for any finite subset of $S$.

A variety of different families of valid kernels are in common use. Some popular choices include squared exponential (SE), polynomial, neural network, Ornstein-Uhlenbeck (OU), Matern, etc. Each of these families typically has a number of free hyper-parameters. Choosing a covariance function for a particular application thus comprises both, the setting of hyper-parameters within a family, and sometimes the comparison across different families through model-selection techniques. Alternatively, flexible and non-parametric covariance functions can be built by exploiting the spectral representation of a GP. Chapter 17 (Reich and Fuentes 2015) introduces such general priors for spatial covariances by applying the DP and the DPM priors to the coefficients of the spectral density.

In general, all covariance functions formally encode some notion of similarity between a pair of random observations based on the distance between corresponding elements of $S$. Consider, for example, the SE kernel given by $k(s,t) = exp(-||s-t||^2/2\tau^2)$. The functional form suggests that observations corresponding to proximal points are highly correlated, with the correlation dropping off exponentially with the distance between the points.

Posterior inference and prediction with GP priors is made immensely easy by using the analytical results for multivariate Gaussians. For this, it is enough to observe that the collection of new and observed variables is a finite subset of the GP and their joint density is a multivariate Gaussian. Hence, the posterior predictive distribution, obtained by conditioning on the observed data, appears as another multivariate normal. The infinite dimension of the prior, while providing substantial modeling flexibility, poses no concern for inference and computation. These properties turn out to be critical for several analytical manipulations with the GP prior.

However a known computational bottleneck is the inversion of $(n \times n)$ matrices that appear in the analytical results, thus making the computational complexity cubic in the number of data points. For large datasets (n > 10,000) this is prohibitive (in both time and space) for any inference, Bayesian or otherwise. So a number of computational methods [e.g., reduced rank matrix approximations (Fine et al. 2001; Smola and Schökopf 2000)] have been developed. Another approach is to exploit structures of special classes of covariance functions for exact computation. These methods are iterative and the computation scales linearly with the size of the data (Johannesson and Cressie 2004). Cressie and Johannesson (2008) extended this approach to a flexible class of covariance functions. The computational complexity also increases drastically in multivariate settings with several spatially dependent response variables. Banerjee et al. (2008) used induced predictive process models as a clever strategy for dimension reduction and to reduce computational cost in this context. An alternative solution to the computational problem is the treed Gaussian process of Gramacy and Lee (2008). The approach proceeds by first partitioning the covariate space into a number of smaller regions, similar to a classification and regression tree (CART). Next, independent GP's are fit to each subregion. The overall inversion of a large matrix is replaced by a number of smaller, computationally feasible inversions. Posterior inference is efficiently handled in the `tgp` package for R.

An excellent reference on Gaussian process models for regression is Rasmussen and Williams (2005).

## 1.6 Conclusion

In this brief review we only introduced some of the most popular BNP models and variations. Some of the chapters use models beyond this selection. Chapter 4 (Ji et al. 2015) uses an Indian buffet process as a prior probability model for a feature allocation problem. Feature allocation generalizes random clusters, that is, non-overlapping subsets, to families of possibly overlapping subsets. Chapter 10 (Nieto-Barajas 2015) introduces several alternative models, including, for example, the normalized generalized gamma (NGG) process. The same NGG process appears in Chap. 6. Some chapters define random functions based on spline bases, including Chap. 14 (Telesca 2015) and Chap. 11 (Zhou and Hanson 2015). Finally, Chap. 8 (Ni et al. 2015) discusses prior probability models for random networks.

The next chapter, Chap. 2 continues this review by discussing some typical applications of basic BNP models.

## References


Banerjee, S., Gelfand, A. E., Finley, A. O., and Sang, H. (2008). Gaussian predictive process models for large spatial data sets. *Journal of the Royal Statistical Society: Series B (Statistical Methodology)*, **70**(4), 825–848.

Bassetti, F., Leisen, F., Airoldi, E., and Guindani, M. (2015). Species sampling priors for modeling dependence: an application to the detection of chromosomal aberrations. In Mitra and Müller (2015).

Cressie, N. and Johannesson, G. (2008). Fixed rank kriging for very large spatial data sets. *Journal of the Royal Statistical Society: Series B (Statistical Methodology)*, **70**(1), 209–226.

Daniels, M. J. and Linero, A. R. (2015). Bayesian nonparametrics for missing data in longitudinal clinical trials. In Mitra and Müller (2015).

DeIorio, M., Müller, P., Rosner, G. L., and MacEachern, S. N. (2002). ANOVA DDP models: A review. In D. D. Denison, M. H. Hansen, C. C. Holmes, B. Mallick, and B. Yu, editors, *Nonlinear Estimation and Classification*, page 467. Springer-Verlag.

Ferguson, T. S. (1973). A Bayesian analysis of some nonparametric problems. *The Annals of Statistics*, **1**, 209–230.

Fine, S., Scheinberg, K., Cristianini, N., Shawe-taylor, J., and Williamson, B. (2001). Efficient svm training using low-rank kernel representations. *Journal of Machine Learning Research*, **2**, 243–264.

Ghosal, S. (2010). The Dirichlet process, related priors and posterior asymptotics. In Hjort et al. (2010), pages 22–34.

Gramacy, R. B. and Lee, H. K. H. (2008). Bayesian treed Gaussian process models with an application to computer modeling. *Journal of the American Statistical Association*, **103**, 1119—1130.

Guha, S., Banerjee, S., Gu, C., and Baladandayuthapani, V. (2015). Nonparametric variable selection, clustering and prediction for large biological datasets. In Mitra and Müller (2015).

Hjort, N. L., Holmes, C., Müller, P., and Walker, S. G. (2010). *Bayesian Nonparametrics*. Cambridge University Press.

Inácio de Carvalho, V., Jara, A., and de Carvalho, M. (2015). Bayesian nonparametric approaches for ROC curve inference. In Mitra and Müller (2015).

Iorio, M. D., Favaro, S., and Teh, Y. W. (2015). Bayesian inference on population structure: from parametric to nonparametric modeling. In Mitra and Müller (2015).

Jara, A., Lesaffre, E., De Iorio, M., and Quintana, F. A. (2010). Bayesian semiparametric inference for multivariate doubly-interval-censored data. *The Annals of Applied Statistics*, **4**, 2126–2149.

Jara, A., García-Zattera, M. J., and st Komárek, A. (2015). Fully nonparametric regression modelling of misclassified censored time-to-event data. In Mitra and Müller (2015).

Ji, Y., Sengupta, S., Lee, J., Müller, P., and Gulutoka, K. (2015). Estimating latent cell subpopulations with Bayesian feature allocation models. In Mitra and Müller (2015).

Johannesson, G. and Cressie, N. (2004). Variance-covariance modeling and estimation for multi-resolution spatial models. In *geoENV IV – Geostatistics for Environmental Applications*, pages 319–330. Springer.

Karabatsos, G. and Walker, S. G. (2015). A Bayesian nonparametric causal model for regression discontinuity designs. In Mitra and Müller (2015).

Lavine, M. (1992). Some aspects of Polya tree distributions for statistical modelling. *The Annals of Statistics*, **20**, 1222–1235.

Lavine, M. (1994). More aspects of Polya tree distributions for statistical modelling. *The Annals of Statistics*, **22**, 1161–1176.

MacEachern, S. (1999). Dependent nonparametric processes. In *ASA Proceedings of the Section on Bayesian Statistical Science*, Alexandria, VA. American Statistical Association.

Mitra, R. and Müller, P., editors (2015). *Nonparametric Bayesian Methods in Biostatistics and Bioinformatics*. Springer-Verlag.

Müller, P. and Quintana, F. A. (2004). Nonparametric Bayesian data analysis. *Statistical Science*, **19**, 95–110.

Müller, P. and Rodríguez, A. (2013). *Nonparametric Bayesian Inference*. IMS-CBMS Lecture Notes. IMS.

Müller, P., Quintana, F., Jara, A., and Hanson, T. (2015). *Nonparametric Bayesian Data Analysis*. Springer Verlag.

Neal, R. M. (2000). Markov chain sampling methods for Dirichlet process mixture models. *Journal of Computational and Graphical Statistics*, **9**, 249–265.

Ni, Y., Marchetti, G. M., Baladandayuthapani, V., and Stingo, F. C. (2015). Bayesian approaches for large biological networks. In Mitra and Müller (2015).

Nieto-Barajas, L. E. (2015). Markov processes in survival analysis. In Mitra and Müller (2015).

Rasmussen, C. E. and Williams, C. K. I. (2005). *Gaussian Processes for Machine Learning (Adaptive Computation and Machine Learning)*. The MIT Press.

Reich, B. J. and Fuentes, M. (2015). Spatial Bayesian nonparametric methods. In Mitra and Müller (2015).

Sethurman, J. (1994). A constructive definition of Dirichlet priors. *Statistica Sinica*, **4**, 639–650.

Shahbaba, B., Behseta, S., and Vandenberg-Rodes, A. (2015). Neuronal spike train analysis using gaussian process models. In Mitra and Müller (2015).

Smola, A. J. and Schökopf, B. (2000). Sparse greedy matrix approximation for machine learning. In *Proceedings of the Seventeenth International Conference on Machine Learning*, ICML '00, pages 911–918, San Francisco, CA, USA. Morgan Kaufmann Publishers Inc.

Telesca, D. (2015). Bayesian analysis of curves shape variation through registration and regression. In Mitra and Müller (2015).

Walker, S. (2013). Bayesian nonparametrics. In P. Damien, P. Dellaportas, N. G. Polson, and D. A. Stephens, editors, *Bayesian Theory and Applications*, pages 249–270. Oxford University Press.

Walker, S., Damien, P., Laud, P., and Smith, A. (1999). Bayesian nonparametric inference for distributions and related functions (with discussion). *Journal of the Royal Statistical Society, Series B*, **61**, 485–527.

Zhang, Z., Pati, D., and Srivastava, A. (2015). Bayesian shape clustering. In Mitra and Müller (2015).

Zhou, H. and Hanson, T. (2015). Bayesian spatial survival models. In Mitra and Müller (2015).

# Chapter 2
# Bayesian Nonparametric Biostatistics

Wesley O. Johnson and Miguel de Carvalho

**Abstract** We discuss some typical applications of Bayesian nonparametrics in biostatistics. The chosen applications highlight how Bayesian nonparametrics can contribute to addressing some fundamental questions that arise in biomedical research. In particular, we review some modern Bayesian semi- and nonparametric approaches for modeling longitudinal, survival, and medical diagnostic outcome data. Our discussion includes methods for longitudinal data analysis, non-proportional hazards survival analysis, joint modeling of longitudinal and survival data, longitudinal diagnostic test outcome data, and receiver operating characteristic curves. Throughout, we make comparisons among competing BNP models for the various data types considered.

## 2.1 Introduction

"Why Bayesian nonparametrics?" Motivation for Bayesian nonparametrics encompasses model flexibility and robustness, as parametric models are often inadequate due to their constraints. Bayesian nonparametric models that embed parametric families of distributions in broader families seem eminently sensible since they allow for flexibility and robustness beyond the constrained parametric family. The models we consider here are in fact richly parametric (formally, using an infinite-dimensional parameter space) rather than nonparametric, which is an unfortunate misnomer

W.O. Johnson (✉)
University of California, Irvine, CA, USA
e-mail: wjohnson@uci.edu

M. de Carvalho
Departamento de Estadística, Pontifica Universidad Católica de Chile, Santiago, Chile
e-mail: mdecarvalho@mat.puc.cl

 
R. Mitra, P. Müller (eds.), *Nonparametric Bayesian Inference in Biostatistics*,
Frontiers in Probability and the Statistical Sciences,
DOI 10.1007/978-3-319-19518-6_2

that we will not attempt to rectify. Bayesian nonparametric models involve placing prior distributions on broad families of probability distributions; examples considered here include Mixtures of Polya trees (MPT) and Dirichlet Processes mixtures (DPM).

The MPT will be seen to be a clear extension of a selected parametric family for data. The DPM is more ambiguous but in some instances could be viewed in the same way. A popular theme in much of the Bayesian nonparametrics literature is to regard a parametric approach as a reference, while allowing data that are modeled nonparametrically to inform a subsequent analysis about the adequacy of the parametric model.

Other Bayesian nonparametric approaches involve the use of Gaussian process priors and consist of probability models over spaces of functions. For these the natural probabilistic concept is that of a random function; conceptually, random functions can be regarded as stochastic processes, and are the subject of Part IV of this volume.

### *2.1.1 Organization of this Chapter*

**Section 2.2 Comments on the DPM and MPT**. In this section we discuss some features of Dirichlet and Polya tree processes; a technical introduction to these and other prior processes can be found in Chap. 1 of this volume (Mitra and Müller 2015).

**Section 2.3 Longitudinal Data: Semiparametric Autoregressive Modeling**. Here we discuss a model that generalizes standard mixed models for longitudinal data, and which includes a functional mean function, and allows for compound symmetry (CS) and autoregressive (AR) covariance structures. The AR structure is specified through a Gaussian process (GP) with an exponential covariance function, which allows observations to be more correlated if they are observed closer in time than if they are observed farther apart. Quintana et al. (2015) generalize this model by considering a DPM of Gaussian processes. In Sect. 2.3.2 we discuss their analysis of data from the Study of Women's Health across the Nation (SWAN) that involves longitudinal outcomes of hormone data for women experiencing the menopausal transition.

**Section. 2.4 Survival Data: Nonparametric and Semiparametric Modeling**. We discuss Bayesian non and semi-parametric modeling for survival regression data; Sect. 2.4 provides some preparation for Part III of this volume, which is entirely dedicated to survival analysis. We first give a selective historical perspective of the development of nonparametric Bayesian survival regression methods (Sect. 2.4.1). We discuss an analysis of time to abortion in dairy cattle with fixed covariates, and then discuss models for time dependent regression survival data, followed by analyses of the Stanford Heart Transplant data and a data set involving the timing of cerebral edema in children diagnosed with ketoacidosis. We end the section with a presentation of a Bayesian nonparametric survival model that allows survival curves to cross, and a subsequent analysis of breast cancer data where survival curves are expected to cross.

**Section 2.5 Joint Modeling of Longitudinal and Survival Data**. We consider the joint modeling of survival data and a longitudinal process. In Sect. 2.4, we discussed a number of survival regression models with time dependent covariates where we fixed the time dependent covariates (TDC) in the same sense that we fix covariates in regression. However, Prentice (1982) pointed out that fixing the TDCs rather than modeling them could bias final estimates. The general rule has been to use the last observation carried forward (LOCF) in the TDC process, despite the fact that the last observation might have occurred some time ago, suggesting that it may not well represent the current value of the process. In Sect. 2.5 we discuss a data analysis performed by Hanson et al. (2011b), which uses the models and methods in Hanson et al. (2009) in conjunction with longitudinal modeling to develop joint models for longitudinal-survival data.

**Section 2.6 Medical Diagnostic Data**. In Sect. 2.6.1 we discuss the subject of Receiver Operating Characteristic (ROC) curve regression, and in Sect. 2.6.2 we consider the issue of Bayesian semi-parametric estimation in ROC regression settings that lack availability of a gold standard test, i.e., when there is no available test that could perfectly classify subjects as diseased and non-diseased. Related literature is reviewed in detail in Chap. 16 (Inácio de Carvalho et al. 2015). We illustrate methods by assessing the potential of a soluble isoform of the epidermal growth factor receptor (sEGFR) for use as a diagnostic biomarker for lung cancer in men, and we assess the effect of age on the discriminatory ability of sEGFR to classify diseased and non-diseased individuals. In Sect. 2.6.3 we discuss joint longitudinal diagnostic outcome modeling and analysis, and we illustrate with longitudinal cow serology and fecal culture data.

In Sect. 2.7 we briefly comment on other types of data that are of interest in biomedical research, and on some current Bayesian nonparametric approaches for modeling.

## 2.2 Comments on the DPM and MPT

We briefly comment on two mainstream prior processes for data analysis: The Dirichlet and Polya tree processes. By themselves, they are perhaps not practical models for data analysis, but it is their mixture forms that are. The Dirichlet Process Mixture (DPM) and the Mixture of Polya Trees (MPT) have been established to be practical tools for data analysis. Models that employ the DPM in various forms are by far the most popular for a variety of reasons including the fact that the DP has been in the literature since at least Ferguson (1973), and DPMs have been developed extensively for use in analyzing data since at least Escobar (1994). Polya Trees have been around since at least Ferguson (1974), but did not seem to be particularly noticed until the 1990s (Mauldin et al. 1992; Lavine 1992, 1994), and were not given a lot of attention until Berger and Guglielmi (2001), and Hanson and Johnson (2002) and Hanson (2006), who developed MPTs for survival analysis and beyond.

A key property of the DPM is that any inferential object that is modeled as a DPM of continuous parametric densities is smooth. Moreover, under some conditions, the DPM of location-scale normal densities has been shown to have strong posterior consistency for the true density (Tokdar 2006). There are many other theoretical works of this type, including Amewou-Atisso et al. (2003), who established large sample consistency properties for semiparametric linear regression models with error distributions that are modeled with median zero processes based on both PTs and DPMs. The original and continuing appeal to DPMs was and is at least partly based on the ease of marginalizing over the DP when performing numerical calculations. The marginalization led to computationally straightforward schemes involving the Polya Urn scheme that researchers often describe as a Chinese restaurant process. Neal (2000) improved upon previous computational schemes pioneered by Escobar (1994); Escobar and West (1995), and MacEachern and Müller (1998), among others. In addition, there are many extensions of the DPM, including the Dependent Dirichlet Process (DDP) (MacEachern 2000), the Nested DP (NDP) (Rodriguez et al. 2008), and the Hierarchical DP (HDP) (Tomlinson and Escobar 1999; Teh et al. 2006), among others, many discussed in Chap. 1 of this volume (Mitra and Müller 2015).[1] The Sethuraman (1994) representation of the DP facilitated the development of all of these, and it provided an easy understanding of the precise meaning of the DP and the DPM. In addition, it facilitated the extension to more general stick-breaking processes, for example the Dunson and Park (2008) application to density regression, among others. The point here is that there is now a wealth of papers that have developed, extended, and used various forms of and which stem from the DP, and which have used these tools to analyze data of all complexities. The DPM is clearly here to stay.

The MPT has many positives as well. It can be selected to be absolutely continuous with probability one, so it is possible to use it directly as a model for data. When used as a model for the error distribution in a linear regression, it is easy to specify that the MPT has median zero with probability one, resulting in a semiparametric median regression model. In Sect. 2.4 we discuss such models for survival data. In addition, it is a flexible model, allowing for multimodality, skewness, etc. It is straightforward to perform MCMC computations for many complex models (Hanson 2006), and there is no need to marginalize the process to make computations simpler. From our point of view, a major positive feature of the Mixture of Finite Polya Tree (MFPT) prior is that it not only allows for a broad/flexible class of distributions but that it has a parametric family of distributions for the data embedded in it, and that the embedding is natural. Thus, if a scientist has previous experience or information that suggests that a log normal family of distributions

[1] See also Müller and Mitra (2013) for a recent survey.

might be appropriate for their survival data, they could hedge their bets by embedding that family as the centering family of an MFPT. Moreover, if they also had scientific information about the log normal family parameters, they could construct an informative prior for those parameters. See Bedrick et al. (1996) and Bedrick et al. (2000) for illustrations of informative prior specification for generalized linear models and for survival models. Thus far, we are not aware of any such nice properties for specifying prior distributions on the parameters of the base distribution in the DPM. Berger and Guglielmi (2001) also took advantage of the fact that a parametric family can be embedded in the PT family in developing a method to test the adequacy of the parametric family to fit data.

A possible advantage of the DPM over the MPT is the ease of extending the DPM to multivariate data, which is straightforward for the DPM. Hanson (2006) has developed MPT methods for multivariate data, and Jara et al. (2009) and Hanson et al. (2011a) improved them. While no comparison between the methods has been performed to date, Hanson reports that the MPT-based method would perform well for joint density estimation, and clearly better for "irregular densities" (personal communication). Another advantage is the smoothness of the DPM. When the weight associated with the MPT is small, density estimates can be quite jagged, despite the fact that Hanson and Johnson (2002, Thm. 2) proved that predictive densities in the context of the semiparametric model that they develop are differentiable under some conditions. For applications, an important issue is prior elicitation for the DPM; cf. Hanson et al. (2005).

In the illustrations below, we take examples that use the DPM, DDP, and MPT. For the MPT based models, we always use a truncated version, which is termed an MFPT. The truncation is at some level, usually termed $M$, of the basic tree structure. In addition, MPTs have weights, $c$, just like the DPM, whereas small weight corresponds to the model being 'more nonparametric.' Some models discussed below, e.g. Hanson and Johnson (2002, 2004), and De Iorio et al. (2009), can be fit using the `R` package `DPpackage` (Jara et al. 2011).

## 2.3 Longitudinal Data: Semiparametric Autoregressive Modeling

### *2.3.1 The Semiparametric Model*

Assume that observations are made on individual $i$ at times $\{t_{i1},\ldots,t_{in_i}\}$, namely $Y_i=\{Y_{ij}: j=1,\ldots,n_i\}$. At time $t_{ij}$ we allow for a vector of possibly time-dependent covariates $x_{ij}^{\mathrm{T}}=(1,x_{i1}(t_{ij}),\ldots,x_{ip}(t_{ij}))$, and assume that $\mathrm{E}(Y_{ij})=x_{ij}^{\mathrm{T}}\beta$. Define the $n_i\times(p+1)$ design matrix $X_i=(x_{i1},\ldots,x_{in_i})^{\mathrm{T}}$, leading to an assumed mean vector $\mathrm{E}(Y_i)=X_i\beta$. Then, allow for a corresponding $n_i\times q$ design matrix $Z_i$, with $q\leqslant p$ and with the column space of $Z_i$ restricted to be contained in the column space of $X_i$.

The starting point for the model to be discussed is a well-known linear mixed model (Diggle 1988) that also allows for AR structure, namely

$$Y_i = X_i\beta + f_i(t) + Z_i b_i + w_i + \varepsilon_i, \quad b_i \mid \xi \sim \mathrm{N}_r(0, D(\xi)), \quad w_i \mid \phi \sim \mathrm{N}_{n_i}(0, H_i(\phi)), \tag{2.1}$$

Here, $H_i(\phi)$ is $n_i \times n_i$ and has a structural form, $\varepsilon_i \sim \mathrm{N}(0, \sigma^2 I_{n_i})$, and $f_i(t)$ is a function evaluated at subject-specific times $t_{ij}$ for individual $i$; in addition, $\xi$ and $\phi$ contain variance–covariance parameters for $b_i$ and $w_i$, respectively.

The $w_i$ are generated by zero-mean Gaussian processes, $\{w_i(t) : t > 0\}$. If $\mathrm{Cov}(w_i(t+s), w_i(t)) = \sigma_w^2 \rho(s)$, with $\rho(s) = \rho^s$, the resulting stationary process is an Ornstein–Uhlenbeck process (Rasmussen and Williams 2006), which yields an exponential covariance function and induces AR structure.[2] The combination of choosing which terms to include in (2.1)—and making particular choices for $H(\phi)$ and $D(\xi)$—when the corresponding effects are included in the model, determines the covariance structure for the data.

The semiparametric autoregressive model extends (2.1) by introducing flexibility beyond the exponential covariance structure. Consider first the GP, $w_i$, for the $i$th subject, with covariance matrix of the form $H_i(\phi) = \sigma_w^2 \tilde{H}_i(\rho)$, where $\phi = (\sigma_w^2, \rho)$ and $\{\tilde{H}_i(\rho)\}_{k,\ell} = \rho^{|t_{i\ell} - t_{ik}|}$. Let $\phi \mid G \sim G$ with $G \sim \mathrm{DP}(\alpha, G_0)$ so that

$$f(w_i \mid G) = \int \mathrm{N}(w_i \mid 0, \sigma_w^2 \tilde{H}_i(\rho))\, \mathrm{d}G(\phi) = \sum_{k=1}^{\infty} \pi_k \mathrm{N}_{n_i}(w_i \mid 0, \tilde{\sigma}_{wk}^2 \tilde{H}_i(\tilde{\rho}_k)), \tag{2.2}$$

is an infinite mixture of multivariate normal densities, where $(\tilde{\sigma}_{wk}^2, \tilde{\rho}_k) \overset{\text{iid}}{\sim} G_0$, and the $\pi_k = V_k \prod_{l<k}(1 - V_l)$, where $V_k \overset{\text{iid}}{\sim} \mathrm{Be}(1, \alpha)$; here, $G_0$ is the centering distribution and $\alpha > 0$ is the so-called precision parameter. A related spatial DP with exponential covariance function in the base distribution was developed by Gelfand et al. (2005).

Model (2.2) implies clustering on autocorrelation structure across subjects, and using the Sethuraman representation, it can be noticed that

$$\mathrm{Cov}(w_i(t+s), w_i(t) \mid G) = \sum_{k-1}^{\infty} \pi_k \tilde{\sigma}_{wk}^2 \tilde{\rho}_k^s.$$

Hence, if the $i$th subject has equally spaced times between observations, the corresponding covariance matrix has equal diagonals with decreasing correlations as $s$ increases, but not necessarily at a geometric rate.

[2] Zeger and Diggle (1994) used $\rho(s) = \alpha + (1 - \alpha)\rho^s$. There are additional choices, including the possibility that $\sigma_w^2$ could depend on $t$, resulting in a nonhomogeneous Ornstein–Uhlenbeck process (Zhang et al. 1998). Taylor et al. (1994) used an integrated Ornstein–Uhlenbeck process (integrating over an Ornstein–Uhlenbeck with exponential covariance function) that results in a covariance function that depends on both $t$ and $s$. With structured covariance functions, the marginal covariance matrix for $Y_i$ is $\mathrm{Cov}(Y_i) = \Sigma_i(\xi, \phi, \sigma^2) = Z_i D(\xi) Z_i^{\mathrm{T}} + H_i(\phi) + \sigma^2 I_{n_i}$.

It is useful to re-write the semiparametric autoregressive model (2.2) hierarchically based on latent parameters $\phi_1,\ldots,\phi_n$, i.e.

$$\begin{aligned}
Y_i \mid \beta, b_i, w_i, \sigma^2 &\overset{\text{ind}}{\sim} \mathrm{N}_{n_i}(X_i\beta + f_i(t) + Z_i b_i + w_i, \sigma^2 I),\\
w_i \mid \phi_i = (\sigma^2_{wi}, \rho_i) &\overset{\text{ind}}{\sim} \mathrm{N}_{n_i}(0, \sigma^2_{wi}\tilde{H}_i(\rho_i)),\\
\phi_1,\ldots,\phi_n \mid G &\overset{\text{iid}}{\sim} G,\\
G &\sim \mathrm{DP}(\alpha, G_0),\\
b_i &\overset{\text{iid}}{\sim} \mathrm{N}(0, D(\xi)),\\
\sigma, \beta, \xi &\sim \mathrm{U}(0,A) \times \mathrm{N}(\beta_0, B) \times p(\xi),
\end{aligned} \tag{2.3}$$

where $w_i$ and $b_i$ are assumed independent for $i = 1,\ldots,n$.

What about posterior sampling? It can be shown that $f(w_i \mid \phi_i)$ is easily obtained, by noting that $w_i \sim \mathrm{N}_{n_i}(0, \sigma^2_{wi}\tilde{H}(\rho_i))$. Then, with

$$r_{ik} = \rho_i^{|t_{i,k+1} - t_{ik}|}, \quad k = 1,\ldots,n_i - 1,$$

Quintana et al. show that

$$\begin{cases} w_{i1} \sim \mathrm{N}_1(0, \sigma_i^2), \\ w_{ik} \mid w_{i1} = \tilde{w}_1, \ldots, w_{ik-1} = \tilde{w}_{k-1} \sim \mathrm{N}_1\left(\tilde{w}_{k-1} r_{ik-1}, \sigma_i^2(1 - r_{ik-1}^2)\right). \end{cases}$$

Thus, $f(w_i \mid \phi_i)$ is obtained as the product of $n_i$ univariate normal probability densities, making it simple to obtain the full conditional distribution of $w_i$ in a Gibbs sampling algorithm.

### *2.3.2 Model Specification for Hormone Data*

Quintana et al. (2015) considered a small subset of data that were obtained from SWAN (Study of Woman Across the Nation, *www.swanstudy.org*). The data included 9 observations for each of 162 women, and contained no missing observations. The data were grouped according to age at the beginning of the study (under 46 and over 46 years), and according to four racial/ethnic groups (African American, Caucasian, Chinese, and Japanese).

The main interest was to model the annual follicle stimulating hormone (FSH) concentrations through the menopausal transition. Concentrations of FSH and other hormones had been modeled to increase according to a (four parameter) sigmoidal shape (Dennerstein et al. 2007). FSH concentrations were measured annually from serum samples in days two through five of the menstrual cycle for women who were still menstruating or on any day that women came in for their annual visit if they were postmenopausal. Times of observation were centered on the year of final menstrual period (FMP), namely $t_i = 0$ corresponds to the year in which the final menses occurred, which is defined to be the actual time of last menses before a 12-month period in which there were none. Thus, year $-3$ is 3 years prior to the FMP, and year $+3$ is 3 years after. The data included women who started at year $-8$ continuing through year 0, and women starting at year $-2$ and continuing through year 6 (after FMP).

The functional part of the model involves a generalized sigmoid function allowing for greater flexibility than the Dennerstein et al. model. Each of the eight age by race-ethnicity groups was modeled with its own generalized sigmoid function. Let $c(i) \in \{1,\ldots,8\}$ be an indicator variable describing the particular combination of four races and two ages corresponding to subject $i$. Here, we set $\beta = (\beta_1,\ldots,\beta_8)$, where $\beta_l$ is the vector of fixed parameters associated with combination $l$.

Quintana et al. used the five parameter generalized sigmoid curve that was discussed in Ricketts and Head (1999):

$$S(t \mid \beta) = \beta_1 + \frac{\beta_2}{1 + f_t \exp\{\beta_3(\beta_4 - t)\} + (1 - f_t)\exp\{\beta_5(\beta_4 - t)\}}, \tag{2.4}$$

where

$$f_t = \frac{1}{1 + \exp\{-C(\beta_4 - t)\}}, \qquad C = \frac{2\beta_3\beta_5}{|\beta_3 + \beta_5|},$$

in which case the fixed effects become $f_i(t_{ij}) = S(t_{ij} \mid \beta_{c(i)})$. The parameters now five-dimensional and the curves defined by (2.4) are not restricted to be monotone, as would be the case of a pure sigmoidal curve. If $\beta_3$ and $\beta_5$ are however both positive, then (2.4) is monotone and increasing, and if both are negative, then it is decreasing. Using a model with fixed effects specified through (2.4), estimated mean profiles can be compared for the eight groups.

The data analysis just below is based on the specification:

$$Y_i = S(t_i \mid \beta_{c(i)}) + b_i\mathbf{1} + w_i + \varepsilon_i, \tag{2.5}$$

where $t_i^{\mathrm{T}} = (t_{i1},\ldots,t_{i9})$, $b_i \overset{\text{ind}}{\sim} \mathrm{N}(0,\sigma_b^2)$ are individual-specific random effects, $\mathbf{1}$ is a vector of ones, $w_i$ is distributed as a DPM of Ornstein–Uhlenbeck (OU) processes, as specified in (2.3), and where $S(t_i \mid \beta_{c(i)})$ is a vector with entries $S(t_{ij} \mid \beta_{c(i)})$, for $j = 1,\ldots,9$.

### Hormone Data Analysis

Quintana et al. (2015) fitted a total of six models to the data, including (2.5) above. The models considered included a parametric version of (2.5) without the OU process, model (2.5) with mixed and fixed linear terms replacing the sigmoid function, a model just like this one, except setting $\rho = 0$, model (2.5) again, but with $\rho = 0$, and finally model (2.5) without OU structure and with a general nonparametric Bayes mixture for the random effects.

They calculated log pseudo-marginal likelihood (LPML) statistics for each model; see Christensen et al. (2010, Sect. 4.9.2), or Gelfand and Dey (1994). This criterion for model selection was first introduced by Geisser and Eddy (1979) and has been used extensively for model selection in recent years; see, for example, Hanson, Branscum, and Johnson (2011b). The pseudo-marginal likelihood used was defined as $\prod_{i=1}^{n} \prod_{j=1}^{n_i} f(y_{ij} \mid y_{(ij)}, X_i, \mathscr{M})$, where $f(y_{ij} \mid y_{(ij)}, X_i, \mathscr{M})$ is the predictive density, under model $\mathscr{M}$, corresponding to individual $i$ at time $j$ based on the data minus $y_{ij}$. LPML value for model (2.5) was $-5966$, and the range for the other five models was $-6673$ to $-6986$; thus the sigmoid function with NP autoregressive structure was the clear winner. Leaving out the AR part of the model was simply not an option.

Plots of fitted values and corresponding probability bands (not shown) were virtually identical for (2.5) and its linear counterpart was virtually identical. The model with linear structure would have however been useless for prediction or for characterizing mean curves as can be seen in Fig. 2.1.

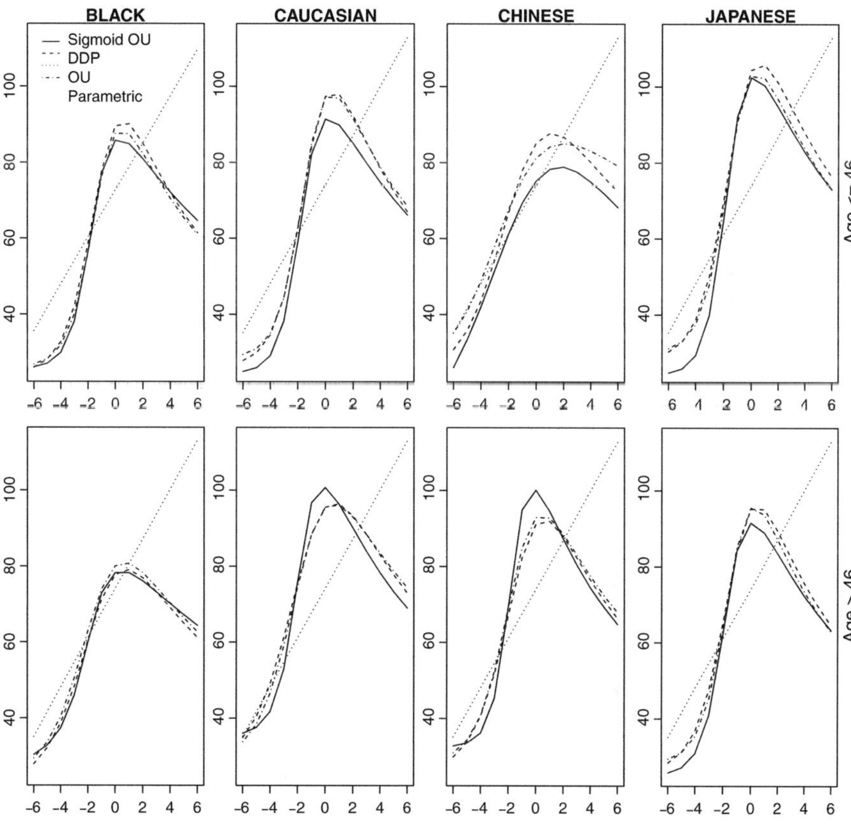


**Fig. 2.1** Predictions of future hormone concentrations ($y$ axis) for eight types of women, using (2.3) (*solid curve*), linear version of (2.3) (*dotted*), parametric sigmoid (*dot dash*), nonparametric random effects with sigmoid (*dashed*). Times of observation ($x$ axis) are centered on the year of final menstrual period (FMP) ($t_i = 0$), so that year $-3$ is 3 years prior to the FMP, and year $+3$ is 3 years after

Figure 2.1 shows model-based future predictions (posterior mean curves) for the eight different types of patient, all on the same time scale. It thus makes sense to compare shapes and levels across race/ethnicity for the same age group, and between age groups for the same race/ethnicity. Generally speaking, all models that include sigmoid mean functions predict that women's FSH hormones will go up sigmoidally, and then curve downwards toward the end of the time frame, regardless of age-race/ethnicity category. On the other hand, the linear effects model, labeled as OU on the graph, predicts a simple linear increase in FSH hormone values in contrast to the others.

Quintana et al. also made inferences comparing the maximum level achieved, the timing of the maximum level achieved and the overall slope of increase in the 4 years before FMP. The most dramatic inference is that Chinese women who are 46 years old and under at baseline achieve their maximum approximately between 1 and 3 years after FMP with 95 % posterior probability, while corresponding intervals for younger women in the other race/ethnic groups are below this interval. Among older women at baseline, there is a 0.95 posterior probability that timing for African Americans is greater than for Caucasians. The posterior probability that the difference in timing comparing younger to older Chinese women is positive is one to four decimal places. There is a clear statistical difference in timing comparing age groups for Chinese women but not for the other groups.

Finally, they estimated correlations among repeated responses on a new patient with *equally spaced times* of observation based on the joint predictive distribution under (2.5). The estimated correlations for these times that were 1–8 years apart were respectively $\{0.43, 0.27, 0.21, 0.17, 0.15, 0.14, 0.14, 0.13\}$, which is quite distinct from an AR structure. Quintana et al. observe that, after about 4 years, the correlations flatten out around 0.14. With a typical AR structure, the estimated correlations would continue to decrease across time.

## 2.4 Survival Data: Nonparametric and Semiparametric Modeling

### *2.4.1 Nonparametric and Semiparametric Survival Regression: A Selective Historical Perspective*

Survival modeling has a long and enduring history that continues. The field took its initial directions from the landmark papers by Kaplan and Meier (1958) (KM) and by Cox (1972).[3] The former paper developed the most famous nonparametric estimator of a survival function for time to event data with censoring called the product limit estimator. The second paper extended the field of survival analysis to semiparametric regression modeling of survival data; the model introduced there

[3] According to Ryan and Woodall (2005); Cox (1972) and Kaplan and Meier (1958) are the two most-cited statistical papers.

is termed the Cox proportional hazards (PH) model and is ubiquitous in medical research. There have literally been hundreds if not thousands of papers addressing various models and methods for performing survival analysis.

The main goal of a large proportion of these papers is to examine the relationship between the time to event, say $T$, and covariate information, say $x$, through the survivor function $S(t \mid x) \equiv \Pr(T > t \mid x)$. This is often done by starting with a model for $T$, like $\log(T) = x\beta + W$ where $\beta$ is a vector of regression coefficients and $W$ is modeled to have a mean zero error distribution.[4] Parametric models have $W$ distributed as normal, or extreme value or logistic, resulting in parametric log normal, Weibull and log logistic survival models. These models are termed parametric accelerated failure time (AFT) models (Kalbfleisch and Prentice 2002, Sect. 2.3.3). If the distribution of $W$ is parameterized to have median zero, which is automatic for the normal and the logistic and involves a slight modification for the extreme value distribution, then the median time to event is $\text{med}(T \mid x) = \mathrm{e}^{x\beta}$.

Models that allow for flexible distributions for $W$ are termed semiparametric. Specifically, the AFT model with fixed covariates $x$ discussed in Hanson and Johnson (2002) asserts $\log(T) = -x\beta + W$ with $\mathrm{e}^{W} \sim \text{MFPT}(M, c, F_\theta)$ and $\theta \sim p(\theta)$, where $M$ is the truncation level for the tree structure and $c$ is the weight that is associated with how much flexibility there will be about the parametric centering model, $F_\theta$. The nonparametric model embeds the family of distributions $\{F_\theta : \theta \in \Theta\}$ in it, in the sense that $\mathrm{E}\{F_W(t) \mid \theta\} = F_\theta(t)$ for all $\theta$ and $t$. Here, for example $F_\theta$ could be a log normal distribution. The survivor function for this model is $S(t \mid x, \beta, S_0) = S_0(t\mathrm{e}^{x\beta})$ and the hazard is $h(t \mid x, \beta, h_0) - \mathrm{e}^{x\beta} h_0(t\mathrm{e}^{x\beta})$.

Alternatively, models can be constructed by considering hazard functions, which can be regarded as instantaneous failure rates, formally defined as $h(t \mid x) = \lim_{\Delta s \to 0} \Pr(T \in (t, t+\Delta s] \mid T > t, x)/\Delta s = f(t \mid x)/S(t \mid x)$, where $f(t \mid x)$ is the density for $T$. The Cox (1972) PH model is $h(t \mid x) = h_0(t)\mathrm{e}^{x\beta}$ where $h_0$ is an arbitrary baseline hazard function. For two distinct individuals, it follows that the ratio of their hazards involves the cancellation of the common baseline hazard and what remains is a constant (in $t$) that only depends on their covariate vectors and the regression coefficients, hence the PH model. The survival function can be written as $S(t \mid x, \beta, H_0) = \exp\{-\mathrm{e}^{x\beta} H_0(t)\}$, where $H_0(t) = \int_0^t h_0(s)\,\mathrm{d}s$, which is termed the baseline cumulative hazard. Defining $S_0(t) = \exp\{-H_0(t)\}$, the survival function can be expressed as $S(t \mid x, \beta, S_0) = S_0(t)^{\mathrm{e}^{x\beta}}$, where $S_0$ is termed the baseline survival function. Under the PH model, survival curves for individuals with distinct covariate values cannot cross. We see that there is a parametric part to the PH model involving $\beta$, and a nonparametric part involving the unknown baseline hazard function, or equivalently the corresponding cumulative hazard, or baseline survival distribution. Bayesian approaches place parametric priors on the former, and nonparametric priors on the latter.

Bayesian methods for survival analysis were somewhat constrained until the advent of modern MCMC methods. Susarla and Van Ryzin (1976) placed a DP prior on $S$, and derived the posterior mean with censored survival data resulting in

[4] For ease of notation, we often write $x\beta$ to denote of $x^{\boldsymbol{T}}\beta$.

the Bayesian analogue to the KM estimator in the no covariate case. There were many ensuing papers, including a paper by Johnson and Christensen (1986) who again placed a DP prior on $S$ and provided analogous results for interval censored data. Kalbfleisch (1978) placed a gamma process prior distribution on $H_0$ in the PH model, and derived empirical Bayes (EB) results for that model by marginalizing over the gamma process and using the marginal likelihood to obtain estimates of $\beta$. Christensen and Johnson (1988) considered the AFT model, placed a DP prior on $e^W$, marginalized over this distribution and maximized the marginal likelihood to obtain EB estimates of regression parameters. Finally, Johnson and Christensen (1989) established the analytical intractability of a fully Bayesian approach to that model.

Subsequently, Kuo and Mallick (1997) developed a Bayesian semiparametric model for AFT data by modeling $W$ with a DP mixture of normal distributions. They performed numerical approximations to posterior inferences using the basic ideas presented in Escobar (1994). Kottas and Gelfand (2001) then developed an AFT model with error distribution modeled as a DPM of split normals that was designed to have median zero and thus resulted in a regression model with $\text{med}(T \mid x, \beta) = e^{x\beta}$, a semiparametric median regression model. Then, Hanson and Johnson (2004) developed a fully Bayesian AFT model for interval censored regression data by placing a mixture of DP priors on $e^W$. While this model is analytically intractable, Hanson and Johnson were able to develop an MCMC algorithm for numerically approximating posterior distributions for all parameters of interest, including survival functions and regression coefficients. Hanson and Johnson (2002) modeled $e^W$ with a mixture of finite Polya trees (MFPT).

### Time-to-Abortion in Dairy Cattle Data Analysis

We illustrate the semiparametric AFT regression model with MFPT model for the error distribution. The model and analysis of these data were presented in Hanson and Johnson (2002). The data included $n = 1344$ dairy cattle that were observed to naturally abort their fetus prematurely. Nine herds from the central valley of California had been monitored and it was of interest to assess the relationship between two characteristics of the dam: Days open (DO), the number of days between the most recent previous birth and conception, and gravidity (GR), the number of previous pregnancies that the dam has had, and the timing to abortion. The herds were followed for 260 days; 16 dams aborted after the 260 days, and hence were right-censored. Hanson et al. (2003) also analyzed these data and determined that it was likely that the baseline densities and hazard functions were bimodal thus ruling out a standard parametric model.

The model used was:

$$\log T_{ij} = -\beta_0 - \beta_1 \text{DO}_{ij} - \beta_2 \text{GR}_{ij} - \gamma_i + W_{ij}, \quad W_{ij} \mid G \overset{\text{iid}}{\sim} G,$$

where $T_{ij}$ is the fetal lifetime of the 1344 fetuses that aborted in each of the $i = 1, \ldots, 9$ herds, with $j = 1, \ldots, h_i$ dams observed to have aborted in herd $i$.

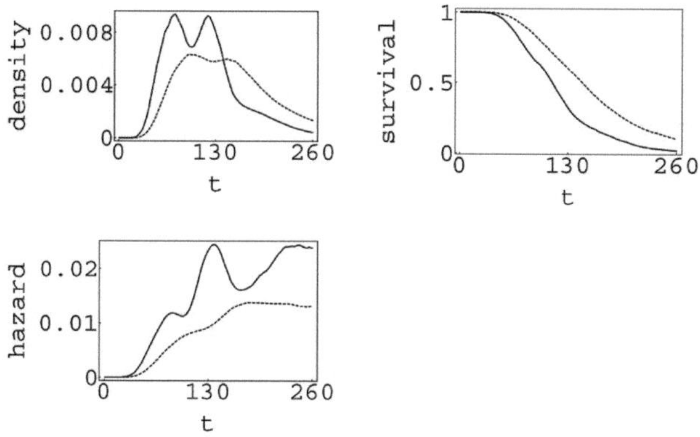


**Fig. 2.2** Predictive densities, survival curves, and hazard curves for herds 4 (*solid*) and 9 (*dashed*); here $t$ denotes time in days

The baseline $G$ was modeled as a mixture of finite ($M = 10 \doteq \log_2 1344$) Polya trees. The fixed effect for herd 1, $\gamma_1$, was fixed at zero and hence herd 1 has the baseline survival distribution. The mixture of Polya trees was centered about the family $G_\theta = \text{N}(0, \theta^2)$ and $p(\beta) \propto 1$ and the prior for $\theta$ was taken to be $\propto \theta^{-2}$. The parameter $w$ was fixed at 10, signifying relative comfort in the parametric log normal family, but small enough to allow for deviations from it. Table 2.1 displays the posterior regression effects. All probability intervals include zero, however there are herd differences. For example, fixing DO and GR, $\exp(\gamma_i - \gamma_j)$, with $j \neq i$, is the ratio of median survival times for herds $j$ and $i$. The median and 95 % probability interval for $\exp(\gamma_4 - \gamma_9)$ is 1.3 (0.9, 2.0), that is, the median time-to-abortion of herd 9 is estimated to be 1.3 times that of herd 4, with a plausible range of 0.9 to 2.0.

**Table 2.1** Posterior inference (posterior medians and 95 % probability intervals) for cow abortion data

| Parameter | Posterior median | 95 % Probability intervals |
|---|---|---|
| Intercept | $-4.79$ | $(-4.89, -4.70)$ |
| DO | $-1.1 \times 10^{-4}$ | $(-6.4 \times 10^{-4}, 3.3 \times 10^{-4})$ |
| GR | 0.01 | $(-0.01, 0.03)$ |
| $\gamma_2$ | $-0.01$ | $(-0.08, 0.05)$ |
| $\gamma_3$ | 0.00 | $(-0.12, 0.10)$ |
| $\gamma_4$ | 0.09 | $(-0.02, 0.21)$ |
| $\gamma_5$ | $-0.03$ | $(-0.14, 0.07)$ |
| $\gamma_6$ | 0.02 | $(-0.16, 0.15)$ |
| $\gamma_7$ | 0.05 | $(-0.02, 0.14)$ |
| $\gamma_8$ | $-0.01$ | $(-0.08, 0.06)$ |
| $\gamma_9$ | $-0.20$ | $(-0.56, 0.16)$ |

Figure 2.2 compares the predictive densities, survival, and hazard functions for herds 4 and 9 evaluated at the population mean values of DO and GR. The predictive survival densities are both clearly bimodal as suggested by Hanson et al. (2003). The herd 4 hazard curve peaks at 86 days and 138 days. Hanson et al. (2003) described these peaks as possibly being related to difficulty in previous calving (the first peak) and the effect of leptospirosis infection (the second peak).

### 2.4.2 Semiparametric Models for Survival Data with Time-Dependent Covariates

A number of semiparametric regression models associating survival time with time-dependent covariates (TDC), have been proposed in the literature, including models due to Cox (1972), Prentice and Kalbfleisch (1979), Aalen (1980), Cox and Oakes (1984), and Sundaram (2006), among many others. In this section, we discuss the extension of the Hanson and Johnson (2002) model, the Sundaram (2006) proportional odds model, and the Cox PH model, to include TDCs, and we discuss the Cox and Oakes (1984, Chap. 8) model—to which we refer as the COTD model—which was designed to incorporate TDCs. This work is discussed in detail in Hanson et al. (2009).

Consider the time-dependent covariate process $\{x(t) : t \in (t_1,\dots,t_k)\}$ where $t_i$s are times of observation, and $x(t)$ is the possibly vector valued observation on the TDC process. Also define $h_0$ to be an arbitrary baseline hazard, and in particular, let it correspond to an individual with constant covariate process values of zero for all times. Let $S_0(t) = \exp\{-\int_0^t h_0(s)\,\mathrm{d}s\}$ be the corresponding baseline survivor function. Prentice and Kalbfleisch (1979) extended the AFT model to TDCs as

$$h(t \mid x(t),\beta,h_0) = \mathrm{e}^{x(t)\beta} h_0(t\mathrm{e}^{x(t)\beta}), \tag{2.6}$$

and Hanson et al. (2009) termed it as the PKTD model. The TD Cox model has hazard function

$$h(t \mid h_0(t),x(t),\beta) = \mathrm{e}^{x(t)\beta} h_0(t), \tag{2.7}$$

and we will call it the CTD model. The TD covariate version of the Sundaram (2006) proportional odds model is

$$\frac{\mathrm{d}}{\mathrm{d}t}\left\{\frac{1-S(t \mid X_t)}{S(t \mid X_t)}\right\} = \mathrm{e}^{x(t)\beta}\frac{\mathrm{d}}{\mathrm{d}t}\left\{\frac{1-S_0(t)}{S_0(t)}\right\} \quad X_t = \{x(s) : s \leqslant t\}, \tag{2.8}$$

and we will call it the POTD model. A generalization of the AFT model due to Cox and Oakes (1984) is

$$S(t \mid x_t,\beta,S_0) = S_0\left(\int_0^t \mathrm{e}^{x(s)\beta}\,\mathrm{d}s\right).$$

Hanson et al. show that $S(t \mid x(t),\beta,S_0)$ for all of these models can be written as easily computable functions of $S_0$ and $\beta$.

Hanson et al. (2009, 2011b) place the same MFPT prior on $S_0$ for all of these models and their model assumes independence of $\beta$ and $S_0$; they use an improper uniform prior for $\beta$. It is however straightforward to incorporate the informative priors for $\beta$ that are discussed in Bedrick et al. (2000) for fixed covariates. This is another nice feature of this semiparametric model.

Hanson et al. (2009) analyzed the classic Stanford Heart Transplant data (Crowley and Hu 1977), and data involving cerebral edema in children with diabetic ketoacidosis. We present parts of their analyses below.

### Stanford Heart Transplant Data Analysis

These data involve the time to death from after entry into the study, which was designed to assess the effect of heart transplant on survival. Individuals entered the study and either received a donor heart at some point according to availability of an appropriate heart and a prioritization scheme, or they left the study and possibly died before a suitable heart was found. The main TDC considered was an indicator of having received a heart, yes or no, at each time $t$. The second and third TDCs were a mismatch score that indicated the quality of the match between donor and recipient hearts, which was centered at 0.5, and age at transplant (AT), which was centered at 35 years. These TDCs switched on when a heart was transplanted.

Crowley and Hu (1977) and Lin and Ying (1995) analyzed these data using the CTD and COTD models, respectively. Hanson et al. (2009) fit these models and the PKTD model using the same MFPT prior on the baseline survivor function with a log logistic base-measure. They truncated the trees at $M = 5$ levels, fixed the PT weight at one, and placed an improper constant prior on $\beta$.

Patients not receiving a new heart have TDC process for the heart transplant, age and mismatch score (MS) that are all zero for all $t$. Let $z_i$ denote the time of transplant for individual $i$ if they did receive a transplant, and define the TDCs

$$x_{i1}(t) = \begin{cases} 0, & \text{if } t < z_i, \\ 1, & \text{if } t \geqslant z_i, \end{cases}$$

and

$$x_{i2}(t) = \begin{cases} 0, & \text{if } t < z_i, \\ \text{AT} - 35, & \text{if } t \geqslant z_i, \end{cases} \quad x_{i3}(t) = \begin{cases} 0, & \text{if } t < z_i, \\ \text{MS} - 0.5, & \text{if } t \geqslant z_i. \end{cases}$$

Let $x_i(t) = (x_{i1}(t), x_{i2}(t), x_{i3}(t))^{\mathrm{T}}$. Results from the three posterior distributions are displayed in Table 2.2.

The models are decisively ranked in the order CTD, COTD, and PKTD, using the LPML criterion. The integrated Cox–Snell residual plots (not shown) were consistent with this ranking and showed nothing that could be construed as extreme lack of fit for any of the models. The CTD model shows statistical importance for status and age but not for mismatch, while the other models do not indicate the statistical importance of status.

**Table 2.2** Posterior inference (posterior medians and 95 % probability intervals) for Stanford Heart Transplant data; the PKTD and CTD models are, respectively, based on (2.6) and (2.7)

| | Model | | |
|---|---|---|---|
| Parameter | PKTD | COTD | CTD |
| Status | −1.76 (−3.86, 1.57) | −1.10 (−2.70, 0.50) | −1.04 (−1.99, −0.17) |
| AT−35 | 0.10 (−0.02, 0.26) | 0.05 (−.004, 0.13) | 0.06 ( .015, 0.11) |
| MS−0.5 | 1.63 (−0.38, 3.89) | 0.64 (−0.30, 1.52) | 0.49 (−0.09, 1.03) |
| LPML | −468.0 | −467.0 | −464.1 |

AT denotes age at transplant while MS denotes mismatch score

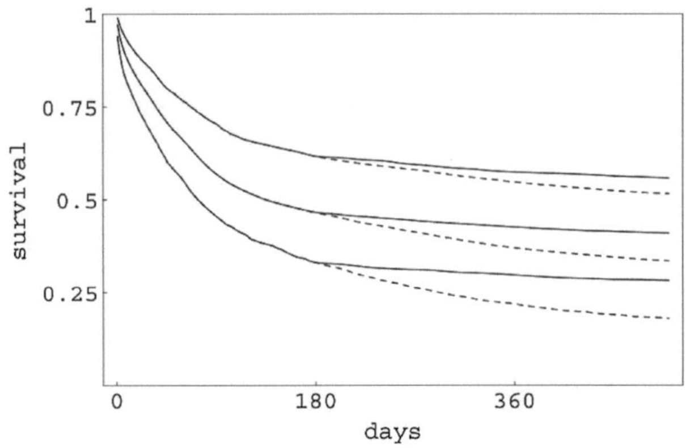


**Fig. 2.3** Estimated survival curves and 95 % probability intervals for individuals with mismatch score 0.5 and age 35. *Solid line* is for individual with a heart transplant at 6 months and *dashed line* is for an individual with no heart transplant

Under the CTD model, Hanson et al. (2009) considered two individuals aged 35 years with mismatch scores of 0.5. The first individual did not receive an HTP while the second did after 6 months. The relative hazard comparing the individual with the no heart transplant to the one with the heart transplant is of course one from time zero to 6 months, and is $e^{-\beta_1}$ from that time on. A 95 % posterior probability interval for the relative hazard after 6 months is (1.19, 7.31), and the posterior median is 2.83. Figure 2.3 displays estimated survivor curves for these two individuals, and their 95 % limits. They also fitted the MFPT with a parametric exponential base that resulted in quite different estimates of regression coefficients. The LPML for this model was −486.3, much smaller than any value in Table 2.2. Chen et al. (2014) later found an AFT model that fit the Stanford data better.

### Cerebral Edema Data Analysis

The data analyzed here were collected by Glaser et al. (2001), who assessed risk factors associated with the onset of cerebral edema (CE) in children with diabetic ketoacidosis. The description that follows is taken from Hanson et al. (2009):

> Cerebral edema is a dangerous complication associated with emergency department and in-patient hospital care of children with diabetic ketoacidosis. Children with symptoms of diabetic ketoacidosis are initially treated in the emergency department, then moved to the hospital, typically the pediatric intensive care unit, over the course of 24 h. The main purpose of treatment is to normalize blood serum chemistry and acid-base abnormalities. A major, but infrequent complication of children associated with diabetic ketoacidosis and its treatment is CE, or swelling in the brain, which may result in death or permanent neurological damage.

Hanson et al. consider only the children in that study who developed CE ($n = 58$). Their goal was to ascertain the effect of treatment procedures in time and fixed covariates on the timing of CE.

Upon admission, various treatments were recorded hourly for up to 24 h, and several initial measurements taken. The only fixed variable considered was age. Two types of TDCs are considered, the first involving the monitoring of biochemical variables over time; Hanson et al. considered serum bicarbonate (BIC) (concentration in the blood measured in mmol per liter) and blood urea nitrogen (BUN) (mg/deciliter). The second type involved actions by physicians; Hanson et al. used fluids administered (FL) (volume of fluids in ml/Kg/hour) and sodium administered (NA) (mEq/Kg/hour). None of the event times are censored. They again used the log logistic family to center the three MFPT survival models, and they set the number of levels for the finite tree to be $M = 4$ and the weight to be one. Table 2.3 gives posterior summaries of the analysis of all three models.

**Table 2.3** Posterior inference (posterior medians and 95 % probability intervals) for cerebral edema data

| | Model | | |
|---|---|---|---|
| Parameter | PKTD | COTD | CTD |
| Age (Fixed) | 0.028 $(-0.01, 0.08)$ | 0.021 $(-0.02, 0.07)$ | 0.044$(-0.02, 0.11)$ |
| Serum-BUN (TD) | $-0.005$ $(-0.02, 0.01)$ | $-0.01$ $(-0.022, 0.005)$ | 0.00 $(-0.03, 0.03)$ |
| Serum-BIC (TD) | $0.04^*$ $(-0.01, 0.13)$ | $0.05^*$ $(-0.02, 0.12)$ | $0.06^*$ $(-0.05, 0.17)$ |
| Serum-BIC$^2$ (TD) | $-0.005$ $(-0.01, 0.006)$ | $-0.006$ $(-0.02, 0.003)$ | $-0.007$ $(-0.02, 0.005)$ |
| Adm-FL (TD) | $-0.03$ $(-0.09, 0.03)$ | $-0.05$ $(-0.10, 0.02)$ | $-0.05$ $(-0.15, 0.04)$ |
| Adm-NA (TD) | $0.60^*$ $(0.16, 0.93)$ | $0.74^*$ $(0.18, 1.2)$ | $0.90^*$ $(0.19, 1.57)$ |
| FL$\times$NA (TD) | $-0.011^*$ $(-0.03, -0.00)$ | $-0.013^*$ $(-0.03, 0.001)$ | $-0.014^*$ $(-0.04, 0.003)$ |
| LPML | $-176$ | $-176$ | $-175$ |

BUN denotes blood urea nitrogen, BIC denotes bicarbonate, while NA denotes sodium administered; the PKTD and CTD models are, respectively, based on (2.6) and (2.7)

Integrated Cox–Snell residual plots did not show radical departures from the assumption of a correct model for any of the three models. Table 2.3 gives LPML values for each model, and there is no obvious distinction among the models according to this criterion. Estimates of regression coefficients for all variables in the models have the same sign and general magnitude across models. Under all models, there is a 99 % posterior probability that the coefficient for Admin-NA is positive and at least a 96 % posterior probability that the coefficient for the interaction is negative. The Serum-BIC variable has at least a 94 % probability of being positive across models; thus the effect of sodium administration appears to be modified by fluids administration. However, the estimated relative hazard under the CTD model, comparing two patients identical in all respects, including the administration of $k$ units of fluids and with the numerator patient having an increase of one unit in NA administration over the patient in the denominator, would be $\exp(0.9-0.014k)$. The effect modification of fluids is thus demonstrated. For small values of $k$, there would be little practical import.

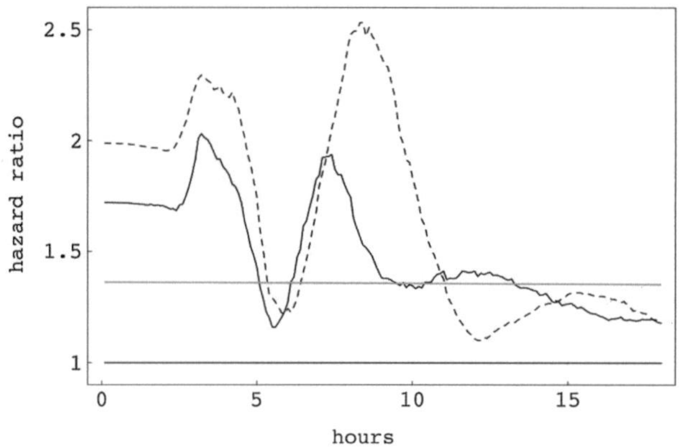


**Fig. 2.4** Cerebral edema hazard ratio for subject with NA $= 0.7$ versus NA $= 0.35$; the *black dashed*, and *solid lines* correspond, respectively, to COTD and PKTD, whereas the *solid gray line* corresponds to CTD; the PKTD and CTD models are, respectively, based on (2.6) and (2.7)

Hence, according to all models, larger values of Serum-BIC are associated with earlier diagnosis of CE. For example, under the CTD model, comparing two children that are otherwise being treated the same over a period of time and who are of the same age, the hazard of cerebral edema for a child with a larger value for BIC will be greater than for one with a lower value.

The posterior density estimates and hazard functions for time to CE corresponding to patients with specified TDC profiles are simple to obtain. Consider hypothetical patients 1 and 2 of age 10, BUN $= 35$, fluids constant at 3.6, and BIC increasing from 5 to 22, as was the case for patient 5 in the data. Figure 2.4 presents an estimated relative hazard comparing hypothetical subject 1, who has NA constant at 0.7, to hypothetical subject 2, who has NA constant at 0.35. Observe that the CTD model gives a constant relative hazard since the only difference in the two subjects is a TDC that is remaining constant over time for both subjects. According to this model, subject 1 is estimated to be about 1.35 times as much at risk of CE as subject 2 for all times. Under the PKTD and COTD models, subject 1 is usually at higher risk of CE, but the estimated relative risk varies considerably over the first 18 h. Observe the similarity of shapes of these two relative hazards, with both peaking twice.

### 2.4.3 A Nonparametric Survival Regression Model

We now discuss the approach by De Iorio et al. (2009) who model censored survival data using a DPM of linear regression models, and which can be shown to be a DDP model. We discuss their analysis of breast cancer data from a cancer clinical trial after describing the model. The model was developed because it was anticipated that survival curves for different treatments would cross each other, which would contraindicate the use of PH, AFT, and PO models.

If we were to posit a parametric survival regression model for the data, we could use the log normal, log logistic, or log extreme value families, among others. These models can be expressed as

$$\log(T) = x^{\mathrm{T}}\beta + \sigma W,$$

where $x$ is a vector of covariates with a one in the first slot for the intercept. We could let $W$ have an $\mathrm{N}(0,1)$, or $\mathrm{Logistic}(0,1)$, or an Extreme-Value(0,1) (re-parameterized to have median zero) distribution. Let $f(t \mid x, \beta, \sigma)$ be the density for an individual with covariate $x$ from one of these models, and let

$$f(t \mid x, G) = \int f(t \mid x^{\mathrm{T}}\beta, \sigma)\,\mathrm{d}G(\beta, \sigma),$$

with $G \sim \mathrm{DP}(\alpha, G_\theta)$ and $\theta \sim p(\theta)$. This is a DPM of regression models where the base of the DP can possibly have unknown parameters and where a further distribution is placed on them.

For simplicity, consider the case with a simple binary covariate, $v$, and a single continuous covariate, $z$. Then $x^{\mathrm{T}} = (1, v, z)$ takes on the values $(1, 0, z)$ or $(1, 1, z)$. So the parametric version of this model would be an analysis of covariance model in the log of the response. Let $x_i$ denote the covariate for individual $i$, for $i = 1, \ldots, n$. Then $x_i^{\mathrm{T}}\beta = \beta_0 + z_i\beta_2$ or $\beta_0 + \beta_1 + z_i\beta_2$. Let $\mathscr{X} = \{x_i : i = 1, \ldots, n\}$ and let $G_{x_i}$ be the induced distribution on $x_i^{\mathrm{T}}\beta$ that is derived from the DP distribution on $G$. The collection $\{G_{x_i} : i = 1, \ldots, n\}$ is a DDP for which the DPM distributions corresponding to the $n$ observations in the data are dependent. The model is termed a linear DDP by De Iorio et al. (2009), and interested readers can find details about the choice of $G_\theta$ and $p(\theta)$ there. Another nice feature of this model is that it can be fit in `DPpackage` (Jara et al. 2011).

### Breast Cancer Data Analysis

De Iorio et al. (2009) illustrate the proposed approach using data on 761 women from a breast cancer clinical trial. Survival times in months are the times until death, relapse, or treatment-related cancer, or censoring. Fifty three percent of the 761 observations are censored. Interest lies in determining whether a high dose of the treatment is more effective overall for treating cancer compared with lower doses. High doses of the treatment are known to be more toxic. It was hoped that the initial risk associated with toxicity would be offset by a subsequent improvement in survival prospects. The main goal of the clinical trial was to compare high versus low dose survival rates.

Two categorical covariates were considered; treatment dose ($-1 = \text{low}$, $1 = \text{high}$) and estrogen receptor (ER) status ($-1 = \text{negative}$, $1 = \text{positive or unknown}$); standardized tumor size was also considered as a continuous covariate, and an interaction between treatment and ER was also included in the model. The centering distribution was log normal.

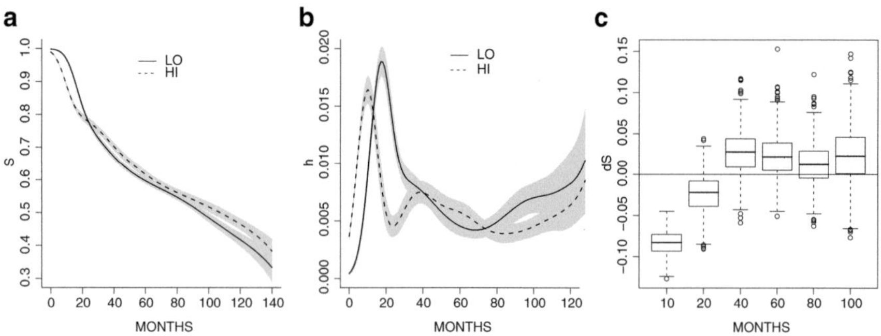


**Fig. 2.5** Inference for high versus low dose. (**a**) Estimated survivor functions (*solid lines*) along with pointwise 50 % probability intervals (*grey bands*). (**b**) Estimated hazard functions (*solid lines*) along with pointwise 50 % probability intervals (*grey bands*). (**c**) Box-plots for posterior distribution of the difference in survival rates at 10, 20, 40, 60, 80, and 100 months between a patient who receives high treatment dose versus a patient who receives the low dose. Remark: (**a**), (**b**), and (**c**) correspond to positive ER status and tumour size equal 2.0

Figure 2.5a,b show the posterior survival and hazard function estimates with their corresponding posterior uncertainty for ER positive patients with tumor size 2.0 cm (equal to the first quartile). As expected, the survivor functions corresponding to the two treatment groups cross, showing a higher level of risk associated with high treatment dose in the first 20 months. Figure 2.5c shows box plots corresponding to posteriors for the difference in survival rates between the two treatment groups for positive ER status and tumour size equal 2.0 cm, across a range of times. There is a statistically important negative effect of high dose due to toxicity early in the study, and a non-statistically important positive effect later in the study. Ultimately, the high dose treatment was abandoned as a result of the study.

## 2.5 Joint Modeling of Longitudinal and Survival Data

Many studies entail an event/survival time of interest and measurements on longitudinal processes that might be associated with patient prognosis. Examples include:

- Blood pressure measurements in dialysis patients (event: Death).
- Daily fertility counts in Mediterranean fruit flies (event: Death).

In the former case, maintaining blood pressure to be sufficiently high plays a key role in long-term prognosis for dialysis patients. In the latter case, it has been argued in the literature that life span of fruit flies might be related to fertility (see Hanson et al. 2011b, for references).

Hanson et al. (2011b) developed a general Bayesian semiparametric methodology for joint analysis. They illustrated and compared Bayesian joint models in which the survival component was taken to be the POTD, CTD, or COTD models that were discussed in Sect. 2.4.2. Comparisons were made using the LPML criterion for model selection. In each instance, baseline survival functions were modeled, as in Sect. 2.4.2, with an MFPT prior.

Two-stage procedures involve modeling the observed longitudinal processes, as, for example, was done in Sect. 2.3.1. That model is then used to predict the 'true' underlying processes, namely the process without measurement error. The predicted processes are then used as if they were the observed TDCs in fitting the time to event data with the TDC survival models discussed in Sect. 2.4.2. Subsequently, we term analyses that condition on the observed processes using LOCF (last observation carried forward) as 'raw' analyses.

Drawbacks of raw and two-stage methods motivated a considerable flourish of research on joint models for longitudinal and survival data (see Tsiatis and Davidian 2004, for a review up to that time). Bayesian approaches to joint analysis include Faucett and Thomas (1996), Wang and Taylor (2001), and Brown and Ibrahim (2003), among others. Joint modeling would appear to be a good idea since one would expect potential benefits from modeling all of the stochastic data, especially when there is the possibility of considerable measurement error, which would be the case when measuring blood pressure, and also beneficial when observations on the process are spaced out in time.

A joint analysis, on the other hand, involves simultaneously modeling longitudinal and survival data and making inferences about the effect of the true process on survival in a single stage of analysis. Let $y(t)$ be the observed vector process. This can be regarded as the vector TDC process discussed in Sect. 2.4.2, only now we consider modeling it rather than simply conditioning on it. Since we expect most processes to be observed with error, let $x(t)$ denote the 'true' (vector) process. In the absence of measurement error $y(t) = x(t)$.

A joint model involving a single process proceeds as follows. All of the models considered involve a baseline survivor curve, $S_0$, and a regression coefficient vector, $\beta$. In each instance, we specify

$$S_0 \mid \theta \sim \text{MFPT}(M, c, G_\theta), \quad \theta \sim p(\theta), \quad p(\beta) \propto \text{const},$$

namely the baseline survivor function has an MFPT prior and the regression coefficients have an improper constant prior distribution. The data consist of $\{(T_i, y_i, t_i) : i = 1, \ldots, n\}$ where $T_i$ is the minimum of the event time and the censoring time, $t_i$ is the vector of observed times, and $y_i$ is the corresponding vector of observations on the process $y_i(t)$, for individual $i$, in a sample of size $n$. We assume that $x_i(t)$ is the 'true' process and that

$$y_i(t) = x_i(t) + \varepsilon_i, \quad \varepsilon_i \sim F_\lambda.$$

If we let $F_\lambda$ be the distribution function of an $\mathrm{N}(0, \sigma^2)$ distribution, one can glean the particular $x_i(t)$ for model (2.1) in Sect. 2.3.1.

Survival modeling is conditional on the longitudinal process. We model the survivor function for individual $i$, $S_i(t \mid x(t_i), S_0, \beta)$, using the POTD, CTD, and COTD models discussed in Sect. 2.4.2, and where $x(t_i) = (x(t_{i1}), \ldots, x(t_{ih_i}))^{\mathrm{T}}$. From this, we know the form of the hazard function and the density. Assuming a parametric model of the form $f(x_i \mid \Delta)$, then the full joint model for a non-censored observation is expressed as

$$f(T_i, y_i \mid x_i, \lambda, S_0, \beta) = f(T_i \mid x_i, S_0, \beta) f(y_i \mid x_i, \lambda) f(x_i \mid \Delta).$$

If an observation is censored, replace $f(T_i \mid x_i, S_0, \beta)$ with $S(T_i \mid x_i, S_0, \beta)$, making the usual assumption that event times and censoring times are independent. We have made the assumption that $T_i$ is conditionally independent of the observed process given the true process and the parameters. Details on inference can be found in Hanson et al. (2011b).

### 2.5.1 Medfly Data Analysis

The data used for illustration came from a study reported in Carey et al. (1998) and further analyzed by Chiou et al. (2003) and Tseng et al. (2005). Tseng et al. (2005) analyzed a sample size of 251 Mediterranean fruit flies with lifetimes ranging from 22 to 99 days. The number of eggs produced per day was recorded throughout their lifespan. We removed the first 2 days from each trajectory since all flies have zero counts on those days.

We present some of the analysis presented in Hanson et al. (2011b). Our case study makes the point that joint or two-stage modeling may not predict as well as simply conditioning on the 'raw' process, for these data. For comparison with the analysis by Tseng et al. (2005), Hanson et al. used the same longitudinal model as they did, as well as some additional more flexible alternatives. Tseng et al. let $y_i(t) = \log\{N_i(t) + 1\}$, the natural log of one plus the number of eggs laid on day $t$, and modelled trajectories as

$$y_i(t) | (b_{i1}, b_{i2}), \tau \sim \mathrm{N}(b_{i1} \log(t) + b_{i2}(t-1), \tau^{-1}), \;\; (b_{i1}, b_{i2}) \mid \mu, \Sigma \overset{\text{iid}}{\sim} \mathrm{N}_2(\mu, \Sigma).$$

where the mean is a log gamma function. Since there are no additional covariates for survival, a single regression coefficient $\beta$ connects the survival model to the longitudinal process $x_i(t) = b_{i1}\log(t) + b_{i2}(t-1)$. The MFPT models used here set $M = 4$ and $c = 1$, with flat priors otherwise. About 16 observations fall into each of the 16 sets at level $M = 4$ if the log logistic family is approximately correct. They also considered the prior $c \sim \text{Gamma}(5,1)$ for a subset of models, obtaining LPML values slightly smaller than with fixed $c = 1$.

All models were fitted with both the MFPT with weight $c = 1$, and parametric log logistic model, corresponding to a weight that grows without bound. According to the LPML statistics presented in Table 2.4, the COTD model performs the worst in this data analysis, regardless of the method used to incorporate the longitudinal predictor (e.g., raw versus modeled) or whether parametric versus MFPT for $S_0$ was assumed. For the two types of raw analysis, the flexibility obtained from an MFPT generalization of the log logistic model improves predictive performance, though not dramatically so. Moreover, it is also clear that two-stage and joint methods predict almost identically but are inferior to simple raw analysis in this setting. Observe from Table 2.5 that point estimates of $\beta$ under the POTD model are similar across types of analysis and that they are different for the COTD model.

From Table 2.4, the general conclusions about predictive model comparison are that a raw LOCF analysis is preferred to two-stage or joint methods, the POTD model is preferred over the COTD and CTD models, and that the COTD model might be excluded from further consideration. On the other hand, Tseng et al. (2005) rejected the CTD model based on a test involving Schoenfeld residuals and proposed the COTD model as a plausible alternative. Hanson et al. (2011b) discuss why these data might not be ideal for joint or two-stage modeling beyond the analysis performed here.

**Table 2.4** LPML across models (larger is better) for medfly data; the POTD and CTD are, respectively, based on (2.8) and (2.7)

| | | Model | | |
|---|---|---|---|---|
| Inference | Method | POTD | CTD | COTD |
| Parametric | Raw | −867 | −870 | −937 |
| MFPT | Raw | −865 | −866 | −938 |
| MFPT | Two-stage | −947 | −959 | −973 |
| Parametric | Joint | −947 | −959 | −973 |
| MFPT | Joint | −945 | −956 | −973 |

Hanson et al. (2011b) also pointed out that not all of the egg count trajectories fit the log gamma structure that is posited for these data. Consequently, they considered a more flexible longitudinal model that represents a compromise between the Tseng et al. approach and using the empirical egg counts (LOCF). They considered a B-spline longitudinal model in conjunction with the POTD model, which resulted in the largest LPML among all models considered, namely LPML $= -879$ for the parametric joint model, worse than parametric raw but much better than using the basis $\{\log(t), t-1\}$. Hanson et al. also argue that preference for the POTD model over the CTD model in our analysis is tantamount to an acceptance that a change in egg laying behavior at a particular time is eventually forgotten.

**Table 2.5** Posterior inference (posterior medians and 95 % probability intervals) across models for medfly data; the POTD and CTD are, respectively, based on (2.8) and (2.7)

| | Model | | |
|---|---|---|---|
| Method | POTD | CTD | COTD |
| Par/Raw | −0.75 (−1.02,−0.53) | −0.65 (−0.74,−0.56) | −0.36 (−0.44,−0.27) |
| MFPT/Raw | −0.74 (−0.85,−0.64) | −0.64 (−0.73,−0.55) | −0.37 (−0.45,−0.29) |
| MFPT/Two-Stage | −0.74 (−0.97,−0.52) | −0.37 (−0.52,−0.24) | 0.16 (−0.01, 0.30) |
| Par/Joint | −0.78 (−1.02,−0.53) | −0.39 (−0.54,−0.25) | 0.19 ( 0.01, 0.33) |
| MFPT/Joint | −0.79 (−1.00,−0.52) | −0.40 (−0.54,−0.24) | 0.19 ( 0.01, 0.32) |

## 2.6 Medical Diagnostic Data

### *2.6.1 ROC Regression*

We consider the quality of a medical diagnostic test for its ability to discriminate between alternative states of health, generally referred to diseased/infected $(D+)$ and non-diseased/infected $(D-)$ states. In many settings of clinical interest, covariates can be used to supplement the information provided by a biomarker, and thus can help to discriminate between $D+$ and $D-$. For example, consider diabetes testing, where blood glucose levels are used to diagnose individuals with diabetes. The covariate, age, plays a key role as older subjects tend to have higher levels of glucose, without that necessarily meaning that there is a higher incidence of diabetes at greater ages. However, since the aging process is believed to be associated with relative insulin deficiency or resistance among the $D-$ individuals, it is relevant to adjust for age in the analysis; see Inácio de Carvalho et al. (2013) and the references therein. The general area we now discuss is called ROC regression.

But first briefly consider the no covariate case using a diagnostic marker $T$. It might be continuous, or dichotomous. If it is dichotomous, the marker outcomes are $T+$, or yes, the individual tested has the infection/disease, or $T-$, or no, they don't. In the case of a continuous marker, a cutoff, $c$, is selected and, without loss of generality, if the marker value exceeds the cutoff, the outcome is $T+$, and is $T-$ otherwise. In either case, observing the yes/no outcome is called a diagnostic test. The quality of the test is determined by considering two types of test accuracy. The sensitivity of the test is defined to be $\text{Se} = \Pr(T+ \mid D+)$, the proportion of the time that the test says yes when it should, and the specificity, $\text{Sp} = \Pr(T- \mid D-)$, the proportion of time the test says no when it should. In the continuous case, we write

Se$(c)$, and Sp$(c)$, and in this case, it is common to plot the false positive rate versus the true positive rate across all possible cutoffs. The ROC curve for a continuous biomarker is thus the plot $\{(1-\text{Sp}(c), \text{Se}(c)) : \text{for all } c\}$. It is possible to re-write this plot as $\{\text{ROC}(t) : t \in [0,1]\}$, where $\text{ROC}(t) = 1 - F_{D+}\{F_{D-}^{-1}(1-t)\}$, (Pepe 2003, Chap. 4), where $F_{D+}(\cdot)$ and $F_{D-}(\cdot)$ are the distribution functions for $D+$ and $D-$ individuals. We now extend this to include adjustment for covariates, $x$.

The key object of interest for modeling in Sect. 2.6.2 is the covariate-adjusted ROC curve, which can be defined just as in the no covariate case, only now Se$(c)$ and Sp$(c)$ are allowed to depend on covariates, $x$. So for every $x$, we have an ROC curve. Here, we define the three-dimensional ROC surface:

$$\{(t, x, \text{ROC}(t \mid x)) : t \in [0,1], x \in \mathbb{R}^p\},$$

where

$$\text{ROC}(t \mid x) = 1 - F_{D+}\{F_{D-}^{-1}(1-t \mid x) \mid x\}. \tag{2.9}$$

We now have two conditional distributions that are allowed to depend on covariates. They may depend on distinct covariates, or one may depend on covariates and the other not. The covariate-adjusted AUC is defined as

$$\text{AUC}(x) = \int_0^1 \text{ROC}(u \mid x)\,\text{d}u,$$

and will be used as our preferred summary measure of covariate-adjusted discriminative power.

In some cases a 'perfect' or gold-standard (GS) test exists, i.e., a test that correctly classifies the subjects as $D+$ and $D-$. In this case, data consist of two samples, one known to be $D+$ and the other known to be $D-$. Observed outcomes for each unit consist of the pair

$$\{\text{Test Covariates}, \text{Test Scores}\};$$

we denote test covariates as $x$. A test score is a continuous diagnostic marker outcome, and a test covariate is simply a covariate that is, at least believed to be, related to a test score. With GS data, the model is identifiable regardless of the amount of separation between $F_{D+}$ and $F_{D-}$; the case where a gold standard test exists is considered in detail in Chap. 16 (Inácio de Carvalho et al. 2015).

Section 2.6.2 focuses on ROC regression for the no gold-standard (NGS) case, thus there is no direct information on whether individual subjects in a study are $D+$ or $D-$. The data consist of a single mixed sample with disease status unknown. The NGS setting typically involves identification issues. However, if there are covariates that allow us to learn about the probability of disease, the model is identifiable under mild assumptions (see Branscum et al. 2015, Appendix 1). We refer to these as disease covariates, and denote them as $x^*$. Hence in this setting we assume that data consist of the triple,

$$\{\text{Disease Covariates}, \text{Test Covariates}, \text{Test Scores}\}.$$

The model discussed in Sect. 2.6.2 was proposed by Branscum et al. (2015), and it was built on the principle that Disease Covariates can be used to mitigate identification issues in the NGS setting. See Branscum et al. (2013) for another GS approach to this problem, and also see Branscum et al. (2008) for an approach that develops much of the machinery used here.

### 2.6.2 A Semiparametric ROC Regression Model in the Absence of a Gold Standard Test

Here we assume there are no test covariates available for $D-$ subjects. For $D+$ subjects we specify the model $Y_{D+} = x^{\mathrm{T}}\beta + \varepsilon_{D+}$, where $x$ is a test covariate, $\beta$ is a coefficient vector, and $\varepsilon_{D+} \sim F_{\varepsilon_{D+}}(\cdot)$. With this specification, (2.9) can be rewritten as

$$\mathrm{ROC}(t \mid x) = 1 - F_{\varepsilon_{D+}}\{F_{D-}^{-1}(1-t) - x^{\mathrm{T}}\beta \mid x\},$$

by noting that $F_{\varepsilon_{D+}}(y - x^{\mathrm{T}}\beta) = F_{D+}(y \mid x)$.

Suppose continuous marker scores $(y_i)$ are obtained on $n$ randomly sampled individuals from a population. Then let $x_i^*$ denote the disease covariate outcome, and let $z_i$ denote latent disease status for subject $i$, with $z_i = 1$ if they are $D+$, and $z_i = 0$ otherwise. Define $\pi_i$ as the probability that subject $i$ is $D+$, for $i = 1, \ldots, n$. The latent $z_i$s are independent and $\mathrm{Bern}(\pi_i)$, with $\pi_i = G_0(x_i^{*\mathrm{T}}\alpha)$, with $\alpha = (\alpha_0, \ldots, \alpha_s)^{\mathrm{T}}$ and where $G_0$ is a standard distribution function, like normal, or logistic. These choices result in probit and logistic regression models for the $z_i$s. Test scores are modeled according to a mixture distribution with conditional density,

$$f(y_i \mid z_i, x_i) = z_i f_{\varepsilon_{D+}}(y_i - x_i^{\mathrm{T}}\beta) + (1 - z_i) f_{D-}(y_i),$$

where $\beta = (\beta_0, \ldots, \beta_p)^{\mathrm{T}}$, $f_{\varepsilon_{D+}}$ is the density associated with $F_{\varepsilon_{D+}}$, and $f_{D-}$ is the density associated with $F_{D-}$. The model for $D-$ subjects can also depend on covariates; test and disease covariates may overlap.

The nonparametric part of the model involves placing independent MFPT priors on $F_{\varepsilon_{D+}}$ and $F_{D-}$; here, $F_{\varepsilon_{D+}}$ is constrained to have median zero to alleviate confounding between $\beta_0$ and the location of $F_{\varepsilon_{D+}}$ (Hanson and Johnson 2002). Since the marker was log transformed, the MFPTs were centered on normal families, the former family having mean zero and the latter having an arbitrary mean. Weights for the PTs were either specified to be one, or given a diffuse gamma distribution. Parametric priors were placed on all hyperparameters. See Branscum et al. (2015) for further details.

### Lung Cancer Data Analysis

Branscum et al. (2015) investigated the potential of a soluble isoform of the epidermal growth factor receptor (sEGFR) to be considered as a diagnostic biomarker for lung cancer in men. The data were gathered a case-control study that was conducted at the Mayo Clinic. The data included 88 controls and 139 lung cancer cases; see Baron et al. (1999, 2003) for further details. Branscum et al. (2015) analyzed the data as if disease status was unknown and used these data to assess the impact of age on the discriminatory ability of sEGFR to distinguish cases and controls. Age was used as a test covariate for controls, and as a disease covariate. They also analyzed the data using known disease status in a GS analysis of the same data for comparative purposes.

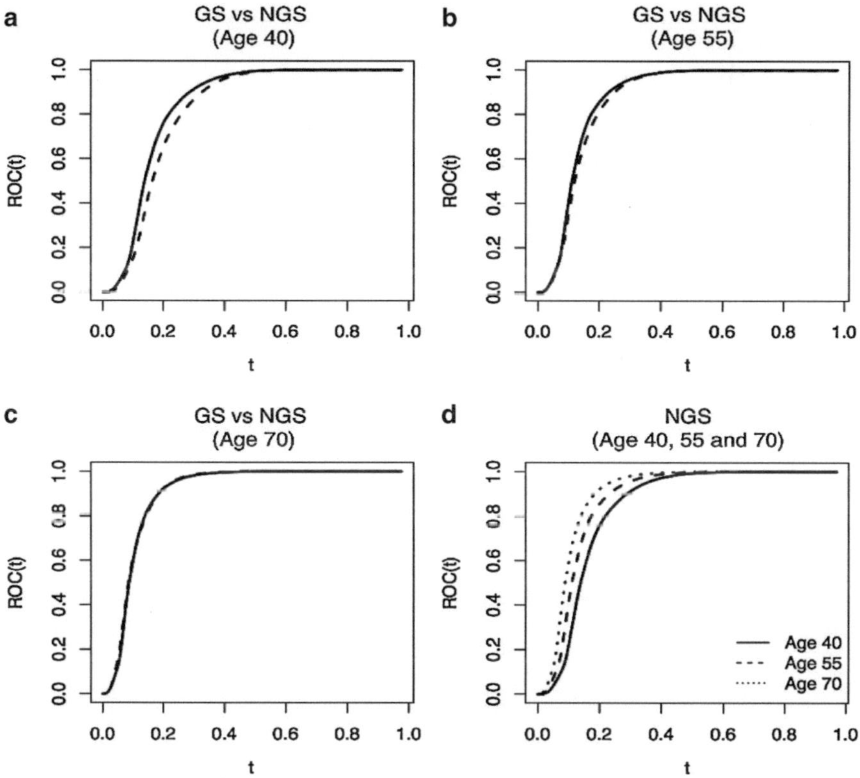


**Fig. 2.6** GS and NGS semiparametric estimates of covariate adjusted ROC curves for ages 40, 55, and 70

The sampling model for the natural log transformed test scores and latent diseased status was:

$$z_i \sim \text{Bern}(\pi_i), \quad \log\left(\frac{\pi_i}{1-\pi_i}\right) = \alpha_0 + \alpha_1 x_i^*,$$
$$f(y_i \mid z_i, x_i) = z_i f_{\varepsilon_{D+}}(y_i - \beta_0 - \beta_1 x_i) + (1 - z_i) f_{D-}(y_i).$$

In Fig. 2.6 we plot semiparametric estimates of the covariate-adjusted ROC curves corresponding to ages 40, 55, and 70. Posterior inferences for covariate-adjusted AUCs for the same ages are displayed in Table 2.6. It is clear that it is easier to diagnose lung cancer in older men than in younger men, and that the NGS analysis provides a reasonable approximation to the GS analysis for these data. As expected, interval inferences are less certain in the NGS case than in the GS case.

LPML and corresponding pseudo Bayes factors were used to compare parametric and semi-parametric models. In the NGS setting, the LPML for the parametric normal model was −439, which was larger than the values for all semi-parametric models considered. The largest LPML statistic for all models considered was −422,

**Table 2.6** Posterior inference (posterior medians and 95 % probability intervals) for the covariate-adjusted AUCs corresponding to ages 40, 55, and 70 based on GS and NGS analyses of the lung cancer data

| | Analysis | |
|---|---|---|
| Parameter | GS | NGS |
| $AUC_{40}$ | 0.78 (0.72, 0.84) | 0.79 (0.71, 0.86) |
| $AUC_{55}$ | 0.83 (0.77, 0.88) | 0.83 (0.75, 0.89) |
| $AUC_{70}$ | 0.87 (0.81, 0.92) | 0.86 (0.77, 0.92) |

for a model with the two MFPTs truncated at four levels and with both weights equal to one. Compared to the parametric model, the pseudo Bayes factor of $e^{17}$ provides strong evidence in favor of the selected semi-parametric model.

### 2.6.3 *Joint Longitudinal Diagnostic Outcome Modeling and Analysis*

Most diagnostic outcome data are cross-sectional, as was the case in the previous section. A main goal in those studies was to estimate sensitivity and specificity of one or more biomarker outcomes over a range of cutoffs, resulting in an estimate of the ROC curve. With cross-sectional data, by definition, sampled individuals include a cross-section of the population. Individuals in this population are either diseased/infected, $D+$, or not, and if they are $D+$, there will be a range of times at which the disease/infection was acquired. For many such maladies, the ability to detect will very much depend on the time of acquisition. For example, it is practically impossible to detect HIV infection in the near term after infection. However, after some time has passed, ELISA and Western Blot tests are able to detect it. If the cross-sectional sample happened to include only newly infected individuals, the estimated sensitivity of the test would be quite low. The purpose of developing the model discussed below was to consider longitudinal or prospective diagnostic outcome data so that it would be possible to estimate the sensitivity of a dichotomous outcome test as a function of time from infection. A major difficulty faced in this endeavor is that it would rarely be known precisely when individuals in a population or sample become infected, or even in many instances if they had become infected. If a perfect/gold standard test is applied, the actual disease status could be known, but not the exact timing. The model developed below does not assume a gold standard and as a result, the latent status and timing of infection/disease are modeled.

Norris et al. (2009, 2014) developed a model for repeated observations in time on a yes/no diagnostic test outcome and a continuous biomarker for a disease. They analyzed longitudinal fecal culture and continuous serum ELISA outcomes for mycobacterium avium paratuberculosis (MAP), the causal agent for Johne's disease in dairy cattle. We discuss their model and analysis in the context of the cow data, but the model would apply to many other data sets as suggested by Norris et al. (2009, 2014).

Once an animal is infected, it is expected that, after some delay, serum antibody outcomes will increase. If animals are being monitored in time, as they are in the cow data set, antibodies should increase to a point that the ELISA outcome exceeds a cutoff, and thus becomes positive for MAP. If an animal is not infected during the study, their ELISA outcomes should remain steady but variable around some baseline value that depends on the cow. The model includes a latent disease status indicator for all cows, and a change point corresponding to time of infection, $t^*$, for animals with a positive disease indicator. The probability of a positive fecal test changes at the time of infection, but the rise in serology score occurs some time later. Norris et al. noted that there was literature that pointed to a 1 year lag after infection. Nonetheless, they modeled lag as an unknown parameter. After the lag, increase in antibodies was modeled to be linear. They also assumed that fecal and serology results are independent for several reasons discussed in their paper. The model takes account of the fact that the fecal test is viable soon after infection whereas the production of detectable serum antibodies involves a lag.

The model incorporates three latent states: (1) no infection during the entire screening period, (2) infection, but insufficient time to mount an antibody reaction during screening period (since "lag" has not elapsed when screening ends), and (3) infection with antibody reaction within screening period (since "lag" elapses before the end of screening period). They define the variable, $k_i \in \{1,2,3\}$, to denote the latent disease state of cow $i$, and they define $t_{ij}$ to be the time of the $j$th screening for the $i$th subject; $(S_{ij}, F_{ij})$ are the serology and fecal culture outcomes of the $i$th subject at time $t_{ij}$; $\mathrm{Se}_F$ is the sensitivity of fecal culture; $\mathrm{Sp}_F$ is the specificity of fecal culture;lag is the time interval between infection and serology reaction, $\Theta$ denotes vector of all model parameters, and $U$ is the vector of all model latents. Figure 2.7 describes the model, discussed below, for a cow with $k_i = 3$.

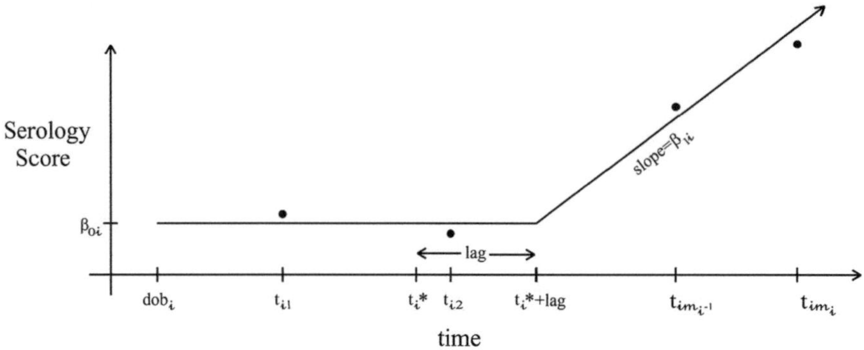


**Fig. 2.7** Serology trajectory with data for cow with $k_i = 3$

The models for cows in latent states 1 and 2 are:

$$S_{ij} \mid \Theta, U, k_i = 1 \sim \beta_{0i} + \varepsilon_{ij}, \quad \perp \quad F_{ij} \mid \Theta, U, k_i = 1 \sim \text{Bern}(1 - \text{Sp}_F),$$
$$S_{ij} \mid \Theta, U, k_i = 2 \sim \beta_{0i} + \varepsilon_{ij} \quad \perp \quad F_{ij} \mid \Theta, U, k_i = 2 \sim \text{Bern}(\pi_{ij}),$$

where $\beta_{0i} \overset{\perp}{\sim} \text{N}(\beta_0, \tau_{\beta_0})$, $\varepsilon_{ij} \overset{\perp}{\sim} \text{N}(0, \tau_e)$, $\beta_{0i} \perp \varepsilon_{ij}$, and $\pi_{ij} = I(t_{ij} \geqslant t_i^*)\text{Se}_F + I(t_{ij} < t_i^*)(1 - \text{Sp}_F)$ for all $i, j$. The model for cows in latent state 3 incorporates a random cow-specific slope for the post-lag serology trajectory, allowing for differing rates of antibody production among infected cows. The function $z^+$ equals $z$ if $z > 0$ and 0 otherwise. The model is:

$$S_{ij} \mid \Theta, U, k_i = 3 \sim \beta_{0i} + \beta_{1i}(t_{ij} - t_i^* - \text{lag})^+ + \varepsilon_{ij}, \ \perp \ F_{ij} \mid \Theta, U, k_i = 3 \sim \text{Bern}(\pi_{ij}),$$

with $\beta_{1i}$,$\beta_{0i}$, and $\varepsilon_{ij}$ pairwise independent; $\beta_{1i}$ is zero until $t_{ij} = t_i^* +$ lag. Hence, the mean serology trajectory is a flat line until $t_i^*$+lag, then it increases linearly with slope $\beta_{1i}$ as shown in Fig. 2.7. We refer the interested reader to Norris et al. for details about the change points, which were modeled with uniform distributions over appropriate ranges, and the disease status variable, which is a simple multinomial for each cow but requires reversible jump methodology to handle the fact that, from one iteration to the next of the Gibbs sampler, the dimension of the parameter space changes according to the (latent status) multinomial outcomes for all $n$ cows.

Norris et al. (2009) analyzed the cow data using the above parametric model, and Norris et al. (2014) extended this model to allow for a DPM of slopes for $k_i = 3$ type cows. The scientific motivation for this was because it was believed that some infected cows may have a more gradual slope, while others a steeper slope after the infection time plus lag. Thus a DPM of slopes will allow for groups of cows with different slopes. Since biology also dictates that antibody slopes must be *non-decreasing* after infection slopes were constrained to be positive by modeling the log-slope as a DPM of normals as follows:

$$\begin{aligned}
\log \beta_{1i} = \gamma_i \mid \mu_i, \tau_i &\overset{\perp}{\sim} \mathrm{N}(\mu_i, \tau_i), \quad \text{for } i : k_i = 3, \\
(\mu_i, \tau_i) \mid G &\overset{\perp}{\sim} G, \\
G \mid \alpha, G_0 &\sim \mathrm{DP}(\alpha, G_0),
\end{aligned}$$

which can also be expressed as

$$\gamma_i \mid G \overset{\perp}{\sim} \int \mathrm{N}(\cdot|\mu_i, \tau_i) G(\mathrm{d}\mu_i, \mathrm{d}\tau_i), \quad G \mid \alpha, G_0 \sim \mathrm{DP}(\alpha, G_0).$$

Let $(n_1, n_2, n_3)$ be the latent numbers of cows in each of the three latent states. Since $G$ is discrete with probability one, at any given iteration of the Gibbs sampler, there will be, say $r$, clusters of distinct values among the $n_3$ realizations of $\theta_i = (\mu_i, \tau_i)$. Cows associated with each of these clusters will have different slopes. At the end of an MCMC run, cows will be belonged to different clusters and corresponding slopes will have changed from iteration to iteration. It is possible to monitor the number of clusters, and the number of modes, at each iteration of the Gibbs sampler and Norris et al. report those results, some of them reproduced below. However, it is impossible to define particular clusters precisely over the entire MCMC sample, due to lack of identifiability of the individual components in the DPM. Nonetheless, through post processing of output, it is possible to allocate cows to clusters that are associated with particular modes in the slope distribution for infected cows using ad hoc methods. The data analysis discussed below uses such a method to make inferences about the sensitivity of the ELISA test, with a particular cutoff, as a function of time since infection for groups of cows deemed to have distinct slopes.

### Analysis of Longitudinal Cow Serology and Fecal Culture Data

The estimated sensitivity and specificity of the FC test were 0.57 (0.52, 0.63) and 0.976 (0.955, 0.990), respectively. The FC test is known to be highly specific. The estimated proportions of animals falling into the three latent status groups is (0.048, 0.25, 0.26), thus the estimated prevalence of MAP in the population sampled at the end of the study is 0.52. The estimated lag is 1.60 (1.32, 1.85), in years.

Figure 2.8 shows some iterates from the posterior log slope distribution; some are bimodal with global maximum near zero and a smaller mode less than zero. The posterior distribution of the number of modes showed a 0.62 probability of one and 0.30 of two modes.

ROC curves at selected times past infection for estimated high and low serology reaction groups are displayed in Fig. 2.9a. By analyzing the posterior iterates of the log-slope distribution shown in Fig. 2.8, Norris et al. obtained rough estimates of the mean and standard deviation of the high and low clusters. Many of the iterates suggest the low cluster is centered around $-1.6$ with a standard deviation of about 0.4 and the higher cluster is centered at about 0.6 with standard deviation of 0.9. The curves depicted in Fig. 2.9 show that discriminatory ability is very poor in the

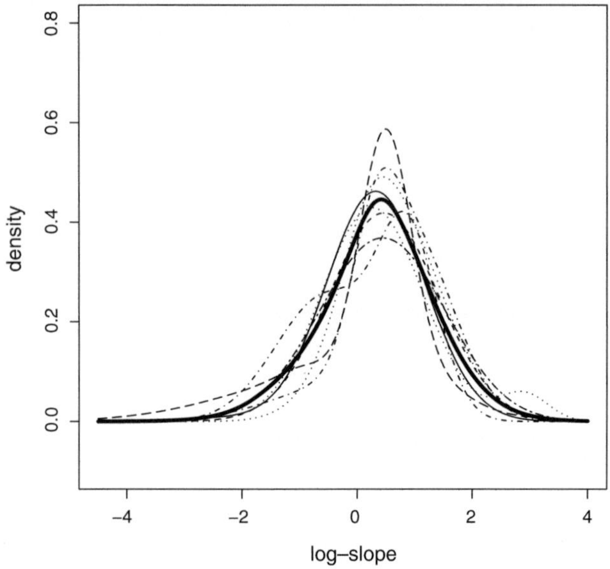


**Fig. 2.8** Posterior iterates of log-slope distribution, with posterior mean in *bold*, for cow serology and fecal culture data

hypothetical low group, and can be very good in the hypothetical high group, and is especially so the longer it has been since infection.

The corresponding graph for low and high groups depicting estimated sensitivity of the dichotomized ELISA as a function of time past infection is shown in Fig. 2.9b. There is a large difference in performance of the ELISA between these two groups.

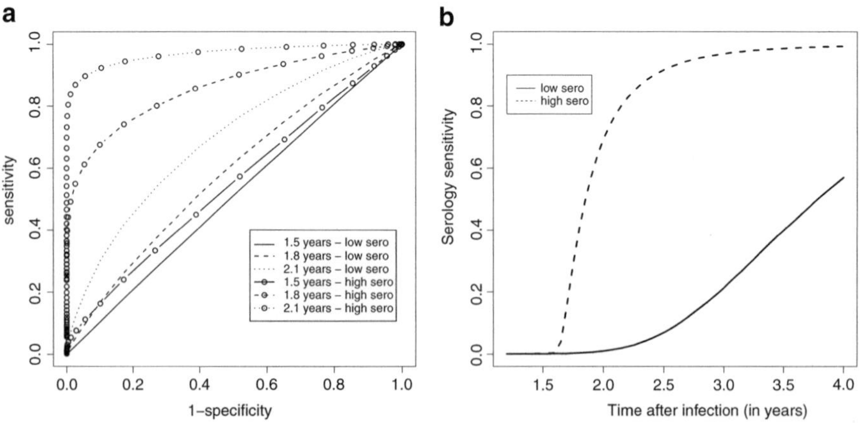


**Fig. 2.9** (**a**) Estimated ROC curves for Johne's disease data for hypothetical groups at selected values of time past infection. (**b**) Estimated sensitivity as a function of time for hypothetical high and low serology groups, with a cutoff level of −1.29

At 3 years past infection, the ELISA applied to the ‘low’ group has estimated sensitivity less than 0.20, while it is one in the ‘high’ group.

More sophisticated methods of post processing allocation to clusters have been developed by Dahl (2006) and Bigelow and Dunson (2009).

## 2.7 Final Remarks

What is statistics all about? As put simply by A. Wald:

> The purpose of statistics,. . . , is to describe certain real phenomena.
> Wald (1952)

Different real phenomena lead us to different types of data, and beyond the ones we have seen above (survival, longitudinal, and medical diagnostic data) there is a wealth of other options arising naturally in biostatistics. These include, for instance:

- **Binary Diagnostic Outcome Data**: Binary diagnostic outcome data are ubiquitous in human and veterinary medicine. While many Bayesian parametric models have been developed, there appears to be a paucity of Bayesian nonparametric approaches in this setting.
- **Compositional Data**: Nonnegative-valued variables constrained to satisfy a unit-sum constraint also find their application in biostatistics. This type of data is known as compositional data; for an application in biostatistics, see Faes et al. (2011), who analyze the composition of outpatient antibiotic use through statistical methods for unit simplex data. Bernstein polynomial-based approaches are tailored for this setting; see, for instance, Petrone (1999) and Barrientos et al. (2015), and the references therein.
- **Functional Data**: Recent advances in technology have led to the development of more sophisticated medical diagnostic data, and, nowadays, applications where measurements are curves or images are becoming commonplace. Dunson (2010, Sect. 3) overviews some recent Bayesian nonparametric approaches for modeling functional data.
- **Missing Data**: In a recent paper at the *The New England Journal of Medicine*, Little et al. (2012) discuss how missing data can compromise inferences from clinical trials. In Chap. 21 (Daniels and Linero 2015) this important subject is considered in detail. An important question that remains after our chapter is: Can we conduct reliable inferences based on the prior processes discussed above, if we have missing data? In terms of Polya trees, Paddock (2002) provides an approach for multiple imputation of partially observed data. Imputation via the Bayesian bootstrap—which can be regarded as a non-informative version of the DP (Gasparini 1995, Theorem 2)—has also been widely applied; more details on the Bayesian bootstrap can be found in Chap. 16 (Inácio de Carvalho et al. 2015).
- **Spatial Data**: This is the subject of Part V of this volume.

- **Time Series Data**: Connected with the topic of longitudinal data is also that of time series data. In this direction some recently proposed models include Nieto-Barajas et al. (2012), Jara et al. (2013), and Nieto-Barajas et al. (2014).

This list continues with multivariate data, shape data, and many more topics, including combinations of the different types of data; see, for example, Chap. 11 (Zhou and Hanson 2015), where models for spatial-survival data are discussed.

We close this introductory part of *Nonparametric Bayesian Methods in Biostatistics and Bioinformatics* with the hope that the next chapters stimulate interaction between experts in Bayesian nonparametric biostatistics and bioinformatics, and that they are useful for those entering this important field of research.

**Acknowledgements** We thank the Editors for the invitation, and we are indebted to our 'partners in crime,' including Adam Branscum, Ron Christensen, Ian Gardner, Maria De Iorio, Alejandro Jara, Prakash Laud, Michelle Norris, Fernando Quintana, Gary Rosner, and Mark Thurmond. Special thanks go to Vanda Inácio de Carvalho, Tim Hanson, and Peter Müller, who made substantive contributions to the penultimate draft of this paper, in addition to their contributions to the work presented. M. de. C was supported by Fondecyt grant 11121186.

## References

Aalen, O. (1980). A model for nonparametric regression analysis of counting processes. In: *Mathematical Statistics and Probability Theory, Lecture Notes in Statistics*, vol. 2, pp. 1–25. New York: Springer.

Amewou-Atisso, M., Ghosal, S., Ghosh, J. K., and Ramamoorthi, R. (2003). Posterior consistency for semi-parametric regression problems. *Bernoulli*, **9**, 291–312.

Baron, A. T., Lafky, J. M., Boardman, C. H., Balasubramaniam, S., Suman, V. J., Podratz, K. C., and Maihle, N. J. (1999). Serum sErbB1 and epidermal growth factor levels as tumor biomarkers in women with stage III or IV epithelial ovarian cancer. *Cancer Epidemiology Biomarkers and Prevention*, **8**, 129–137.

Baron, A. T., Cora, E. M., Lafky, J. M., Boardman, C. H., Buenafe, M. C., Rademaker, A., Liu, D., Fishman, D. A., Podratz, K. C., and Maihle, N. J. (2003). Soluble epidermal growth factor receptor (sEGFR/sErbB1) as a potential risk, screening, and diagnostic serum biomarker of epithelial ovarian cancer. *Cancer Epidemiology Biomarkers and Prevention*, **12**, 103–113.

Barrientos, A. F., Jara, A., and Quintana, F. A. (2015). Bayesian density estimation for compositional data using random Bernstein polynomials. *Journal of Statistical Planning and Inference (DOI: 10.1016/j.jspi.2015.01.006).*

Bedrick, E. J., Christensen, R., and Johnson, W. (1996). A new perspective on priors for generalized linear models. *Journal of the American Statistical Association*, **91**, 1450–1460.

Bedrick, E. J., Christensen, R., and Johnson, W. O. (2000). Bayesian accelerated failure time analysis with application to veterinary epidemiology. *Statistics in Medicine*, **19**, 221–237.

Berger, J. O. and Guglielmi, A. (2001). Bayesian and conditional frequentist testing of a parametric model versus nonparametric alternatives. *Journal of the American Statistical Association*, **96**, 174–184.

Bigelow, J. L. and Dunson, D. B. (2009). Bayesian semiparametric joint models for functional predictors. *Journal of the American Statistical Association*, **104**, 26–36.

Branscum, A. J., Johnson, W. O., Hanson, T. E., and Gardner, I. A. (2008). Bayesian semiparametric ROC curve estimation and disease diagnosis. *Statistics in Medicine*, **27**, 2474–2496.

Branscum, A. J., Johnson, W. O., and Baron, A. T. (2013). Robust medical test evaluation using flexible Bayesian semiparametric regression models. *Epidemiology Research International*, **ID 131232**, 1–8.

Branscum, A. J., Johnson, W. O., Hanson, T. E., and Baron, A. T. (2015). Flexible regression models for ROC and risk analysis with or without a gold standard. *Submitted*.

Brown, E. R. and Ibrahim, J. G. (2003). A Bayesian semiparametric joint hierarchical model for longitudinal and survival data. *Biometrics*, **59**, 221–228.

Carey, J. R., Liedo, P., Müller, H.-G., Wang, J.-L., and Chiou, J.-M. (1998). Relationship of age patterns of fecundity to mortality, longevity, and lifetime reproduction in a large cohort of mediterranean fruit fly females. *The Journals of Gerontology, Ser. A: Biological Sciences and Medical Sciences*, **53**, 245–251.

Chen, Y., Hanson, T., and Zhang, J. (2014). Accelerated hazards model based on parametric families generalized with Bernstein polynomials. *Biometrics*, **70**, 192–201.

Chiou, J.-M., Müller, H.-G., Wang, J.-L., and Carey, J. R. (2003). A functional multiplicative effects model for longitudinal data, with application to reproductive histories of female medflies. *Statistica Sinica*, **13**, 1119–1133.

Christensen, R. and Johnson, W. (1988). Modelling accelerated failure time with a Dirichlet process. *Biometrika*, **75**, 693–704.

Christensen, R., Johnson, W., Branscum, A., and Hanson, T. E. (2010). *Bayesian Ideas and Data Analysis*. Boca Raton, FL: Chapman & Hall/CRC.

Cox, D. R. (1972). Regression models and life-tables (with discussion). *Journal of the Royal Statistical Society, Ser. B*, **34**, 187–220.

Cox, D. R. and Oakes, D. (1984). *Analysis of Survival Data*. Boca Raton, FL: Chapman & Hall/CRC.

Crowley, J. and Hu, M. (1977). Covariance analysis of heart transplant survival data. *Journal of the American Statistical Association*, **72**, 27–36.

Dahl, D. B. (2006). Model-based clustering for expression data via a Dirichlet process mixture model. In: Bayesian Inference for Gene Expression and Proteomics, Eds: Kim-Anh Do, Peter Müller & Marina Vannucci, New York: Springer, pp. 201–218.

Daniels, M. J. and Linero, A. R. (2015). Bayesian nonparametrics for missing data in longitudinal clinical trials. In: *Nonparametric Bayesian Methods in Biostatistics and Bioinformatics,* Eds: R. Mitra & P. Müller, New York: Springer.

De Iorio, M., Johnson, W. O., Müller, P., and Rosner, G. L. (2009). Bayesian nonparametric non-proportional hazards survival modelling. *Biometrics*, **65**, 762–771.

Dennerstein, L., Lehert, P., Burger, H., and Guthrie1, J. (2007). New findings from non-linear longitudinal modelling of menopausal hormone changes. *Human Reproduction Update*, **13**, 551–557.

Diggle, P. J. (1988). An approach to the analysis of repeated measurements. *Biometrics*, **44**, 959–971.

Dunson, D. B. (2010). Nonparametric Bayes applications to biostatistics. In: *Bayesian Nonparametrics*, Eds: N. L. Hjort et al., Cambridge UK: Cambridge University Press, pp. 223–273.

Dunson, D. B. and Park, J.-H. (2008). Kernel stick-breaking processes. *Biometrika*, **95**, 307–323.

Escobar, M. D. (1994). Estimating normal means with a Dirichlet process prior. *Journal of the American Statistical Association*, **89**, 268–277.

Escobar, M. D. and West, M. (1995). Bayesian density estimation and inference using mixtures. *Journal of the American Statistical Association*, **90**, 577–588.

Faes, C., Molenberghs, G., Hens, N., Muller, A., Goossens, H., and Coenen, S. (2011). Analysing the composition of outpatient antibiotic use: A tutorial on compositional data analysis. *Journal of Antimicrobial Chemotherapy*, **66**, 89–94.

Faucett, C. L. and Thomas, D. C. (1996). Simultaneously modelling censored survival data and repeatedly measured covariates: A Gibbs sampling approach. *Statistics in Medicine*, **15**, 1663–1685.

Ferguson, T. S. (1973). A Bayesian analysis of some nonparametric problems. *The Annals of Statistics*, **1**, 209–230.

Ferguson, T. S. (1974). Prior distribution on the spaces of probability measures. *The Annals of Statistics*, **2**, 615–629.

Gasparini, M. (1995). Exact multivariate Bayesian bootstrap distributions of moments. *The Annals of Statistics*, **23**, 762–768.

Geisser, S. and Eddy, W. F. (1979). A predictive approach to model selection. *Journal of the American Statistical Association*, **74**, 153–160.

Gelfand, A. E. and Dey, D. K. (1994). Bayesian model choice: Asymptotics and exact calculations. *Journal of the Royal Statistical Society, Ser. B*, **56**, 501–514.

Gelfand, A. E., Kottas, A., and MacEachern, S. N. (2005). Bayesian nonparametric spatial modeling with Dirichlet process mixing. *Journal of the American Statistical Association*, **100**, 1021–1035.

Glaser, N., Barnett, P., McCaslin, I., Nelson, D., Trainor, J., Louie, J., Kaufman, F., Quayle, K., Roback, M., Malley, R., *et al.* (2001). Risk factors for cerebral edema in children with diabetic ketoacidosis. *The New England Journal of Medicine*, **344**, 264–269.

Hanson, T. and Johnson, W. O. (2002). Modeling regression error with a mixture of Polya trees. *Journal of the American Statistical Association*, **97**, 1020–1033.

Hanson, T. and Johnson, W. O. (2004). A Bayesian semiparametric AFT model for interval-censored data. *Journal of Computational and Graphical Statistics*, **13**, 341–361.

Hanson, T., Bedrick, E. J., Johnson, W. O., and Thurmond, M. C. (2003). A mixture model for bovine abortion and foetal survival. *Statistics in Medicine*, **22**, 1725–1739.

Hanson, T., Sethuraman, J., and Xu, L. (2005). On choosing the centering distribution in Dirichlet process mixture models. *Statistics & Probability Letters*, **72**, 153–162.

Hanson, T., Johnson, W., and Laud, P. (2009). Semiparametric inference for survival models with step process covariates. *Canadian Journal of Statistics*, **37**, 60–79.

Hanson, T. E. (2006). Inference for mixtures of finite Polya tree models. *Journal of the American Statistical Association*, **101**, 1548–1565.

Hanson, T. E., Monteiro, J. V., and Jara, A. (2011a). The Polya tree sampler: Toward efficient and automatic independent Metropolis–Hastings proposals. *Journal of Computational and Graphical Statistics*, **20**, 41–62.

Hanson, T. E., Branscum, A. J., and Johnson, W. O. (2011b). Predictive comparison of joint longitudinal-survival modeling: A case study illustrating competing approaches (with discussion). *Lifetime Data Analysis*, **17**, 3–28.

Inácio de Carvalho, V., Jara, A., Hanson, T. E., and de Carvalho, M. (2013). Bayesian nonparametric ROC regression modeling. *Bayesian Analysis*, **8**, 623–646.

Inácio de Carvalho, V., Jara, A., and de Carvalho, M. (2015). Bayesian nonparametric approaches for ROC curve inference. In: *Nonparametric Bayesian Methods in Biostatistics and Bioinformatics*, Eds: R. Mitra & P. Müller, New York: Springer.

Jara, A., Hanson, T. E., and Lesaffre, E. (2009). Robustifying generalized linear mixed models using a new class of mixtures of multivariate Polya trees. *Journal of Computational and Graphical Statistics*, **18**, 838–860.

Jara, A., Hanson, T., Quintana, F., Müller, P., and Rosner, G. L. (2011). DPpackage: Bayesian semi- and nonparametric modeling in R. *Journal of Statistical Software*, **40**, 1–30.

Jara, A., Nieto-Barajas, L., and Quintana, F. (2013). A time series model for responses on the unit interval. *Bayesian Analysis*, **8**, 723–740.

Johnson, W. and Christensen, R. (1986). Bayesian nonparametric survival analysis for grouped data. *Canadian Journal of Statistics*, **14**, 307–314.

Johnson, W. and Christensen, R. (1989). Nonparametric Bayesian analysis of the accelerated failure time model. *Statistics & Probability Letters*, **8**, 179–184.

Kalbfleisch, J. D. (1978). Non-parametric Bayesian analysis of survival time data. *Journal of the Royal Statistical Society, Ser. B*, **40**, 214–221.

Kalbfleisch, J. D. and Prentice, R. L. (2002). *The Statistical Analysis of Failure Time Data*. New York: Wiley.

Kaplan, E. L. and Meier, P. (1958). Nonparametric estimation from incomplete observations. *Journal of the American Statistical Association*, **53**, 457–481.

Kottas, A. and Gelfand, A. E. (2001). Bayesian semiparametric median regression modeling. *Journal of the American Statistical Association*, **96**, 1458–1468.

Kuo, L. and Mallick, B. (1997). Bayesian semiparametric inference for the accelerated failure-time model. *Canadian Journal of Statistics*, **25**, 457–472.

Lavine, M. (1992). Some aspects of Polya tree distributions for statistical modelling. *The Annals of Statistics*, **20**, 1222–1235.

Lavine, M. (1994). More aspects of Polya tree distributions for statistical modelling. *The Annals of Statistics*, **22**, 1161–1176.

Lin, D. and Ying, Z. (1995). Semiparametric inference for the accelerated life model with time-dependent covariates. *Journal of Statistical Planning and Inference*, **44**, 47–63.

Little, R. J., D'Agostino, R., Cohen, M. L., Dickersin, K., Emerson, S. S., Farrar, J. T., Frangakis, C., Hogan, J. W., Molenberghs, G., Murphy, S. A., et al. (2012). The prevention and treatment of missing data in clinical trials. *The New England Journal of Medicine*, **367**, 1355–1360.

MacEachern, S. N. (2000). Dependent Dirichlet processes. Technical report, Department of Statistics, The Ohio State University.

MacEachern, S. N. and Müller, P. (1998). Estimating mixture of Dirichlet process models. *Journal of Computational and Graphical Statistics*, **7**, 223–238.

Mauldin, R. D., Sudderth, W. D., and Williams, S. (1992). Polya trees and random distributions. *The Annals of Statistics*, **20**, 1203–1221.

Mitra, R. and Müller, P. (2015). Bayesian nonparametric models. In: *Nonparametric Bayesian Methods in Biostatistics and Bioinformatics*, Eds: R. Mitra & P. Müller, New York: Springer.

Müller, P. and Mitra, R. (2013). Bayesian nonparametric inference—Why and how (with discussion). *Bayesian Analysis*, **8**, 269–302.

Neal, R. (2000). Markov chain sampling methods for Dirichlet process mixture models. *Journal of Computational and Graphical Statistics*, **9**, 249–265.

Nieto-Barajas, L. E., Müller, P., Ji, Y., Lu, Y., and Mills, G. B. (2012). A time-series DDP for functional proteomics profiles. *Biometrics*, **68**, 859–868.

Nieto-Barajas, L. E., Contreras-Cristán, A., *et al.* (2014). A Bayesian nonparametric approach for time series clustering. *Bayesian Analysis*, **9**, 147–170.

Norris, M., Johnson, W. O., and Gardner, I. A. (2009). Modeling bivariate longitudinal diagnostic outcome data in the absence of a gold standard. *Statistics and its Interface*, **2**, 171–185.

Norris, M., Johnson, W. O., and Gardner, I. A. (2014). Bayesian semi-parametric joint modeling of biomarker data with a latent changepoint: Assessing the temporal performance of Enzyme-Linked Immunosorbent Assay (ELISA) testing for paratuberculosis. *Statistics and its Interface*, **7**, 417–438.

Paddock, S. M. (2002). Bayesian nonparametric multiple imputation of partially observed data with ignorable nonresponse. *Biometrika*, **89**, 529–538.

Pepe, M. S. (2003). *The Statistical Evaluation of Medical Tests for Classification and Prediction*. Oxford, UK: Oxford University Press.

Petrone, S. (1999). Random Bernstein polynomials. *Scandinavian Journal of Statistics*, **26**, 373–393.

Prentice, R. (1982). Covariate measurement errors and parameter estimation in a failure time regression model. *Biometrika*, **69**, 331–342.

Prentice, R. L. and Kalbfleisch, J. D. (1979). Hazard rate models with covariates. *Biometrics*, pages 25–39.

Quintana, F., Johnson, W. O., Waetjen, E., and Gold, E. (2015). Bayesian nonparametric longitudinal data analysis. *Submitted*.

Rasmussen, C. E. and Williams, C. K. I. (2006). *Gaussian Processes for Machine Learning*. Cambridge, MA: MIT Press.

Ricketts, J. and Head, G. (1999). A five-parameter logistic equation for investigating asymmetry of curvature in baroreflex studies. *American Journal of Physiology—Regulatory, Integrative and Comparative Physiology*, **277**, R441–R454.

Rodriguez, A., Dunson, D. B., and Gelfand, A. E. (2008). The nested Dirichlet process (with discussion). *Journal of the American Statistical Association*, **103**, 1131–1154.

Ryan, T. P. and Woodall, W. H. (2005). The most-cited statistical papers. *Journal of Applied Statistics*, **32**, 461–474.

Sethuraman, J. (1994). A constructive definition of Dirichlet priors. *Statistica Sinica*, **2**, 639–650.

Sundaram, R. (2006). Semiparametric inference for the proportional odds model with time-dependent covariates. *Journal of Statistical Planning and Inference*, **136**, 320–334.

Susarla, V. and Van Ryzin, J. (1976). Nonparametric Bayesian estimation of survival curves from incomplete observations. *Journal of the American Statistical Association*, **71**, 897–902.

Taylor, J., Cumberland, W., and Sy, J. (1994). A stochastic model for analysis of longitudinal AIDS data. *Journal of the American Statistical Association*, **89**, 727–736.

Teh, Y. W., Jordan, M. I., Beal, M. J., and Blei, D. M. (2006). Hierarchical Dirichlet processes. *Journal of the American Statistical Association*, **101**, 1566–1581.

Tokdar, S. T. (2006). Posterior consistency of Dirichlet location-scale mixture of normals in density estimation and regression. *Sankhyā: The Indian Journal of Statistics*, **68**, 90–110.

Tomlinson, G. and Escobar, M. (1999). Analysis of densities. Technical report, University of Toronto.

Tseng, Y.-K., Hsieh, F., and Wang, J.-L. (2005). Joint modelling of accelerated failure time and longitudinal data. *Biometrika*, **92**, 587–603.

Tsiatis, A. A. and Davidian, M. (2004). Joint modeling of longitudinal and time-to-event data: An overview. *Statistica Sinica*, **14**, 809–834.

Wald, A. (1952). On the principles of statistical inference. *Notre Dame Mathematical Lectures*, No. 1, Notre Dame, Ind.

Wang, Y. and Taylor, J. M. G. (2001). Jointly modeling longitudinal and event time data with application to acquired immunodeficiency syndrome. *Journal of the American Statistical Association*, **96**, 895–905.

Zeger, S. L. and Diggle, P. J. (1994). Semiparametric models for longitudinal data with application to CD4 cell numbers in HIV seroconverters. *Biometrics*, **50**, 689–699.

Zhang, D., Lin, X., Raz, J., and Sowers, M. (1998). Semiparametric stochastic mixed models for longitudinal data. *Journal of the American Statistical Association*, **93**, 710–719.

Zhou, H. and Hanson, T. (2015). Bayesian spatial survival models. In: *Nonparametric Bayesian Methods in Biostatistics and Bioinformatics*, Eds: R. Mitra & P. Müller, New York: Springer.

# Part II
# Genomics and Proteomics

# Chapter 3
# Bayesian Shape Clustering

Zhengwu Zhang, Debdeep Pati, and Anuj Srivastava

**Abstract** Curve clustering is an important fundamental problem in biomedical applications involving clustering protein sequences or cell shapes in microscopy images. Existing model-based clustering techniques rely on simple probability models that are not generally valid for analyzing shapes of curves. In this chapter, we talk about an efficient Bayesian method to cluster curve data using a carefully chosen metric on the shape space. Rather than modeling the infinite-dimensional curves, we focus on modeling a summary statistic which is the inner product matrix obtained from the data. The inner-product matrix is modeled using a Wishart with parameters with carefully chosen hyperparameters which induce clustering and allow for automatic inference on the number of clusters. Posterior is sampled through an efficient Markov chain Monte Carlo procedure based on the Chinese restaurant process. This method is demonstrated on a variety of synthetic data and real data examples on protein structure analysis.

## 3.1 Introduction

The chapter provides a review of the paper "Bayesian Clustering of Shapes of Curves Using Dirichlet-Wishart Prior' (Zhang et al. 2015). The work evolved from our investigation of a long-standing problem of clustering protein sequences. Protein structure analysis is an outstanding scientific problem in structural biology. A large number of new proteins are regularly discovered and scientists are interested in learning about their functions in larger biological systems. Since protein functions are closely related to their folding patterns and structures in native states, the task of

Z. Zhang • D. Pati (✉) • A. Srivastava
Florida State University, Tallahassee, FL, USA
e-mail: zhengwu@stat.fsu.edu; debdeep@stat.fsu.edu; anuj@stat.fsu.edu

 
R. Mitra, P. Müller (eds.), *Nonparametric Bayesian Inference in Biostatistics*, Frontiers in Probability and the Statistical Sciences,
DOI 10.1007/978-3-319-19518-6_3

structural analysis of proteins becomes important. In terms of evolutionary origins, proteins with similar structures are considered to have common evolutionary origin. The structural classification of proteins (SCOP) database (Murzin et al. 1995) provides a manual classification of protein structural domains based on similarities of their structures and amino acid sequences. Refer to Fig. 3.1 for a snapshot of the proteins in $\mathbb{R}^3$ and the three-coordinates of the protein sequences. Clustering

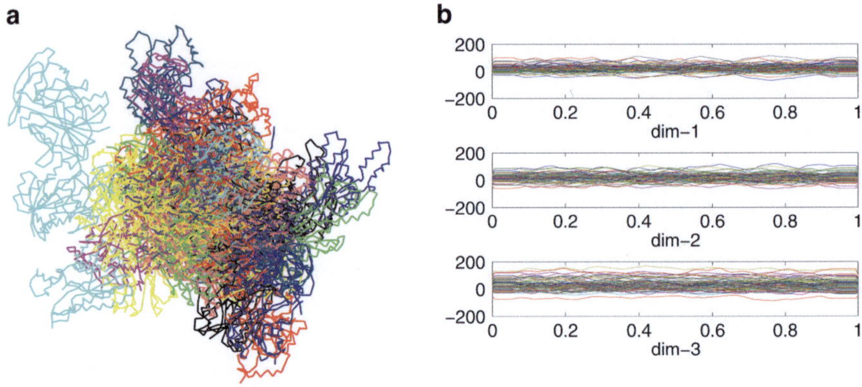


**Fig. 3.1** Protein sequences. (**a**) Raw protein structure data in $\mathbb{R}^3$. (**b**) Three-dimensional components of the protein sequences, where the $x$-axis indicates the length of each sequence

protein sequences based on their shapes is extremely important to trace the evolutionary relationship between proteins and for detecting conserved structural motifs. Since protein sequences are non-Euclidean objects, clustering poses several challenges. To address this problem, we first need to clarify the term "shape" of an object living in a possibly non-Euclidean space. Shape is defined to be a property of an object which are invariant to rotations, translations, and scaling. There are many difficulties when analyzing shapes. Firstly, it is important to develop representations and metrics such that the analysis is invariant to parameterization in addition to the standard transformations (rigid motion and scaling). Furthermore, under the chosen representations and metrics, the analysis must be performed on infinite-dimensional and sometimes nonlinear spaces, which poses an additional difficulty. Before discussing these issues in detail, we start with a brief review of clustering.

Clustering is an important area of research in unsupervised classification of large object databases. The general goal here is to choose groups of objects so as to maximize homogeneity within clusters and minimize homogeneity across clusters. The clustering problem has been addressed by researchers in many disciplines. A few well-known methods are metric based, e.g. K-means (MacQueen 1967), hierarchical clustering (Ward 1963), clustering based on principal components, spectral clustering (Ng et al. 2002), and so on (Jain and Dubes 1988; Ozawa 1985). Traditional clustering methods are complemented by methods based on a probability model where one assumes a data generating distribution (e.g., Gaussian) and infers clustering configurations that maximize certain objective function (Banfield and Raftery 1993;

Fraley and Raftery 1998, 2002, 2006; MacCullagh and Yang 2008). A model-based clustering can be useful in addressing challenges posed by traditional clustering methods. This is because a probability model allows the number of clusters to be treated as a parameter in the model, and can be embedded in a Bayesian framework providing quantification of uncertainty in the number of clusters and clustering configurations.

A popular probability model is obtained by considering that the population of interest consists of $K$ different sub-populations and the density of the observation $y$ from the $k$th sub-population is $f_k$. Given observations $y_1, \ldots, y_n$, we introduce indicator random variables $(c_1, \ldots, c_n)$ such that $c_i = k$ if $y_i$ comes from the $k$th sub-population. The maximum likelihood inference is based on finding the value of $(c, f_1, \ldots, f_k)$ that maximizes the likelihood $\prod_{i=1}^n f_{c_i}(y_i)$. Typically $K$ is assumed to be known or a suitable upper bound is assumed for convenience. When $y_i \in \mathbb{R}^p$, $f_k$ is commonly parametrized by a multivariate Gaussian density with mean vector $\mu_k$ and covariance matrix $\Sigma_k$. An alternative is to use a nonparametric Bayesian approach which has an appealing advantage of allowing $K$ to be unknown and inferring it from the data. A further bonus of a Bayesian approach is the easy availability of an entire distribution of the inferred number of clusters.

The vast majority of the literature on model-based clustering is almost exclusively focused on Euclidean data. This is primarily due to the easy availability of parametric distributions on the Euclidean space as well as computational tractability of estimating the cluster centers. For clustering functional data, e.g. shapes of proteins, one encounters several challenges. Unlike Euclidean data, where the notions of cluster centers and cluster variance are standard, these quantities and the resulting quantification of homogeneity within clusters is not obvious for shape spaces. Moreover, it is important to use representations and metrics for clustering objects that are invariant to shape-preserving transformations (rigid motions, scaling, and re-parametrization). Such a metric will be referred to as the elastic metric. For example, Kurtek et al. (2012) take a model-based approach for clustering of curves using an elastic metric that has proper invariances. However, under the chosen representations and metrics, even simple summary statistics of the observed data are difficult to compute. They presented a special representation of curves, called the square-root velocity function (SRVF), under which a specific elastic metric becomes an $L^2$ metric and simplifies the shape analysis. Other existing shapes clustering methods (Belongie et al. 2002; Liu et al. 2012) either extract finite-dimensional features to represent the shapes or project the high-dimensional shape space to a low-dimensional space (Yankov and Keogh 2006; Auder and Fischer 2012), and then apply clustering methods for Euclidean data; these are not generally valid in all applications. Also, several methods (Srivastava et al. 2005; Gaffney and Smyth 2005) have been proposed to cluster non-Euclidean data based on a distance-based notion of dispersion, thus, avoiding the computation of shape means (e.g., Karcher means), but they all assume a given number of clusters.

In this chapter we develop a model-based clustering method for non-Euclidean curve data that does not require the knowledge of cluster number $K$ apriori. This approach is based on modeling a summary statistic that encodes the clustering information, namely the inner product matrix. The salient points of this approach are: (1) The comparison of curves is based on the inner product matrix under elastic shape analysis (ESA), so that the analysis is invariant to all desired shape-preserving transformations. (2) The inner product matrix is modeled using a Wishart distributions with priors induced by the Chinese restaurant process (CRP) (Vogt et al. 2010). A model directly on the inner-product matrix has an appealing advantage of reducing computational cost substantially by avoiding computation of the Karcher means. (3) We formulate and sample from a posterior on the number of clusters, and use the mode of this distribution for final clustering. We illustrate our ideas through several synthetic and real data examples. The results show that our model on the inner product matrix leads to a more accurate estimate of the number of clusters as well as the clustering configurations compared to a Bayesian nonparametric model directly on the data, even in the Euclidean case.

This chapter is organized as follows. The mathematical details of the metric used for computing the inner product and the model specifications are presented in Sect. 3.2. In Sect. 3.3, we illustrate our methodology on several synthetic data examples and on clustering protein sequences.

## 3.2 Methodology

We develop a model based on the Wishart distribution for the inner product matrices to cluster shapes of curves. Since we model a summary statistic of the data as in Adametz and Roth (2011) and Vogt et al. (2010) instead of the infinite dimensional data points, our method is computationally efficient. However, unlike Adametz and Roth (2011) and Vogt et al. (2010) which consider a standard $\mathbb{L}^2$ metric to calculate the distance matrices, the inner product matrix is calculated using a specific representation of curves called SRVF (Srivastava et al. 2011). This along with some registration techniques makes the inner product invariant to the shape preserving transformations, thus eliminating the drawback of Adametz and Roth (2011) and Vogt et al. (2010). Moreover, a Bayesian nonparametric approach allows us to do automatic inference on the number of clusters. Below, we describe the mathematical framework for computing the inner product matrix.

### *3.2.1 Inner Product Matrix Using Elastic Shape Analysis*

We adapt the ESA introduced in Srivastava et al. (2011) to calculate the inner product matrix in the SRVF space for the non-Euclidean functional data. Let $\beta : D \to \mathbb{R}^p$ be a parameterized curve in $\mathbb{R}^p$ with domain $D$. We restrict our attention to those $\beta$ which are absolutely continuous on $D$. Usually $D = [0,1]$ for open curves and $D = \mathbb{S}^1$ for closed curves. Define $\mathscr{F} = \{\beta : D \to \mathbb{R}^p : \beta \text{ is absolutely continuous on } D\}$ and a continuous mapping: $Q : \mathbb{R}^p \to \mathbb{R}^p$ as

$$Q(x) \equiv \begin{cases} x/\sqrt{|x|} & \text{if } |x| \neq 0 \\ 0 & \text{otherwise.} \end{cases}$$

Here, $|\cdot|$ is the Euclidean two-norm in $\mathbb{R}^p$. For the purpose of studying the shape of a curve $\beta$, we will represent it as: $q : D \to \mathbb{R}^p$, where $q(t) \equiv Q(\dot{\beta}(t))$. The function $q : D \to \mathbb{R}^p$ is called SRVF. It can be shown that for any $\beta \in \mathscr{F}$, the resulting SRVF is square integrable. Hence, we will define $\mathbb{L}^2(D,\mathbb{R}^p)$ to be the set of all SRVFs. For every $q \in \mathbb{L}^2(D,\mathbb{R}^p)$ there exists a curve $\beta$ (unique up to a constant, or a translation) such that the given $q$ is the SRVF of that $\beta$.

There are several motivations for using SRVF for functional data analysis. First, an elastic metric becomes the standard $\mathbb{L}^2$ metric under the SRVF representation (Srivastava et al. 2011). This elastic metric is invariant to the re-parameterization of curves and provides nice physical interpretations. Although the original elastic metric has a complicated expression, the SRVF transforms it into the $\mathbb{L}^2$ metric, thus providing a substantial simplification in terms of computing the metric.

By representing a parameterized curve $\beta$ by its SRVF $q$, we have taken care of the translation variability, but the scaling, rotation, and the re-parameterization variabilities still remain. In some applications like clustering of protein sequences, it is not advisable to remove the scaling variabilities as the length can be a predictor of its biological functions. On the contrary, in applications like clustering images with the camera placed at variable distances, it is necessary to remove the scales by rescaling all curves to be of unit length, i.e., $\int_D |\dot{\beta}(t)|dt = \int_D |q(t)|^2 dt = 1$. The set of all SRVFs representing unit-length curves is a unit hypersphere in the Hilbert manifold $\mathbb{L}^2(D,\mathbb{R}^p)$. We will use $\mathscr{C}^o$ to denote this hypersphere, i.e., $\mathscr{C}^o = \{q \in \mathbb{L}^2(D,\mathbb{R}^p) | \int_D |q(t)|^2 dt = 1\}$. A rigid rotation in $\mathbb{R}^p$ is represented as an element of $SO(p)$, the special orthogonal group of $p \times p$ matrices. The rotation action is defined to be $SO(p) \times \mathscr{C}^o \to \mathscr{C}^o$ as follows. If a curve is rotated by a rotation matrix $O \in SO(p)$, then its SRVF is also rotated by the same matrix, i.e. the SRVF of $O\beta(t)$ is $Oq(t)$, where $q$ is the SRVF of $\beta$. A re-parameterization function is an element of $\Gamma$, the set of all orientation-preserving diffeomorphisms of $D$. For any $\beta \subset \mathscr{F}$ and $\gamma \subset \Gamma$, the composition $\beta \circ \gamma$ denotes the re-parameterization of $\beta$ by $\gamma$. The SRVF of $\beta \circ \gamma$ is given by: $\tilde{q}(t) = q(\gamma(t))\sqrt{\dot{\gamma}(t)}$. We will use $(q,\gamma)$ to denote $q(\gamma(t))\sqrt{\dot{\gamma}}$ in the following.

It is easy to show that the actions of $SO(p)$ and $\Gamma$ on $C^o$ commute each other, thus we can form a join action of the product group $SO(p) \times \Gamma$ on $\mathscr{C}^o$ according to $((O,\gamma),q) = O(q \circ \gamma)\sqrt{\dot{\gamma}}$. The action of the product group $\Gamma \times SO(p)$ is by isometries under the chosen Riemannian metric. The orbit of an SRVF $q \in \mathscr{C}^o$ is the set of SRVFs associated with all the reparameterizations and rotations of a given curve and is given by: $[q] = \text{closure}\{(q,(O,\gamma)) | (O,\gamma) \in SO(p) \times \Gamma\}$. The specification of orbits is important because each orbit uniquely represents a shape and, therefore, analyzing the shapes is equivalent to the analysis of orbits. The set of all such orbits is denoted by $\mathscr{S}$ and termed the *shape space*. $\mathscr{S}$ is actually a quotient space given by $\mathscr{S} = \mathscr{C}^o/(SO(p) \times \Gamma)$. Now we can define an inner product on the space $\mathscr{S}$ which is invariant to translation, scaling, rotation, and reparameterization of curves.

**Definition 3.1 (Inner Product on Shape Space of Curves).** For given curves $\beta_1, \beta_2 \in \mathscr{F}$ and the corresponding SRVFs, $q_1, q_2$, we define the inner product, $s_{\beta_1,\beta_2}$ or $\langle [q_1],[q_2] \rangle_{\mathbb{H}}$, to be:

$$s_{\beta_1,\beta_2} = \sup_{\gamma \in \Gamma, O \in SO(d)} \langle q_1, (q_2, (O, \gamma)) \rangle .$$

Note that this inner product is well defined because the action on $SO(p) \times \Gamma$ is by isometries.

**Optimization over $SO(p)$ and $\Gamma$:** The maximization over $SO(p)$ and $\Gamma$ can be performed iteratively as in Srivastava et al. (2011). In our case, we use Dynamic Programming algorithm to solve for an optimal $\gamma$ first. Then we fix $\gamma$, and search for the optimal rotation $O$ in $SO(p)$ using a rotational Procrustes algorithm (Kurtek et al. 2012).

### *3.2.2 Likelihood Specification for the Inner Product Matrix*

Let $S_{+n}(\mathbb{R})$ denote the set of all $n \times n$ symmetric non-negative definite matrices over $\mathbb{R}$. Depending on whether we rescale the curves to have unit length or not, we define two classes of inner product matrices: (1) $U_{+n}(\mathbb{R}) = \{A \in S_{+n}(\mathbb{R}) : a_{ii} = 1, |a_{ij}| \leq 1, 1 \leq i \neq j \leq n\}$ and (2) $S_{+n}(\mathbb{R})$. In this chapter, we do not make a distinction between these two cases and specify our model for the larger subspace $S_{+n}(\mathbb{R})$ irrespective of whether we rescale the curves or not. As illustrated using the experimental results in Sect. 3.3, having a probability model on a slightly larger space does not pose any practical issues when we actually rescale the curves.

For a scaled inner product matrix $S \in S_{+n}(\mathbb{R})$, let $S \sim W_n(\Sigma, d)$, the Wishart distribution with degrees of freedom $d$ and parameter $\Sigma = \mathbb{E}(S)$ of rank $n$ ($d > n$). To allow rank-deficient $S$, a *generalized Wishart distribution* with degrees of freedom $d$ ($d < n$) can be defined as

$$p(S \mid \Sigma, d) \propto |S|^{(d-n-1)/2} |\Sigma^{-1}|^{d/2} \exp\left\{ -\frac{d}{2}\mathrm{tr}(\Sigma^{-1}S) \right\}, \tag{3.1}$$

where $|\cdot|$ implies the product of non-zero eigenvalues and $\mathrm{tr}(\cdot)$ is the sum of the diagonal elements.

For an observation of $S$, the log-likelihood function is

$$l(\Sigma; S, d) \propto -\frac{d}{2}\log(|\Sigma|) - \frac{d}{2}\mathrm{tr}(\Sigma^{-1}S) \tag{3.2}$$

for $\Sigma \in S_{+n}(\mathbb{R})$. One can easily identify this as an exponential family distribution with canonical parameter $W = \Sigma^{-1}$, and the deviance is minimized at $\Sigma = S$ (McCullagh 2009). Therefore, $\Sigma$ encodes the similarity between the observed shapes measured by the inner product matrix $S$. For instance, $\Sigma_{jk}$ encodes the similarity between $y_i$ and $y_j$ as measured by the inner product $\langle [q_i], [q_j] \rangle_{\mathbb{H}}$, where $q_i$ and $q_j$ are the SRVFs of $y_i$ and $y_j$, respectively.

Clustering is equivalent to finding an optimal partition of the data. We use $P = \{P_1, P_2, \ldots, P_K\} \in \mathscr{P}$ to denote a partition of set $\{1, 2, \ldots, n\}$ into $K$ classes, where $\mathscr{P}$ denotes the set of all partitions of $\{1, 2, \ldots, n\}$. A partition $P$ can also be represented by *membership indicators* $\{c_i, i = 1, \ldots, n\}$, where $c_i = j$ if $i \in P_j, j = 1, \ldots, K$, or a *membership matrix* $B \in \mathbb{R}^{n \times n}$, defined as $B_{ij} = \begin{cases} 1 & \text{if } c_i = c_j \\ 0 & \text{otherwise} . \end{cases}$ If we assume: (1) observed shapes $\{y_i \in \mathscr{F}, i = 1, \ldots, n\}$ come from several sub-populations, and (2) observed shapes from the same population are placed next to each other; one would expect to observe a block pattern in the inner product matrix $S$ because the observations from the same cluster will have similar inner product. Figure 3.2 on the left panel shows one example of such inner product matrix, which is calculated from simulated Euclidean data with three clusters. One can observe three large-value-blocks along the diagonal.

To perform Bayesian inference on the clustering configuration, we define the following prior on $\Sigma$ that enables clustering of the observations. Motivated by MacCullagh and Yang (2008), Adametz and Roth (2011), and Vogt et al. (2010), consider the following decomposition of $\Sigma$.

Let

$$\Sigma = \alpha I + \beta B \,, \tag{3.3}$$

where $\alpha, \beta \in \mathbb{R}$, $I$ is the identity matrix and $B \in \mathbb{R}^{n \times n}$ is the membership matrix. Equation (3.3) decomposes the scalar matrix $\Sigma$ into a sparse matrix $\alpha I$ and a low-rank matrix $\beta B$, where $B$ encodes the clustering information. For convenience of introducing a conjugate prior for $\alpha$ (Vogt et al. 2010), we re-parameterize this model into $\Sigma = \alpha(I + \theta B)$, where $\theta = \beta / \alpha$. Intuitively, the parameter $\theta$ controls the strength of similarity between two observations measured by their inner product—a large $\theta$ indicates a strong association, and vice versa. Refer to Fig. 3.2 for an illustration of the membership matrix $B$ and the corresponding $\Sigma$ matrix.

In a clustering based on the inner product matrix $S$ via Bayesian inference, the primary goal is to infer the posterior distribution on the membership matrix $B$. To clarify terms of model, likelihood and priors in the Bayesian framework, we refer $S \sim W_n(\Sigma, d)$ as our model on the inner product matrix $S \in \mathbb{R}^{n \times n}$, $p(S|\Sigma, d)$ as the likelihood, and we put priors on $\Sigma$ and $d$. The prior on $\Sigma$ is induced by first letting $\Sigma = \alpha(I + \theta B)$ and then put priors on $\alpha$, $\theta$, and $B$. Below, we discuss the specification of prior distributions for those parameters.

### 3.2.3 Priors and Hyperpriors

A popular method of inducing a prior distribution on the space of partitions $\mathscr{P}$ is the CRP (Pitman 2006) induced by a Dirichlet process (Ferguson 1973, 1974). Since a prior on $\{c_i, i = 1, \ldots, n\}$ induces a prior on $\mathscr{P}$ and, hence, on the space of membership matrices $B$, it is enough to specify a prior on $\{c_i, i = 1, \ldots, n\}$. We assume

$$P(c_n = j \mid c_1, \ldots, c_{n-1}) = \begin{cases} \frac{n_j}{n-1+\xi} & \text{if } c_n = j \text{ for some } 1 \leq j \leq K \\ \frac{\xi}{n-1+\xi} & \text{otherwise,} \end{cases} \tag{3.4}$$

where $n_j = \#\{i : 1 \leq i < n, c_i = j\}$ and $\xi > 0$ is the precision parameter which controls the prior probability of introducing new clusters. The expected cluster size under CRP is given by $\sum_{i=1}^{n} \frac{\xi}{\xi+i-1} \sim \xi \log(\frac{\xi+n}{\xi})$.

#### 3.2.3.1 Hyperpriors

We need to choose hyperpriors for parameters associated with the prior distributions.

**Priors on $\alpha$ and $\theta$ :** $\alpha$ is assigned an inverse Gamma distribution, denoted $\alpha \sim$ Inv-Gamma$(r,s)$ for constants $r,s > 0$. An inverse Gamma distribution for $\alpha$ allows us to marginalize out $\alpha$ in the posterior distribution, thus obviating the need to sample from its conditional posterior distribution in the Gibbs sampler (refer to Sect. 3.2.3.2). Recall that $\theta$ controls the strength of similarity within cluster. Thus a large $\theta$ will encourage tight clusters (elements in each cluster are very similar). We will explore the sensitivity of the final clustering to $\theta$ in Sect. 3.3. We assume a discrete uniform distribution for $\theta$ on the set $\{\theta_1, \ldots, \theta_m\}$, with $P(\theta = \theta_i) = \frac{1}{m}$, $i = 1, \ldots, m$.

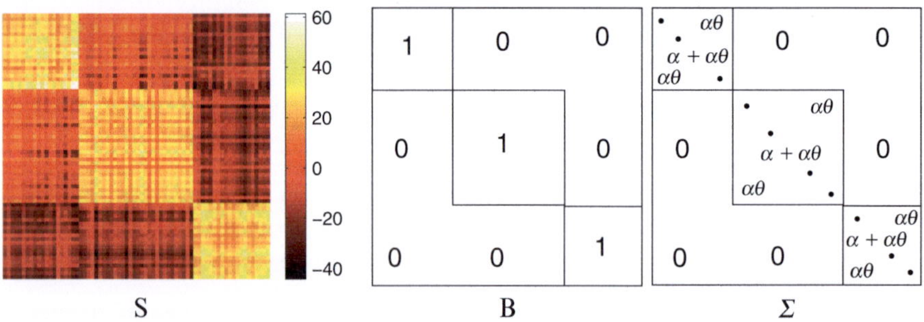


**Fig. 3.2** From the *left* to *right*: an inner product matrix $S$, a partition matrix $B$, and a scale matrix $\Sigma$

**Choice of $\xi$ and $d$:** Recall that the $\xi$ controls the prior probability of introduction of new clusters in the CRP (3.4). We start with an initial guess of the number of clusters $C_0$ using standard algorithms for shape clustering (Yankov and Keogh 2006; Auder and Fischer 2012). In our experience, $C_0/\log n$ provides reasonable choice for $\xi$.

Also, recall that $d$ is the degrees of freedom for the Wishart distribution. Since $d$ represents the rank of the inner product matrix $S$, it is natural to estimate $d$ using the number of largest eigenvalues of $S$ which explains 95 % of the total variation. This forms an empirical Bayes estimate of $d$, denoted $d_{EB}$. Let the eigenvalues of $S$ be $\{\lambda_1,\ldots,\lambda_m\}$, where $m \leq n$ and $\lambda_1 \geq \lambda_2 \geq \ldots \geq \lambda_m$. $d_{EB}$ is taken to be the smallest integer such that $\frac{\sum_{j=1}^{d_{EB}} \lambda_j}{\sum_{i=1}^{m} \lambda_i} \geq 0.95$.

#### 3.2.3.2 Posterior Computation and Final Selection of Clusters

Next, we develop a Gibbs sampling algorithm to sample from the posterior distribution of the unknown parameters. To that end, we propose the following simplifications to the likelihood. The trace and determinant that involve the $\alpha$ and $\theta$ in Eq. (3.1) can be computed analytically (Vogt et al. 2010; Adametz and Roth 2011). Observe that

$$\left|\Sigma^{-1}\right| = \alpha^{-n} \prod_{j=1}^{J} (1+\theta n_j)^{-1}, \tag{3.5}$$

where $n_j$ is the number of elements in $j$th cluster. Clearly, the $j$th cluster corresponds to the $j$th diagonal block in $B$; refer to Fig. 3.2a. Let $S_{jj}, j = 1,\ldots J$ be a *sub-square-matrix* in $S$ corresponding to the $j$th diagonal block in $B$, and $\bar{S}_{jj} = \mathbf{I}_j^t S \mathbf{I}_j$. Let $\mathbf{I}_j \in \mathbb{R}^{n\times 1}$ be such that the $i$th element is $\mathbf{1}(c_i = j)$ for $i = 1,\ldots,n$. Then

$$\mathrm{tr}(\Sigma^{-1}S) = \sum_{j=1}^{J} \frac{1}{\alpha}\left\{\mathrm{tr}(S_{jj}) - \frac{\theta}{1+n_j\theta}\bar{S}_{jj}\right\} = \frac{1}{\alpha}\left\{\mathrm{tr}(S) - \sum_{j=1}^{J} \frac{\theta}{1+n_j\theta}\bar{S}_{jj}\right\}. \tag{3.6}$$

Substituting (3.5) and (3.6) in (3.1) with (3.3), we obtain

$$P(S \mid B,\alpha,\theta,d) \propto$$
$$\alpha^{-nd/2} \prod_{j=1}^{J} (1+\theta n_j)^{-d/2} \exp\left[-\frac{d}{2\alpha}\left\{\mathrm{tr}(S) - \sum_{j=1}^{J} \frac{\theta}{1+n_j\theta} S_{jj}\right\}\right]. \tag{3.7}$$

If $\alpha \sim$ Inv-Gamma$(r_0 d/2, s_0 d/2)$, it is possible to integrate out $\alpha$ analytically in (3.7) as $P(S \mid B,\theta,d) = \int P(\alpha)P(S \mid B,\alpha,\theta,d)d\alpha$ yielding

$$P(S \mid B,\theta,d) \propto \prod_{j=1}^{J} (1+\theta n_j)^{-d/2} \left[\frac{d}{2}\left\{\mathrm{tr}(S) - \sum_{j=1}^{J} \frac{\theta}{1+n_b\theta}\bar{S}_{jj} + s_0\right\}\right]^{-(n+r_0)d/2}. \tag{3.8}$$

Using the prior distributions for $\theta, B$ with the $d_{EB}$ plugged in the likelihood (3.7), we get the posterior distribution of the membership matrix $B$:

$$P(B|S,\theta,d,\xi) \propto P(S|B,\theta,d_{EB})P(B|\xi)P(\theta). \tag{3.9}$$

Figure 3.3 shows the graphical model representation of our Bayesian model. Some suggestions of specifying hyper-priors are summarized in Table 3.1. We use Markov chain Monte Carlo (MCMC) algorithm to obtain posterior samples $B^{(1)},\ldots,B^{(M)}$ for a suitable large integer $M > 0$ using (3.9). The detailed algorithm is described in the following.

**Table 3.1** Suggestions for specifying hyper-priors in our model

| Hyper-parameters | Description | Suggested values |
|---|---|---|
| $\theta$ | Parameter for $\Sigma$, $\Sigma = \alpha(I+\theta B)$ | $\theta \sim$ Uniform$(\theta_1,\ldots,\theta_n)$ large $\theta$—tight clusters, small $\theta$—loose clusters |
| $\alpha$ | Parameter for $\Sigma$ | $\alpha \sim$ Inv-Gamma$(r,s)$, where $r,s$ are constants |
| $d$ | Degrees of freedom of Wishart | Estimated using the rank of $S$ |
| $\xi$ | Parameter for CRP | $\xi = K/\log(n)$, $K$ is initial estimated # of clusters |

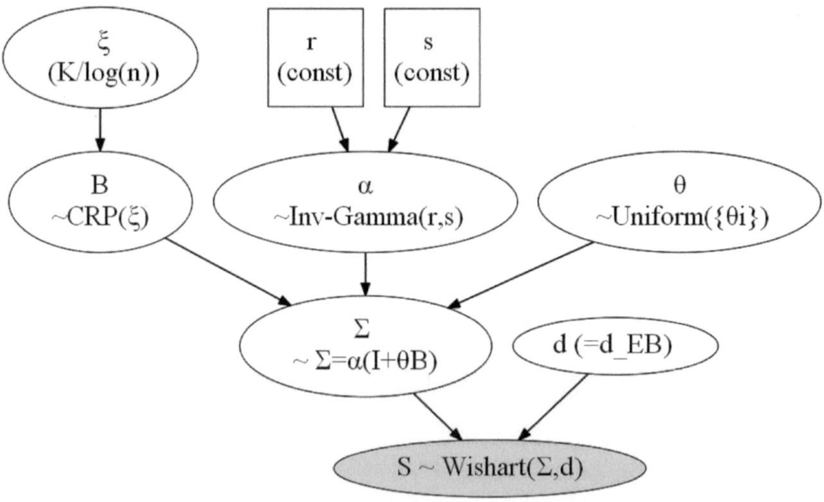


**Fig. 3.3** Graphical model representation of our Bayesian model. *Squares* indicate fix parameters and *circles* indicate random variables

**Algorithm 1 Posterior sampling using the MCMC**
*Given the prior parameters $d_{EB}$, $r$, $s$, $\xi$, $\theta$, and the inner product matrix $S$ from the $n$ observations, and let $N_\theta = length(\theta)$, we want to sample $Iter$ number of posterior samples of membership matrices $B$:*

1. *Initialize the cluster number $K$ (a large integer), the cluster indices $\{c_i, i = 1,\ldots,n\}$ and obtain the initial membership matrix $B^{(0)}$;*
2. **For** *each sweep of the MCMC ($it = 0$ to $Iter$)*
   a. **For** *each $\theta_i, i = 1,\ldots,N_\theta$ , obtain posteriors $P(\theta_i|\cdot) \propto P(S|\theta_i, B^{(it)}, d_{EB})p(\theta_i)$ using (3.8). Normalize $\{P(\theta_i|\cdot)\}$ and sample $\theta^{it}$ from the discrete distribution on the support points $\theta_i, i = 1,\ldots,N_\theta$ with probabilities $\{P(\theta_1|\cdot), \ldots, P(\theta_{N_\theta}|\cdot)\}$. The complexity for this step is $O(N_\theta * k_{B^{(it)}})$, where $k_{B^{(it)}}$ is the number of clusters obtained from $B^{(it)}$.*
   b. **For** *each observation ($i = 1$ to $n$)*
      i. **For** *each cluster ($j = 1$ to $k_{B^{(it)}} + 1$)*
         A. *Assign current observation ($y_i$) to the $j$-th cluster, update the membership matrix $B^{(it)}$ to $B'_{it}$, and calculate the posterior $\pi_j = P(B'_{it}|S, \theta^{it}, d_{EB}, \xi)$ using (3.9). The complexity for this step is $O(1)$[1].*
      ii. *Normalize $\{\pi_1, \ldots, \pi_{k_{B^{(it)}}+1}\}$ and sample $c_i$ from a discrete distribution on support points $\{1, 2, \ldots, k_{B^{(it)}} + 1\}$ with probabilities $(\pi_1, \ldots, \pi_{k_{B^{(it)}}+1})$. Update $B^{(it)}$. Complexity for this step is $O\{\log(k_{B^{(it)}})\}$ (Bringmann and Panagiotou 2012).*
   c. *After completing Step 2b, we obtain one MCMC sample of $B^{(it)}$.*
3. *Repeat step 2 so that we have $Iter$ many samples. Discard the first few samples (burn-in), and relabel the remaining $B^{(it)}$s as $B^{(1)}, \ldots, B^{(M)}$.*

From Algorithm 1, the complexity of each sweep of the MCMC is $O\{N_\theta K + nK \log K\}$. Usually $N_\theta \leq n$, leading to an overall complexity of $O(nK \log(K))$.

Once we obtain the posterior samples $\{B^{(i)}, i = 1,\ldots,M\}$, our goal is to estimate the clustering configuration. However, the space of membership matrices $B$ is huge, and we would expect the posterior to explore only an insignificant fraction of the space based on a moderate values of $M$. Therefore, instead of using the mode of $\{B^{(i)}, i = 1,\ldots,M\}$, we devise the following alternate strategy to estimate the clustering configuration more accurately. We treat the set of the membership matrices, denoted as $\mathscr{F}_B$, as a subset of symmetric $n \times n$ matrices with restrictions: (1) $B(i,j) = \{0,1\}$ for all $i, j = 1,\ldots,n$; (2) $B(i,\cdot) = B(j,\cdot)$ and $B(\cdot,i) = B(\cdot,j)$ if $i$th observation and $j$th observation are in the same cluster. The final matrix $B^*$ is obtained by calculating the **extrinsic mean** of the posterior samples defined as follows.

[1] Since only one observation changes the cluster index, one can explicitly calculate the difference between the old values of (3.5) and (3.6) and new values in O(1) steps.

**Algorithm 2 Calculating extrinsic mean of membership matrices**
*Given the samples* $B^{(1)},\dots,B^{(M)}$*, the extrinsic mean* $B^*$ *is calculated as the following:*

1. *Find the mode of the number of clusters* $k_0$ *based on the samples* $B^{(1)},\dots,B^{(M)}$.
2. *Calculate the Euclidean mean and threshold it onto the set of membership matrices* ($\mathscr{F}_B$)*:*
   a. **Euclidean mean***: Let* $\bar{B} = \frac{1}{M}\sum_{t=1}^{M} B^{(t)}$.
   b. **Thresholding***: threshold the Euclidean mean onto* $\mathscr{F}_B$*:* $B^* = threshold(\bar{B},t^*)$*, where* $t^*$ *is the largest threshold such that* $B^*$ *has* $k_0$ *clusters. Setting* $k = N$ *and* $iter = M$*, the thresholding procedure is described below:*
   **While** ($k \neq k_0$)*, do*
      i. *Set* $J_{array} = \{1,\dots,N\}$, $B^* = zeros(N,N)$*. Also set* $iter = iter - 1$*, let* $t^* = iter/M$.
      ii. **For** $j$ *in* $J_{array}$*, calculate*
         A. $\mathbf{v} = \mathbf{1}(\bar{B}(j,\cdot) > t^*)$*; record the index of elements in* $\mathbf{v}$ *equal to* 1*, denoted as set* $C$*. Let* $J_{array} = J_{array} - C$*, which means remove elements in* $C$ *from* $J_{array}$.
         B. **For** $i$ *in set* $C$*, set* $B^*(i,\cdot) = \mathbf{v}$, $B^*(\cdot,i) = \mathbf{v}^t$ *and* $\bar{B}(i,\cdot) = \mathbf{0}, \bar{B}(\cdot,i) = \mathbf{0}^t$.
      iii. *Set* $k = \#B^*$*, which is number of clusters in* $B^*$.

## 3.3 Experimental Results

We demonstrate the performance of our model (Wishart-CRP, denoted by W-CRP) both on synthetic data (in Sect. 3.3.1) and the protein dataset discussed in the introduction (in Sect. 3.3.2). For the Euclidean datasets, we generated 8000 samples from the posterior distribution and discarded a burn-in of 1000, whereas those numbers for the non-Euclidean data are 4000 and 1000, respectively. Convergence was monitored using trace plots of the deviance as well as several parameters. The high effective sample size of the main parameters of interest shows good mixing of the Markov chain. Also we get essentially identical posterior modes with different starting points and moderate changes to hyperparameters.

### 3.3.1 Synthetic Examples

Zhang et al. (2015) consider both Euclidean and non-Euclidean synthetic datasets. In this review, we will mainly focus on the non-Euclidean examples since they form a central part of our motivating study. Our goal is to cluster synthetic shapes taken from the MPEG-7 database (Jeannin and Bober 1999). The full database has 1400 shape samples, 20 shapes for each class. We first choose 100 shapes to form a subset of 10 classes with 10 shapes from each class. The observations are randomly permuted and the inner product matrix $S$ is calculated using Definition 3.1. Then we perform our clustering method on $S$ (note that $S \in U_+(\mathbb{R})$) with parameters $d_{EB} = 32$, $r = 3$, $s = 4$, $\theta = \{0.1, 0.2, 0.5, 1, 5, 10, 20, 50, 100, 200, 500, 1000\}$, and $\xi = 3$. We impose a prior on $\alpha$ with Inv-Gamma$(3,4)$, and $\xi$ is estimated by $\tilde{K}/\log(100)$, where $\tilde{K}$ is an estimate of number of clusters ($\tilde{K} = 15$ in this case). The clustering result is shown in Fig. 3.4, where (a) and (b) show the inner product (I-P) matrix before and after clustering, (c) shows the final clustering result, and (d) shows the histogram of cluster number $K$ obtained from 4000 MCMC samples of $B$. From the result, one can see that our algorithm clusters these 100 shapes well other than splitting one class.

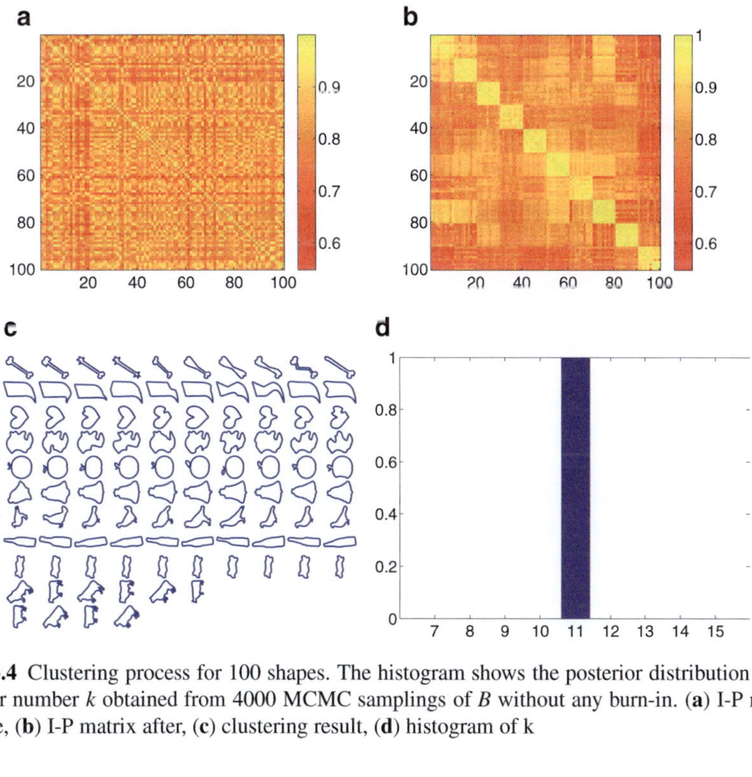


**Fig. 3.4** Clustering process for 100 shapes. The histogram shows the posterior distribution of the cluster number $k$ obtained from 4000 MCMC samplings of $B$ without any burn-in. (**a**) I-P matrix before, (**b**) I-P matrix after, (**c**) clustering result, (**d**) histogram of k

Next, we study the sensitivity of the cluster number $K$ to the parameter $d$, degrees of freedom of the Wishart distribution. Note that in the Euclidean case, $d$ can be easily estimated since $d$ (in the case of $d < n$) is the dimension of the data. Figure 3.5 shows the estimated cluster number $K$ versus the value of parameter $d$ in the dataset shown in Fig. 3.4. It is evident that the estimates of $K$ are robust to different choices of $d$.

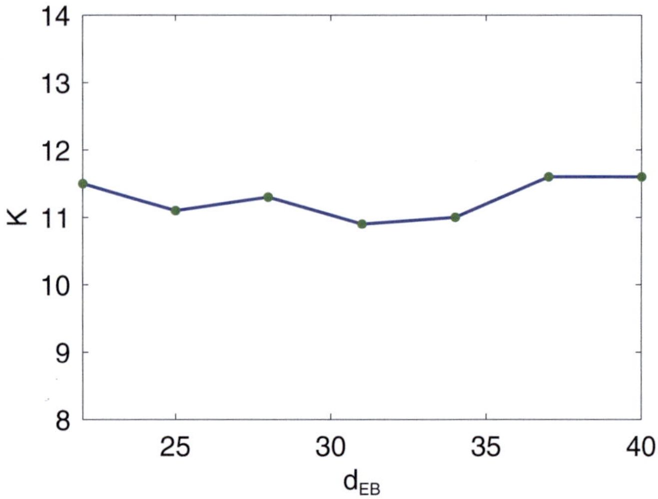


**Fig. 3.5** Clustering sensitivity analysis of parameter $d$ in the dataset shown in Fig. 3.4

To compare with the existing methods in shape analysis, we test our method on another subset of MPEG-7 dataset that was used in Bicego and Murino (2004), Bicego et al. (2004), and Bicego and Murino (2007). The dataset contains six classes of shapes with 20 shapes per class. To quantify the clustering result, we use the "classification rate" defined in Jain and Dubes (1988). For each cluster, we note the predominant shape class, and for those shapes assigned to the cluster which do not belong to the dominant class are recognized to be misclassified. The classification rate is the total number of dominant shapes for all classes divided by the total number of shapes. However, this measure is known to be sensitive towards larger clusters.

The Rand index (Torsello et al. 2007) is an alternative measure of the quality of classification which measures the similarity between the clustering result and the ground truth, defined as $RI = a/\binom{n}{2}$. Here $a$ is the number of the "agreements" between the clustering and the ground truth, which is defined as the sum of two quantities: (1) the number of pairs of elements belonging to the same class that are assigned to the same cluster; (2) the number of pairs of elements belonging to different sets that are assigned to different classes. If the clustering result is the same as the ground truth, $RI = 1$, otherwise $RI < 1$. The Rand index penalizes the over-segmentation while the classification rate does not. Table 3.2 compares the overall classification rate and Rand index of our method with other methods, such as Fourier descriptor combined with support vector machine based classification (FD + SVM), hidden Markov model (HMM + Wtl) with weighted likelihood classification (Bicego and Murino 2007), HMM with OPC approach (HMM + OPC) (Bicego et al. 2004), ESA (Srivastava et al. 2011) with k-medians (K-medians), ESA with pairwise clustering method (ESA + PW) (Srivastava et al. 2005). Our model, with Wishart-CRP applied on the elastic inner product (EIP) matrix is denoted by EIP + W-CRP. The classification rate, Rand index and the computational time of K-medians, ESA + PW and our method are obtained based on the average of 5 runs on a laptop with an i5-2450M CPU and 8GB memory. The computational time of our approach (EIP + DW) includes the cost of calculating the inner product matrix $S$ (642.6 s) and generating the 4000 MCMC samples (131.6 s). A faster approach for calculating the elastic inner product matrix defined in our chapter is available in Huang et al. (2014). For ESA + PW and K-medians method, we set $K = 6$ since we know the true $K$ in this case. The classification rates for FD+ SVM, HMM + Wtl, and HMM + OPC are reported from Bicego and Murino (2007), and these rates are based on the 1-nearest neighbor classification. As evident from the results, our model can automatically find the cluster number $K = 6$, and the classification rate is better than the competitors.

**Table 3.2** Comparison of the classification rate on MPEG-7 dataset

| Classifier | FD+ SVM | HMM+ WtL | HMM+ OPC | K-medians | ESA+ PW | EIP+ W-CRP |
|---|---|---|---|---|---|---|
| Classification rate (%) | 94.29 | 96.43 | 97.4 | 81.5 | 96.67 | 100.00 |
| Rand index | – | – | – | 0.91 | 0.98 | 1.00 |
| Time (seconds) | – | – | – | 648.5 | 707.4 | 774.2 |

### *3.3.2 Clustering Real Protein Sequences*

In the following experiments, we will use our model to cluster protein sequences (refer to Fig. 3.1 in the Introduction). We obtained the protein sequences from SCOP database (Murzin et al. 1995) which provides a manual classification of protein structural domains based on similarities of their structures and amino acid sequences.

In the first experiment, we choose a small protein structure dataset obtained from SCOP with only 88 proteins. Based on SCOP, these proteins are from four classes (SCOP provides the ground truth). Those proteins are pre-processed similar to an earlier study (Liu et al. 2011). To have a good estimate of the SRVFs from the raw data, we smooth the protein sequences with a Gaussian kernel. We also added one residue at both N and C terminal of each protein chain by extrapolating from the two terminal residues to allow some degrees of freedom on matching boundary residues. The added residues are removed after matching. Note that these smoothed SRVFs will only be used for searching optimal re-parameterizations $\gamma$ and rotations $SO(3)$ to get the inner product between protein structures. Then we apply our model to the inner product matrix $S \in U_+(\mathbb{R})$ and get the clustering result, where we use parameters $\theta \in \{0.1, 0.2, 0.3, 0.4\}$ and $\xi = 1$. The final clustering results are shown in Fig. 3.6. The clustering rate is 100 % compared with the ground truth provided by SCOP.

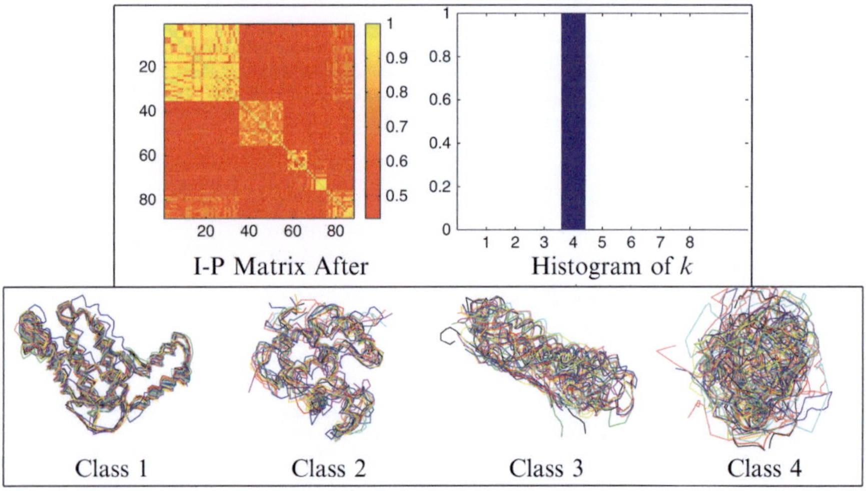


**Fig. 3.6** Protein structure classification on a SCOP subset containing 88 proteins. The *first row* shows the inner product matrix between protein structures after clustering and the histogram of cluster number $k$. The *second row* shows the four clusters

In the next experiment, we choose 20 classes with at least 10 elements in each class from SCOP dataset to form a subset with 602 proteins. The final clustering result shows some clusters with only a few elements which we consider as *outliers*. In this experiment, our model identifies 17 outliers (seven small clusters). After removing these outliers, the remaining 585 proteins are clustered into 38 classes. The clustering rate is 84.1 %. The first row in Fig. 3.7 shows the inner product matrix corresponding to the 585 protein structures (after putting elements in the same cluster together), and the posterior estimate of the partition matrix $B$. The second row shows first four clusters of the clustering result after the alignment (removing shape-preserving transformations). One can see that inside each cluster, the shapes of these protein structures are very similar to each other. As comparisons, we remove the outliers detected by our method, then apply ESA + PW and K-medians method to cluster the left 585 proteins by setting $K = 20$. ESA + PW gets 75.99 % of classification rate and K-medians gets 63.42 %. The Rand indexes for our model, ESA + PW and K-medians are 0.95, 0.93, and 0.91 respectively. As evident, we obtain a good clustering result based on only the shape of the proteins.

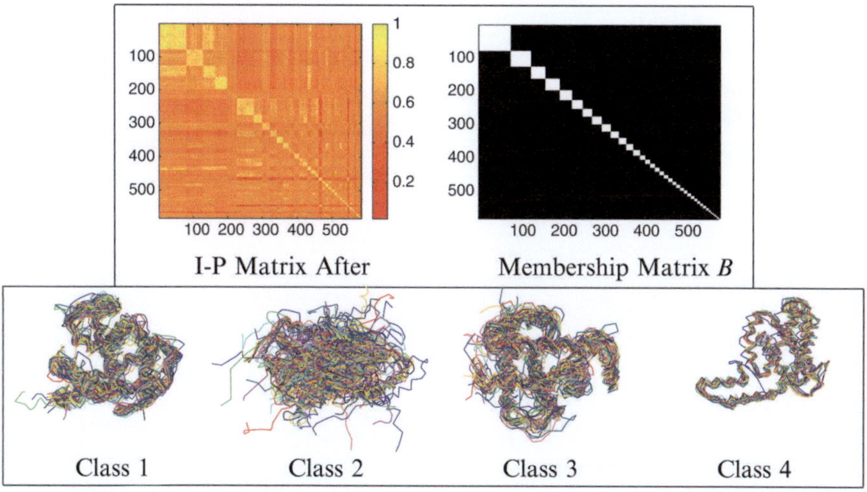


**Fig. 3.7** Protein structure classification on a SCOP subset of 602 proteins. The *first row* shows the inner product matrix between protein structures after clustering and the corresponding inferred $B$, respectively. The *second row* shows the first four clusters

# References


Adametz, D. and Roth, V. (2011). Bayesian partitioning of large-scale distance data. In *Neural Information Processing Systems (NIPS)*, pages 1368–1376.

Auder, B. and Fischer, A. (2012). Projection-based curve clustering. *Journal of Statistical Computation and Simulation*, **82**(8), 1145–1168.

Banfield, J. D. and Raftery, A. E. (1993). Model-based Gaussian and non-Gaussian clustering. *Biometrics*, pages 803–821.

Belongie, S., Malik, J., and Puzicha, J. (2002). Shape matching and object recognition using shape contexts. *IEEE Transactions on Pattern Analysis and Machine Intelligence*, **24**(4), 509–522.

Bicego, M. and Murino, V. (2004). Investigating hidden Markov models' capabilities in 2D shape classification. *IEEE Trans. Pattern Anal. Mach. Intell*, **26**, 281–286.

Bicego, M. and Murino, V. (2007). Hidden Markov model-based weighted likelihood discriminant for 2D shape classification. *IEEE Trans. Pattern Anal. Mach. Intell*, **16**, 2707–2719.

Bicego, M., Murino, V., and Figueiredo, M. A. (2004). Similarity-based classification of sequences using hidden Markov models. *Pattern Recognition*, **37**(12), 2281–2291.

Bringmann, K. and Panagiotou, K. (2012). Efficient sampling methods for discrete distributions. In *In Proc. 39th International Colloquium on Automata, Languages, and Programming (ICALP'12*, pages 133–144. Springer.

Ferguson, T. S. (1973). A Bayesian analysis of some nonparametric problems. *The Annals of Statistics*, pages 209–230.

Ferguson, T. S. (1974). Prior distributions on spaces of probability measures. *The Annals of Statistics*, pages 615–629.

Fraley, C. and Raftery, A. E. (1998). How many clusters? Which clustering method? Answers via model-based cluster analysis. *The Computer Journal*, **41**(8), 578–588.

Fraley, C. and Raftery, A. E. (2002). Model-based clustering, discriminant analysis, and density estimation. *Journal of the American Statistical Association*, **97**(458), 611–631.

Fraley, C. and Raftery, A. E. (2006). MCLUST version 3: an R package for normal mixture modeling and model-based clustering. Technical report, DTIC Document.

Gaffney, S. and Smyth, P. (2005). Joint probabilistic curve clustering and alignment. In *Neural Information Processing Systems (NIPS)*, pages 473–480. MIT Press.

Huang, W., Gallivan, K., Srivastava, A., and Absil, P.-A. (2014). Riemannian optimization for elastic shape analysis. *Mathematical theory of Networks and Systems*.

Jain, A. K. and Dubes, R. C. (1988). *Algorithms for Clustering Data*. Prentice-Hall, Inc., Upper Saddle River, NJ, USA.

Jeannin, S. and Bober, M. (1999). Shape data for the MPEG-7 core experiment CE-Shape-1 @ONLINE.

Kurtek, S., Srivastava, A., Klassen, E., and Ding, Z. (2012). Statistical modeling of curves using shapes and related features. *Journal of the American Statistical Association*, **107**(499), 1152–1165.

Liu, M., Vemuri, B. C., Amari, S.-I., and Nielsen, F. (2012). Shape retrieval using hierarchical total Bregman soft clustering. *IEEE Transactions on Pattern Analysis and Machine Intelligence*, **34**(12), 2407–2419.

Liu, W., Srivastava, A., and Zhang, J. (2011). A mathematical framework for protein structure comparison. *PLoS Computational Biology*, **7**(2).

MacCullagh, P. and Yang, J. (2008). How many clusters? *Bayesian Analysis*, **3**(1), 1–19.

MacQueen, J. (1967). Some methods for classification and analysis of multivariate observations. In *Proceedings of the Fifth Berkeley Symposium on Mathematical Statistics and Probability, Volume 1: Statistics*, pages 281–297, Berkeley, Calif. University of California Press.

McCullagh, P. (2009). Marginal likelihood for distance matrices. *Statistica Sinica*, **19**, 631–649.

Murzin, A. G., Brenner, S. E., Hubbard, T., and Chothia, C. (1995). SCOP: A structural classification of proteins database for the investigation of sequences and structures. *Journal of Molecular Biology*, **247**(4), 536–540.

Ng, A. Y., Jordan, M. I., Weiss, Y., *et al.* (2002). On spectral clustering: analysis and an algorithm. *Advances in Neural Information Processing Systems*, **2**, 849–856.

Ozawa, K. (1985). A stratificational overlapping cluster scheme. *Pattern Recognition*, **18**(3–4), 279–286.

Pitman, J. (2006). *Combinatorial stochastic processes*, volume 1875. Springer-Verlag.

Srivastava, A., Joshi, S., Mio, W., and Liu, X. (2005). Statistical shape analysis: clustering, learning, and testing. *IEEE Transactions on Pattern Analysis and Machine Intelligence*, **27**, 590–602.

Srivastava, A., Klassen, E., Joshi, S. H., and Jermyn, I. H. (2011). Shape analysis of elastic curves in Euclidean spaces. *IEEE Transactions on Pattern Analysis and Machine Intelligence*, **33**, 1415–1428.

Torsello, A., Robles-Kelly, A., and Hancock, E. (2007). Discovering shape classes using tree edit-distance and pairwise clustering. *International Journal of Computer Vision*, **72**(3), 259–285.

Vogt, J. E., Prabhakaran, S., Fuchs, T. J., and Roth, V. (2010). The translation-invariant Wishart-Dirichlet process for clustering distance data. In *Proceedings of the 27th International Conference on Machine Learning (ICML-10)*, pages 1111–1118.

Ward, J. H. (1963). Hierarchical grouping to optimize an objective function. *Journal of the American Statistical Association*, **58**(301), 236–244.

Yankov, D. and Keogh, E. (2006). Manifold clustering of shapes. In *Proceedings of ICDM*, pages 1167–1171, Washington, DC, USA.

Zhang, Z., Pati, D., and Srivastava, A. (2015). Bayesian clustering of shapes of curves. *Journal of Statistical Planning and Inference* (to appear).

# Chapter 4
# Estimating Latent Cell Subpopulations with Bayesian Feature Allocation Models

Yuan Ji, Subhajit Sengupta, Juhee Lee, Peter Müller, and Kamalakar Gulukota

**Abstract** Tumor cells are genetically heterogeneous. The collection of the entire tumor cell population consists of different subclones that can be characterized by mutations in sequence and structure at various genomic locations. Using next-generation sequencing data, we characterize tumor heterogeneity using Bayesian nonparametric inference. Specifically, we estimate the number of subclones in a tumor sample, and for each subclone, we estimate the subclonal copy number and single nucleotide mutations at a selected set of loci. Posterior summaries are presented in three matrices, namely, the matrix of subclonal copy numbers ($\boldsymbol{L}$), subclonal variant alleles ($\boldsymbol{Z}$), and the population frequencies of the subclones ($\boldsymbol{w}$). The proposed method can handle a single or multiple tumor samples. Computation via Markov chain Monte Carlo yields posterior Monte Carlo samples of all three matrices, allowing for the assessment of any desired inference summary. Simulation and real-world examples are provided as illustration. An R package is available at http://www.cran.r-project.org/web/packages/BayClone2/index.html.

Y. Ji • S. Sengupta • K. Gulukota
NorthShore University HealthSystem/The University of Chicago, 1001 University Place, Evanston, IL 60201, USA
e-mail: koaeraser@gmail.com;subhajit06@gmail.com; KGulukota@northshore.org

J. Lee (✉)
Department of Statistics, University of California Santa Cruz, 1156 High Street, Santa Cruz, CA 95064, USA
e-mail: juheelee@soe.ucsc.edu

P. Müller
The University of Texas at Austin, 1, University Station, C1200, Austin, TX 78712, USA
e-mail: pmueller@math.utexas.edu

 
R. Mitra, P. Müller (eds.), *Nonparametric Bayesian Inference in Biostatistics*, Frontiers in Probability and the Statistical Sciences,
DOI 10.1007/978-3-319-19518-6_4

# 4.1 Introduction

## *4.1.1 Biological and Statistical Background*

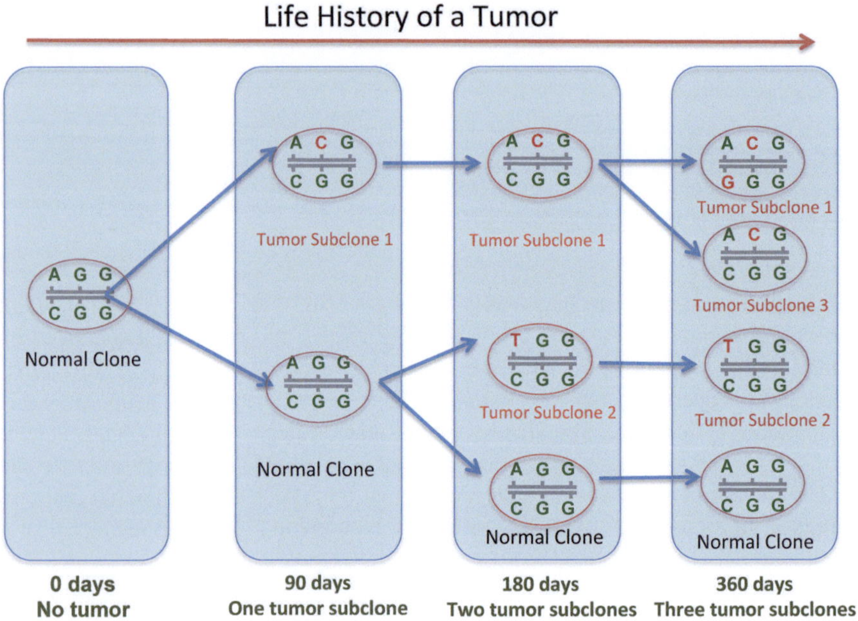


**Fig. 4.1** Evolving subclones over time result in TH. On days 90, 180, and 360, three somatic mutations (represented by branching *arrows*) result in three tumor subclones

Inference for tumor heterogeneity (TH) remains a critical gap in current literature. The ability to precisely break down a tumor into a set of subclones with distinct genetics would provide opportunities for breakthroughs in cancer treatment by potentially facilitating individualized cancer treatment that exploits TH. It would open doors for cocktail-type combinational treatments, with each treatment targeting a specific tumor subclone based on its genetic characteristics. However, many details of TH remain a mystery to scientists.

TH arises as a result of a sequence of somatic mutations over the life history of a tumor. Figure 4.1 illustrates the single progenitor model in Russnes et al. (2011) whereby tumor cells are originated from a single ancestor normal cell (on day 0). Tumor subclones are generated over time as somatic mutations accumulate at various loci. On day 360, four subclones have formed, each possessing a unique but overlapping set of mutations. Other biological models based on multiple progenitors and cancer stem cells have also been proposed (Russnes et al. 2011; Nowell 1976). The overall picture remains the same: as a result of the accumulation of somatic mutations individual tumors harbor multiple subclonal genomes that are spatially and temporally heterogeneous (Navin et al. 2010; Russnes et al. 2011). Numerous recent studies confirm this nature of tumors (Dexter et al. 1978; Weinberg 2007; Stingl and Caldas 2007; Shackleton et al. 2009; Polyak 2011; Marjanovic et al. 2013; Almendro et al. 2013).

While there is broad agreement on the origin and nature of TH (Russnes et al. 2011; Greaves and Maley 2012; Frank and Nowak 2004; Biesecker and NB 2013; Frank and Nowak 2003; De 2011; Bedard et al. 2013; Navin et al. 2011), the main challenge remains to genetically characterize heterogeneous subclones within each tumor sample. The characterization is in terms of sequence and structural variants. Some progress has been made on quantifying structural variants such as subclonal copy number variants (CNVs) (Oesper et al. 2013a). However, DNA nucleotide differences between subclonal genomes have remained undetectable despite the technological breakthrough in short read-based sequencing (Mardis 2008) and more recently single-cell sequencing (Shapiro et al. 2013). Some recent landmark work (Serena et al. 2012; Roth et al. 2014a) has used statistical inference based on DP priors, defining subclones based on non-overlapping clusters of mutations. These methods proceed assuming that the clusters will need to fit into a phylogenetic tree to describe the underlying genetic evolution of the tumors. It is unclear how sensitive the final tree construction is to the estimated clusters. Also, variation of the mutation frequencies within the same cluster is ignored in the phylogenetic tree construction.

### *4.1.2 Bayesian Feature Allocation Models for Tumor Heterogeneity*

Taking a different approach, in Lee et al. (2015a,b) we developed a new class of Bayesian nonparametric (BNP) models for inference on TH that allows for overlapping sets of mutations between subclones. To reflect the underlying biology, we use random feature allocation models such as the Indian buffet process (IBP) to model subclones (or haplotypes) as columns of a random categorical (trinary) (or binary) matrix. In Lee et al. (2015b) we use an extension of the IBP known as the categorical Indian buffet process (cIBP). Theoretically the extension is non-trivial and lays the foundation for new BNP models that are suitable for TH inference and that other researchers can build upon.

In Lee et al. (2015a) we focus on haplotypic inference based on short reads from next-generation sequencing (NGS) data. Figure 4.2 shows a stylized illustration of pair-end reads in hypothetical NGS data. At the two loci, we observe three different haplotypes, GG, GC, and AC. The identification of more than two haplotypes implies the presence of multiple subclones with distinct genomes because human cells are diploid. We infer the exact DNA sequences of each haplotype. In Lee et al. (2015b) we further extend the model by integrating CNVs with single nucleotide variants (SNVs). We provide the desired description of TH based on DNA variations in both, sequence and structure. Such inference will significantly impact downstream treatment of individual tumors, ultimately allowing personalized prognosis. For example, tumor samples with large proportions of cells bearing somatic

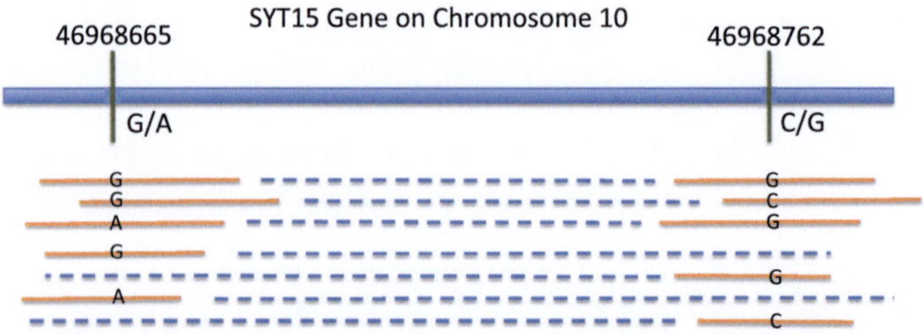


| NT at 46968665 | NT at 46968762 | Number of Reads | Local Haplotype |
|---|---|---|---|
| G | G | 23 | GG |
| G | C | 10 | GC |
| A | G | 9 | AG |
| _ | G | 43 | AG/GG |
| G | _ | 26 | GG/GC |
| A | _ | 13 | AG/AC |
| _ | C | 11 | GC/AC |

**Fig. 4.2** Local haplotype variants (LHV) based on pair-end reads mapped to two proximal nucleotides (NT). Three local haplotypes (GG, GC, AG) are directly observed from the pair-end short reads that are mapped to at least one NT. A "–" in the bottom table indicates that the read is not mapped to the corresponding NT

mutations on tumor suppressor genes should be treated differently than those that have a small proportion or none of such cells. In addition, metastatic or recurrent tumors may possess very different compositions of cellular genomes and should be treated differently.

We study nucleotide differences between cellular genomes that characterize such subclones within a single tumor (intra-tumor TH). Sets of phased SNVs define haplotypes, which in turn we use to characterize TH. For example, in Fig. 4.1, by day 360, the tumor sample possesses five alleles for the haplotype of three loci, which are (ACG, GGG, CGG, TGG, AGG). The presence of more than two alleles indicates TH. We distinguish local and non-local haplotype variants (LHVs and nLHVs).

**LHV** When SNVs are proximal on the genome (say, a few hundred base pairs apart), direct evidence of subclonality (the presence of subclones) can be identified since they may be simultaneously mapped by the same short reads. For example, in Fig. 4.2, an LHV consists of two loci about 100 base pairs apart and exhibit three alleles (GG, GC, AG). After filtering artifacts in the data and ruling out other potential genetic variations (such as copy number gain coupled with a mutation), the only explanation for having three alleles is TH, that is heterogeneous tumor cells possessing different DNA sequences at these loci. For instance, there could be two different subclones, one with alleles (GG, GC) at the two SNVs and the other with alleles (GG, AG). The two subclones could be equally distributed in the sample, which could explain the observed read counts (23, 10, 9) for haplotypes (GG, GC, AG), respectively, in Fig. 4.2.

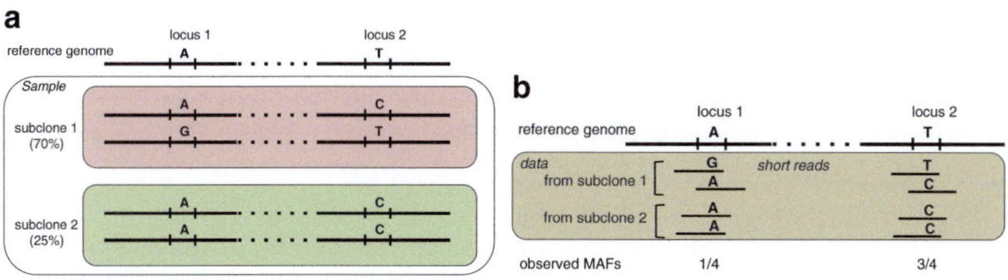


**Fig. 4.3** Non-local haplotype variant (nLHV): (**a**) shows TH with three haplotypes defining two subclones; (**b**) shows hypothetical short reads for this sample

We have developed a computational pipeline for the detection of LHVs across genome using NGS data (http://www.compgenome.org/lochap).

**nLHV** When SNVs are distant, they will not be mapped by the same short reads. Therefore, haplotypes that scaffold these SNVs are not directly observable and need to be estimated. Figure 4.3 illustrates inference for nLHVs. Panel (a) presents a hypothetical tumor sample consisting of two subclones with two long-range loci. One subclone has alleles (AC, GT), and the other subclone has (AC, AC). Together, the sample will present three alleles (AT, GT, AC). Panel (b) shows a stylized example of short reads data if the tissue sample in panel (a) were sequenced. Since two loci are far apart on the genome, short reads are separately mapped to each of them. The observed variant allele fraction (VAF), defined as the fraction of reads bearing the variant sequence, reflects the subclonal sequences and proportions in the original sample. In this example, the observed VAFs are 1/4 and 3/4 for the two loci, respectively. These numbers would change depending on absence/presence of the mutation at the loci and/or different proportions of the subclones in the sample.

In the rest of this chapter we focus on nLHVs as they cannot be directly observed from the NGS data.

### *4.1.3 Existing Methods*

Recent literature introduced several useful tools for subclonal inference. This includes, in particular, ThetA (Oesper et al. 2013b), SciClone (Miller et al. 2014), TrAp (Strino et al. 2013), PhyloSub (Jiao et al. 2014), PhyloWGS (Deshwar et al. 2014), and Clomial (Zare et al. 2014), CloneHD (Fischer et al. 2014). ThetA only considers subclonal copy numbers and is among the earliest methods for subclonal inference. TrAp emphasizes identifiability and sufficient sample size for unique mathematical solutions. SciClone and Clomial assume a binary matrix, focusing on SNVs at copy neutral regions with heterozygous mutations. PhyloSub and PhyloWGS consider possible genotypes at SNVs accounting for potential copy number changes and phylogenetic constraints. CloneHD provides inference similar to our method, but assumes the availability of data from matched normal samples. Also,

CloneHD only provides point estimates of the subclonal copy numbers and subclonal mutations, lacking a description of uncertainty for the inferred subclones. Most recently, CHAT (Li and Li 2014) adjusts the estimation of subclonal cellular fractions for both CNVs and SNVs, but still stops short of directly inferring subclonal copy numbers or variant allele counts.

Most of these methods are based on either finite mixture models or Dirichlet process models and aim to infer subclones based on clusters of mutations. In this chapter we review two recently proposed methods that are described in Lee et al. (2015a,b). Our approach differs from previously proposed methods in key aspects. Based on latent feature models, including the IBP and the novel cIBP (Sengupta 2013; Sengupta et al. 2015), we build hierarchical Bayesian models to facilitate joint inference on subclonal copy number, mutations, and cellular fractions simultaneously. We do not exploit a phylogenetic tree structure for the inferred subclones. We do so assuming that in any given tumor sample not all subclones on the phylogenetic tree may be present in sufficient proportion to allow inference. Instead, existing subclones may only represent nodes on a subset of branches of the phylogenetic tree.

In the rest of this chapter we review the inference methods proposed in Lee et al. (2015a,b). We summarize the underlying BNP models in Sect. 4.2. We also briefly discuss a special Markov chain Monte Carlo (MCMC) scheme based on splitting the data into a small training set and a large test set. The approach allows us to implement practicable transdimensional MCMC posterior simulation. In Sect. 4.3 we describe a simulation study. Section 4.4 reports a data analysis for an in-house data set to illustrate intra-tumor heterogeneity. The last section concludes with a final discussion.

## 4.2 Probability Model

### *4.2.1 Models on SNVs Alone*

Lee et al. (2015a) propose inference for TH using SNV only. The approach is based on nonparametric Bayesian model-based inference. Examining the VAFs of short reads mapped to multiple loci, we propose a binomial sample model and a feature allocation prior linking latent subclones and observed data. We do not directly model subclones. Instead we model haplotypes, with pairs of haplotypes defining a subclone. For a diploid organism, like humans, the existence of $C > 2$ haplotypes is evidence for tumor heterogeneity.

To begin, let $N_{st}$ denote the total number of reads in sample $t$ that are mapped to the genomic location $s = 1,\dots,S$ and $t = 1,\dots,T$. Among $N_{st}$ reads, let $n_{st}$ be the number of variant reads. Here a variant read refers to a read that harbors a variant allele at location $s$. For example, if the reference genome has an "A" at a genomic location and a read bears an "C" at the same location, the read is considered a variant read for that location. We use a binomial sampling model

$$n_{st} \stackrel{indep}{\sim} \text{Bin}(N_{st}, p_{st}) \text{ with } p_{st} = w_{t0}p_0 + \sum_{c=1}^{C} w_{tc}z_{sc} \equiv \varepsilon_{t0} + \sum_{c=1}^{C} w_{tc}z_{sc}. \quad (4.1)$$

The model for the expected allele fraction $p_{st}$ formalizes the assumption that the tumor sample is composed of a mixture of different haplotypes, $c = 1, \ldots, C$, each of which defined as a set of mutational statuses $z_{sc} \in \{0, 1\}$ at locus $s$ (and see below for the interpretation of the first term $w_{t0}p_0$). Here $z_{st} = 1$ (0) indicates that haplotype $c$ includes (does not include) mutation at location $s$. Sample-specific weights $w_{tc}$ record the fraction of haplotype $c$ in sample $t$. The construction of the haplotypes, including the number of haplotypes, $C$ and the indicators $z_{sc}$ are latent. The key term, $\sum_{c=1}^{C} w_{tc}z_{tc}$, indirectly infers haplotypes by explaining $p_{st}$ as arising from sample $t$ being composed of a mix of hypothetical subclones which do ($z_{sc} = 1$) or do not ($z_{sc} = 0$) include a mutation at location $s$. The indicators $z_{sc}$ are collected in a $(S \times C)$ binary matrix $Z$. The number of latent haplotypes, $C$, is unknown. Conditional on $C$, the binary matrix $Z$ describes $C$ latent tumor haplotypes that are present in the observed samples. Joint inference on $C$, $Z$, and $\mathbf{w}_t$ explains tumor heterogeneity.

In addition, we impose a special column in the $Z$ matrix, labeled as $c = 0$, which assumes that all the locations harbor mutations. That is, $z_{s0} = 1, \forall s$. We call this column the "background subclone." It does not represent any real biological subclone in the sample but accounts for experimental noise such as sequencing error. For example, it is known that NGS experiments would produce false base calls in a small proportion of short reads due to technical and computational errors. As such, there is a chance that a mutational base (nucleotide) may be falsely called at each SNV. Subclone $c = 0$ accounts for these false calls. We let $\varepsilon_{t0} = w_{t0}p_0$ in (4.1) to represent the sample-specific ($w_{t0}$) and experiment-specific ($p_0$) background noise.

Model (4.1) defines the sampling model. We complete the Bayesian inference model with a prior on the unknown parameters $(C, \mathbf{Z}, \mathbf{w})$. We use a geometric distribution, $C \sim \text{Geom}(r)$ where $\text{E}(C) = 1/r$. Conditional on $C$, we assume a feature allocation model for the binary matrix $\mathbf{Z}$ using a finite Indian buffet process ($\text{IBP}_C$)

$$\mathbf{Z} \sim \text{IBP}_C(\alpha). \quad (4.2)$$

See below for a definition of the IBP prior. For the moment we only need that $\text{IBP}_C(\alpha)$ defines a prior for the $(S \times C)$ binary matrix $\mathbf{Z}$.

Finally, we complete the model with a prior distribution for the weights $w_{tc}$ in (4.1). The haplotypes are common for all tumor samples, but the relative weights $w_{tc}$ vary across tumor samples. We assume independent Dirichlet priors as follows. Let $\theta_{tc}$ denote an (unscaled) abundance level of subclone $c$ in tissue sample $t$. We assume $\theta_{tc} \mid C \stackrel{iid}{\sim} \text{Gamma}(d, 1)$ for $c = 1, \ldots, C$ and $\theta_{t0} \stackrel{iid}{\sim} \text{Gamma}(d_0, 1)$. We then define

$$w_{tc} = \theta_{tc} / \sum_{c'=0}^{C} \theta_{tc'},$$

as the relative weight of subclone $c$ in sample $t$. This is equivalent to $\mathbf{w}_t \mid C \stackrel{iid}{\sim} \text{Dir}(d_0, d, \ldots, d)$ for $t = 1, \ldots, T$. Using $d_0 < d$ implies that the background subclone takes a smaller proportion in a sample.

In summary, we assume a binomial sampling model (4.1) with success probability, $p_{st}$. Given $C$, $\mathbf{Z}$ and $\mathbf{w}$, we define $p_{st}$ as a mixture over $C$ haplotypes. In specific, $p_{st}$ is determined by $C$, $\mathbf{Z}$ and $\mathbf{w}_t$ with the earlier two describing the latent haplotypes, and the last specifying the relative abundance of each subclone in sample $t$. Based on this model, we carried out posterior inference using MCMC simulations. Details are reported in Lee et al. (2015a).

#### 4.2.1.1 The Finite IBP

We define a model for a random $(S \times C)$ binary matrix. We define the model as a hierarchical model $p(\mathbf{Z} \mid \boldsymbol{\mu}, C)\, p(\boldsymbol{\mu} \mid C)$ with latent column-specific probabilities $\boldsymbol{\mu} = (\mu_1, \ldots, \mu_C)$. Let $m_c = \sum_{s=1}^{S} z_{sc}$ denote the number of 1s in column $c$. We use

$$p(\mathbf{Z} \mid \boldsymbol{\mu}, C) = \prod_{s=1}^{S} \prod_{c=1}^{C} \mu_c^{z_{sc}} (1 - \mu_c)^{(1 - z_{sc})} = \prod_{c=1}^{C} \mu_c^{m_c} (1 - \mu_c)^{S - m_c}, \tag{4.3}$$

$$p(\mu_c \mid C) = \text{Be}(\alpha/C, 1), \quad c = 1, \ldots, C. \tag{4.4}$$

We write $\mathbf{Z} \sim \text{IBP}_C(\alpha)$. The limit of the model, as $C \to \infty$ becomes a constructive definition of the IBP (Griffiths and Ghahramani 2005; Teh et al. 2007). The model is symmetric with respect to arbitrary indexing of the SNVs, simply because of the symmetry in (4.3) and (4.4). Note that $m_c = 0$ is possible with positive prior probability.

### 4.2.2 *Linked Models on SNVs and CNVs*

Lee et al. (2015b) develop an important extension of model (4.1) and (4.2). Recall that the columns of the binary matrix $\mathbf{Z}$ do not represent subclones, but rather haplotypes, with pairs of haplotypes defining subclones. In the previous model we stopped short of inference on actual subclones and characterized TH by describing heterogeneity of haplotypes. Since humans are diploid, up to two copies of the genome can be mutated at each locus. That is, for a given locus, the potential genotypes can be 0, 1, and 2 mutational alleles, which represents the three possible genotypes at a locus, namely, homozygous wild type, heterozygous and homozygous mutations. A binary matrix $\mathbf{Z}$ cannot capture this. In addition, CNVs further complicate the matter as a copy number loss or gain will affect the number of mutational alleles at each locus. To account for a biologically more accurate description, taking these details into account, Lee et al. (2015b) proposed a linked feature allocation model based on cIBP (Sengupta 2013; Sengupta et al. 2015), which we review next.

#### 4.2.2.1 Representing CNV (L) and SNV (Z)

We now define a subclone by two characteristics, subclonal copy numbers and subclonal variant allele counts at a set of loci. For each locus, we want to infer the number of alleles (copy number) and the number of variant alleles, which fully describe the genotype at the locus. To start, we first construct an integer-valued random matrix $\boldsymbol{L}$ to characterize subclonal copy numbers. Each column corresponds to a subclone and each row corresponds to a locus. The number of columns is unknown and random. The $c$−th column $\boldsymbol{\ell}_c = (\ell_{1c}, \ldots, \ell_{Sc})$ are the copy numbers across $S$ loci for subclone $c$. For example, in Fig. 4.4, $\ell_{sc} = 3$ for $s = 1$ and $c = 2$ since subclone 2 has three alleles at locus 1. As a prior distribution for $\boldsymbol{L}$, $p(\boldsymbol{L})$, we use a finite version of cIBP (Sengupta 2013; Sengupta et al. 2015).

**a**

| | Subclone 1 | Subclone 2 | Normal Clone |
|---|---|---|---|
| locus 1 | 2 | 3 | 2 |
| locus 2 | 1 | 2 | 2 |
| locus 3 | 2 | 2 | 2 |

**b**

| | Subclone 1 | Subclone 2 | Normal Clone |
|---|---|---|---|
| locus 1 | 0 | 2 | 0 |
| locus 2 | 1 | 1 | 0 |
| locus 3 | 0 | 0 | 0 |

**c**

| | Subclone 1 | Subclone 2 | Normal Clone |
|---|---|---|---|
| Day 90 | 30% | 0 | 70% |
| Day 180 | 30% | 15% | 55% |
| Day 360 | 20% | 30% | 50% |

**Fig. 4.4** Three matrices describe a subclonal structure. (**a**) $\boldsymbol{L}$ describes the subclonal copy numbers, (**b**) $\boldsymbol{Z}$ describes the numbers of subclonal variant alleles, and (**c**) $\boldsymbol{w}$ describes the cellular fractions of subclones

Next, we introduce a second integer-valued matrix $\boldsymbol{Z}$ with dimensions matching $\boldsymbol{L}$. We use $\boldsymbol{Z}$ to record SNVs. Denote by $\boldsymbol{z}_c$ the $c$-th column of $\boldsymbol{Z}$. The vector $\boldsymbol{z}_c = (z_{1c}, \ldots, z_{Sc})$, $z_{sc} < \ell_{sc}$ records the number of variant alleles, out of the $\boldsymbol{\ell}_c$ copies, that bear a mutant sequence different from the reference sequence at loci $s$, $s = 1, \ldots, S$ in subclone $c$. For example, in Fig. 4.4, $z_{sc} = 1$ for $s = 2$ and $c = 1$, indicating that one allele bears a variant sequence. By definition, the number of variant alleles $z_{sc}$ in a subclone cannot be larger than the copy number $\ell_{sc}$ of the subclone, i.e., $z_{sc} \leq \ell_{sc}$. Jointly, the two random integer vectors $\boldsymbol{\ell}_c$ and $\boldsymbol{z}_c$ describe a subclone and its genetic architecture at the corresponding loci. Lastly, we introduce the $\boldsymbol{w}$ matrix. Each row $\boldsymbol{w}_t = (w_{t1}, \ldots, w_{tC})$ represents the cellular fractions of the $C$ subclones in each sample (for example, see the leftmost matrix in Fig. 4.4).

#### 4.2.2.2 Sampling Model and Prior

The total number of reads mapped to locus $s$ in sample $t$, $N_{st}$ can be used to infer CNVs at locus $s$ in that copy number gain (loss) in subclones may lead to large (small) value of $N_{st}$ compared to those at loci without any CNV. We augment the binomial sampling model (4.1) with a second model for $N_{st}$, to account for subclonal copy numbers. Following Klambauer et al. (2012), we assume a Poisson sampling model for $N_{st}$. Conditional on $N_{st}$ we continue to use the same binomial model for $n_{st}$ as in (4.1). In summary,

$$N_{st} \mid \phi_t, M_{st} \overset{indep}{\sim} \text{Poi}(\phi_t M_{st}/2) \text{ and } n_{st} \mid N_{st}, p_{st} \overset{indep}{\sim} \text{Bin}(N_{st}, p_{st}). \tag{4.5}$$

Here, $M_{st}$ is the sample copy number that represents an average copy number across subclones. We will formally define and model $M_{st}$ using subclonal copy numbers ($\boldsymbol{L}$) next. The factor $\phi_t$ is the expected number of reads in sample $t$ if there were no CNV, that is, copy number equals 2. In other words, when $M_{st} = 2$, the Poisson mean becomes $\phi_t$. The interpretation of $p_{st}$ in (4.5) remains unchanged as the expected VAF for mutation $s$ in sample $t$. However, the representation and prior model for $p_{st}$ is changed. In the following discussion we will represent $p_{st}$ in terms of the underlying matrices $\boldsymbol{L}$ and $\boldsymbol{Z}$.

Recall that $C$ and $w_{tc}$ denote the unknown number of subclones in $T$ samples and the proportion of subclone $c$, $c = 1, \ldots, C$, in sample $t$, respectively. We first relate $M_{st}$ to CNV at locus $s$ in sample $t$. Let $\ell_{sc} \in \{0, 1, 2, \ldots, Q\}$ denote the number of copies at SNV $s$ in subclone $c$ where $Q$ is a prespecified maximum number of copies. The event $\ell_{sc} = 2$ means no CNVs at SNV $s$ in subclone $c$, $\ell_{sc} = 1$ indicates one copy loss, and $\ell_{sc} = 3$ indicates one copy gain. Then the mean number of copies for sample $t$ can be expressed as the weighted sum of the number of copies over $C$ latent subclones where the weight $w_{tc}$ denotes the cellular fractions of subclone $c$ in sample $t$. The expected VAF $p_{st}$ is written as the total number of recorded variant alleles which is a mixture over the $C$ latent subclones with the same weights $w_{tc}$, relative to the total number $M_{st}$. In summary, we assume

$$M_{st} = \ell_{s0} w_{t0} + \sum_{c=1}^{C} w_{tc} \ell_{sc}, \text{ and } p_{st} = \frac{p_0 z_{s0} w_{t0} + \sum_{c=1}^{C} w_{tc} z_{sc}}{M_{st}}. \tag{4.6}$$

The mixtures $\sum w_{tc} \ell_{sc}$ and $\sum w_{tc} z_{sc}$ reflect the key assumption of decomposing the sample copy number into a weighted average of subclonal copy numbers. The first terms $\ell_{s0} w_{t0}$ and $p_0 z_{s0} w_{t0}$ account for noise that can arise from upstream bioinformatics analysis. For example, a small number of short reads may be erroneously mapped to locus $s$ due to ambiguity in human reference genome or due to base calling error. As such, we use $\ell_{s0} w_{t0}$ to denote the expected copy number from a hypothetical background subclone to account for potential noise and artifacts in the data, labeled as subclone $c = 0$. Arbitrarily we assume no CNVs at any locations for the background subclone, that is, $\ell_{s0} = 2$ for all $s$. This use of $c = 0$ is also introduced in the previous section.

Next we complete the sampling model (4.6) with a prior probability model on the unknown parameters $(\boldsymbol{L}, \boldsymbol{Z}, \boldsymbol{w})$. We assume a feature-allocation prior for a latent random matrix of copy numbers, $\boldsymbol{L} = [\ell_{sc}]$, $c = 1, \ldots, C$ and $s = 1, \ldots, S$, using a finite cIBP

$$\boldsymbol{L} \sim \text{cIBP}_C(C, \alpha, Q). \tag{4.7}$$

See below for a definition of the cIBP. For the moment we only need that it defines a random $(n \times C)$ categorical matrix with $Q+1$ levels $\ell_{sc} \in \{0,\ldots,Q\}$.

Recall that the $S \times C$ matrix $\boldsymbol{Z}$ with entries, $z_{sc} \in \{0,\ldots,\ell_{sc}\}$ reports the number $z_{sc} \leq \ell_{sc}$ of alleles bearing a variant sequence among the total of $\ell_{sc}$ copies at locus $s$ in subclone $c$. Assume $z_{sc} = 0$ if $\ell_{sc} = 0$, and given $\ell_{sc} > 0$ we assume a uniform distribution

$$z_{sc} \mid \ell_{sc} \sim \text{Unif}(0,1,\ldots,\ell_{sc}), \tag{4.8}$$

where Unif$(\cdot)$ indicates a discrete uniform distribution.

The priors for $\boldsymbol{w}$ and $C$ are unchanged from the previous model. Note that we have accounted for different average read counts in $T$ samples through $\phi_t$, where $\phi_t$ represents the expected read count with two copies in sample $t$ and assume $\phi_t \overset{indep}{\sim}$ Gamma$(a_t,b_t)$ where $\text{E}(\phi_t) = a_t/b_t$. This completes the model construction.

#### 4.2.2.3 The Categorical Indian Buffet Process

We define a prior probability model for a random $(n \times C)$ categorical matrix $\boldsymbol{L}$ with entries $\ell_{sc} \in \{0,\ldots,Q\}$. Let $\boldsymbol{\pi}_c = (\pi_{c0},\pi_{c1},\ldots,\pi_{cQ})$ where $p(\ell_{sc} = q) = \pi_{cq}$ and $\sum_{q=0}^{Q} \pi_{cq} = 1$. As a prior distribution of $\boldsymbol{\pi}_c$, we use a beta-Dirichlet distribution developed in Kim et al. (2012). Conditional on $C$, $p(\ell_{sc} \neq 2) = (1-\pi_{c2})$ follows a beta distribution with parameters, $\alpha/C$ and $\beta$ and $\widetilde{\boldsymbol{\pi}} = (\widetilde{\pi}_{c0},\widetilde{\pi}_{c1},\widetilde{\pi}_{c3},\ldots,\widetilde{\pi}_{cQ})$, where $\widetilde{\pi}_{cq} = \pi_{cq}/(1-\pi_{c2})$ with $q \neq 2$, follows a Dirichlet distribution with parameters, $(\gamma_0,\gamma_1,\gamma_3,\ldots,\gamma_Q)$. Assuming a priori independence among subclones, we write $\boldsymbol{\pi}_c \overset{iid}{\sim}$ Be-Dir$(\alpha/C,\beta,\gamma_0,\gamma_1,\gamma_3,\ldots,\gamma_Q)$. For $\beta = 1$, the marginal limiting distribution of $\boldsymbol{L}$ can be shown to define a cIBP as $C \to \infty$ (Sengupta 2013; Sengupta et al. 2015).

### 4.2.3 *Posterior Simulation*

Let $\boldsymbol{x} = (\boldsymbol{L},\boldsymbol{Z},\boldsymbol{\theta},\boldsymbol{\phi},\boldsymbol{\pi},p_0)$ denote all unknown parameters, where $\boldsymbol{\theta} = \{\theta_{tc}\}$ and $\boldsymbol{\pi} = \{\pi_{cq}\}$. In Lee et al. (2015b), we implement inference via posterior MCMC simulation. That is, by generating a Monte Carlo sample of $\boldsymbol{x}_i \sim p(\boldsymbol{x} \mid \boldsymbol{n},\boldsymbol{N})$, $i = 1,\ldots,I$. MCMC posterior simulation proceeds by repeatedly using transition probabilities that update a subset of parameters at a time. See, for example, Brooks et al. (2011) for a review.

For fixed $C$ such MCMC simulation is straightforward. Gibbs sampling transition probabilities are used to update $\ell_{sc}$, $z_{sc}$, $\pi_{cq}$ and $\phi_t$ and Metropolis-Hastings transition probabilities are used to update $\boldsymbol{\theta}$ and $p_0$.

The construction of transition probabilities that involves a change of $C$ is more difficult, since the dimension of $\boldsymbol{L}$, $\boldsymbol{Z}$, $\boldsymbol{\pi}$, and $\boldsymbol{\theta}$ changes as $C$ varies. Lee et al. (2015a) introduce a clever trick for transdimensional posterior MCMC simulation. The method is not universally applicable, but works for inference in the earlier discussed models. We split the data into a small training set $(\boldsymbol{n}',\boldsymbol{N}')$ with $n'_{st} = bn_{st}$, $N'_{st} = bN_{st}$, and a test data set, $(\boldsymbol{n}'',\boldsymbol{N}'')$ with $n''_{st} = (1-b)n_{st}$, etc. In the implementation we actually use random $b_{st} \sim \text{Be}(a,b)$, finding that a random $b_{st}$ worked better than a fixed fraction $b$ across all samples and loci. Let $p_1(\boldsymbol{x} \mid C) = p(\boldsymbol{x} \mid \boldsymbol{N}',\boldsymbol{n}',C)$ denote the posterior distribution under $C$ using the training sample. We use $p_1$ in two instances. First, we replace the original prior $p(\boldsymbol{x} \mid C)$ by $p_1(\boldsymbol{x} \mid C)$ and, second, we use $p_1(\cdot)$ as proposal distribution $q(\widetilde{\boldsymbol{x}} \mid \widetilde{C}) = p_1(\widetilde{\boldsymbol{x}} \mid \widetilde{C})$ in a reversible jump (RJ) style transition probability where $\widetilde{C}$ is a proposed value of $C$. The test data is then used to evaluate the acceptance probability. The critical advantage of using the same $p_1(\cdot)$ as prior and proposal distribution is that the normalization constant cancels out in the Metropolis-Hastings acceptance probability.

To summarize the joint posterior distribution, we first find the maximum a posteriori (MAP) estimate $C^\star$ from its marginal posterior distribution and then obtain posterior point estimates $\boldsymbol{L}^\star$, $\boldsymbol{Z}^\star$, $\boldsymbol{w}^\star$, $\boldsymbol{\phi}^\star$, and $\boldsymbol{\pi}^\star$ conditional on $C^\star$. See Lee et al. (2015a,b) for more detailed discussion.

## 4.3 Simulation

We assess the proposed model in a simulation study. In short, we generate random $n_{st}$ and $N_{st}$ for a set of $S = 100$ loci in $T = 25$ hypothetical samples based on simulated truth about subclones. We consider four subclones ($C^{\text{TRUE}} = 4$) as well as a background subclone ($c = 0$) in the simulation truth. We use the true $\boldsymbol{L}$ shown in Fig. 4.5a where green color (light grey) in the panels implies a copy number gain ($\ell_{sc} = 3$) and red color (dark grey) shows two copy loss ($\ell_{sc} = 0$), for $c = 1,\ldots,4$ and $s = 1,\ldots,100$. Panel (b) shows the simulation truth $\boldsymbol{Z}$. Similar to $\boldsymbol{L}^{\text{TRUE}}$, green color indicates three variant alleles and red color zero. We generate $\phi_t^{\text{TRUE}}$ from Gamma$(600,3)$ for $t = 1,\ldots,25$. We then generate $\boldsymbol{w}_t^{\text{TRUE}}$ as follows. We let $\boldsymbol{a}^{\text{TRUE}} = (12,9,4.5,1.5)$ and for each $t$ randomly permute $\boldsymbol{a}^{\text{TRUE}}$. Let $\boldsymbol{a}_\pi^{\text{TRUE}}$ denote a random permutation of $\boldsymbol{a}^{\text{TRUE}}$. We generate $\boldsymbol{w}^{\text{TRUE}} \sim \text{Dir}(0.3,\boldsymbol{a}_\pi^{\text{TRUE}})$. That is, the first parameter of the Dirichlet prior for the $(C^{\text{TRUE}}+1)$-dimensional weight vector is 0.3, and the remaining parameters are a permutation of $\boldsymbol{a}^{\text{TRUE}}$. Using the assumed $\boldsymbol{L}^{\text{TRUE}}$, $\boldsymbol{Z}^{\text{TRUE}}$ and $\boldsymbol{w}^{\text{TRUE}}$ and letting $p_0^{\text{TRUE}} = 0.05$, we generate $N_{st}$ from $\text{Poi}(\phi_t^{\text{TRUE}} M_{st}^{\text{TRUE}}/2)$ and $n_{st}$ from $\text{Bin}(N_{st}, p_{st}^{\text{TRUE}})$. The weights $\boldsymbol{w}^{\text{TRUE}}$ are shown in Fig. 4.5c. Similar to the other heatmaps, green color (light grey) in panel (c) represents high abundance of a subclone in a sample, red color (dark grey) low abundance for $c = 0,\ldots,4$ and $t = 1,\ldots,30$. The samples in rows are rearranged for better display.

To fit the proposed model in Sect. 4.2.2, we fix the hyperparameters as $r = 0.2$, $\alpha = 2$, $\gamma_q = 0.5$ for $q = 0,1,3(=Q)$, $d_0 = 0.5$, $d = 1$, $a_{00} = 0.3$ and $b_{00} = 5$. For the prior of $\phi_t$, we let $b = 3$ and $a$ to be the median of the observed $N_{st}$. For each

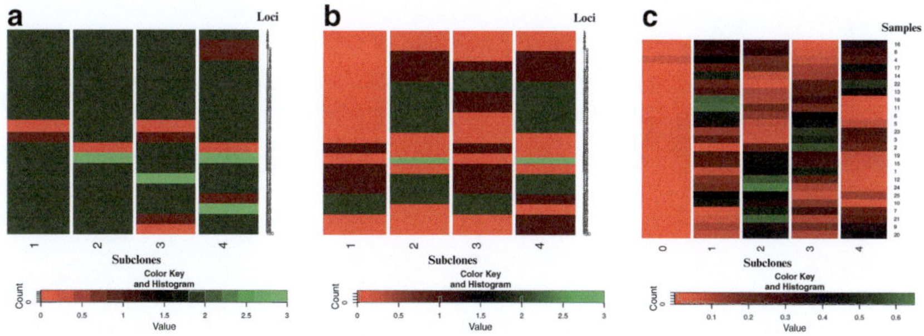


**Fig. 4.5** Heatmaps of (**a**) $\boldsymbol{L}^{\text{TRUE}}$, (**b**) $\boldsymbol{Z}^{\text{TRUE}}$, and (**c**) $\boldsymbol{w}^{\text{TRUE}}$ in the simulation study

value of $C$, we initialize $\boldsymbol{Z}$ using the observed sample proportions and $\boldsymbol{L}$ using the initial $\boldsymbol{Z}$. We generate initial values for $\boldsymbol{\theta}_{tc}$ and $p_0$ by prior draws. We generate $b_{st} \stackrel{iid}{\sim} \text{Be}(25, 975)$ to construct the training set and run the MCMC simulation over 16,000 iterations, discarding the first 6,000 iterations as initial burn-in.

Figure 4.6a reports the posterior distribution of $C$ in which the dashed vertical line represents the true value $C^{\text{TRUE}} = 4$. The posterior mode $C^\star = 4$ recovers the truth. Panels (d) through (f) illustrate the posterior point estimates, $\boldsymbol{L}^\star$, $\boldsymbol{Z}^\star$, and $\boldsymbol{w}^\star$. Compared to the simulation truth in Fig. 4.5, the posterior estimate recovers subclones 1 through 3 with high accuracy. $\boldsymbol{\ell}_c^\star$ for subclone 4 has large discrepancy. We suspect that this is due to small $w_{tc}$ with $c = 4$ for almost all samples seen in the last column of Fig. 4.5c. The discrepancy between $\boldsymbol{\ell}_4^\star$ and $\boldsymbol{\ell}_4^{\text{TRUE}}$ is related to the underestimation of $\boldsymbol{w}_{tc}^\star$ under $c = 4$. More importantly, there is ambiguity about the true latent structure, as is seen by the still excellent fit of the data. Conditional on $C^\star$, we computed $\hat{M}_{st}$ and $\hat{p}_{st}$ and compared to the true values. Figure 4.6b,c show that the fit under the model is great, except that the histograms have a slightly thicker left tail, also possibly due to the misspecification of $\boldsymbol{\ell}_4$.

For comparison, we implement PyClone (Roth et al. 2014b) with the same simulated data. We let the normal copy number, the minor parental copy number, and the major parental copy number to be 2, 0, and 3, respectively, at each locus. PyClone considers copy number changes and estimates the variant allelic prevalence (fraction of clonal population having a mutation) at a locus in a sample. The formulation of their variant allelic prevalences (named as "cellular prevalences") is similar to that of our $p_{st}$. It uses a Dirichlet process model to identify a non-overlapping clustering of the loci based on their cellular prevalences. Cellular prevalences over loci and samples may vary but the clustering of loci is shared by samples. Figure 4.7a shows a heatmap of posterior estimates of the cellular prevalences (by color) and mutational clustering (by separations with white horizontal lines) under PyClone. Panel (b) of the figure shows a heatmap of $p_{st}^{\text{TRUE}}$. The loci (rows) of the two heatmaps are re-arranged in the same order for easy comparison. By comparing the two heatmaps, the cellular prevalence estimates under PyClone are close to $p_{st}^{\text{TRUE}}$ and yields a reasonable estimates of a clustering of the loci. However, PyClone does not attempt to construct a description of subclones with genomic variants.

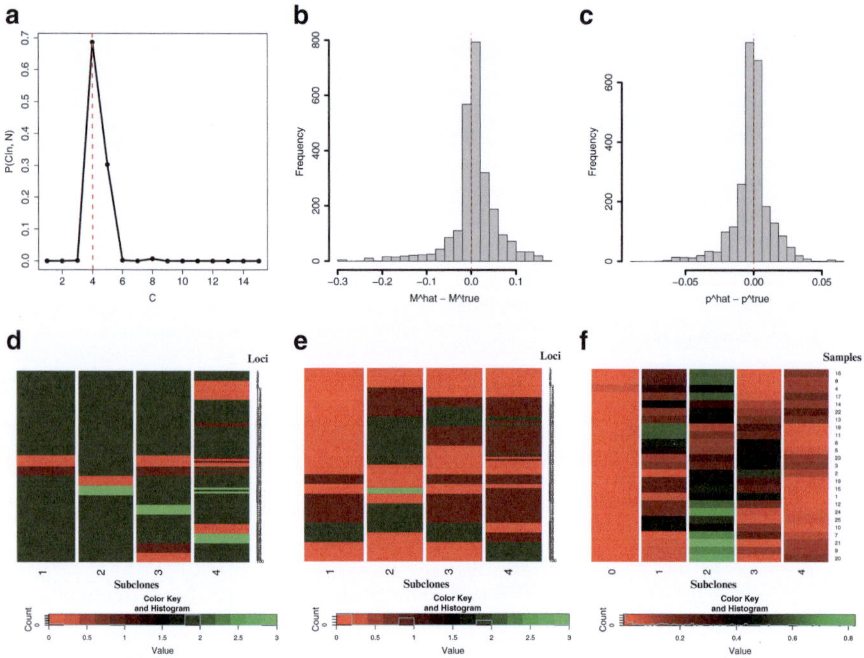


**Fig. 4.6** Posterior inference for the simulation study. (**a**) Posterior distribution of $C$, (**b**) $\hat{M}_{st} - M_{st}^{\text{TRUE}}$, (**c**) $\hat{p}_{st} - p_{st}^{\text{TRUE}}$, (**d**) $\boldsymbol{L}^{\star}$, (**e**) $\boldsymbol{Z}^{\star}$, (**f**) $\boldsymbol{w}^{\star}$

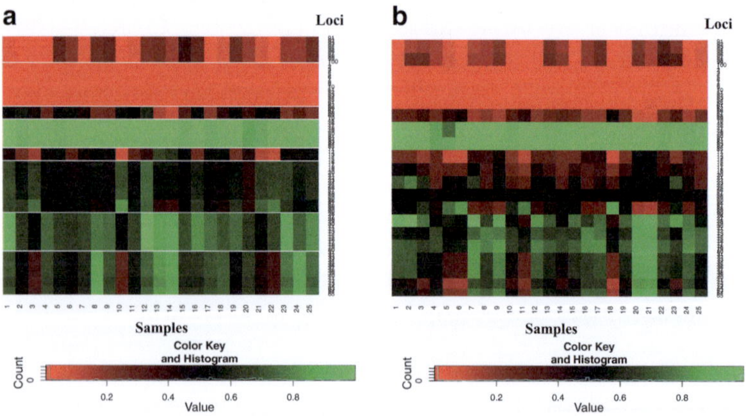


**Fig. 4.7** Heatmaps of estimated cellular prevalences from PyClone (**a**) and $p_{st}^{\text{TRUE}}$ (**b**) for the simulation study

## 4.4 Lung Cancer Data

We record whole-exome sequencing for $T = 4$ surgically dissected tumor samples taken from the same lung cancer patient. The four samples are spatially close to each other. It is of interest to see if spatially proximal tumor samples are genetically homogeneous. To this end, we extract genomic DNA from each tissue and construct an exome library from these DNA using Agilent SureSelect capture probes. The exome library is then sequenced in paired-end fashion on an Illumina HiSeq 2000 platform. About 60 million reads—each 100 bases long—are produced. We map the reads to the human genome (version HG19) (Church et al. 2011) using BWA (Li and Durbin 2009) and call variants using GATK (McKenna et al. 2010). Post-mapping, the mean coverage of the samples is between 60 and 70 fold.

A total of nearly 115,000 loci and small indels are called within the exome coordinates. We restrict our attention to loci that (1) make a difference to the protein translated from the gene, and (2) that exhibit significant coverage in all samples with $n_{st}/N_{st}$ not being too close to 0 or 1; and (3) we use expert judgment to some more loci. The described filter rules leave in the end $S = 101$ loci for the four intra-tumor samples.

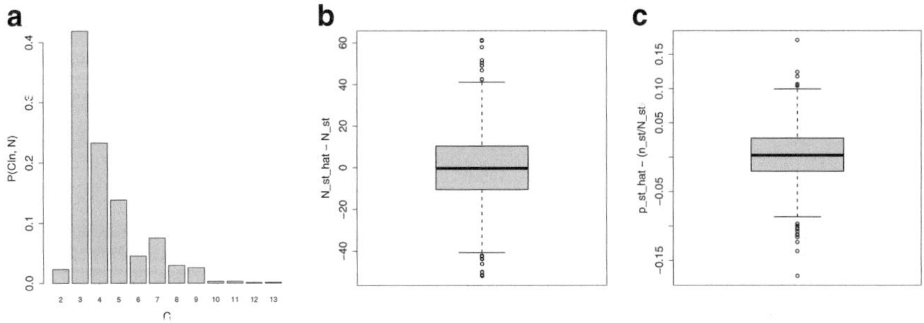


**Fig. 4.8** Posterior inference for the lung cancer dataset. (**a**) $p(C \mid \boldsymbol{n}, \boldsymbol{N})$, (**b**) $\hat{N}_{st} - N_{st}$, (**c**) $\hat{p}_{st} - (n_{st}/N_{st})$

We use hyperparameters similar to those in the simulation study. Figure 4.8 summarizes posterior inference under the proposed model. Panel (a) shows $C^\star = 2$, i.e., two estimated subclones. Using posterior samples with $C = C^\star$, we computed $\hat{N}_{st}$ and $\hat{p}_{st}$ and compared them to the observed data. The differences are centered at 0, implying a good fit to the data. Conditional on $C^\star = 2$, we found $\boldsymbol{L}^\star$, $\boldsymbol{Z}^\star$, and $\boldsymbol{w}^\star$ (see Fig. 4.9). The loci in $\boldsymbol{L}^\star$ and $\boldsymbol{Z}^\star$ are re-arranged in the same order for better illustration. The estimated weights $\boldsymbol{w}^\star$ in Fig. 4.9 show a great similarity across the four samples. This lack of heterogeneity across samples suggests that for this tumor, spatial proximity is implicative of genetic homogeneity.

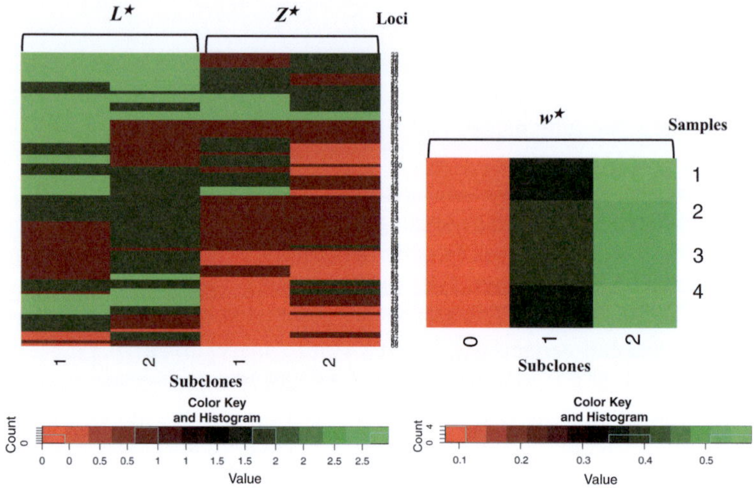


**Fig. 4.9** Posterior point estimates for the lung cancer dataset

## 4.5 Conclusions

We review Bayesian feature allocation models for the estimation of tumor heterogeneity, in terms of subclonal copy numbers, subclonal variant allele counts, and cellular fractions. By jointly modeling CNV and SNV, we characterize the genetic landscape of a tumor sample based on both sequence and structure variations. Such inference impacts downstream treatment of individual tumors, ultimately allowing precise prognosis. For example, a tumor with large proportions of cells bearing somatic mutations on oncogenes would respond more favorably to treatments that suppress the expression of the oncogenes. Also, temporal TH can be inferred if tumor samples from different time points are available from the same patient. Treatment options can be adaptively modified over time based on the heterogeneity among the samples. Indeed, large effort has already been invested in clinic regarding treating cancer based on the genetic contents instead of tumor origin (http://www.cancer.gov/clinicaltrials/noteworthy-trials/match), and trials regarding TH are being conducted (Catenacci 2014).

Many extensions are still needed to improve TH calling. For example, sometimes additional sources of information on CNVs such as an SNP array may be available. We then extend the proposed model to incorporate this information into the modeling of $\boldsymbol{L}$. Another extension is to cluster patients on the basis of the imputed TH. This extension may help clinicians assign different treatment strategies, and be a natural basis of adaptive clinical trial designs.

**Acknowledgements** Yuan Ji and Peter Müller's research is partially supported by NIH R01 CA132897.

## References


Almendro, V., Marusyk, A., and Polyak, K. (2013). Cellular heterogeneity and molecular evolution in cancer. *Annual Review of Pathology: Mechanisms of Disease*, **8**, 277–302.

Bedard, P., Hansen, A., Ratain, M., and Siu, L. (2013). Tumour heterogeneity in the clinic. *Nature*, **501**.

Biesecker, L. and NB, S. (2013). A genomic view of mosaicism and human disease. *Nature Reviews Genetics*, **14**.

Brooks, S., Gelman, A., Jones, G., and Meng, X.-L. (2011). *Handbook of Markov Chain Monte Carlo*. CRC Press.

Catenacci, D. V. (2014). Next-generation clinical trials: Novel strategies to address the challenge of tumor molecular heterogeneity. *Molecular oncology*.

Church, D. M., Schneider, V. A., Graves, T., Auger, K., Cunningham, F., Bouk, N., Chen, H.-C., Agarwala, R., McLaren, W. M., Ritchie, G. R., *et al.* (2011). Modernizing reference genome assemblies. *PLoS biology*, **9**(7), e1001091.

De, S. (2011). Somatic mosaicism in healthy human tissues. *Trends in genetics : TIG*, **27**(6), 217–223.

Deshwar, A., Vembu, S., Yung, C. K., Jang, G. H., Stein, L., and Morris, Q. (2014). Reconstructing subclonal composition and evolution from whole genome sequencing of tumors. *ArXiv 1406.7250*, page ArXiv Tech. Report.

Dexter, D. L., Kowalski, H. M., Blazar, B. A., Fligiel, Z., Vogel, R., and Heppner, G. H. (1978). Heterogeneity of tumor cells from a single mouse mammary tumor. *Cancer Research*, **38**(10), 3174–3181.

Fischer, A., Vázquez-García, I., Illingworth, C. J., and Mustonen, V. (2014). High-Definition Reconstruction of Clonal Composition in Cancer. *Cell Reports*, **7**, 1740–1752.

Frank, S. and Nowak, M. (2003). Cell biology: Developmental predisposition to cancer. *Nature*, **422**.

Frank, S. and Nowak, M. (2004). Problems of somatic mutation and cancer. *BioEssays*, **26**, 291–299.

Greaves, M. and Maley, C. C. (2012). Clonal evolution in cancer. *Nature*, **481**(7381), 306–313.

Griffiths, T. L. and Ghahramani, Z. (2005). Infinite latent feature models and the indian buffet process. In *In NIPS*, pages 475–482. MIT Press.

Jiao, W., Vembu, S., Deshwar, A., Stein, L., and Morris, Q. (2014). Inferring clonal evolution of tumors from single nucleotide somatic mutations. *BMC Bioinformatics*, **15**(1), 35.

Kim, Y., James, L., and Weissbach, R. (2012). Bayesian analysis of multistate event history data: beta-Dirichlet process prior. *Biometrika*, **99**(1), 127–140.

Klambauer, G., Schwarzbauer, K., Mayr, A., Clevert, D.-A., Mitterecker, A., Bodenhofer, U., and Hochreiter, S. (2012). MOPS: mixture of Poissons for discovering copy number variations in next-generation sequencing data with a low false discovery rate. *Nucleic Acids Research*, **40**(9), e69–e69.

Lee, J., Müller, P., Gulukota, K., and Ji, Y. (2015a). A Bayesian feature allocation model for tumor heterogeneity. *Annals of Applied Statistics*, page In press.

Lee, J., Müller, P., Sengupta, S., Gulukota, K., and Ji, Y. (2015b). Bayesian inference for tumor subclones accounting for sequencing and structural variants. *arXiv:1409.7158v1*.

Li, B. and Li, J. Z. (2014). A general framework for analyzing tumor subclonality using SNP array and DNA sequencing data. *Genome Biology*, **15**, 473.

Li, H. and Durbin, R. (2009). Fast and accurate short read alignment with burrows–wheeler transform. *Bioinformatics*, **25**(14), 1754–1760.

Mardis, E. R. (2008). Next-generation DNA sequencing methods. *Annu. Rev. Genomics Hum. Genet.*, **9**, 387–402.

Marjanovic, N. D., Weinberg, R. A., and Chaffer, C. L. (2013). Cell plasticity and heterogeneity in cancer. *Clinical chemistry*, **59**(1), 168–179.

McKenna, A., Hanna, M., Banks, E., Sivachenko, A., Cibulskis, K., Kernytsky, A., Garimella, K., Altshuler, D., Gabriel, S., Daly, M., *et al.* (2010). The Genome Analysis Toolkit: a MapReduce framework for analyzing next-generation DNA sequencing data. *Genome research*, **20**(9), 1297–1303.

Miller, C. A., White, B. S., Dees, N. D., Griffith, M., Welch, J. S., Griffith, O. L., Vij, R., Tomasson, M. H., Graubert, T. A., Walter, M. J., *et al.* (2014). Sciclone: Inferring clonal architecture and tracking the spatial and temporal patterns of tumor evolution. *PLoS computational biology*, **10**(8), e1003665.

Navin, N., Krasnitz, A., Rodgers, L., Cook, K., Meth, J., Kendall, J., Riggs, M., Eberling, Y., Troge, J., Grubor, V., *et al.* (2010). Inferring tumor progression from genomic heterogeneity. *Genome research*, **20**(1), 68–80.

Navin, N., Kendall, J., Troge, J., Andrews, P., and Rodgers, L. (2011). Tumour evolution inferred by single-cell sequencing. *Nature*, **472**.

Nowell, P. (1976). The clonal evolution of tumor cell populations. *Science*, **195**, 23–28.

Oesper, L., Mahmoody, A., and Raphael, B. J. (2013a). THetA: inferring intra-tumor heterogeneity from high-throughput DNA sequencing data. *Genome biology*, **14**(7).

Oesper, L., Mahmoody, A., and Raphael, B. J. (2013b). Theta: inferring intra-tumor heterogeneity from high-throughput dna sequencing data. *Genome Biol*, **14**(7), R80.

Polyak, K. (2011). Heterogeneity in breast cancer. *The Journal of clinical investigation*, **121**(10), 3786.

Roth, A., Khattra, J., Yap, D., Wan, A., Laks, E., Biele, J., Ha, G., Aparicio, S., Alexandre, B., and Shah, S. P. (2014a). PyClone: statistical inference of clonal population structure in cancer. *Nature methods*, **11**(4), 396–398.

Roth, A., Khattra, J., Yap, D., Wan, A., Laks, E., Biele, J., Ha, G., Aparicio, S., Bouchard-Côté, A., and Shah, S. P. (2014b). PyClone: statistical inference of clonal population structure in cancer. *Nature methods*.

Russnes, H. G., Navin, N., Hicks, J., and Borresen-Dale, A.-L. (2011a). Insight into the heterogeneity of breast cancer through next-generation sequencing. *The Journal of Clinical Investigation*, **121**(10), 3810–3818.

Sengupta, S. (2013). *Two Models Involving Bayesian Nonparametric Techniques (Ph.D thesis)*. Ph.d thesis, University of Florida.

Sengupta, S., Guluokta, K., Lee, J., Müller, P., and Ji, Y. (2015). BayClone: Bayesian nonparametric inference of tumor subclones using NGS data. In *Proceedings of The Pacific Symposium on Biocomputing (PSB) 2015*, pages 20: 467–478.

Serena, N., Van Loo, P., Wedge, D., Alexandrov, L., Greenman, C., Lau, K., Raine, K., Jones, D., Marshall, J., Ramakrishna, M., Shlien, A., Cooke, S., Hinton, J., Menzies, A., Stebbings, L., Leroy, C., Jia, M., Rance, R., Mudie, L., Gamble, S., Stephens, P., Stuart, M., Tarpey, P., Papaemmanuil, E., Davies, H., Varela, I., David, M., Bignell, G., Leung, K., Butler, A., Teague, J., Martin, S., Jönsson, G., Mariani, O., Boyault, S., Miron, P., Fatima, A., Langerød, A., Aparicio, S., Tutt, A., Sieuwerts, A., Borg, r., Thomas, G., Salomon, A., Richardson, A., Anne-Lise, B., Futreal, P., Stratton, M., Campbell, P., and of the International Cancer Genome Consortium, B. C. W. G. (2012). The life history of 21 breast cancers. *Cell*, **149**(5), 994–991007.

Shackleton, M., Quintana, E., Fearon, E. R., and Morrison, S. J. (2009). Heterogeneity in cancer: cancer stem cells versus clonal evolution. *Cell*, **138**(5), 822–829.

Shapiro, E., Biezuner, T., and Linnarsson, S. (2013). Single-cell sequencing-based technologies will revolutionize whole-organism science. *Nature reviews. Genetics*, **14**(9), 618–630.

Stingl, J. and Caldas, C. (2007). Molecular heterogeneity of breast carcinomas and the cancer stem cell hypothesis. *Nature Reviews Cancer*, **7**(10), 791–799.

Strino, F., Parisi, F., Micsinai, M., and Kluger, Y. (2013). Trap: a tree approach for fingerprinting subclonal tumor composition. *Nucleic Acids Research*, **41**(17), e165.

Teh, Y. W., Görür, D., and Ghahramani, Z. (2007). Stick-breaking construction for the Indian buffet process. In *Proceedings of the International Conference on Artificial Intelligence and Statistics*, volume 11.

Weinberg, R. A. (2007). *The biology of cancer*, volume 255. Garland Science New York.

Zare, H., Wang, J., Hu, A., Weber, K., Smith, J., Nickerson, D., Song, C., Witten, D., Blau, C. A., and Noble, W. S. (2014). Inferring clonal composition from multiple sections of a breast cancer. *PLoS computational biology*, **10**(7), e1003703.

# Chapter 5
# Species Sampling Priors for Modeling Dependence: An Application to the Detection of Chromosomal Aberrations

Federico Bassetti, Fabrizio Leisen, Edoardo Airoldi, and Michele Guindani

**Abstract** We discuss a class of Bayesian nonparametric priors that can be used to model local dependence in a sequence of observations. Many popular Bayesian nonparametric priors can be characterized in terms of exchangeable species sampling sequences. However, in some applications, common exchangeability assumptions may not be appropriate. We discuss a generalization of species sampling sequences, where the weights in the predictive probability functions are allowed to depend on a sequence of independent (not necessarily identically distributed) latent random variables. More specifically, we consider conditionally identically distributed (CID) Pitman-Yor sequences and the Beta-GOS sequences recently introduced by Airoldi et al. (*Journal of the American Statistical Association*, **109**, 1466–1480, 2014). We show how those processes can be used as a prior distribution in a hierarchical Bayes modeling framework, and, in particular, how the Beta-GOS can provide a reasonable alternative to the use of non-homogenous Hidden Markov models, further allowing

F. Bassetti
Dipartimento di Matematica, Universitá di Pavia, via Ferrata, 1, 27100 Pavia, Italy
e-mail: federico.bassetti@unipv.it

F. Leisen
School of Mathematics, Statistics and Actuarial Sciences, University of Kent, Cornwallis Building, CT2 7NF Canterbury, Kent, UK
e-mail: fabrizio.leisen@gmail.com

E. Airoldi
Department of Statistics, Harvard University, 1 Oxford Street, Cambridge, MA 02138, USA
e-mail: airoldi@fas.harvard.edu

M. Guindani (✉)
Department of Biostatistics, University of Texas MD Anderson Cancer Center, Houston, TX, USA
e-mail: mguindani@mdanderson.org

 
R. Mitra, P. Müller (eds.), *Nonparametric Bayesian Inference in Biostatistics*, Frontiers in Probability and the Statistical Sciences,
DOI 10.1007/978-3-319-19518-6_5

unsupervised clustering of the observations in an unknown number of states. The usefulness of the approach in biostatistical applications is discussed and explicitly shown for the detection of chromosomal aberrations in breast cancer.

## 5.1 Introduction

Due to their clustering properties, Bayesian nonparametric methods have been widely employed for the analysis of various types of data in genetics, e.g. for identifying disease subtypes and isolating discriminating genes, proteins or samples (see, e.g., Kim et al. 2006; Guindani et al. 2009; Lee et al. 2013). In order to take into account measurement characteristics (e.g., continuous support, long tails, skewness, multimodality or overdispersion of the frequency distribution), it is often convenient to employ a hierarchical model specification. At the top level of the hierarchy, observations are assumed to be conditionally independent given some "latent" process, i.e. the sampling distribution is

$$y_i|\theta_i \overset{ind}{\sim} p(y_i|\theta_i) \quad i = 1, 2, \ldots \tag{5.1}$$

where $p(\cdot|\theta_i)$ denotes a probability density function or probability mass function, dependent on the values of a set of parameters $\theta_i$. The distribution of the $\theta_i$'s is then assumed to follow a process that captures relevant features of the data. Let $p$ denote the unknown distribution of the model parameters, and $Q$ be a prior probability measure for $p$. Then, the hierarchical model specification can be concisely described as follows:

$$\begin{aligned} \theta_1, \theta_2 \ldots | p &\sim p \\ p &\sim Q. \end{aligned} \tag{5.2}$$

Model (5.1), (5.2) schematically encompasses both popular Dirichlet Process mixtures (Lo 1984) and Dependent Dirichlet Process mixtures (MacEachern 1999). The prior process $p$ can often be represented by means of a sequence of predictive distributions that typically encode exchangeability assumptions on the model parameters and the data (see Sect. 5.2). In some applications, however, the usual exchangeability assumptions may be hardly justified. For example, if $\theta_1, \theta_2, \ldots$ represent a process in time (space), then the model should properly account for the dependence relations among nearby time points (neighboring locations).

To illustrate the point, in Fig. 5.1a we consider the frequency of genome copy number abnormalities, as estimated from data obtained in a classical study of the genetic determinants of breast cancer pathophysiologies (Chin et al. 2006). The raw data measure genome copy number gains and losses over 145 primary breast tumor samples, across the 23 chromosomes, obtained using BAC array comparative genomic hybridization (CGH). Regions of relative gains or losses are identified by measuring the fluorescence ratio of cancer and normal female genomic DNA, la-

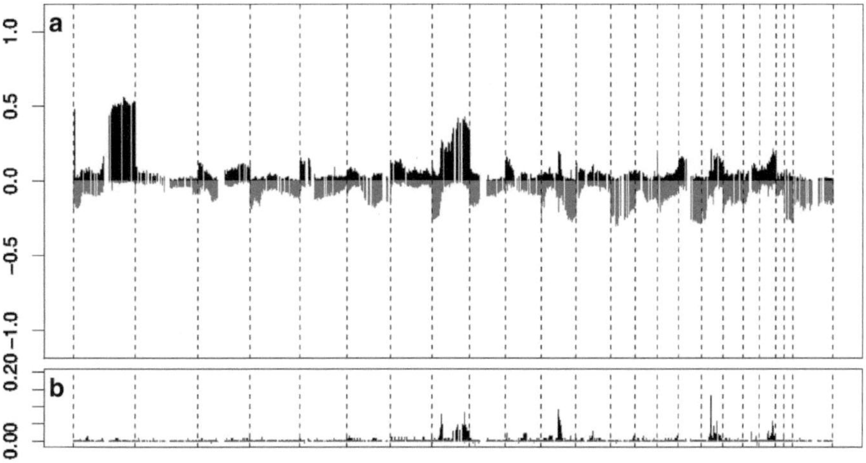


**Fig. 5.1** (**a**) Frequencies of genome copy number gains and losses plotted as a function of genomic location. (**b**) Frequency of tumors showing high-level amplification. The *dashed vertical lines* separate the 23 chromosomes

beled with distinct fluorescent dyes and co-hybridized on a microarray in the presence of Cot-1 DNA to suppress unspecific hybridization of repeat sequences (see Redon et al. 2009). The reference DNA is assumed to have two copies of each chromosome. If the test sample has no copy number aberrations, the log2 of the intensity ratio is theoretically equal to zero.

Array CGH data are typically very noisy and spatially correlated. More specifically, copy number gains or losses at a region are often associated with an increased probability of gains and losses at a neighboring region. Bayesian models for array CGH data have been recently investigated by Guha et al. (2008), DeSantis et al. (2009), Baladandayuthapani et al. (2010), Du et al. (2010), Cardin et al. (2011), and Yau et al. (2011), among others. Guha et al. propose a four state homogenous Bayesian HMM to detect copy number amplifications and deletions and partition tumor DNA into regions (clones) of relatively stable copy number. DeSantis et al. extend this approach and propose a supervised Bayesian latent class approach for classification of the clones, which relies on a heterogenous hidden Markov model to account for local dependence in the intensity ratios. In a heterogeneous hidden Markov model, the transition probabilities between states depend on each single clone or the distance between adjacent clones (Marioni et al. 2006). Using a Bayesian nonparametric approach, Du et al. propose a sticky Hierarchical DP-HMM (Fox et al. 2011; Teh et al. 2006) to infer the number of states in an HMM, while also imposing state persistence. Yau et al. (2011) also propose a nonparametric Bayesian HMM, but use instead a DP mixture to model the likelihood in each state.

In this chapter, we present an alternative approach, which flexibly models the evolution of the parameters $\theta_1, \theta_2, \ldots$ by means of a general class of non-exchangeable species sampling sequences. As it is typical in a Bayesian non-parametric setting, we allow clustering of the observations, further assuming that the number of states is unknown and can be inferred from the data. Furthermore, in finite HMMs, the distribution of state durations is necessarily restricted to a geometric form, so that departures from this assumption, e.g. state persistence, must be appropriately accounted for in the modeling (Yu 2010; Fox et al. 2011; Johnson and Willsky 2013). The species sampling priors, which we discuss in the next section, model "non-homogenous" assumptions in the state durations more flexibly, since the weights in the species sampling rule can adapt to take into account local dependences in the data.

## 5.2 Species Sampling Sequences: Basics and Extensions

In this section, we review basic definitions and properties of species sampling (SS) sequences, and also discuss their generalizations to a class of random sequences that are appealing for modeling non-exchangeable observations.
More specifically, for defining SS-sequences, we refer to the hierarchical formulation (5.2), and characterize the sequence of random variables $\theta_1, \theta_2, \dots$ by means of the sequence of predictive probability functions,

$$P\{\theta_{n+1} \in \cdot \,|\theta_1, \dots, \theta_n\} = \sum_{i=1}^{n} q_{n,i}\delta_{\theta_i}(\cdot) + q_{n,n+1}G_0(\cdot), \tag{5.3}$$

where $\delta_x(\cdot)$ denotes a point mass at $x$, and $G_0$ is a non-atomic probability measure (base measure, Pitman 1996). The weights $q_{n,i}$, $i = 1, \dots, n+1$, are non-negative functions of $(\theta_1, \dots, \theta_n)$, such that $\sum_{i=1}^{n+1} q_{n,i} = 1$, and define the probability that the sampled value of $\theta_{n+1}$ coincides with one of the previous values in the sequence or is a new draw from the base measure. In (5.3), it is implicitly assumed that $\theta_1 \sim G_0$. If $q_{n,n+1} < 1$, there's a positive probability of ties among the $\theta_i$'s, that is some of the $\theta_i$'s will share a common value. We can collect the unique values in a vector $(\theta_1^*, \dots \theta_{K_n}^*)$, where $K_n$ indicates the (random) number of distinct values in the sub-sequence $\theta(n) = (\theta_1, \dots, \theta_n)$. Alternatively, we can say that (5.3) implicitly defines a random partition $\Pi^{(n)} = \{\Pi_1^{(n)}, \dots, \Pi_{K_n}^{(n)}\}$ of the set $\{1, \dots, n\}$ into $K_n$ blocks, where $i \in \Pi_j^{(n)}$ if and only if $\theta_i = \theta_j^*$.

If the probability of a tie, $P(\theta_{n+1} = \theta_j^* | \theta(n))$, depends only on the cardinality of each block, i.e. the frequency $n_{jn} = |\Pi_j^{(n)}|$ of each value $\theta_j^*$ in $\theta(n)$, $j = 1, \dots, K_n$, then the sequence $\theta_1, \theta_2, \dots$ is exchangeable. The result characterizes all exchangeable SS-sequences (see Fortini et al. 2000; Hansen and Pitman 2000; Lee et al. 2008, for more details). The most notable example of exchangeable SS-sequences is the Blackwell MacQueen sampling rule, which defines a Dirichlet Process (see Blackwell and MacQueen 1973; Ishwaran and Zarepour 2003). Let $p$ be a DP with mass parameter $\gamma$ and base measure $G_0(\cdot)$, denoted as $p \sim DP(\gamma, G_0)$. Then, the corresponding sequence of predictive probability function is the well-known Blackwell MacQueen sampling rule, which sets $q_{n,i} = \frac{1}{n+\gamma}$ and $q_{n,n+1} = \frac{\gamma}{n+\gamma}$ in (5.3).

The dependence of the weights only on the sequence $\theta(n)$ may be seen as a limiting feature in some applications, e.g. whenever one could contemplate that additional covariate information might affect the clustering of the observations. For example, Park and Dunson (2010) propose a generalized product partition model (GPPM) in which the clustering process is predictor-dependent. Their GPPM relax the exchangeability assumption through the incorporation of predictors, implicitly defining a generalized Pólya urn scheme. Similarly, Müller and Quintana (2010) define a product partition model that includes a regression on covariates, which allows units with similar covariates to have greater probability of being clustered together.

Here, we consider a generalization of the predictive rule (5.3), where the weights are allowed to depend on a sequence of independent (not necessarily identically distributed) latent random variables $W_1, W_2, \dots$ More specifically, we consider a sequence $(\theta_n)_{n\geq 1}$ characterized by the following predictive distributions,

$$P\{\theta_{n+1} \in \cdot\, | \theta(n), W(n)\} = \sum_{i=1}^{n} p_{n,i}\delta_{\theta_i}(\cdot) + r_n G_0(\cdot), \tag{5.4}$$

where $W(n) = (W_1, \dots, W_n)$ and the weights $p_{n,i}$ are strictly positive functions of the partitions $\Pi^{(n)}$ and the random variables $W(n)$, i.e. $p_{n,i} = p_{n,i}(\Pi^{(n)}, W(n)) > 0$, with $\sum_{i=1}^{n} p_{n,i} < 1$ and $r_n := 1 - \sum_{i=1}^{n} p_{n,i}$.

The specific choice of the weights $p_{n,i}$'s determines the clustering behavior of the sequence $(\theta_n)_n$. In this chapter, we focus on the general class of *conditionally identically distributed* (CID) sequences (Berti et al. 2004). This class generalizes the notion of exchangeable sequences, while still preserving some of their important characteristics. Formally, a sequence $(\theta_n)_{n\geq 1}$ is CID with respect to a filtration $\mathscr{G} = (\mathscr{G}_n)_{n\geq 0}$, whenever for each $n \geq 0$ all the random variables $\theta_{n+i}$, with $i \geq 1$, are identically distributed conditionally on $\mathscr{G}_n$. In the definition it is assumed that $\mathscr{G}$ contains the natural filtration of $(\theta_i)_{i\geq 1}$. It is clear that every exchangeable sequence is a CID sequence with respect to its natural filtration, but a CID sequence does not necessarily need to be exchangeable nor stationary. Indeed, if a CID sequence is stationary, then it is also exchangeable. A remarkable property of CID sequences is that the $\theta_i$'s are marginally identically distributed. No representation theorem is known for CID sequences. However, it can be shown that given any bounded and measurable function $f$, the predictive mean $E[f(\theta_{n+1})|\theta_1, ..., \theta_n]$ and the empirical mean $\frac{1}{n}\sum_{i=1}^{n} f(\theta_i)$ converge to the same limit as $n$ goes to infinity. For details, we refer to Berti et al. (2004). Finally, if the sequence of observations $(Y_1, Y_2, \dots)$ follows the hierarchical model (5.1) and the latent process $(\theta_1, \theta_2, \dots)$ is a CID sequence, then it can be shown that the sequence of observations $Y_i$'s also forms a CID sequence. This result has been proved in Airoldi et al. (2014) specifically for the Beta-GOS prior (see below); however, the proof can be easily extended to a general CID sequence.

Two interesting types of CID sampling sequences are the following:

(a) **CID Pitman-Yor sequences.** A Pitman-Yor process (Pitman 2006) is an exchangeable sequence characterized by the following predictive probability functions,

$$P\{\theta_{n+1} \in \cdot\, | \theta_1, \ldots, \theta_n\} = \Sigma_{j=1}^{K_n} \frac{n_{jn} - \alpha}{\gamma + n} \delta_{\theta_j^*}(\cdot) + \frac{\gamma + \alpha K_n}{\gamma + n} G_0(\cdot), \tag{5.5}$$

for $\gamma > 0$ and $\alpha \in [0,1]$, as a function of the partition $\Pi^{(n)} = \{\Pi_1^{(n)}, \ldots, \Pi_{K_n}^{(n)}\}$ of the set $\{1, \ldots, n\}$ into $K_n$ blocks. When $\alpha = 0$, the sequence (5.5) defines a Dirichlet Process, $DP(\theta, G_0)$. There exists a generalization of the classical Pitman-Yor process (5.5) as a CID sequence. More specifically, the CID generalization assumes that the weights in (5.4) are functions of a sequence of random variables $W(n)$, with weights $p_{n,i}(\Pi^{(n)}, W(n)) = (W_i - \alpha/n_{k_i n})/(\gamma + \Sigma_{j=1}^n W_j)$ and $r_n(\Pi^{(n)}, W(n)) = (\gamma + \alpha K_n)/(\gamma + \Sigma_{j=1}^n W_j)$ where $n_{k_i n}$ denotes the cardinality of the block in $\Pi^{(n)}$ that contains observation $i$. Then,

$$P\{\theta_{n+1} \in \cdot\, | \theta(n), W(n)\} = \sum_{j=1}^{K_n} \frac{\left(\Sigma_{i \in \Pi_j^{(n)}} W_i\right) - \alpha}{\gamma + \Sigma_{i=1}^n W_i} \delta_{\theta_j^*}(\cdot) + \frac{\gamma + \alpha K_n}{\gamma + \Sigma_{i=1}^n W_i} G_0(\cdot), \tag{5.6}$$

which reduces to (5.5) if $W_n = 1$. Similarly to the Chinese Restaurant Process (CRP) representation of the Dirichlet Process, Eq. (5.6) has an intuitive illustration in terms of the seating allocation at a restaurant. In this representation, each customer enters the restaurant with a distinctive "mark" (the random variables $W_i$'s). When customers enter the restaurant, they have the possibility to start a new table (with probability dependent on the parameter $\gamma$) or join a table already occupied by other customers. In the CID version, the "attractiveness" of a table depends on $\Sigma_{i \in \Pi_j^{(n)}} W_i$ in (5.6), i.e. the sum of the individual marks for each customer already seating at the table. In other words, the process takes into account possible additional variability in the "seating plan" due to individual random effects.

In terms of clustering, the asymptotic behavior of the CID version of the DP, obtained by setting $\alpha = 0$ in (5.6), is similar to that of the classical DP: if the $W_i$'s are i.i.d. with finite variance and mean $E[W_i] = m$, then $K_n/\log(n)$ converges almost surely to $\gamma/m$ (see Bassetti et al. 2010, Example 5.8). The situation is less simple for the case in which $\alpha \neq 0$.

(b) **Beta-GOS sequences.** An alternative specification of (5.4) considers weights obtained as a product of independent Beta random variables. More specifically, Airoldi et al. (2014) assume that the random variables $(W_i)_{i \geq 1}$ are draws from independent Beta$(\alpha_i, \beta_i)$ distributions, and then set $p_{n,i} = (1 - W_i)\prod_{j=i+1}^n W_j$ and $r_n = \prod_{j=1}^n W_j$ in (5.4). The resulting sequence $(\theta_1, \theta_2, \ldots)$ defines the so-called *Beta-GOS* sequence, a particular case of a generalized ottawa sequence

(GOS) in the class of CID sequences (see, for details, Bassetti et al. 2010). The choice of Beta latent variables allows for a flexible specification of the species sampling weights, while it still retains simplicity and interpretability of the sequence allocation scheme. As a matter of fact, this allocation rule can also be described in terms of a preferential attachment scheme, similarly to the CID Pitman-Yor sequences. Also in this scheme, each customer, $\theta_i$, is characterized by a random weight (or "mark"), $1-W_i$, and can join the table where any of the previous customer is sitting by means of a "geometric-type" assignment scheme. More precisely, suppose we have customers $\theta_1,\ldots,\theta_n$ together with their marks up to time $n$, $(1-W_1,\ldots,1-W_n)$. Then, the $(n+1)$-th individual will be assigned to the same table as the previous customer, $\theta_n$, with probability $1-W_n$; the probability of pairing $\theta_{n+1}$ to $\theta_{n-1}$ will be $W_n(1-W_{n-1})$, and so forth. In general, in this representation, $W_i$ will represent a "repulsion" score associated with customer $i$. Thus, each weight $p_{n,i}$ will be represented by the product of the $W_j$'s associated with the latest $n-j$ subjects and the "mark" or "attractiveness" score of customer $i$, $1-W_i$. Summarizing, customer $\theta_{n+1}$ will occupy a new table (i.e., $\theta_{n+1}\sim G_0$) with probability $r_n$, or instead they will join one of the previously occupied tables, say table $j$, with probability $\sum_{i:\theta_i=\theta_j^*} p_{n,i}$. Of course, the seating assignment and the clustering behavior of the sequence is determined by the specification of the parameters $\alpha_i$ and $\beta_i$ in the distribution of the $W_i$'s. We briefly discuss the issue in the next section, where we review some asymptotic results and their interpretation in terms of clustering of the sequence, for a set of parameter specifications.

In the next sections, we will focus specifically on the use of the Beta-GOS sequences for modeling latent dependence in Bayesian hierarchical models and we will discuss their application to the detection of chromosomal aberrations in array CGH data.

## 5.3 A Beta-GOS Hierarchical Model

In this section, we focus on the Beta-GOS sequences and discuss how they can be used to define a prior in a hierarchical model (for a broader discussion, see Airoldi et al. 2014). Although the discussion pertains specifically to the Beta-GOS process, the basic modeling idea naturally extends to the CID Pitman-Yor sequences and the general CID sequences. We then discuss the prior specification of the parameters of the Beta random variables in the Beta-GOS. Finally, we briefly present the MCMC sampling algorithm for conducting posterior inference with this type of models.

Similarly to the hierarchical Bayesian specification in (5.1), (5.2), we can assume that at the highest level of the hierarchy the sampling distribution is specified as

$$Y_i|\theta_i \overset{ind.}{\sim} f(y_i|\theta_i), \quad i=1,\ldots,n, \tag{5.7}$$

where the vector $(\theta_1,\ldots,\theta_n)^T$ is a realization of a Beta-GOS process characterized by auxiliary random variables $W_i \sim Be(\alpha_i,\beta_i)$, $i=1,\ldots,n$, and base measure $G_0$. We can succinctly denote the Beta-GOS prior as

$$\theta_1,\ldots,\theta_m \sim \text{Beta-GOS}(\boldsymbol{\alpha}_n,\boldsymbol{\beta}_n,G_0), \tag{5.8}$$

where $\boldsymbol{\alpha}_n = (\alpha_1,\ldots\alpha_n)$ and $\boldsymbol{\beta}_n = (\beta_1,\ldots,\beta_n)$. As discussed in Sect. 5.2, the Beta-GOS is a particular case of a CID sequence. Hence, in particular, marginally $\theta_i \sim G_0$, $i=1,\ldots,n$. Therefore, the base $G_0$ can be regarded as a centering distribution, as it is typical in DP mixture models: $G_0$ represents a vague parametric prior assumption on the distribution of the parameters of interest. The hierarchical model may be extended by putting hyper-priors on the remaining parameters of the model, including the hyper-parameters of the base measure $G_0$ as well as the vectors $\boldsymbol{\alpha}_n$ and $\boldsymbol{\beta}_n$.

The parameters of the Beta random variables control the asymptotic behavior of the sequence, and the clustering properties of the prior. For example, if we set $\alpha_i = i+\gamma-1, \beta_i = 1$, for given $\gamma > 0$, then $K_n/\log(n)$ converges in distribution to a $Gamma(\gamma,1)$ random variable. As a comparison, for a $DP(\gamma,G_0)$, it is well known that $K_n/\log(n)$ converges almost surely to $\gamma$. If we set $\alpha_i = a, \beta_i = b$, for some $a,b>0$, then $K_n$ converges almost surely to a finite random variable. This result naturally implies that the resulting partition is characterized by a few big clusters, as $n$ increases. We refer to Airoldi et al. (2014) for further details and proofs. In addition, the parameters $\alpha_i$ and $\beta_i$ implicitly model the autocorrelation expected a priori in the dynamics of the sequence. The probability of a tie may decrease with $n$ and atoms that have been observed at farthest times may have a greater probability to be selected if they have also been observed more recently. More specifically, setting $\alpha_i = \gamma-1+j$ ($\gamma>0$) and $\beta_i = 1$ implies that $E[r_n] = \gamma/(\gamma+n)$ and $E[p_{n,i}] = 1/(\gamma+n)$, $i=1,\ldots,n$. This specification can be seen as a feature of a process with a long memory, since all the previous observations have the same weight on average. For $\alpha_i = a, \beta_i = b$, $E[r_n] = (a/(a+b))^n$ and $E[p_{n,i}] = (a/(a+b))^{n-i}(b/(a+b))$. Hence, the probabilities of ties d decrease exponentially as a function of the lag $n-i$, describing a short memory process. In practice, the determination of the parameters of the Beta distributions is not trivial, and may be problem dependent, especially given the sensitivity of the clustering behavior to the values of $\alpha_i$ and $\beta_i$. As a general rule, following what it is usually done with Dirichlet processes priors, we suggest to elicit the parameters on the basis of the expected number of clusters a priori, i.e. $E(K_n) = 1+\sum_{j=1}^{n-1} E[r_j]$. For example, one could set $\alpha_i = a$ and $\beta_i = b$ to represent a short memory process, and the values of $a,b$ can be chosen based on the asymptotic relationship $E(K_n) \approx \frac{a+b}{b}$. We further suggest to choose $b=1$, or anyway $b<a$, to encourage a priori low autocorrelation of the sequence, since then $E(p_{n,n}) < 0.5$. As a matter of fact, in Sect. 5.5 we will follow the previous guidelines in the application to the detection of chromosomal aberrations, since biological considerations lead to expect the true number of states to be around 4. On the other hand, the single-parameter specification $\alpha_j = j+\gamma-1$, $\beta_j = 1$ should be the default choice in those applications where prior information on the expected number of clusters is more vague, and the choice of the parameter $\gamma$ should be based on $E(K_n) = \sum_{j=0}^{n-1} \frac{\gamma}{\gamma+j} \sim \gamma\log(n)$, for large $n$.

### 5.3.1 MCMC Posterior Sampling

Posterior inference for the model (5.7) and (5.8) entails learning about the clustering and corresponding estimates of the parameters $\theta_i$. In this section, we describe a Gibbs sampler scheme. The basic idea is to describe the partition $\Pi^{(n)}$ by introducing a sequence of labels $C_i$, $i = 1, \ldots, n$ which record the pairing of observation $i$ with one of the previous observations, $j < i$. Hence, here the label $C_i$ is not a simple indicator of the cluster membership, as it is typical in most MCMC algorithms devised for the Dirichlet process, although cluster membership can be easily retrieved by analyzing the sequence of pairings. In what follows, $C_i$ will be sometimes referred to as the $i$-th pairing label. In particular, if the $i$-th observation is not paired to any of the preceding ones, we set $C_i = i$. Then, $\theta_i$ is a draw from the base distribution $G_0$, and thus it generates a new cluster. This slightly different representation of data points in terms of data-pairing labels, instead of cluster-assignment labels, turns useful to develop an MCMC sampling scheme for non-exchangeable processes, as described in Blei and Frazier (2011) and Airoldi et al. (2014). It is easy to see that the pairing sequence $(C_n)_{n\geq 1}$ assigns $C_1 = 1$ and has full conditional distribution

$$\begin{aligned} P\{C_n = i|C_1, \ldots, C_{n-1}, W\} &= P\{C_n = i|W_1, \ldots, W_{n-1}\} \\ &= r_{n-1}\mathbb{I}\{i = n\} + p_{n-1,i}\mathbb{I}\{i \neq n\}, \end{aligned} \tag{5.9}$$

for $i = 1, \ldots, n$, where $\mathbb{I}(\cdot)$ denotes the indicator function, such that, given a set A, $\mathbb{I}(A) = 1$ if $A$ is true and 0 otherwise. The clustering configuration is a by-product of the representation in terms of data-pairing labels. If two observations are connected by a sequence of interim pairings, then they are in the same cluster. Given $C(n) = (C_1, \ldots, C_n)$, then we denote by $\Pi(C(n))$ the partition generated by the pairings $C(n)$, i.e. $\Pi^{(n)}$. For any $n$ and any $i \leq n$, let $C_{-i} = (C_1, \ldots, C_{i-1}, C_{i+1}, \ldots, C_n)$; analogously, let $W(n) = (W_1, \ldots, W_n)$, and $W_{-i} = (W_1, \ldots, W_{i-1}, W_{i+1}, \ldots, W_n)$. Then, the full conditional for the pairing indicators $C_i$'s is

$$\begin{aligned} P\{C_i = j|C_{-i}, Y(n), W(n)\} &\propto P\{C_i = j, Y(n)|C_{-i}, W(n)\} \\ &= P\{Y(n)|C_i = j, C_{-i}, W(n)\}P\{C_i = j|C_{-i}, W(n)\}. \end{aligned} \tag{5.10}$$

The second term in (5.10) is the prior predictive rule (5.9), whereas

$$P\{Y(n)|C_i = j, C_{-i}, W(n)\} = \prod_{k=1}^{|\Pi(C_{-i},j)|} \int \prod_{l \in \Pi(C_{-i},j)_k} f(Y_l|\theta_j^*)\, G_0(d\theta_j^*),$$

where $\Pi(C_{-i}, j)$ denotes the partition generated by $(C_1, \ldots, C_{i-1}, j, C_{i+1}, \ldots, C_n)$. If $G_0$ and $f(y|\theta)$ are conjugate, the latter integral has a closed form solution. The non-conjugate case could be handled by appropriately adapting the algorithms of MacEachern and Müller (1998) and Neal (2000). As far as the full conditional for the latent variables $W_i$'s, we can show that $W_i|C(n), W_{-i}, Y(n) \sim \text{Beta}(A_i, B_i)$, where $A_i = \alpha_i + \sum_{j=i+1}^{n} \mathbb{I}\{C_j < i \text{ or } C_j = j\}$, and $B_i = \beta_i + \sum_{j=i+1}^{n} \mathbb{I}\{C_j = i\}$; hence, they depend only on the clustering configurations and not on the values of $W_{-i}$.

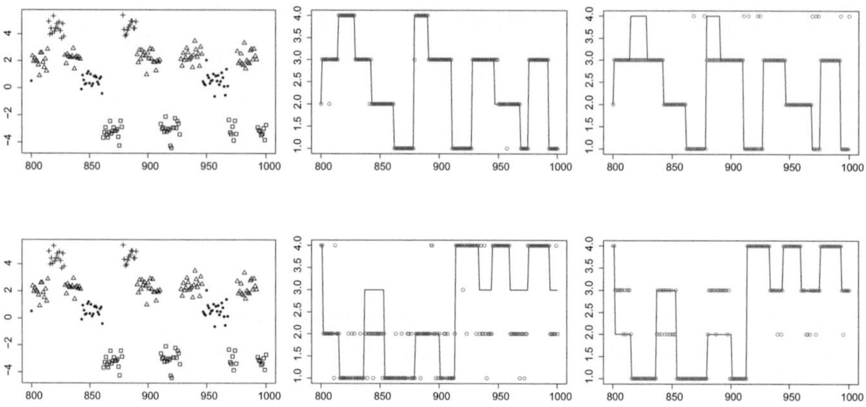

**Fig. 5.2** Illustrative segmentation-type plots for the simulation study in Sect. 5.4. *Right column*: subset of data for two replicates. *Center column top*: an example of allocation for a Beta-GOS($\alpha_i = 1, \beta_i = 1$) plotted *vs* the truth (*black line*); *bottom* considers a Beta-GOS($\alpha_i = i, \beta_i = 1$). *Left column* illustrates the fitting by a HMM with 4 states

Then, let's consider the set of cluster centroids $\theta_i^*$'s. The algorithm above allows faster mixing of the chain by integrating over the distribution of the $\theta_i^*$. However, in case inference on the vector $(\theta_1, \dots, \theta_m)$ is of interest, it is possible to sample the unique cluster values at each iteration, as

$$\theta_j^* | C(n), W(n), Y(n) \propto \prod_{i \in \Pi_j(n)} p(Y_i | \theta_j^*) G_0(d\theta_j^*), \tag{5.11}$$

where $\Pi_j(n)$ denotes the partition set of those observations with $\theta_i = \theta_j^*, i = 1, \dots, n$. Again, if $f(y|\theta)$ and $G_0$ are conjugate, the full conditional of $\theta_j^*$ is available in closed form, otherwise we can update $\theta_j^*$ by standard Metropolis Hastings algorithms (Neal 2000).

Finally, we note that if $\pi(\boldsymbol{\alpha}_n, \boldsymbol{\beta}_n)$ is a prior distribution for the Beta hyper-parameters $\boldsymbol{\alpha}_n$ and $\boldsymbol{\beta}_n$, one could implement a Metropolis Hasting scheme to learn about their posterior distribution, since

$$\boldsymbol{\alpha}_n, \boldsymbol{\beta}_n | C(n), Y(n) \propto \pi(\boldsymbol{\alpha}_n, \boldsymbol{\beta}_n) \prod_{i=1}^{n} \frac{B(A_i, B_i)}{B(\alpha_i, \beta_i)}, \tag{5.12}$$

where $A_i$ and $B_i$ are defined as above and $B(x,y) = \Gamma(x)\Gamma(y)/\Gamma(x+y)$ denotes the Beta function. Equation (5.12) is an adaptation of well-known results for the Dirichlet Process (Escobar and West 1995).

## 5.4 A Comparison with Hidden Semi-Markov Models

In many problems (e.g., change point detection), hidden Markov Models are used as computationally convenient substitutes for temporal processes that are known to be more complex than what could be implied by first order Markovian dynamics. Here, we generate non-exchangeable sequences from a hidden semi-Markov process (HSMM; Ferguson 1980; Yu 2010) and study how the Beta-GOS process performs in fitting this type of data. Hidden semi-Markov processes are an extension of the popular hidden Markov model where the time spent in each state (state occupancy or sojourn time) is given by an explicit (discrete) distribution. A geometric state occupancy distribution characterizes ordinary hidden Markov models. Therefore, hidden semi-Markov process have also been referred to as "hidden Markov Models with explicit duration" (Mitchell et al. 1995; Dewar et al. 2012) or "variable-duration hidden Markov Models" (Rabiner 1989).

We generate 1000 datasets (1000 observations each) using a hidden semi-Markov process with four states and a negative binomial distribution for the state occupancy distribution. More specifically, we parametrize the negative binomial in terms of its mean and an ancillary parameter, which is directly related to the amount of overdispersion of the distribution (Hilbe 2011; Airoldi et al. 2006). If the data are not overdispersed, the Negative Binomial reduces to the Poisson, and the ancillary parameter is zero. For the simulations presented here, we consider a $\text{NegBin}(15, 0.15)$, which corresponds to assuming a large overdispersion (17.25). We also consider $\tau = 0.5$ for the noise. We fit the data by means of a Beta-GOS model with Beta hyper-parameters defined by: (a) $\alpha_i = i, \beta_i = 1$; (b) $\alpha_i = 5, \beta_i = 1$; (c) $\alpha_i = 1, \beta_i = 1$, $i = 1, \ldots, n$. Those choices correspond to assuming different clustering behaviors; in particular, different expected number of clusters a priori. We then compare the Beta-GOS with the fit resulting from hidden Markov models, assuming 3, 4, and 5 states, respectively. Results from the simulations are reported in Fig. 5.3, where the HMM was implemented using the R package "RHmm" (Taramasco and Bauer 2012). Figure 5.3 shows that the Beta-GOS is a viable alternative to HMM, as it can provide more accurate inference than a single hidden Markov model where the number of states is fixed a priori. The fit obtained with the Beta-GOS appears quite robust to the different choices of the hyper-parameters. Figure 5.2 illustrates the clustering induced by the Beta-GOS and a 4-state HMM for a subset of the data generated in two specific simulation replicates. The middle column illustrates the allocation, respectively, from a Beta-GOS($\alpha_i = 1, \beta_i = 1$) (*top*) and a Beta-GOS($\alpha_i = i, \beta_i = 1$) (*bottom*), whereas column (c) illustrates the clustering attained by the HMM. Overall, the segmentation-plots suggest similarity in the allocations induced by the Beta-GOS and the HMM. In some instances, the Beta-GOS fit seems to allow shorter stretches of contiguous identical states, as illustrated in the top row of Fig. 5.2. On the other hand, when data are characterized by elevated intra-claster variability, as in the bottom row of Fig. 5.2, both the Beta-GOS and the HMM could fail to attain a fair representation of the true clustering structure of the data. Our practical experience suggests that the issue is more prominent for the "default" Beta-GOS($\alpha_i = i, \beta_i = 1$) than for the "informative" Beta-GOS($\alpha_i = a, \beta_i = b$)

| i) Data Generating Process: Hidden Semi Markov Model (HSMM) with 4 states and NegBin(15,0.15) | | | | | | |
|---|---|---|---|---|---|---|
| **Model Fitting Method** | **Beta-GOS** | | | **HMM** | | |
| | $\alpha_n = n; \beta_n = 1$ | $\alpha_n = 5; \beta_n = 1$ | $\alpha_n = 1; \beta_n = 1$ | 3 States | 4 States | 5 States |
| Estimated Number of Clusters | 3.69±0.81 | 3.89±0.96 | 4.06±0.97 | 2.99±0.12 | 3.96±0.25 | 4.90±0.48 |
| Accuracy of Cluster Assignment | 0.86±0.14 | 0.90±0.12 | 0.90±0.12 | 0.71±0.11 | 0.83±0.12 | 0.88±0.13 |

**Fig. 5.3** Summary statistics for the simulation studies described in Sect. 5.4. The table compares the Beta-GOS and a hidden Markov model under different specifications of hyper-parameters. The data generating process assumes a hidden semi-Markov with state occupancy distribution NegBin(15,0.15) and two levels of the sampling noise $\tau = 0.25$ and $\tau = 0.5$

formulations. This is in accordance with the discussion in Sect. 5.3 and, in particular, with the consideration that a Beta-GOS($\alpha_i = i, \beta_i = 1$) should represent a long memory process.

## 5.5 Application to the Analysis of Array CGH Data

We apply the Beta-GOS model (5.7) and (5.8) to the analysis of the array CGH data from Chin et al. (2006) which we presented in Sect. 5.1. More specifically, we consider the raw log2 intensity ratio measurements and seek to identify and cluster clones with similar levels of amplification/deletion for each breast tumor sample and each chromosome in the dataset. For array CGH data, it is typical to distinguish regions with a normal amount of chromosomal material, from regions with single copy loss (deletion), single copy gain and amplifications (multiple copy gains). Therefore, we present here the results of the analysis where the latent Beta hyper-parameters are set to $\alpha_i = 3$ and $\beta_i = 1$, corresponding to $E(K_n) = 4$ states for large $n$. We have also considered $\alpha_n = n$ and $\beta_n = 1$, with no remarkable differences in the results. We complete the specification of model (5.7) and (5.8) with a vague base distribution, Normal$(0,10)$, and a vague inverse gamma distribution for $\tau$ centered around $\tau = 0.1$. This choice of $\tau$ is motivated by the typical scale of array CGH data and is in accordance with similar choices in the literature (see, for example, Guha et al. 2008).

Figure 5.4 exemplifies the fit to chromosome 8 on two tumor samples. The model is able to identify regions of reduced copy number variation and high amplification. Note how contiguous clones tend to be clustered together, in a pattern typical of these chromosomal aberrations. Figure 5.1 shows the frequencies of genome copy number gains and losses among all 145 samples plotted as a function of genome location. In order to identify a copy number aberration for this plot, for each chromosome and sample, at each iteration we consider the cluster with lowest absolute mean and order the other clusters accordingly. The lowest absolute mean is chosen to identify the copy neutral state. Following Guha et al. (2008) any other cluster is identified as a copy number gain or loss if its mean, say $\hat{\mu}_{(j)}$, is farther than a specified threshold from the minimum absolute mean, say $\hat{\mu}_{(1)}$, i.e. if $\hat{\mu}_{(j)} - \hat{\mu}_{(1)} > \varepsilon$. We experimented with choices of $\varepsilon$ in the range $[0.05, 0.15]$, but we report here only

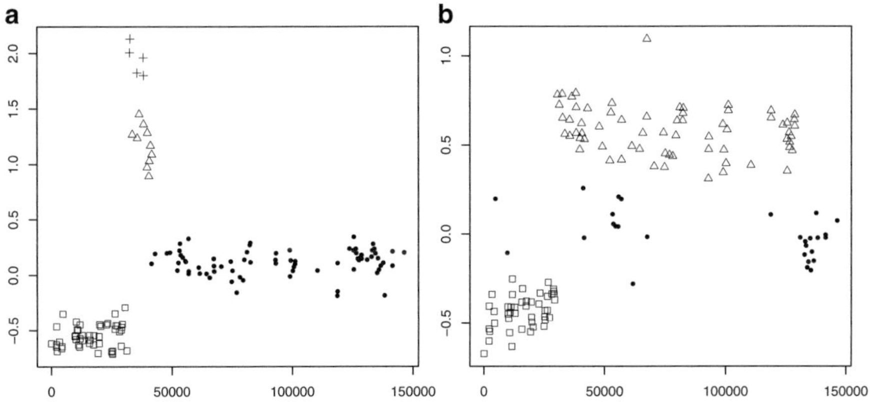


**Fig. 5.4** Model fit overview: Array CGH gains and losses on chromosome 8 for two samples of breast tumors in the dataset in Chin et al. (2006). Points with different shapes denote different clusters

the results for $\varepsilon = 0.1$. Furthermore, if the mean of a cluster is above the mean of all declared gains plus two standard deviations, all genes in that cluster are considered high level amplifications. We identify a clone with an aberration (or high level amplification) if it is such in more than 70 % of the MCMC iterations; then, we compute the frequency of aberrations and high level amplifications among all 145 samples, which are the values reported, respectively, at the top and bottom of Fig. 5.1. As expected, the clusters identified by the model tend to be localized in space all over the genome. This feature may be facilitated by the increasingly low reinforcement of far away clones embedded in the Beta-GOS, and corresponds to the understanding that clones that live at adjacent locations on a chromosome can be either amplified or deleted together due to the recombination process.

**Table 5.1** False discovery rate analysis for clones with high-level amplification previously identified by Chin et al. (2006)

| Amplicon | Flanking clone (left) | Flanking clone (right) | Kb start | Kb end | FDR q-value |
|---|---|---|---|---|---|
| 8p11-12 | RP11-258M15 | RP11-73M19 | 33579 | 43001 | 0.021 |
| 8q24 | RP11-65D17 | RP11-94M13 | 127186 | 132829 | 0.021 |
| 11q13-14 | CTD-2080I19 | RP11-256P19 | 68482 | 71659 | 0.022 |
| 11q13-14 | RP11-102M18 | RP11-215H8 | 73337 | 78686 | 0.024 |
| 12q13-14 | BAL12B2624 | RP11-92P22 | 67191 | 74053 | 0.011 |
| 17q11-12 | RP11-58O8 | RP11-87N6 | 34027 | 38681 | 0.017 |
| 17q21-24 | RP11-234J24 | RP11-84E24 | 45775 | 70598 | 0.017 |
| 20q13 | RMC20B4135 | RP11-278I13 | 51669 | 53455 | 0.021 |
| 20q13 | GS-32I19 | RP11-94A18 | 55630 | 59444 | 0.017 |

The individual amplicons are reported together with the locations of the flanking clones on the array platform

Finally, we considered some regions of chromosomes 8, 11, 17, and 20 that have been identified by Chin et al. (2006) and have been shown to correlate to increased gene expression in their analysis. We adapt the procedure described in Newton et al. (2004) to compute a region-based measure of the false discovery rate (FDR) and determine the $q$-values for the neutral-state and aberration regions estimated from our model. The $q$-value is the FDR analogue of the $p$-value, as it measures the minimum FDR threshold at which we may determine that a region corresponds to significant copy number gains or losses (Storey 2003, 2007). More specifically, after conducting a clone based test as described in the previous paragraph, we identify regions of interest by taking into account the strings of consecutive calls. These regions then constitute the units of the subsequent cluster based FDR analysis. Alternatively, the regions of interest could be pre-specified on the basis of the information available in the literature. The optimality of the type of procedures here described for cluster based FDR is discussed in Sun et al. 2015. See also Heller et al. 2006, Müller et al. 2007 and Ji et al. 2008. In Table 5.1 we report the $q$-values from a set of candidate oncogenes in well-known regions of recurrent amplification (notably, 8p12, 8q24, 11q13–14, 12q13–14, 17q21–24, and 20q13). Our findings also lead to detect chromosomal aberrations in the same locations reported by Chin et al. (2006).

## 5.6 Final Remarks

We have discussed a set of generalizations of the predictive rules that characterize the species sampling mechanism underlying many commonly used Bayesian Nonparametric priors, such as the Dirichlet process and the Pitman Yor process. Those generalizations allow the clustering of the observations in the sequence to depend on latent random variables or "marks", which are associated with each observation. Although the resulting sequence is in general not exchangeable, the framework provides a flexible way to model latent and local dependence in the observations.

We illustrated this feature in an application to a study of chromosomal aberrations in breast cancer. Although it's known that copy number gains and losses are spatially correlated, the extent of such correlation varies along the genome. Homogeneous Hidden Markov models have been widely employed to model copy number data (Guha et al. 2008), but it's been recognized that such models may not completely capture local dependence in the intensity ratios, which results in location-dependent transition probabilities and corresponding locally varying state persistence properties of the aberrations (DeSantis et al. 2009; Du et al. 2010; Fox et al. 2011). By considering species sampling sequences where the weights are modeled as functions of latent Beta random variables, we have defined a Beta-GOS process prior that provides an alternative Bayesian nonparametric formalism to model heterogeneity and local spatial dependence across observations that are sequentially ordered. In particular, since the Beta-GOS model does not rely on the estimation of a single transition matrix across time points, as in a homogenous HMM, we do not need to consider an explicit parameter to account for state persistence, as in Fox et al. (2011), or assume a distribution for the sojourn times, as assumed in Hidden Semi-Markov models. Indeed, since the predictive weights depend on the sequence of observations itself, the use of such prior appears to be particularly convenient when the underlying generative process is non-stationary, e.g. as a possible alternative to more complicated non-homogeneous HMMs. In addition, our modeling approach enables unsupervised clustering of the observations in an unknown number of states, as it is typical of Bayesian nonparametric priors.

The previous considerations remain valid also for the CID Pitman-Yor sequences we presented in Sect. 5.2 and can be extended to other types of conditionally identically distributed sequences characterized by the predictive rule (5.4). We believe that the flexibility of the latent specification and the possibility to tie the clustering implied by the Generalized Pólya Urn scheme directly to a set of latent random variables provides an opportunity to flexibly model the complex relationships typical of many heterogenous datasets encountered in biostatistics. For example, the approach may be helpful for modeling individual random effects in longitudinal studies. In functional data analysis, these priors could be used to detect change points in a curve. Further developments may substitute the latent variable specification with a probit/logistic specification, and define a generalized Pólya Urn scheme that allows the clustering at each observation to be dependent on a set of individual covariates, possibly varying with time.

## References

Airoldi, E., Costa, T., Bassetti, F., Leisen, F., and Guindani, M. (2014). Generalized Species Sampling Priors With Latent Beta Reinforcements. *Journal of the American Statistical Association*, **109**, 1466–1480.

Airoldi, E. M., Anderson, A., Fienberg, S., and Skinner, K. (2006). Who wrote Ronald Reagan's radio addresses? *Bayesian Anal.*, **1**, 289–320.

Baladandayuthapani, V., Ji, Y., Talluri, R., Nieto-Barajas, L. E., and Morris, J. S. (2010). Bayesian random segmentation models to identify shared copy number aberrations for array cgh data. *Journal of the American Statistical Association*, **105**(492), 1358–1375.

Bassetti, F., Crimaldi, I., and Leisen, F. (2010). Conditionally identically distributed species sampling sequences. *Adv. in Appl. Probab*, **42**, 433–459.

Berti, P., Pratelli, L., and P., R. (2004). Limit Theorems for a Class of Identically Distributed Random Variables. *Ann. Probab.*, **32**(3), 2029–2052.

Blackwell, D. and MacQueen, J. (1973). Ferguson distributions via Pólya urn schemes. *Ann. Statist.*, **1**(353–355).

Blei, D. and Frazier, P. (2011). Distance dependent Chinese restaurant processes. *Journal of Machine Learning Reseach*, **12**, 2461–2488.

Cardin, N., Holmes, C., Consortium, T. W. T. C. C., Donnelly, P., and Marchini, J. (2011). Bayesian hierarchical mixture modeling to assign copy number from a targeted cnv array. *Genetic Epidemiology*, **35**(6), 536–548.

Chin, K., DeVries, S., Fridlyand, J., Spellman, P. T., Roydasgupta, R., Kuo, W.-L., Lapuk, A., Neve, R. M., Qian, Z., Ryder, T., Chen, F., Feiler, H., Tokuyasu, T., Kingsley, C., Dairkee, S., Meng, Z., Chew, K., Pinkel, D., Jain, A., Ljung, B. M., Esserman, L., Albertson, D. G., Waldman, F. M., and Gray, J. W. (2006). Genomic and transcriptional aberrations linked to breast cancer pathophysiologies. *Cancer Cell*, **10**(6), 529–541.

DeSantis, S. M., Houseman, E. A., Coull, B. A., Louis, D. N., Mohapatra, G., and Betensky, R. A. (2009). A latent class model with hidden markov dependence for array cgh data. *Biometrics*, **65**(4), 1296–1305.

Dewar, M., Wiggins, C., and Wood, F. (2012). Inference in Hidden Markov Models with Explicit State Duration Distributions. *Signal Processing Letters, IEEE*, **19**(4), 235–238.

Du, L., Chen, M., Lucas, J., and Carlin, L. (2010). Sticky hidden Markov modelling of comparative genomic hybridization. *IEEE TRANSACTIONS ON SIGNAL PROCESSING*, **58(10)**, 5353–5368.

Escobar, M. D. and West, M. (1995). Bayesian density estimation and inference using mixtures. *Journal of the American Statistical Association*, **90**, 577–588.

Ferguson, J. D. (1980). Variable duration models for speech. In *Proceedings of the Symposium on the Applications of Hidden Markov Models to Text and Speech*, pages 143–179.

Fortini, S., Ladelli, L., and Regazzini, E. (2000). Exchangeability, predictive distributions and parametric models. *Sankhya*, **62**(1), 86–109.

Fox, E., Sudderth, E., Jordan, M., and Willsky, A. (2011). A sticky HDP-HMM with application to speaker diarization. *Annals of Applied Statistics*, **5(2A)**, 1020–1056.

Guha, S., Li, Y., and Neuberg, D. (2008). Bayesian hidden Markov modelling of array cgh data. *JASA*, **103**, 485–497.

Guindani, M., Müller, P., and Zhang, S. (2009). A Bayesian discovery procedure. *Journal of the Royal Statistical Society: Series B (Statistical Methodology)*, **71**(5), 905–925.

Hansen, B. and Pitman, J. (2000). Prediction rules for exchangeable sequences related to species sampling. *Statist. Probab. Lett.*, **46**(251–256).

Heller, R., Stanley, D., Yekutieli, D., Rubin, N., and Benjamini, Y. (2006). Cluster-based analysis of fMRI data. *Neuroimage*, **33**, 599–608.

Hilbe, J. M. (2011). *Negative Binomial Regression*. Cambridge University Press.

Ishwaran, H. and Zarepour, M. (2003). Random probability measures via Pólya sequences: revisiting the Blackwell-MacQueen urn scheme. Technical report, Arxiv.org.

Ji, Y., Lu, Y., and Mills, G. (2008). Bayesian models based on test statistics for multiple hypothesis testing problems. *Bioinformatics*, **24**, 943–949.

Johnson, M. J. and Willsky, A. S. (2013). Bayesian nonparametric hidden semi-Markov models. *J. Mach. Learn. Res.*, **14**(1), 673–701.

Kim, S., Tadesse, M. G., and Vannucci, M. (2006). Variable selection in clustering via dirichlet process mixture models. *Biometrika*, **93**(4), 877–893.

Lee, J., Quintana, F., Müller, P., and Trippa, L. (2008). Defining Predictive Probability Functions for Species Sampling Models. . *Statist.Sci.*, **2**(209–222).

Lee, J., Müller, P., Zhu, Y., and Ji, Y. (2013). A nonparametric Bayesian model for local clustering with application to proteomics. *Journal of the American Statistical Association*, **108**(503), 775–788.

Lo, A. (1984). On a class of Bayesian nonparametric estimates: I density estimates. *Ann. Statist.*, **12 (1)**, 351–357.

MacEachern, S. N. (1999). Dependent nonparametric processes. In *Proceedings of the Section on Bayesian Statistical Science*.

MacEachern, S. N. and Müller, P. (1998). Estimating mixtures of Dirichlet process models. *Journal of Computational and Graphical Statistics*, **7**, 223–238.

Marioni, J. C., Thorne, N. P., and Tavaré, S. (2006). Biohmm: a heterogeneous hidden Markov model for segmenting array CGH data. *Bioinformatics*, **22**(9), 1144–1146.

Mitchell, C., Harper, M., and Jamieson, L. (1995). On the complexity of explicit duration hmm's. *Speech and Audio Processing, IEEE Transactions on*, **3**(3), 213–217.

Müller, P. and Quintana, F. (2010). Random partition models with regression on covariates. *Journal of Statistical Planning and Inference*, **140**(10), 2801–2808.

Müller, P., Parmigiani, G., and Rice, K. (2007). FDR and Bayesian multiple comparisons rules. In J. Bernardo, M. Bayarri, J. Berger, A. Dawid, D. Heckerman, A. Smith, and M. West, editors, *Bayesian Statistics 8*. Oxford, UK: Oxford University Press.

Neal, R. M. (2000). Markov Chain Sampling Methods for Dirichlet Process Mixture Models. *Journal of Computational and Graphical Statistics*, **9**, 249–265.

Newton, M. A., Noueiry, A., Sarkar, D., and Ahlquist, P. (2004). Detecting differential gene expression with a semiparametric hierarchical mixture method. *Biostatistics*, **5**, 155—176.

Park, J. and Dunson, D. (2010). Bayesian generalized product partition model. *Statistica Sinica*, **20**(1203–1226).

Pitman, J. (1996). *Some developments of the Blackwell-MacQueen urn scheme*, volume 30, pages 245–267. Lecture Notes-Monograph Series, Institute of Mathematical Statistics, Hayward, California.

Pitman, J. (2006). *Combinatorial Stochastic Processes*. Lecture Notes in Mathematics. Springer:Berlin / Heidelberg.

Rabiner, L. (1989). A tutorial on hidden Markov models and selected applications in speech recognition. *Proceedings of the IEEE*, **77**(2), 257–286.

Redon, R., Fitzgerald, T., and Carter, N. (2009). Comparative genomic hybridization: DNA labeling, hybridization and detection. In M. Dufva, editor, *DNA Microarrays for Biomedical Research*, volume 529 of *Methods in Molecular Biology*, pages 267–278. Humana Press.

Storey, J. D. (2003). The positive false discovery rate: a Bayesian interpretation and the q-value. *The Annals of Statistics*, **31**, 2013–2035.

Storey, J. D. (2007). The optimal discovery procedure for large-scale significance testing, with applications to comparative microarray experiments. *Biostatistics*, **8**, 414–432.

Sun, W., Reich, B. J., Tony Cai, T., Guindani, M., and Schwartzman, A. (2015). False discovery control in large-scale spatial multiple testing. *Journal of the Royal Statistical Society Series B*, **77**, 59–83.

Taramasco, O. and Bauer, S. (2012). RHMM: Hidden Markov models simulations and estimations. Technical report, CRAN.

Teh, Y. W., Jordan, M. I., Beal, M. J., and Blei, D. M. (2006). Hierarchical Dirichlet processes. *Journal of the American Statistical Association*, **101**(476), 1566–1581.

Yau, C., Papaspiliopoulos, O., Roberts, G. O., and Holmes, C. (2011). Bayesian non-parametric hidden Markov models with applications in genomics. *Journal of the Royal Statistical Society: Series B (Statistical Methodology)*, **73**(1), 37–57.

Yu, S.-Z. (2010). Hidden semi-markov models. *Artificial Intelligence*, **174**(2), 215–243. Special Review Issue.

# Chapter 6
# Modeling the Association Between Clusters of SNPs and Disease Responses

Raffaele Argiento, Alessandra Guglielmi, Chuhsing Kate Hsiao, Fabrizio Ruggeri, and Charlotte Wang

**Abstract** The aim of the paper is to discuss the association between SNP genotype data and a disease. For genetic association studies, the statistical analyses with multiple markers have been shown to be more powerful, efficient, and biologically meaningful than single marker association tests. As the number of genetic markers considered is typically large, here we cluster them and then study the association between groups of markers and disease. We propose a two-step procedure: first a Bayesian nonparametric cluster estimate under normalized generalized gamma process mixture models is introduced, so that we are able to incorporate the information from a large-scale SNP data with a much smaller number of explanatory variables. Then, thanks to the introduction of a genetic score, we study the association between the relevant disease response and groups of markers using a logit model. Inference is obtained via an MCMC truncation method recently introduced in the literature. We also provide a review of the state of art of Bayesian nonparametric cluster models and algorithms for the class of mixtures adopted here. Finally, the model is applied to genome-wide association study of Crohn's disease in a case-control setting.

R. Argiento (✉) • F. Ruggeri
CNR-IMATI, Via Bassini 15, 20133 Milano, Italy
e-mail: raffaele@mi.imati.cnr.it

A. Guglielmi
Dipartimento di Matematica, Politecnico di Milano, piazza Leonardo 32, 20133 Milano, Italy
e-mail: alessandra.guglielmi@polimi.it

C.K. Hsiao • C. Wang
Institute of Epidemiology and Preventive Medicine, National Taiwan University,
Taipei 100, Taiwan

 
R. Mitra, P. Müller (eds.), *Nonparametric Bayesian Inference in Biostatistics*,
Frontiers in Probability and the Statistical Sciences,
DOI 10.1007/978-3-319-19518-6_6

## 6.1 Introduction

In recent years, researchers have been trying to identify genetic variants influencing complex diseases through genetic association studies. For genome-wide association studies, single marker tests are common approaches to figure out possible disease-associated markers; however, these methods are often criticized for the problem of multiple testing and low power (Asimit and Zeggini 2010; Bansal et al. 2010). Dealing with these problems, multiple-marker tests, such as candidate multiple-marker tests, haplotype analysis, SNP-set analysis, gene-set analysis, and pathway analysis, have become popular solutions. SNP-set analysis is more flexible in terms of defining an analytic genomic region based on investigators' prior knowledge and available biological information (Huang et al. 2011; Nguyen et al. 2011). Such methods are beneficial to evaluate the joint effects of grouped variants in a pre-specified region. For instance, at the genome-wide level, the information about the relation among an enormous set of genes may not be complete, and thus it can be difficult to decipher the association between the genome-wide markers and the disease phenotypes. In that case, scientists resort to methodologies that cluster or categorize the genomic markers into several relatively smaller and manageable sets before performing multiple-marker association tests. Therefore, clustering methods are usually considered as the first step in the analysis of SNP genotypes. Algorithms utilizing mathematical formulations of similarity include principal component analysis, k-means, and Hamming distance metric. Even if the actual computation is easy, these tools do not call for any statistical model for the data, and hence inference is poor. In addition, such clustering algorithms present one more limitation, i.e., the determination of the number of clusters a priori.

From the statistical perspective, Bayesian nonparametrics can handle the problems of clustering in a natural way through species sampling mixtures, where the mixing distribution represents the proportion of (possibly infinite) groups, and the estimation of its number is a by-product of the model. There is a quite lively literature on the topic. We refer here to the early works by Quintana and Iglesias (2003), Medvedovic et al. (2004), and Dahl (2006) for the Dirichlet process mixture DPM. However, none of these papers considered categorical genotype data, while a DPM model for clustering single nucleotide polymorphisms (SNPs) is described in Onogi et al. (2011).

When data collection aims at studying a disease, even when a cluster procedure has been selected and performed, the impact of its uncertainty on subsequent association analysis is rarely assessed. Previous genome association studies include Wei et al. (2010), Wakefield (2007), Wakefield (2009). More recently, Molitor et al. (2010) and Papathomas et al. (2012) propose a Bayesian nonparametric model for grouping patients according to the clustering of categorical covariates, then associated with a relevant outcome through a regression model. See Liverani et al. (2015) for an R package to perform inference under these models. On the other hand, Müller et al. (2011) achieve a covariate-driven clustering including the covariates directly into the prior of the partition induced by the infinite species-sampling mixture model and modeling the likelihood at the cluster level. Since in genomics the covariate matrix usually includes a very large number of columns (i.e., $p >> n$), another stream of research within Bayesian Nonparametrics aims at identifying the relevant covariates for clustering patients; see Tadesse et al. (2005), Yau and Holmes (2011), and Chung and Dunson (2009).

In the paper we propose a two-step procedure: we first cluster SNPs, i.e. columns of the covariate matrix, where the rows identify patients, and then, thanks to the introduction of a genetic score, we study association between the relevant disease response and selected groups of markers. In particular, we induce clustering among SNPs through a nonparametric model, based on normalized generalized gamma mixtures. The class of normalized generalized gamma processes has been recently introduced in the statistical literature by Lijoi et al. (2007). This class encompasses the Dirichlet process, and has been proved to be very flexible in its clustering ability. With this flexible Bayesian nonparametric clustering model, the number of genome-wide markers can be reduced effectively to a manageable number of marker-sets (clusters) for association studies, and then markers in the same cluster can be investigated simultaneously to retain their possible, though unknown, interacting relation.

It is worth mentioning that our approach is similar to the one in Papathomas et al. (2012). The latter authors cluster individuals in groups (e.g., high risk, average risk and low risk for a certain disease) and then evaluate which covariates are influent in clustering, using DPMs. In our approach the procedure is reversed: we first cluster the SNPs according to a normalized generalized gamma mixture model with multinomial kernels, and then investigate which groups of SNPs affect the risk of disease of an individual via the logit model.

The paper is organized as follows. Section 6.2 introduces a quite general formulation for clustering in Bayesian Nonparametrics, using the normalized generalized gamma process as species sampling mixing measure. Moreover, we revise the approach to the definition of point estimates of the partition parameter assigning the clustering structure, and the MCMC algorithms under normalized generalized gamma process mixtures. Section 6.3 details the clustering model for general SNP data; in particular, we describe SNP data for non-experts in Sect. 6.3.1, the model in Sect. 6.3.2 and introduce the genetic score in Sect. 6.3.3. Section 6.4 provides the application to genotyping of Crohn's disease patients and Sect. 6.5 concludes.

## 6.2 Clustering Through Bayesian Nonparametric Models

A standard approach in Bayesian nonparametrics dealing with clustering can be described as follows. If $(X_1,\ldots,X_n)$ represents the data, its conditional distribution can be assigned as:

$$(X_1,\ldots,X_n)|S_1,\ldots,S_k,\phi_1,\ldots,\phi_k \sim \prod_{j=1}^{k}\left\{\prod_{i\in S_j} f(x_i|\phi_j)\right\}, \tag{6.1}$$

where $\rho := \{S_1, \dots, S_k\}$ is a partition of the data label set $\{1, \dots, n\}$ and $\{f(\cdot|\phi), \phi \in \Theta\}$ is a parametric family of densities on $\mathbb{X}$, the space of data values. Observe that here $k$ is the number of clusters in the partition $\rho$. From (6.1), it is clear that, conditionally on $\rho$, data are independent between different clusters and independent and identically distributed (i.i.d.) within each cluster. To complete the Bayesian model we need to assign a prior for $(\rho, \boldsymbol{\phi})$, with $\boldsymbol{\phi} := (\phi_1, \dots, \phi_k)$. We assume that

$$\pi(\rho) = \mathbb{P}(\rho = \{S_1, \dots, S_k\}) = p(|S_1|, \dots, |S_k|), \tag{6.2}$$

where $p(\cdot)$ is an *infinite* exchangeable partition probability function. Moreover, conditionally on $\rho$, we assume that the parameters $(\phi_1, \dots, \phi_k)$ in (6.1) are i.i.d. from some fixed distribution $P_0$ on $\Theta$. According to Pitman (1996), for any distribution $P_0$ and any exchangeable partition probability function $p(\cdot)$, there exists a unique species sampling prior $\Pi(\cdot; p, P_0)$ on the space of all probabilities on $\Theta$, such that model (6.1) under the specified prior is equivalent to

$$\begin{aligned} X_i|\theta_i &\overset{\text{ind.}}{\sim} f(\cdot|\theta_i) \quad i = 1, \dots, n \\ \theta_i|P &\overset{\text{i.i.d.}}{\sim} P \quad i = 1, \dots, n \qquad P \sim \Pi(\cdot; p, P_0), \end{aligned} \tag{6.3}$$

where $P_0$ represents the expectation of $P$. In this case, we say that $\theta_i$ is the latent variable corresponding to $X_i$ in the mixture model (6.3).

In particular, in this work, $P$ is the normalized generalized gamma (NGG) process prior, introduced in Regazzini et al. (2003). It is well known that such a process $P$ can be represented as

$$P = \sum_{i=1}^{+\infty} \xi_i \delta_{\tau_i} = \sum_{i=1}^{+\infty} \frac{J_i}{T} \delta_{\tau_i} \tag{6.4}$$

where $\xi_i := J_i/T$, $(J_i)_i$ are the points of a Poisson process on $\mathbb{R}^+$ with mean intensity $(\kappa/\Gamma(1-\sigma))s^{-1-\sigma}e^{-s}\mathbb{I}_{(0,+\infty)}(s)$, and $T = \sum_i J_i$; the random variables $\tau_i$ are independent from $\{J_i\}$, and $\tau_i$'s are i.i.d. from $P_0$. Here $0 \le \sigma < 1$, $\kappa \ge 0$. We write $P \sim NGG(\sigma, \kappa, P_0)$, where $(\sigma, \kappa, P_0)$ are the parameters of the NGG-process. See Lijoi et al. (2007) and Argiento et al. (2010) for more details. This class encompasses the Dirichlet processes: when $\sigma = 0$ and $\kappa > 0$, $P$ is the Dirichlet process (Ferguson 1973) with measure parameter $\kappa P_0(\cdot)$. On the other hand, when $\sigma = 1/2$, $P$ reduces to the normalized inverse-Gaussian process.

One of the main arguments in favor of NGG processes, when compared with DPs, is its higher flexibility in clustering. For instance, when considering a sample of size $n$ from an NGG process, the distribution of the number $K_n$ of distinct values in the sample has a further degree of freedom, $\sigma$, which tunes its variance, unlike the DP case where the distribution of $K_n$ can be highly peaked. The parameter $\sigma$ also drives a richer reinforcement mechanism in the predictive distributions of the sample. Moreover, NGG processes are of Gibbs-type, a class of random probabilities which stands out for their mathematical tractability (see De Blasi et al. 2014). Figure 6.1 shows the prior induced on $K_n$ when we consider a sample of size $n = 150$

from an NGG-process, under different values of $(\sigma, \kappa)$, with $\mathbb{E}(K_n)$ set to 27. As $\sigma$ increases, the prior becomes vaguer and vaguer.

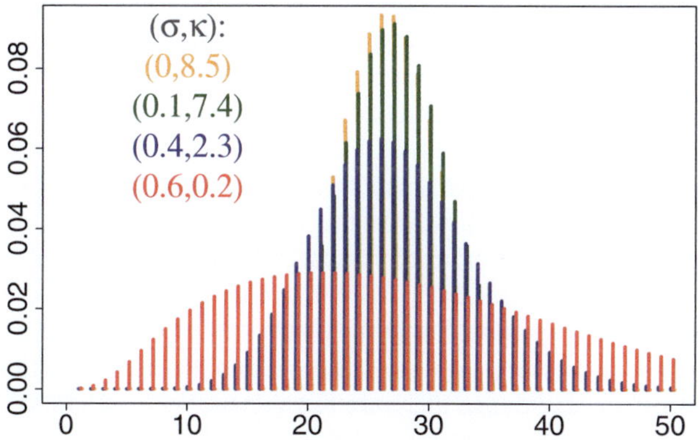


**Fig. 6.1** Prior distribution of $K_n$, the number of distinct values in a sample of size $n = 150$ from an NGG process. All the couples $(\sigma, \kappa)$ here yield $\mathbb{E}(K_n) = 27$

Observe that the equivalence between models (6.1), (6.2), on the one hand, and (6.3), on the other, holds thanks to the natural clustering rule and identifiability of the likelihood. By *natural clustering rule* we mean the following: given $\theta_1, \ldots, \theta_n$, $X_l$ and $X_j$ belong to the same cluster if, and only if, $\theta_l = \theta_j$. The partition $\rho = \{S_1, \ldots, S_k\}$ of the data label set $\{1, \ldots, n\}$ is induced by the natural clustering rule, and $\boldsymbol{\phi} = (\phi_1, \ldots, \phi_k)$ are the distinct values among the $\theta_i$'s. Under both model formulations, cluster estimates are based on the posterior distribution of $\rho$, given the data, i.e. $\mathscr{L}(\rho|data)$.

In the Bayesian nonparametric model-based context, the choice of a suitable point estimate $\hat{\rho}$ of the random partition $\rho$ is a key point. From a computational point of view, once we have obtained an MCMC sample from the posterior law $\mathscr{L}(\rho|data)$, a Bayesian estimate of $\rho$ is computed via a summary of the latter sample. Nevertheless, in general, the search for such a posterior estimate can be a difficult task because of the large support of the prior, and consequently of the posterior, of $\rho$. All the methods that we are going to shortly revise here were originally proposed for DPMs, but, of course, they can be extended to any species sampling model mixtures, since they are all based on the marginal posterior law $\mathscr{L}(\rho|data)$ only; see Argiento et al. (2014). Medvedovic et al. (2004) estimate the pairwise similarity matrix induced by $\mathscr{L}(\rho|data)$, and then use a hierarchical clustering algorithm to estimate $\rho$. A different approach (Quintana and Iglesias 2003; Dahl 2006; Lau and Green 2007; Fritsch and Ickstadt 2009) follows a theoretical-decision perspective: a suitable loss function $L(\rho, \hat{\rho})$ is fixed, giving the cost of estimating the "true" $\rho$ by $\hat{\rho}$. The Bayesian estimate is therefore given by any $\hat{\rho}$ which minimizes the posterior expectation of the loss function, i.e.

$$\hat{\rho} \in \arg\min_{y} \mathbb{E}[L(\rho, y)|data].$$

However, the "exact" computation of such an estimate is generally difficult; Quintana and Iglesias (2003) provide an algorithm tailored for the loss function they use, while Lau and Green (2007) adopt Binder's loss function. The latter authors formulate an equivalent binary integer programming problem, which can be solved exactly; unfortunately, this is only computationally feasible for very small sample sizes, so that the same authors resort to a Monte Carlo plug-in estimate, like the one we are using here. We consider Binder's function as in Lau and Green (2007) assigning cost $b$ when two elements are wrongly clustered together and cost $a$ when two elements are erroneously assigned to different clusters. If no information is available, we set $a = b$. Indeed, we run the MCMC algorithm, approximating the posterior $\mathscr{L}(\rho|data)$ once in order to estimate the loss function, and then we plug this estimate in and run the MCMC a second time, obtaining a posterior sample of configurations. Finally we choose as $\hat{\rho}$ the configuration, among the latter sampled ones, that minimizes the sampled values of the (approximated) loss function.

To the best of our knowledge there are few papers, with an applied perspective, adopting NGG-processes as an ingredient. This probably happens because NGG processes yield inherent computational difficulties. Recent works that include NGG processes as an ingredient in their models are Caron (2012) and Caron and Fox (2014), both on statistical networks: the former for bipartite random graphs, and the latter for sparse and exchangeable random graphs. See also Chen et al. (2012) for an application of such multivariate priors in a dynamic topic modeling context. However, there is a recent and very lively literature on algorithms to draw inference for NGG-mixtures, which has resulted into a number of efficient algorithms. Here we refer to *marginal* and *conditional* Gibbs samplers. The former integrate out the infinite dimensional parameter (i.e., the random probability), resorting to generalized Polya urn schemes; see Favaro and Teh (2013). On the other hand, by a conditional algorithm we mean a Gibbs sampler imputing the nonparametric mixing measure and updating it as a component of the algorithm itself. This latter group includes the slice sampler, which has been extended also to NGG-mixtures in Griffin and Walker (2011).

Conditional algorithms are called truncation methods if the infinite parameter (i.e., the mixing measure) is approximated by truncation of the infinite sums defining the process. Truncation can be achieved a-posteriori, when one approximates the infinite parameter P given the data, or a-priori to approximate the nonparametric mixing distribution with a finite dimensional random probability measure, so that a simpler mixture model need to be implemented. In the latter framework, the best known method for DPM models is the blocked Gibbs sampler in Ishwaran and James (2001). Barrios et al. (2013) propose an a-posteriori truncation algorithm for mixtures of normalized completely random measures (and therefore for NGG-mixtures) using the so-called Ferguson–Klass representation of completely random measures. Argiento et al. (2010) propose a simple adaptive truncation method evaluating an upper bound in probability for the jumps excluded from the summation. Recently, two a-priori truncation methods have been introduced by Griffin (2014), who proposes an adaptive truncation algorithm for posterior inference with priors either of stick-breaking or normalized random measure with independent increments type, and Argiento et al. (2015). As in Muliere and Tardella (1998), Argiento et al. (2015) build a finite-dimensional approximation of the NGG process, so that a blocked Gibbs sampler can be implemented as in Ishwaran and James (2001).

Finally, we highlight the difference between the approaches in Papathomas et al. (2012) and Müller et al. (2011) on the one hand, and ours on the other. First of all, note that the aims behind the product partition model (PPMx) in Müller et al. (2011) and Papathomas et al. (2012) are similar. The latter is based on Dirichlet processes, while the former is more general. However, the PPMx model is built by considering the prior of the partition given the covariates, while the likelihood is cluster-specific; Papathomas et al. (2012) consider the likelihood of the covariates given the partition and variable selection parameters. Both papers provide clustering of patients, i.e. of the rows of the covariate matrix as shown by gray boxes in right panel of Fig. 6.2; then, the association with the relevant responses is assessed through a cluster-specific model. On the contrary, we group columns of the covariate matrix (left panel of Fig. 6.2), so that, for each patient, the genotype of multiple markers in each group is summarized through the genetic score (see (6.7)) and the association can be studied by a regression model with a few covariates.

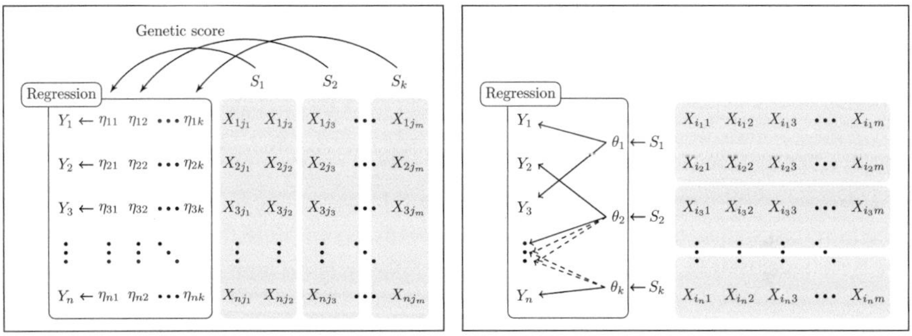


**Fig. 6.2** Scheme of our (*left*) and Papathomas et al. (2012) (*right*) models. In both graphs, the first cluster contains only two elements for ease of notation

## 6.3 Application to SNPs: Data Description and Model Specification

In this section we describe the model illustrated above when applied to general SNP data. In the first subsection we shortly describe what usually is meant by SNP data. In particular, we revise elementary definitions of the human genome, SNP data, and minor alleles frequencies. For a deeper understanding of this notions, see, for instance, Reece et al. (2014) and Mooney (2005). In the second subsection we detail the model for clustering SNP covariates, while the third introduces a suitable genetic score to incorporate either the posterior distribution or the posterior estimate of the random partition into a logistic model; this will allow us to perform an association study between groups of SNPs and the disease.

### 6.3.1 SNP Data for Beginners

The *human genome* is an ordered sequence of 22+1 pairs of chromosomes. The members of each pair are nearly (but not exactly) identical to each other. Chromosomes are very long molecules of a double-stranded chemical known as DNA. The two strands are linked by a long sequence of units called *nucleotides pairs*, or *base pairs*. Each nucleotide can assume four base specifications (A,T,G, and C), which only pair as A-T, and G-C. Genetic information is stored in the exact list, or sequence of nucleotides pairs. It is worth mentioning that, when a DNA sequence is recorded, the genotype of only one strand is registered, since the nucleotides on the other strand are determined. The human genome sequence is 99.6 % identical in all people. This is true for everyone, regardless of race or heritage. However, there are certain positions where people might have different nucleotide pairs. These positions are known as SNPs. It seems there is no consistent conclusion about the frequency of SNPs occurring in human genome. Actually, the probability to find an SNP on the DNA sequence depends on the density of the genes, i.e. the probability of SNPs to appear is higher for a region with more genes than a region with less genes. According to the dbSNP database, 112,743,739 SNPs were found in humans as of October 14, 2014 (NCBI dbSNP Build 142 http://www.ncbi.nlm.nih.gov/projects/SNP/snp_summary.cgi ). The total length of the human genome is over 3 billion base pairs. So, the average may roughly be 112,743,739/3 billion = 0.038 per base pair; that is, 38 SNPs per 1 kb. An SNP may be thought of as an address:

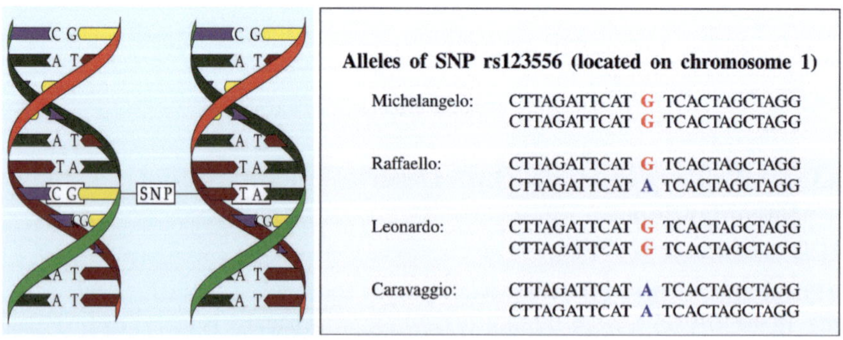


**Fig. 6.3** *Left*: a segment of a chromosome pair where the box identifies an SNP address. *Right*: illustrative table reporting chromosome pairs for four individuals

it is a physical location on a particular chromosome, and may have various "occupants," A, T, G, or C. The term *allele* is used to identify the nucleotide which may occupy an SNP; see left panel of Fig. 6.3 for a graphical illustration. As chromosomes are paired, an SNP genotype consists of two alleles, one from each member of the paired chromosome. A minor allele frequency is defined as the frequency at which the least common allele of an SNP is present in a given population.

As an illustrative example, Fig. 6.3 (right) represents a possible segment of the two sequences of DNA from chromosome pair 1 at the rs123556 SNP for four individuals. It is known that A is the minor allele in rs123556 SNP; from the figure, Michelangelo and Leonardo have the G allele on both of their copies of this chromosome (no minor allele). However, Raffaello has one G and one A allele (one minor allele), and Caravaggio has two A alleles (two minor alleles). Therefore, to code the categorical random variable $X$ representing the minor allele configuration of an individual at a specific SNP (genotyping), one has simply to count the number of minor alleles assuming values in $\{0,1,2\}$. Referring to the right panel of Fig. 6.3, we have:

| Categorical variable | | X |
|---|---|---|
| Michelangelo | $\stackrel{\text{genotype}}{\rightarrow}$ | 0 |
| Raffaello | $\stackrel{\text{genotype}}{\rightarrow}$ | 1 |
| Leonardo | $\stackrel{\text{genotype}}{\rightarrow}$ | 0 |
| Caravaggio | $\stackrel{\text{genotype}}{\rightarrow}$ | 2 |

The data that we are going to analyze in this paper are of the same kind we have discussed so far. Since our final goal is the detection of the association between SNPs and a disease, we will consider data from $n$ patients: a binary response (experiencing the disease or not) and a long list of $m$ categorical covariates reporting the SNP genotype of the patient. Summing up, we will analyze data of this kind:

- a vector $\boldsymbol{Y} = (Y_1, \ldots, Y_n)$ of the disease indicators;
- a matrix $\boldsymbol{X} = \{X_{ij}\}$ $i = 1, \ldots, n$ and $j = 1, \ldots, m$, where $X_{ij} \in \{0,1,2\}$ is the genotype for the $j$-th SNP of patient $i$.

Generally the SNPs are located in different regions of many chromosomes. To fix notation, suppose that we observe $m_1$ SNPs from region 1, $m_2$ SNPs from region 2, ..., $m_L$ SNPs from region $L$, where $m = m_1 + m_2 + \cdots + m_L$. So, our SNPs can be arranged into a large matrix $\boldsymbol{X}$, where its rows are described by the vectors $\boldsymbol{X}_i = (\boldsymbol{X}_{i,1}, \ldots, \boldsymbol{X}_{i,L})$, $i = 1, \ldots, n$; here, $\boldsymbol{X}_{i,l}$ is the subvector containing the SNPs of the $i$-th patient from region $l$, $l = 1, \ldots, L$, that is

$$\boldsymbol{X}_i = (\overbrace{X_{i,1}, \ldots, X_{i,m_1}}^{\boldsymbol{X}_{i,1}}, \overbrace{X_{i,m_1+1}, \ldots, X_{i,m_1+m_2}}^{\boldsymbol{X}_{i,2}}, \ldots, \overbrace{X_{i,m_1+\cdots+m_{L-1}+1}, \ldots, X_{i,m}}^{\boldsymbol{X}_{i,L}})$$

where $m_l$ is the number of SNPs in region $l = 1, \ldots, L$.

### 6.3.2 Cluster Model Specification

Let $N_{jls} = \sum_{i=1}^{n} I(X_{i\, m_0+\cdots+m_{l-1}+j} = s)$ be the total number of subjects whose genotypes on the $j$-th SNP (in region $l$) are recorded as $s$, where $s \in \{0,1,2\}$, and $m_0 = 0$. We model $\boldsymbol{N}_{jl} = (N_{jl0}, N_{jl1}, N_{jl2})$, for $j = 1,\ldots,m_l$, $l = 1,\ldots,L$, as conditionally independent multinomial distributed random variables, given $\boldsymbol{\theta}_{jl} = (\theta_{jl0}, \theta_{jl1}, \theta_{jl2})$. Specifically, data, both within and between regions, are conditionally independent, according to the following likelihood

$$\boldsymbol{N}_{1l},\ldots,\boldsymbol{N}_{m_l l} | \boldsymbol{\theta}_{1l},\ldots,\boldsymbol{\theta}_{m_l l} \sim \prod_{j=1}^{m_l} Mult(n, \boldsymbol{\theta}_{jl}). \tag{6.5}$$

All the blocks $\boldsymbol{\theta}_{1l},\ldots,\boldsymbol{\theta}_{m_l l}$, $l = 1,\ldots,L$, of latent parameters are a priori independent, and distributed as follows:

$$\begin{aligned} \boldsymbol{\theta}_{1l},\ldots,\boldsymbol{\theta}_{m_l l} | P_l &\overset{\text{i.i.d.}}{\sim} P_l \quad l = 1,\ldots,L \\ P_1,\ldots,P_L &\overset{\text{i.i.d.}}{\sim} NGG(\sigma, \kappa, P_{0l}), \end{aligned} \tag{6.6}$$

where $\sigma \in [0,1)$, $\kappa \in \mathbb{R}^+$ and $P_{0l}(\cdot)$ are Dirichlet distributions $Dir(a_{0l}, a_{1l}, a_{2l})$ on the simplex $\{(x_1,x_2,x_3) \in \mathbb{R}^3 : 0 < x_1,x_2,x_3 < 1, x_1 + x_2 + x_3 = 1\}$, and $a_{0l}, a_{1l}, a_{2l} \in \mathbb{R}^+$.

Observe that, since $P_1,\ldots,P_L$ are a priori independent, and the likelihood can be factorized, they will be independent a posteriori as well. This is a precise choice that we are going to make: in fact, we do not want to share information across different regions of the chromosomes; it is well recognized in the geneticist community that different chromosome regions, not close to each others, may not be passed from parents to offspring together due to the so-called random crossover, so that independence among different regions may be assumed. On the contrary, if we had wanted to share information among the regions, we could have resorted to a dependent prior for the vector of the random probabilities $(P_1,\ldots,P_L)$, as, for instance, the nested Dirichlet process (see Rodriguez et al. 2008).

Model (6.5) and (6.6) induces a partition on the indices $\{1,\ldots,m_l\}$ of $\boldsymbol{N}_{1l},\ldots,\boldsymbol{N}_{m_l l}$ through the natural clustering rule described in Sect. 6.2, for each $l = 1,\ldots,L$. The SNPs in the same cluster within each region share the same allele probabilities $\boldsymbol{\phi}$'s, which are the unique values of a sample of size $m_l$ from the same $P_l$. Therefore, our model specifies the overall fraction of minor alleles across all SNPs in the cluster.

The natural clustering rule also induces a partition of indices of the columns of the data matrix $\boldsymbol{X}$, which varies in $\{1,\ldots,m_1,\ldots,m_1+\cdots+m_{L-1}+1,\ldots,m_1+\cdots+m_L\}$ $(m_1+\cdots+m_L = m)$. We will adopt the same notation $\boldsymbol{\rho} = (\rho_1,\ldots,\rho_L)$, where $\rho_l = (S_{1l},\ldots,S_{k_l l})$, $l = 1,\ldots,L$, denotes both the partition of $\{1,\ldots,n_l\}$ (i.e., the index of the observed $\boldsymbol{N}_{jl}$'s) and a partition of the column indices of $\boldsymbol{X}$.

### 6.3.3 Association Study Between SNP Clusters and Disease

Our goal is the identification of groups of important SNPs within each region $l$, $l = 1,\ldots,L$, in genomic association studies.

Let us denote by $\boldsymbol{\phi}_{hl}$ the parameter vector of the likelihood of all the $\boldsymbol{N}_{jl}$'s belonging to the same group $S_{hl}$, where $h = 1,\ldots,k_l$. For any vector of cluster configurations $(\rho_1,\ldots,\rho_L)$, and vectors of distinct values in the latent variables $\boldsymbol{\phi}_l = (\boldsymbol{\phi}_{1l},\ldots,\boldsymbol{\phi}_{k_l l})$, $l = 1,\ldots,L$, we construct a genetic score $\eta(S_{hl},i)$ of the $i$–th subject with respect to group $S_{hl}$,

$$\eta(S_{hl},i) = \ln \frac{\mathbb{P}(\mathbf{X}_{i,S_{hl}}|\boldsymbol{\phi}_{hl})}{|S_{hl}|} \tag{6.7}$$

where

$$\mathbb{P}(\mathbf{X}_{i,S_{hl}}|\boldsymbol{\phi}_{hl}) = \prod_{j\in S_{hl}} \mathbb{P}(X_{ij}|S_{hl},\boldsymbol{\phi}_{hl}) = \prod_{j\in S_{hl}} \phi_{hlX_{ij}}.$$

The genetic score $\eta(S_{hl},i)$ can be interpreted as the standardized log-probability of observing the SNP genotype configuration of patient $i$ within cluster $S_{hl}$, given the parameters $\rho_l$'s and $\boldsymbol{\phi}_l$'s. Substantially, conditional to the cluster estimation, $\eta(S_{hl},i)$ quantifies how likely the $i$-th individual genotypes are carried in the group $S_{hl}$. It is essentially the genetic content of this subject in a specific chromosome region. This representation provides the information of the genotype via latent variables $\phi_{hl}$, without using directly the observed SNP genotype vectors so that the group characteristics can be enhanced and the sparseness due to the discrete $X_{i,S_{hl}}$ can be avoided.

Here, the association study can be carried out for the disease status of each patient, fitting the following logit regression model; for each $i = 1,\ldots,n$, conditionally on the parameters, the patients are independent and

$$\begin{aligned}
\text{logit}(P(Y_i = 1)) = \beta_0 + &\overbrace{\beta_1\eta(S_{11},i) + \cdots + \beta_{k_1}\eta(S_{k_1 1},i)}^{k_1 \text{ regressors from region 1}} \\
&\overbrace{+\beta_{k_1+1}\eta(S_{12},i) + \cdots + \beta_{k_1+k_2}\eta(S_{k_2 2},i)}^{k_2 \text{ regressors from region 2}} \\
&\ldots \\
&\overbrace{+\beta_{k_1+\cdots+k_{L-1}+1}\eta(S_{1L},i) + \cdots + \beta_k\eta(S_{k_L L},i)}^{k_L \text{ regressors from region L}}.
\end{aligned} \tag{6.8}$$

Observe that the overall number of groups is $k = k_1 + \cdots + k_L$, that is generally much smaller than $m$, the observed number of SNPs. It is better to consider multiple marker tests instead of a single marker test since for genetic association studies, the statistical analyses with multiple markers have been shown to be more powerful, efficient, and biologically meaningful than single marker association tests. Examples include regularized regression models like lasso or ridge regression (Chen et al.

2010), gene-set enrichment analysis (Hu and Tzeng 2014), pathway (Ramanan et al. 2012), and network analysis (Lee et al. 2011). These analyses are useful for a large number of markers from pre-specified genetic regions in which genes are interacting in the same pathway or network. Such tools, however, may be limited when utilized on a genome scale.

A vague prior is assumed on the regression parameters $(\beta_1, \ldots, \beta_k)$, and the posterior distribution is computed via an MCMC algorithm. We resort to a variable selection procedure (hard shrinkage, Johnstone and Silverman 2004) to study the association between the disease and the clusters of SNPs: a group $S_{hl}$ is not significative if the posterior 90 % credible interval of the corresponding parameter $\beta$ does not contain zero; if this posterior credible interval is entirely contained in $\mathbb{R}^+$ ($\mathbb{R}^-$), then it denotes positive (negative) association between the clusters and the disease phenotype $Y_i$.

The likelihood in (6.8) is conditioned not only on the parameters $\beta_p$'s, but also on the partition $\rho$ and the corresponding $\phi$'s. Therefore, under this model, we may alternatively consider two strategies to perform the association study. On the one hand, we can plug in Bayesian point estimates $\hat{\rho}$ and $\hat{\phi}$ to compute first the genetic score (6.7) and later infer through the regression model. In this way we take into account the collective association effect of multiple markers (SNPs) to the disease. On the other hand, we can incorporate the uncertainty on the clustering in our association study, i.e. we use the information contained in the whole posterior distribution of $\rho$. In this case observe that, given $\rho$ and $\phi$, once we have chosen a rule to classify a cluster (whether associated or not to the disease), for each $j = 1, \ldots, m$, we can define single marker effect indices as

$$A_j := A_j(\boldsymbol{\rho}, \boldsymbol{\phi}) = \begin{cases} 1 \text{ if SNP} j \text{ belongs to a positively associated cluster} \\ 0 \text{ if SNP} j \text{ belongs to a non associated cluster} \\ -1 \text{ if SNP} j \text{ belongs to a negatively associated cluster.} \end{cases}$$

If $(\boldsymbol{\rho}^{(1)}, \boldsymbol{\phi}^{(1)}), \ldots, (\boldsymbol{\rho}^{(G)}, \boldsymbol{\phi}^{(G)})$ is a sample from $\mathscr{L}(\boldsymbol{\rho}, \boldsymbol{\phi} | X)$, we can evaluate the posterior frequencies that $A_j = a$ with $a = -1, 0, 1$. If the mode of these posterior frequencies, for each SNP $j$, occurs at $-1$, 1, and 0, then the SNP is classified as carrying negative, positive, or no association with the disease, respectively.

## 6.4 Application to SNPs: Bayesian Inference

Our data come from the genotyping of Crohn's disease patients from the The Wellcome Trust Case Control Consortium (2007) (WTCCC). Using the same dataset, Wei et al. (2010) concluded that the disease is associated with five chromosome regions:

| | |
|---|---|
| Chro. 10 region q24.1 | Chro. 16 region q12.1 |
| Chro. 1 region p313 | Chro. 5 region p13.1 |

Chro. 2 region q37.1

In this study 1748 patients with Crohn's disease and 2938 share controls were considered, so that the total number of patients is $n = 4686$. After excluding SNPs with minor allele frequency lower than 0.01 or in Hardy-Weinberg disequilibrium, a total number of $m = 3704$ SNPs were left for our analysis; therefore, here $L = 5$, whereas $m_1, \ldots, m_5$ are in Table 6.1.

We assume model (6.5) and (6.6) but resort to the a-priori truncation algorithm in Argiento et al. (2015) in order to deal with the random probability measures $P_l$, which have infinite supports (i.e., the sum in (6.4) is infinite). As mentioned in the Introduction, this algorithm uses an approximation of NGG processes which is a discrete measure where the weights are obtained by normalization of the jumps of a Poisson process, however considering only jumps larger than a threshold $\varepsilon > 0$. The number of jumps of this new process, called $\varepsilon$-NGG process, turns out to be a Poisson random variable. Argiento et al. (2015) prove that, as $\varepsilon$ goes to 0, the $\varepsilon$-NGG process converges in distribution to the NGG process. Here we assume $\varepsilon$ very small, $\varepsilon = 10^{-6}$. The prior specification is completed assuming that, a priori, $\sigma$ and $\kappa$ are independent, and $\sigma \sim$beta$(2,18)$, $\kappa \sim$gamma$(2,0.1)$. It is well known that, when considering parametric mixtures, the hyperparameters of the mean distribution $P_0$ strongly affect the inference on clusters. Here we have fixed them according to the empirical Bayes approach, i.e. $a_l = \frac{1}{m}\sum_{j=1}^{m} S_{jl}$, $l = 0,1,2$. We considered a final sample size of 5000 iterations after 10,000 iterations as burn-in. On average, we got 20-min run times on a laptop with Intel Core i7-2620M processor.

Table 6.1 reports the estimated numbers of clusters for each region. The fourth

**Table 6.1** Descriptive statistics and posterior summaries of SNPs and clusters

| $l$ | Region | $m_l$ | $\hat{k}$ | Cluster size | | | Association | |
|---|---|---|---|---|---|---|---|---|
| | | | | min | median | max | positive | negative |
| 1 | 1p31.3 | 1357 | 38 | 8 | 34.5 | 72 | 6 | 3 |
| 2 | 2q37.1 | 662 | 33 | 6 | 18 | 44 | 2 | 8 |
| 3 | 5p13.1 | 554 | 31 | 2 | 16 | 43 | 3 | 7 |
| 4 | 10q24.2 | 390 | 27 | 2 | 15 | 39 | 0 | 4 |
| 5 | 16q12.1 | 742 | 36 | 1 | 20.5 | 42 | 5 | 8 |

column ($\hat{k}$) shows the estimated number of clusters, corresponding to the sizes of the estimated partitions according to Binder's loss function with equal costs (see Sect. 6.2). The next three columns show the minimum, the median, and the maximum, respectively, of the cluster sizes of the estimated partition of each region.

Following the allocation of SNP clusters prescribed by the estimated partitions, we compute the genetic score, as in (6.7), of each cluster to investigate the cluster effects under the logit linear model (6.8). The regression parameters $\beta_p$'s are a priori independent, all marginally Gaussian distributed with zero mean and variance equal to 1000. As mentioned in Sect. 6.3.3, we may alternatively compute the plug-in estimates of the genetic scores (6.7), or evaluate them at each iteration of the

MCMC chain. Here we report the association results under the former strategy. Of course, this computation requires an estimate of the parameters $\phi$ in each cluster. However, when one is interested in cluster specific inference, label-switching can be a problem. To solve it, a variety of strategies have been proposed (for a review see Jasra et al. 2005), typically using post-processing strategies (see Molitor et al. 2010; Müller et al. 2011).

Here we follow a simple strategy: we take advantage of the conjugacy of $P_0$, so that, if $\hat{\rho}_l = \left\{\hat{S}_{1l}, \ldots, \hat{S}_{\hat{k}_l l}\right\}$ is the estimated partition, $\hat{\boldsymbol{\phi}}_{hl}$ is fixed as the within-cluster posterior mean $\mathbb{E}(\boldsymbol{\phi}_{hl} | X_{\hat{S}_{hl}})$.

Figure 6.4 displays the posterior 90 % credible intervals of all cluster-specific regression parameters for the five different regions; the dashed intervals are those overlapping 0, while the dark gray ones are those significative. The last two columns in Table 6.1 report the numbers of positive and negative associated clusters. On the whole, we found 41 groups of significative SNPs. Observe that, to fix notation in (6.8), within each region, clusters are labeled according to their sampling biased order of appearance, i.e. cluster 1 contains observation $i = 1$, then cluster 2 contains the first observation not in cluster 1, and so on.

Figure 6.5 shows the observed frequencies $S_1$ and $S_2$ for each SNP; observations are depicted as red squares (blue diamonds) if they belong to a cluster that is positively (negatively) associated with the Crohn's disease, while gray points represent the others. We found 46 significant clusters containing 1070 SNPs in total. These SNPs are located in 106 genes: 16 of them have been mentioned to be in relationship to Crohn's disease in literature, including *ATG16L1, IL23R,* and *IL12RB2* (de Paus et al. 2013; Cho 2008; Duerr et al. 2006). Moreover 6 SNPs have been reported in association with Crohn's disease: rs10889675, rs11805303, and rs2201841, located in gene *IL23R* of region 1p31.1, belong to our cluster 26, 16, and 16, respectively. SNP rs6871834 of region 5p13.1 is in cluster 11, and rs1861759 and rs2066843 of region 16q12.1 are in cluster 27 and 11, respectively.

The significant clusters show exciting results. For example, in region 1p31.3, cluster 2, 5, 15, 16, and 26 are all parts of the gene *IL23R*, while cluster 2 and 5 are part of gene *IL12RB2*. On the other hand, cluster 11 and 30 in region 2q37.1 are parts of gene *ATG16L1*. These clusters reveal different effects on Crohn's disease. Cluster 16 in 1p31.3 and cluster 30 in 2q37.1 are protective, while cluster 2, 5, 15, and 26 in 1p31.3 and cluster 11 in 2q37.1 are deleterious. This may be a reason why previous association studies did not have consistent findings about the association between *IL23R* and Crohn's disease (Duerr et al. 2006; Glas et al. 2007; Jostins et al. 2012).

The genetic findings obtained under the NGG mixture model just discussed are similar to those obtained under the DPM model (see Wang et al. 2014). Under both models the significative clusters contain SNPs that have been recognized to be related to Crohn's disease in the genetic literature, and under both models the significative clusters belong to genes related with the disease. However the cluster

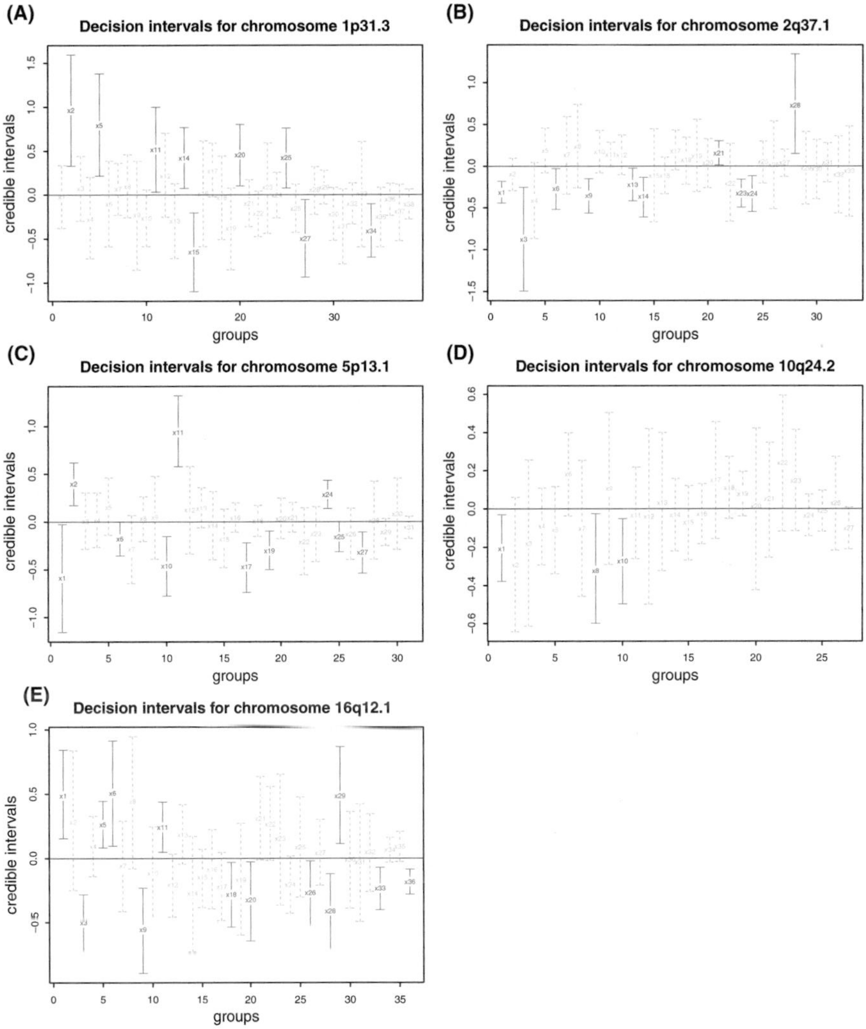


**Fig. 6.4** 90 % posterior credible intervals of $\beta_k$, for each of the five regions. The credible intervals are colored and depicted according to their significance: *solid* (*red*) for deleterious, *solid* (*green*) for protective, and *dashed* (*gray*) for non-associated effects

estimates under the two Bayesian nonparametric models are different. Under the NGG mixture model the median of the cluster sizes is generally higher, and it varies less (among the five regions) with respect to the DMP model. Part of our current research concerns further quantitative comparison of the two models from statistical and genetic point of views.

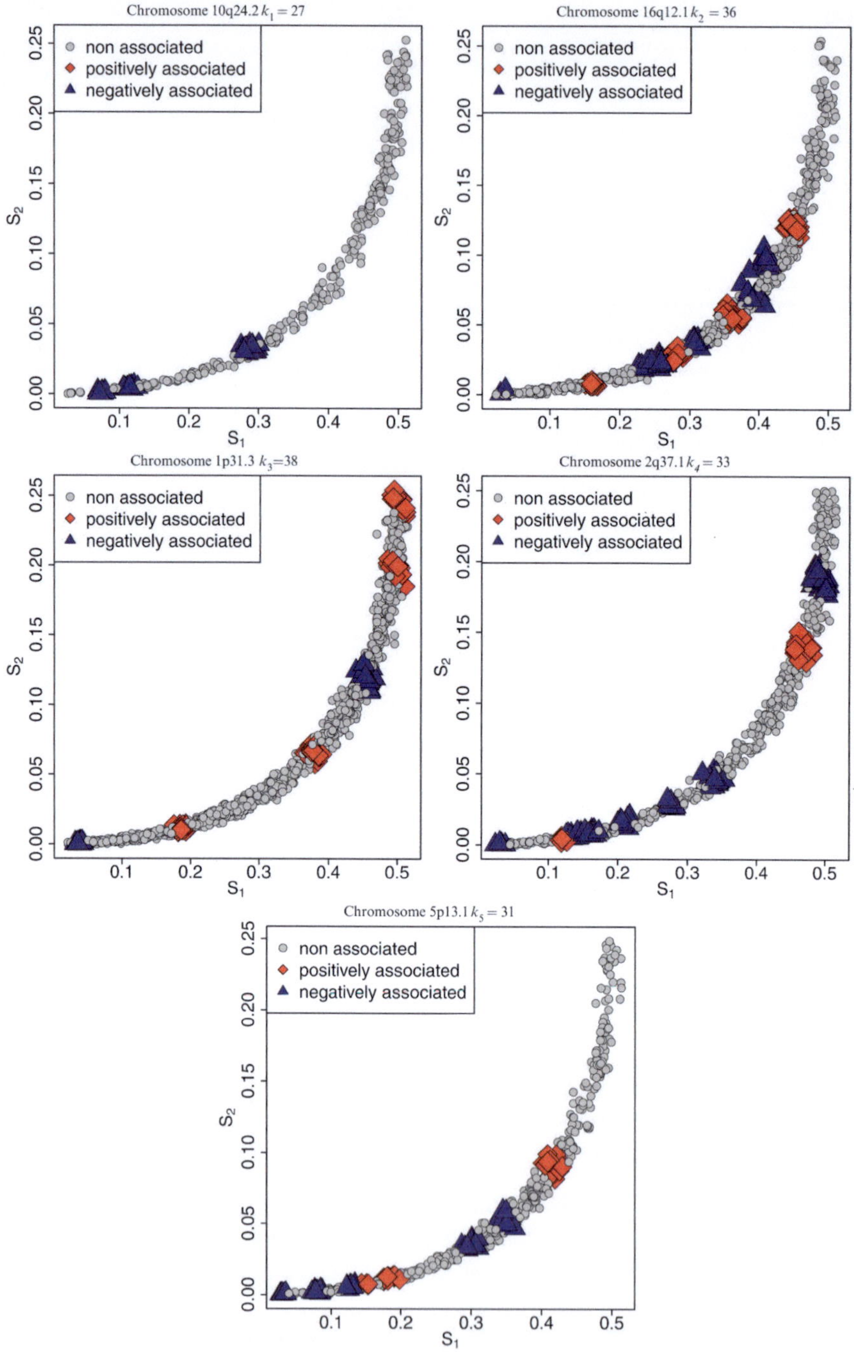


**Fig. 6.5** Observed frequencies $S_1$ and $S_2$ for each SNP in the five regions. Observations are depicted as *red squares* (*blue diamond*) if they belong to a cluster that is positively (negatively) associated with Crohn's disease, while *gray points* represent SNPs in cluster which were not associated with the disease

## 6.5 Conclusions

The paper presents an approach, based on an NGG mixture model, useful to detect association between SNPs and a disease. In particular, we proposed a two-step procedure: we first clustered group of SNPs, i.e. columns of the covariate matrix, and then, studied association between the relevant disease response and groups of markers via a logit model, identifying clusters having positive, negative, or null effect on the disease. From a computational point of view, we used a blocked Gibbs sampler recently introduced in the literature, which approximates the infinite dimensional mixing measure via a finite sum. This yields a direct and efficient algorithm, so that computations are simplified (from infinite to finite dimension). We also provided a review of the state of art of Bayesian nonparametric cluster models and algorithms for NGG mixtures.

Besides the positive aspects discussed in the paper, the procedure could be useful when considering chromosomes which could be associated with more than one disease. In such situation, the proposed method would be computationally more efficient of existing ones based on association first and then clustering; in fact, SNPs could be clustered only once and then just association studies would be performed for each disease.

As discussed in the paper, here, in order to cluster SNPs, an NGG mixture model has been considered, unlike in Wang et al. (2014) who used a Dirichlet process mixture. Since the Dirichlet process is a special case of NGG, we plan to perform comparisons between the findings of the two approaches, in a more formal way with respect to the discussion of the previous section, for instance computing the Bayes factor of $\sigma = 0$ versus $\sigma > 0$.

Empirical Bayes has been used to choose the hyperparameters of the Dirichlet distributions $P_{0l}$, $l = 1, \ldots, L$. Given the large number of SNPs, it is difficult to have experts' opinions on all of them to choose the hyperparameters subjectively but we recommend to use them when available.

Supported by biological reasons, we have assumed that SNPs from different chromosomes are independent but it would be interesting, especially from a statistical viewpoint, to check if there is some significant relation across chromosomes; dependent hierarchical NGG could be a possible model fit for such analysis.

## References

Argiento, R., Guglielmi, A., and Pievatolo, A. (2010). Bayesian density estimation and model selection using nonparametric hierarchical mixtures. *Comput. Stat. Data Anal.*, **54**, 816–832.

Argiento, R., Cremaschi, A., and Guglielmi, A. (2014). A density-based algorithm for cluster analysis using species sampling Gaussian mixture models. *J. Comput. Graph. Stat.*, **23**, 1126–1142.

Argiento, R., Bianchini, I., and Guglielmi, A. (2015). A blocked Gibbs sampler for NGG-mixture models via a priori truncation. *Statist. Comp.*, Online First.

Asimit, J. and Zeggini, E. (2010). Rare variant association analysis methods for complex traits. *Annu. Rev. Genet.*, **44**, 293–308.

Bansal, V., Libiger, O., Torkamani, A., and Schork, N. J. (2010). Statistical analysis strategies for association studies involving rare variants. *Nat. Rev. Genet.*, **11**, 773–785.

Barrios, E., Lijoi, A., Nieto-Barajas, L. E., Prünster, I., *et al.* (2013). Modeling with normalized random measure mixture models. *Stat. Sci.*, **28**, 313–334.

Caron, F. (2012). Bayesian nonparametric models for bipartite graphs. In *Advances in Neural Information Processing Systems*, pages 2051–2059.

Caron, F. and Fox, E. B. (2014). Bayesian nonparametric models of sparse and exchangeable random graphs. *arXiv preprint arXiv:1401.1137*.

Chen, C., Ding, N., and Buntine, W. (2012). Dependent hierarchical normalized random measures for dynamic topic modeling. International conference on machine learning (ICML), Edimburg, UK.

Chen, L. S., Hutter, C. M., Potter, J. D., Liu, Y., Prentice, R. L., Peters, U., and Hsu, L. (2010). Insights into colon cancer etiology via a regularized approach to gene set analysis of GWAS data. *Am. J. Hum. Genet.*, **86**, 860–871.

Cho, J. H. (2008). The genetics and immunopathogenesis of inflammatory bowel disease. *Nat. Rev. Immunol.*, **8**, 458–466.

Chung, Y. and Dunson, D. (2009). Nonparametric Bayes conditional distribution modeling with variable selection. *J. Am. Stat. Assoc.*, **104**, 1646–1660.

Dahl, D. B. (2006). Model-based clustering for expression data via a Dirichlet process mixture model. In V. M. Do K.-A., Müller P., editor, *Bayesian inference for gene expression and proteomics*, pages 201–218. Cambridge: Cambridge University Press.

De Blasi, P., Favaro, S., Lijoi, A., Mena, R., Prünster, I., and Ruggiero, M. (2014). Are Gibbs-type priors the most natural generalization of the Dirichlet process? *IEEE Trans. Pattern Anal. Mach. Intell.*, **37**, 212–229.

de Paus, R. A., Geilenkirchen, M. A., van Riet, S., van Dissel, J. T., and van de Vosse, E. (2013). Differential expression and function of human il-12r$\beta$2 polymorphic variants. *Molecular immunology*, **56**(4), 380–389.

Duerr, R. H., Taylor, K. D., Brant, S. R., Rioux, J. D., Silverberg, M. S., Daly, M. J., Steinhart, A. H., Abraham, C., Regueiro, M., Griffiths, A., *et al.* (2006). A genome-wide association study identifies IL23R as an inflammatory bowel disease gene. *Science*, **314**, 1461–1463.

Favaro, S. and Teh, Y. W. (2013). MCMC for normalized random measure mixture models. *Stat. Sci.*, **28**, 335–359.

Ferguson, T. S. (1973). A Bayesian analysis of some nonparametric problems. *Ann. Statist.*, **1**, 209–230.

Fritsch, A. and Ickstadt, K. (2009). Improved criteria for clustering based on the posterior similarity matrix. *Bayesian Anal.*, **4**, 367–392.

Glas, J., Seiderer, J., Wetzke, M., Konrad, A., Török, H.-P., Schmechel, S., Tonenchi, L., Grassl, C., Dambacher, J., Pfennig, S., *et al.* (2007). rs1004819 is the main disease-associated il23r variant in German Crohn's disease patients: combined analysis of IL23R, CARD15, and OCTN1/2 variants. *PloS One*, **2**, e819.

Griffin, J. E. (2014). An adaptive truncation method for inference in Bayesian nonparametric models. *Statist. Comp.*, Online First, 1–19.

Griffin, J. E. and Walker, S. G. (2011). Posterior simulation of normalized random measure mixtures. *J. Comput. Graph. Stat.*, **20**, 241–259.

Hu, J. and Tzeng, J.-Y. (2014). Integrative gene set analysis of multi-platform data with sample heterogeneity. *Bioinformatics*, **30**, 1501–1507.

Huang, H., Chanda, P., Alonso, A., Bader, J. S., and Arking, D. E. (2011). Gene-based tests of association. *PLoS Genetics*, **7**, e1002177.

Ishwaran, H. and James, L. F. (2001). Gibbs sampling methods for stick-breaking priors. *J. Amer. Statist. Assoc.*, **96**, 161–173.

Jasra, A., Holmes, C., and Stephens, D. (2005). Markov chain Monte Carlo methods and the label switching problem in Bayesian mixture modeling. *Stat. Sci.*, **20**, 50–67.

Johnstone, I. M. and Silverman, B. W. (2004). Needles and straw in haystacks: Empirical Bayes estimates of possibly sparse sequences. *Ann. Statist.*, **32**, 1594–1649.

Jostins, L., Ripke, S., Weersma, R. K., Duerr, R. H., McGovern, D. P., Hui, K. Y., Lee, J. C., Schumm, L. P., Sharma, Y., Anderson, C. A., *et al.* (2012). Host-microbe interactions have shaped the genetic architecture of inflammatory bowel disease. *Nature*, **491**, 119–124.

Lau, J. W. and Green, P. J. (2007). Bayesian model based clustering procedures. *J. Comput. Graph. Stat*, **16**, 526–558.

Lee, I., Blom, U. M., Wang, P. I., Shim, J. E., and Marcotte, E. M. (2011). Prioritizing candidate disease genes by network-based boosting of genome-wide association data. *Genome Res.*, **21**, 1109–1121.

Lijoi, A., Mena, R. H., and Prünster, I. (2007). Controlling the reinforcement in Bayesian nonparametric mixture models. *J. R. Stat. Soc., B*, **69**, 715–740.

Liverani, S., Hastie, D. I., Azizi, L., Papathomas, M., and Richardson, S. (2015). PReMiuM: An R package for profile regression mixture models using Dirichlet processes. *J. Stat. Softw.*, forthcoming.

Medvedovic, M., Yeung, K. Y., and Bumgarner, R. E. (2004). Bayesian mixture model based clustering of replicated microarray data. *Bioinformatics*, **20**, 1222–1232.

Molitor, J., Papathomas, M., Jerrett, M., and Richardson, S. (2010). Bayesian profile regression with an application to the national survey of children's health. *Biostatistics*, **11**, 484–498.

Mooney, S. (2005). Bioinformatics approaches and resources for single nucleotide polymorphism functional analysis. *Brief Bioinform.*, **6**, 44–56.

Muliere, P. and Tardella, L. (1998). Approximating distributions of random functionals of Ferguson-Dirichlet priors. *Can. J. Stat.*, **26**, 283–297.

Müller, P., Quintana, F. A., and Rosner, G. A. (2011). A product partition model with regression on covariates. *J. Comput. Graph. Stat*, **20**, 260–278.

Nguyen, L. B., Diskin, S. J., Capasso, M., Wang, K., Diamond, M. A., Glessner, J., Kim, C., Attiyeh, E. F., Mosse, Y. P., Cole, K., *et al.* (2011). Phenotype restricted genome-wide association study using a gene-centric approach identifies three low-risk neuroblastoma susceptibility loci. *PLoS Genetics*, **7**, e1002026.

Onogi, A., Nurimoto, M., and Morita, M. (2011). Characterization of a Bayesian genetic clustering algorithm based on a Dirichlet process prior and comparison among Bayesian clustering methods. *BMC Bioinformatics*, **12**, 263–278.

Papathomas, M., Molitor, J., Hoggart, C., Hastie, D., and Richardson, S. (2012). Exploring data from genetic association studies using Bayesian variable selection and the Dirichlet process: application to searching for gene × gene patterns. *Genet. Epidemiol.*, **36**, 663–674.

Pitman, J. (1996). Some developments of the Blackwell-MacQueen urn scheme. In B. M. Ferguson TS, Shapley LS, editor, *Statistics, Probability and Game Theory: Papers in Honor of David Blackwell*, pages 245–267. Hayward: Institute of Mathematical Statistics.

Quintana, F. A. and Iglesias, P. L. (2003). Bayesian clustering and product partition models. *J. R. Stat. Soc., B*, **65**, 557–574.

Ramanan, V. K., Shen, L., Moore, J. H., and Saykin, A. J. (2012). Pathway analysis of genomic data: concepts, methods, and prospects for future development. *Trends Genet.*, **28**, 323–332.

Reece, J. B., Urry, L. A., Cain, M. L., Wasserman, S. A., Minorsky, P. V., and Jackson, R. (2014). *Campbell Biology*. Boston: Pearson.

Regazzini, E., Lijoi, A., and Prünster, I. (2003). Distributional results for means of normalized random measures with independent increments. *Ann. Statist.*, **31**, 560–585.

Rodriguez, A., Dunson, D. B., and Gelfand, A. E. (2008). The nested Dirichlet process. *J. Am. Stat. Assoc.*, **103**, 1131–1154.

Tadesse, M. G., Sha, N., and Vannucci, M. (2005). Bayesian variable selection in clustering high-dimensional data. *J. Am. Stat. Assoc.*, **100**, 602–617.

The Wellcome Trust Case Control Consortium (2007). Genome-wide association study of 14,000 cases of seven common diseases and 3,000 shared controls. *Nature*, **447**, 661–678.

Wakefield, J. (2007). A Bayesian measure of the probability of false discovery in genetic epidemiology studies. *Am. J. Hum. Genet.*, **81**, 208–227.

Wakefield, J. (2009). Bayes factors for genome-wide association studies: comparison with p-values. *Genet. Epidemiol.*, **33**, 79–86.

Wang, C., Ruggeri, F., Hsiao, C., and Argiento, R. (2014). Bayesian nonparametric clustering and association studies for large-scale SNP observations. *Submitted.*

Wei, Y. C., Wen, S. H., Chen, P. C., Wang, C. H., and Hsiao, C. K. (2010). A simple Bayesian mixture model with a hybrid procedure for genome-wide association studies. *Eur. J. Hum. Genet.*, **18.8**, 942–947.

Yau, C. and Holmes, C. (2011). Hierarchical Bayesian nonparametric mixture models for clustering with variable relevance determination. *Bayesian Anal.*, **6**, 329–351.

# Chapter 7
# Bayesian Inference on Population Structure: From Parametric to Nonparametric Modeling

Maria De Iorio, Stefano Favaro, and Yee Whye Teh

**Abstract** Making inference on population structure from genotype data requires to identify the actual subpopulations and assign individuals to these populations. The source populations are assumed to be in Hardy-Weinberg equilibrium, but the allelic frequencies of these populations and even the number of populations present in a sample are unknown. In this chapter we present a review of some Bayesian parametric and nonparametric models for making inference on population structure, with emphasis on model-based clustering methods. Our aim is to show how recent developments in Bayesian nonparametrics have been usefully exploited in order to introduce natural nonparametric counterparts of some of the most celebrated parametric approaches for inferring population structure. We use data from the 1000 Genomes project (http://www.1000genomes.org/) to provide a brief illustration of some of these nonparametric approaches.

M. De Iorio (✉)
Department of Statistical Science, University College, London, UK
e-mail: m.deiorio@ucl.ac.uk

S. Favaro
Department of Economics and Statistics, University of Torino
and Collegio Carlo Alberto, Torino, Italy
e-mail: stefano.favaro@unito.it

Y.W. Teh
Department of Statistics, University of Oxford, Oxford, UK
e-mail: y.w.teh@stats.ox.ac.uk

 
R. Mitra, P. Müller (eds.), *Nonparametric Bayesian Inference in Biostatistics*,
Frontiers in Probability and the Statistical Sciences,
DOI 10.1007/978-3-319-19518-6_7

## 7.1 Introduction

Population stratification or structure refers to the presence of a systematic difference in allele frequencies between populations due to the fact that populations are typically heterogeneous in terms of their genetic ancestry. A particular type of population structure is genetic admixtures, which derive from the genetic mixing of two or more previously separated groups in the recent past. A typical example is offered by African-Americans. The analysis of population structure based on genotypes at co-dominant marker loci presents an important problem in population genetics. In particular it is central to the understanding of human migratory history and the genesis of modern populations, while the associated admixture analysis of individuals is important in correcting the confounding effects of population ancestry in gene mapping and association studies. As allele frequencies are known to vary among populations of different genetic ancestry, similarly phenotypic variation, such as disease risk, is observed among group of different genetic ancestry. Population structure is also relevant in the analysis of gene flow in hybridization zones (Field et al. 2011) and invasive species (Ray and Quader 2014), conservation genetics (Wasser et al. 2007), and domestication events (Park et al. 2004).

The advent of high density genotyping arrays and next generation resequencing technologies has led to the production of enormous quantity of data, offering an opportunity to investigate ancestry and genetic relationships among individuals in a population in unprecedented level of details. Nevertheless, this enormous quantity of available data poses new statistical and computational challenges. Making inference on population structure from genotype data requires to identify the actual subpopulations and, in particular, assign individuals to these populations. The source populations are assumed to be in the Hardy-Weinberg equilibrium, namely the likelihood of the genotype of an individual, conditional on its subpopulation of origin, is simply the product of the frequencies of its alleles in that population. The allelic composition of these populations and even the number of populations are unknown and, therefore, object of inference.

A full range of statistical approaches, parametric and nonparametric as well as frequentist and Bayesian, have been proposed for inferring population structure. Two of the prevailing approaches used to infer genetic ancestry from a sample of chromosomes are Principal Components Analysis (PCA) and Structured Association. PCA has been used to infer population structure from genetic data for several decades (Novembre and Stephens 2008) and consists in projecting individuals in a lower dimensional space so that the locations of individuals in the projected space reflect the genetic similarities among them. Clusters of individuals in the projected space can be interpreted as genetic populations, while admixture of two populations results in sets of individuals lying along a line. PCA is a computationally efficient method which can handle large numbers of markers, and is useful for visualizing population structure. The first few principal components are often used to correct for population stratification in genetic association studies. PCA is implemented in EIGENSTRAT (Patterson et al. 2006). In a structured association approach the goal is to explicitly infer genetic ancestry: individuals are assigned to subpopulation clusters, possibly allowing fractional cluster membership in the case of genetic admixtures. Techniques from model based clustering are usually employed.

In this chapter we focus on structured association methods, in particular concentrating on Bayesian approaches which model population structure and admixture using mixture models. An influential early Bayesian parametric mixture model has been proposed by Pritchard et al. (2000). Specifically, assuming that marker loci are unlinked and at linkage equilibrium with one another within populations, each individual is assumed to come from one of $K$ populations and alleles at different loci are modeled conditionally independently given population specific allele frequencies. In the case of genetic admixtures, each individual is associated with proportions of its genome coming from different populations, while alleles at different loci are suitably modeled conditionally independently given the admixture proportions. Independent prior distributions on the allelic profile parameters of each population are introduced and full posterior inference is performed through Markov chain Monte Carlo (MCMC) algorithms. With regard to the determination of the number of extant populations $K$, Pritchard et al. (2000) proposed the use of model selection techniques based on marginal likelihoods, though it has been noted that such estimates are highly sensitive to the prior specification.

Falush et al. (2003) improved the admixture model of Pritchard et al. (2000) by taking into account the correlations among neighboring loci. In particular Falush et al. (2003) model linked loci by using a Markov model which segments each chromosome into contiguous regions with shared genetic ancestry. This Markov model allows for the estimation of local genetic ancestry information from genotype data, as opposed to the global admixture proportions in Pritchard et al. (2000). Such local ancestry estimation gives more fine-grained information about the admixture process. The nonparametric counterpart of the simple population structure model in Pritchard et al. (2000) is described in Huelsenbeck and Andolfatto (2007), while the Hierarchical Dirichlet process of Teh et al. (2006) offers the Bayesian nonparametric extension of the admixture model. Recently, De Iorio et al. (2015) have proposed a Bayesian nonparametric counterpart of the linkage admixture model of Falush et al. (2003). In particular the nonparametric approach provides a methodology for modeling population structure that simultaneously gives estimates of local ancestries and bypasses difficult model selection issues arising in the parametric models by Pritchard et al. (2000) and Falush et al. (2003).

The chapter is structured as follows. In Sect. 7.2 we review the Bayesian parametric approaches introduced by Pritchard et al. (2000) and Falush et al. (2003) for modeling population structure with and without admixture and in presence of linked loci and correlated allele frequencies. In Sect. 7.3 we show how recent developments in Bayesian nonparametrics have been usefully exploited in order to introduce natural nonparametric counterparts of the parametric approaches by Pritchard et al. (2000) and Falush et al. (2003). Some of these Bayesian nonparametric approaches are briefly illustrated using data from the 1000 Genomes project (http://www.1000genomes.org/). The goal of the 1000 Genomes project consists in finding most genetic variants that have frequencies of at least 1 % in the populations under study by sequencing the genomes of a large number of individuals, providing in this way a valuable resource on human genetic variation.

## 7.2 Parametric Modeling

Suppose we sample $N$ haploid individuals at $L$ loci from a population with unknown structure. For simplicity we discuss the haploid case, extension to the diploid case is straightforward. We denote by $X = (X_l^{(i)})_{1 \le i \le N, 1 \le l \le L}$ the observed data, i.e., $x_l^{(i)}$ is the genotype of individual $i$ at locus $l$. Assuming $K$ subpopulations characterized by a set of allele frequencies at each locus, and with $K$ being fixed, in this section we review some Bayesian parametric models for making inference on the unknown population structure.

### 7.2.1 Models with and Without Admixture

We start by assuming that marker loci are unlinked and at linkage equilibrium with one another within populations. Let $Z = (z^{(i)})_{1 \le i \le N}$ denote the unknown allocation vector which assigns each individual to a population of origin, i.e., $z^{(i)}$ denotes the population from which individual $i$ originated. Let $Q = (q_k)_{1 \le k \le K}$ denote the unknown population proportions, i.e., $q_k$ is the proportion of individuals that originated from population $k$. Furthermore, let $J_l$ be the number of distinct alleles observed at locus $l$, and let $P = (p_{klj})_{1 \le k \le K, 1 \le l \le L, 1 \le j \le J_l}$ be the unknown allele frequencies in the populations, i.e., $p_{klj}$ is the frequency of allele $j$ at locus $l$ in population $k$. Throughout this chapter we use "allele copies" to refer to an allele carried at a particular locus by a particular individual.

Under this framework Pritchard et al. (2000) introduced a model without admixture among populations, namely the genome of each individual is assumed to be originated entirely from one of the $K$ populations. Given the population of origin of each individual, the genotype is generated by drawing alleles copies independently from the appropriate population frequency distribution. Formally, the model without admixture is specified as

$$\Pr[z^{(i)} = k \,|\, Q] = q_k \tag{7.1}$$

and

$$\Pr[x_l^{(i)} = j \,|\, Z, P] = p_{z^{(i)} l j} \tag{7.2}$$

independently for each $x_l^{(i)}$. This model can be easily extended to diploid or, in general, to polyploid data. For polyploid data the allocation variables $z^{(i)}$'s along each of the chromosomes of individual $i$ form independent vectors. We refer to Falush et al. (2003) for details.

The model (7.1)–(7.2) is completed by specifying a prior distribution for $Q$ and $P$. As regard to $Q$, Pritchard et al. (2000) assumed that the probability that individual $i$ originated in population $k$ is the same for all $k$. Hence, they proposed to use the uniform distribution $q_k = 1/K$, independently for all individuals. Different distributions for $Q$ have been considered in Anderson and Thompson (2002) to model cases with some populations being more represented in the sample than others. As regard to $P$, Pritchard et al. (2000) followed Balding and Nichols (1995) and Ranalla and Mountain (1997) in using the Dirichlet distribution to model the allele frequencies at each locus within each populations, i.e.,

$$p_{kl\cdot} \sim \text{Dirichlet}(\lambda_1, \ldots, \lambda_{J_l}), \tag{7.3}$$

for the allele frequencies $p_{kl\cdot}$, independently for any $k$ and $l$. Furthermore, they assumed $\lambda_i = 1$ for all $j = 1, \ldots, J_l$, which gives a uniform distribution over the allele frequencies. By means of (7.2) and (7.3) we can use the following MCMC scheme to construct a Markov chain with stationary distribution $\Pr[Z, P \mid X]$. Start with initial values $Z^{(0)}$ for $Z$ and, for $m \geq 1$: i) sample $P^{(m)}$ from $\Pr[P \mid X, Z^{(m-1)}]$ and ii) sample $Z^{(m)}$ from $\Pr[Z \mid X, P^{(m)}]$. For sufficiently large $m$ and $c$, $(Z^{(m)}, P^{(m)})$, $(Z^{(m+c)}, P^{(m+c)})$, $(Z^{(m+2c)}, P^{(m+2c)})$, ... are approximately random samples from the target distribution $\Pr[Z, P \mid X]$.

An obvious limitation of the model without admixture is that, in practice, individuals may have recent ancestors in more than one population. In order to overcome this fundamental drawback, Pritchard et al. (2000) introduced a more flexible model in which only a fraction of the individual's genome is assumed to have originated from one of the $K$ populations. This more general model allows individuals to have mixed ancestry. Let $Z = (z_l^{(i)})_{1 \leq i \leq N, 1 \leq l \leq L}$ be the unknown populations of origin of the allele copies, i.e., $z_l^{(i)}$ is the population from which the allele copies at locus $l$ of individual $i$ are originated. Furthermore, let $Q = (q_k^{(i)})_{1 \leq i \leq N, \leq k \leq K}$ be the unknown admixture individual proportions, i.e., $q_k^{(i)}$ is the proportion of the genome of individual $i$ that originated from population $k$.

Under this more general framework, Pritchard et al. (2000) introduced a model which allows for admixture: given the population of origin of each allele copies, the genotype is generated by drawing allele copies independently from the appropriate population frequency distribution. Formally, the model with admixture is specified as follows:

$$\Pr[z_l^{(i)} = k \mid Q] = q_k^{(i)} \tag{7.4}$$

and

$$\Pr[x_l^{(i)} = j \mid Z, P] = p_{z_l^{(i)} l j} \tag{7.5}$$

independently for each $x_l^{(i)}$. This model can be easily extended to diploid or, in general, to polyploid data. For polyploid data the allocation variables $z^{(i)}$'s along each of the chromosomes of individual $i$ form independent vectors. We refer to Falush et al. (2003) for details.

The admixture model (7.4)–(7.5) is completed by specifying a prior distribution for $Q$ and $P$. Pritchard et al. (2000) proposed the use of the Dirichlet distribution (7.3) for $P$, as for the model without admixture. The specification of a prior distribution for $Q$ depends on the type and amount of mixed ancestry we expect to see. In particular Pritchard et al. (2000) proposed the use of a symmetric Dirichlet distribution to model the admixture proportions of each individual. Specifically, they specified the distribution

$$q^{(i)} \sim \text{Dirichlet}(\alpha, \ldots, \alpha) \tag{7.6}$$

for the admixture proportions $q^{(i)}$, independently for each individual. If $\alpha$ tends to 0, then the admixture model reduces to the model without admixture. Different distributions for $Q$ have been considered in Anderson and Thompson (2002). The following MCMC scheme may be used to sample from $\Pr[Z,P,Q \mid X]$. Start with initial values $Z^{(0)}$ for $Z$ and, for $m \geq 1$: i) sample $P^{(m)}$ and $Q^{(m)}$ from $\Pr[P,Q \mid X, Z^{(m-1)}]$, ii) sample $Z^{(m)}$ from $\Pr[Z \mid X, P^{(m)}, Q^{(m)}]$ and update $\alpha$ using a Metropolis-Hastings step. As before, for sufficiently large $m$ and $c$, note that $(Z^{(m)}, P^{(m)}, Q^{(m)})$, $(Z^{(m+c)}, P^{(m+c)}, Q^{(m+c)})$, $(Z^{(m+2c)}, P^{(m+2c)}, Q^{(m+2c)})$, ... are approximately random samples from the target distribution $\Pr[Z,P,Q \mid X]$.

### *7.2.2 Extensions: Linked Loci and Correlated Allele Frequencies*

Falush et al. (2003) extended the admixture model of Pritchard et al. (2000) to allow for linkage between loci. In particular they considered the correlations in ancestry, which cause linkage disequilibrium between linked loci. This linkage disequilibrium naturally occurs because the chromosome is composed of a set of chunks that are derived, as intact units, from one or another of the ancestral populations. In order to model linked loci, Falush et al. (2003) assumed that the breakpoints between successive segments occur as a Poisson process at a rate $r$ per unit of genetic distance, and that the population of origin of each chunk in individual $i$ is independently drawn according to the vector $q^{(i)}$, which continues to represent the admixture proportions of the $i$-th individual.

More formally the linkage admixture model of Falush et al. (2003) assumes that for each individual $i$ the random variables $z_l^{(i)}$'s are dependent across $l$ and, in particular, they form a reversible Markov chain. Specifically, for any positive $r$, one has the following specification

$$\Pr[z_1^{(i)} = k \mid Q] = q_k^{(i)} \tag{7.7}$$

and

$$\Pr[z_{l+1}^{(i)} = k' \mid z_l^{(i)} = k, Q] = \begin{cases} \mathrm{e}^{-d_l r} + (1 - \mathrm{e}^{-d_l r}) q_{k'}^{(i)} & \text{if } k' = k \\ (1 - \mathrm{e}^{-d_l r}) q_k^{(i)} & \text{if } k \neq k' \end{cases} \tag{7.8}$$

independently for each individual, where $d_l$ denotes the genetic distance from locus $l$ to locus $l+1$, assumed known. The admixture model (7.4)–(7.5) is recovered by letting $r \to +\infty$. We refer to Falush et al. (2003) for details on the MCMC scheme for sampling from $\Pr[Z,P,Q \mid X]$.

Falush et al. (2003) also introduce an extension of the admixture model of Pritchard et al. (2000) in order to allow for correlated allele frequencies, namely the allele frequencies in one population provide information about the allele frequencies in another population. Indeed it is expected that allele frequencies in closely related populations tend to be very similar. In order to model closely related populations, Pritchard et al. (2000) replaced the prior distribution (7.3) with $p_{kl\cdot} \sim \text{Dirichlet}(f^{(l)}J_l\mu_1^{(l)},\ldots,f^{(l)}J_l\mu_{J_l}^{(l)})$, where $\mu_j^{(l)}$ is the mean sample frequency at locus $l$, and $f^{(l)} > 0$ determines the strength of the correlations across populations at locus $l$. Clearly, when $f^{(l)}$ is large, the allele frequencies in all populations tend to be similar to the mean allele frequencies in the sample.

Alternatively, Falush et al. (2003) assume that the populations all diverged from a common ancestral population at the same time, but allow that the populations may have experienced different amounts of drift since the divergence event. Specifically, let $p_{Alj}$ be the frequency of allele $j$ at locus $l$ in a hypothetical ancestral population $A$. The $K$ populations in the sample have each undergone independent drift away from the ancestral allele frequencies, at rates parameterized by $F_1,\ldots,F_K$, respectively. More formally,

$$p_{Al\cdot} \sim \text{Dirichlet}(\lambda_1,\ldots,\lambda_{J_l}) \tag{7.9}$$

independently for each $l$. Note that the prior distribution has the same form as that used in the model with uncorrelated population frequencies. Then, conditionally on $P_A$,

$$p_{kl\cdot} \sim \text{Dirichlet}\left(p_{Al1}\frac{1-F_k}{F_k},\ldots,p_{AlJ_l}\frac{1-F_k}{F_k}\right) \tag{7.10}$$

independently for each population $k$ and for each locus $l$. According to (7.10) the size of the parameter $F_k$ tells us about the effective size of population $k$ during the time since divergence, with large values of $F_k$ indicating a smaller effective population size. We refer to Falush et al. (2003) for details on the MCMC scheme for sampling from $\Pr[Z,P,Q|X]$.

The Bayesian parametric approaches described in this section are implemented in STRUCTURE (http://pritchardlab.stanford.edu/structure.html), which is arguably the most widely used software for estimating genetic ancestry. The reader is referred to Pritchard et al. (2000) and Falush et al. (2007) for a description of the basic algorithms. Extensions can be found in Falush et al. (2007) and Hubisz et al. (2009). ADMIXTURE (http://genetics.ucla.edu/software/admixture/index.html) is an alternative software which provides a faster implementation of a similar model to the one defined in STRUCTURE. In particular ADMIXTURE uses maximum likelihood inference to estimate population allele frequencies and ancestry proportions, rather than sampling from posterior distribution through MCMC algorithms. See, e.g., Alexander et al. (2009) for details.

## 7.3 Nonparametric Modeling

The Bayesian parametric models reviewed in Sect. 7.2 assume the number of populations $K$ to be fixed. In order to deal with an unknown $K$, Pritchard et al. (2000) suggest a method based upon an *ad hoc* approximation of the marginal likelihood to determine the number of populations needed to explain the observations. In particular STRUCTURE is run for different values of $K$, and the number of populations is determined by the value of $K$ which maximises the marginal likelihood of the data. Alternatively, ADMIXTURE uses a cross validation approach to estimate $K$, by fitting the model on a subset of genotype data and then predicting the excluded genotypes. Other parametric approaches have been proposed by Corander et al. (2003), Corander et al. (2004), and Evanno et al. (2005). In this section we review some Bayesian nonparametric models for making inference on population structure. In the nonparametric framework both the allocation vectors $Z$ and the number of ancestral populations $K$ are unknown.

### *7.3.1 Models with and Without Admixture*

A Bayesian nonparametric counterpart of the model without admixture of Pritchard et al. (2000) has been proposed in Huelsenbeck and Andolfatto (2007). This model makes use of the Dirichlet process by Ferguson (1973), which allows both the assignment of individuals to populations and the number $K$ of populations to be random variables. A simple and intuitive definition of the Dirichlet process follows from the stick-breaking construction introduced by Sethuraman (1994). Specifically, let $(v_j)_{j\geq 1}$ be a collection of independent Beta random variables with parameter $(1, \alpha_0)$, and let $(\theta_i)_{i\geq 1}$ be a collection of random variables, independent of $(v_j)_{j\geq 1}$, and independent and identically distributed according to a nonatomic probability measure $G_0$. The discrete random probability measure $Q_0 = \sum_{j\geq 1} q_j \delta_{\theta_j}$, with $q_j = v_j \prod_{1\leq l\leq j-1}(1 - v_l)$, is a Dirichlet process with parameter $\alpha_0 G_0$.

Here and in the following discussion we denote with $\mathrm{DP}(\alpha_0, G_0)$ the distribution of a Dirichlet process with parameter $(\alpha_0 G_0)$. The Bayesian nonparametric model without admixture introduced by Huelsenbeck and Andolfatto (2007) can be specified as follows:

$$z^{(i)} \mid Q_0 \overset{\text{iid}}{\sim} Q_0$$

$$Q_0 \sim \mathrm{DP}(\alpha_0, G_0) \tag{7.11}$$

and

$$x_l^{(i)} \mid Z, P \overset{\text{ind}}{\sim} P_Z$$

$$P_Z \sim \mathrm{Dirichlet}(\lambda_1, \ldots, \lambda_{J_l}), \tag{7.12}$$

for $i = 1,\dots,N$ and $l = 1,\dots,L$. See, e.g., Dawson and Belkhir (2001) and Pella and Masuda (2006) for alternative Bayesian nonparametric models which exploit the Dirichlet process at the allocation level.

The sample $Z$ from $Q_0$ induces a random partition of $\{1,\dots,N\}$ which determines the allocation of individuals into a random number $K$ of populations with random frequencies $(n_1,\dots,n_K)$. The parameter $\alpha_0$ determines the degree to which individuals are grouped together into the same population. Indeed, Blackwell and MacQueen (1973) show that

$$\Pr[z^{(N)} \in \cdot \,|\, z^{(1)},\dots,z^{(N-1)}] = \sum_{i=1}^{K} \frac{n_i}{N-1+\alpha_0}\delta_{\theta_i}(\cdot) + \frac{\alpha_0}{N-1+\alpha_0}G_0(\cdot). \quad (7.13)$$

The allocation directed by the predictive distribution (7.13) can be intuitively described by means of the following Chinese restaurant metaphor. See, e.g., Aldous (1985) for a detailed account. Consider a Chinese restaurant with an unbounded number of tables. Each $z^{(i)}$ corresponds to a customer who enters the restaurant, whereas the distinct values $\theta_j$'s correspond to the tables at which the customers sit. Customer $i$ sits at the table indexed by $\theta_j$ with probability proportional to the number $n_j$ of customers already seated there, in which case we set $z^{(i)} = \theta_j$, and it sits at a new table with probability proportional to $\alpha_0$, in which case we increment $K$ by 1, draw $\theta_K$ from $G_0$ and set $z^{(i)} = \theta_K$.

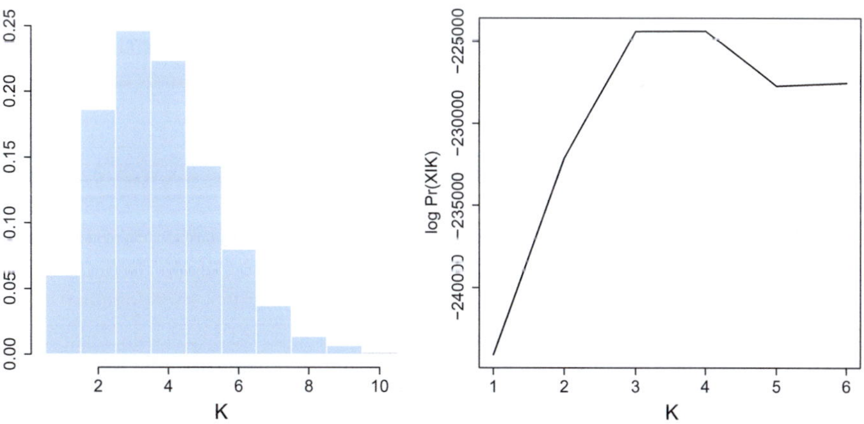


**Fig. 7.1** *Left panel*: posterior distribution of the number $K$ of populations present in the sample. *Right panel*: $\log \Pr(\text{Data} \,|\, K)$ from STRUCTURE

The approach of Huelsenbeck and Andolfatto (2007) is implemented in STRUCTURAMA (http://cteg.berkeley.edu/structurama/). Posterior inference is performed through an MCMC scheme which aims at determining the mean partition, a partitioning of individuals among populations which minimizes the squared distance to the sampled partitions. To illustrate the model (7.11)–(7.12), we consider 305 individuals from the 1000 Genomes project. The sample is composed of 95 chromosomes with European ancestry (CEU), 107 chromosomes of African (YRI) origin, and 103 individuals of East Asian (CHB) ancestry. In order to phase the genotype data we use SHAPEIT2 (http://mathgen.stats.ox.ac.uk/genetics_software/shapeit/shapeit.html), providing a sample of 610 haplotypes. We have analyzed a collection of 1000 bi-allelic loci from chromosome 2. Posterior inference on the number of populations in the sample is shown in the left panel of Fig. 7.1. The posterior distribution of $K$ has its mode in 3, which is the true number of populations present in the sample. We have also run the model without admixture implemented in STRUCTURE for each value of $K$, $K = 1, \ldots, 6$. The right panel of Fig. 7.1 shows the estimated $\log \Pr(\text{Data} \,|\, K)$. Note that the value $K = 3, 4$ seems to maximize the model log-likelihood. With this regard, it is worth pointing out that Pritchard et al. (2000) warn against possible drawbacks of using this criterion and to interpret the results with caution and give suggestions for improvement.

A Bayesian nonparametric counterpart of the linkage admixture model of Falush et al. (2003) has been recently introduced and investigated in De Iorio et al. (2015). This model extends the hierarchical Dirichlet process introduced by Teh et al. (2006). The hierarchical Dirichlet process is defined as a distribution over a collection of discrete random probability measures. Specifically, let $\alpha_0$ and $\alpha$ be positive constants and let $G_0$ be a nonatomic probability measure. The hierarchical Dirichlet process defines a set of local discrete random probability measures $(Q_i)_{i \in \{1,\ldots,N\}}$, for some index $N \geq 1$, and a global discrete random probability measure $Q_0$ such that $Q_0$ is a Dirichlet process with parameter $\alpha_0 G_0$ and, given $Q_0$, $(Q_i)_{i \in \{1,\ldots,N\}}$ is a collection of independent Dirichlet processes, each one with the same parameter $\alpha Q_0$. Because the global $Q_0$ has support at the points $(\theta_i)_{i \geq 1}$, each local random probability measure $Q_i$ necessarily has support at these points as well, and thus can be written as $Q_i = \sum_{j \geq 1} q_{ij} \delta_{\theta_j}$, with $q_{ij} = w_{ij} \prod_{1 \leq l \leq j-1} (1 - w_{il})$, where $(w_{ij} \,|\, v_1, \ldots, v_j)_{j \geq 1}$ are independent random variables from a Beta distribution with parameter $(\alpha v_j, \alpha(1 - \sum_{1 \leq l \leq j} v_l))$. Note that the Dirichlet process with parameter $\alpha_0 G_0$ is recovered by letting $\alpha \to 0$.

Due to the sharing of atoms among discrete random probability measures, the hierarchical Dirichlet process is the natural generalization of the Dirichlet process to model linked sets of admixture proportions and constitutes the Bayesian nonparametric counterpart of the admixture model defined in (7.4)–(7.5). Individual genotypes will have portions that arise from different populations which are shared among individuals. The hierarchical Dirichlet process models the allocation vector $Z = (z_l^{(i)})_{1 \leq i \leq N, 1 \leq l \leq L}$ that specifies the populations of origin of the allele copies. Accordingly, the resulting Bayesian nonparametric "admixture" model can be specified as follows:

$$
\begin{aligned}
z_l^{(i)} \,|\, Q_i &\overset{\text{iid}}{\sim} Q_i \\
Q_i \,|\, Q_0 &\overset{\text{iid}}{\sim} \mathrm{DP}(\alpha, Q_0), \\
Q_0 &\sim \mathrm{DP}(\alpha_0, G_0)
\end{aligned}
\tag{7.14}
$$

and

$$x_l^{(i)} \mid Z, P \overset{\text{ind}}{\sim} P_Z$$

$$P_Z \sim \text{Dirichlet}(\lambda_1, \ldots, \lambda_{J_l}), \tag{7.15}$$

for $i = 1, \ldots, N$ and $l = 1, \ldots, L$. Note that the hierarchical Dirichlet process prior assumption allows to model $Z$ in terms of an unknown number of populations $K$ with unknown frequencies that are shared out across individuals; each individual's genome is then modeled according to an unknown number of populations with unknown allocation frequencies. For polyploid data the $z$'s along each of the chromosomes of individual $i$ are assumed to be independent.

The allocation mechanism induced by the hierarchical Dirichlet process can be intuitively described by the following generalization of the Chinese restaurant metaphor. Consider a finite collection of Chinese restaurants, one for each index $i \in \{1, \ldots, N\}$, with a shared menu. Each $z_l^{(i)}$ corresponds to customer $l$ in restaurant $i$. Let $\psi_t^{(i)}$ be the $t$-th table in restaurant $i$ and let $\theta_k$ denote the $k$-th dish. If $n_{itk}$ is the number of customers in restaurant $i$ seated around the table $t$ and being served dish $k$, $m_{ik}$ is the number of tables in restaurant $i$ serving the dish $k$, and $K$ is the number of unique dishes served in the franchise, then

$$\Pr[z_L^{(i)} \in \cdot \mid z_1^{(i)}, \ldots, z_{L-1}^{(i)}, Q_0] = \sum_{t=1}^{m_{i\cdot}} \frac{n_{it\cdot}}{L-1+\alpha} \delta_{\psi_t^{(i)}}(\cdot) + \frac{\alpha}{L-1+\alpha} Q_0(\cdot). \tag{7.16}$$

where $n_{it\cdot} = \sum_k n_{itk}$ and $m_{i\cdot} = \sum_k m_{ik}$. In other words, the customer $z_l^{(i)}$ sits at the table indexed by $\psi_t^{(i)}$ with probability proportional to the number of customers $n_{it\cdot}$ already seated there, in which case we set $z_l^{(i)} = \psi_t^{(i)}$, and it sits at a new table with probability proportional to $\alpha$, in which case we increment $m_{i\cdot}$, set $n_{im_{i\cdot}\cdot} = 1$, draw $\psi_{m_{i\cdot}}^{(i)}$ from $Q_0$ and set $z_l^{(i)} = \psi_{m_{i\cdot}}^{(i)}$. Note that $\psi_{m_{i\cdot}}^{(i)}$ is drawn from $Q_0$ and this is the only reference to $Q_0$ in the predictive (7.16). In particular, one has

$$\Pr[\psi_t^{(i)} \in \cdot \mid \psi_1^{(1)}, \ldots, \psi_{t-1}^{(i)}] = \sum_{k=1}^{K} \frac{m_{\cdot k}}{m_{\cdot\cdot} + \alpha_0} \delta_{\theta_k}(\cdot) + \frac{\alpha_0}{m_{\cdot\cdot} + \alpha_0} G_0(\cdot). \tag{7.17}$$

where $m_{\cdot k} = \sum_i m_{ik}$ and $m_{\cdot\cdot} = \sum_i \sum_k m_{ik}$. In other words, to table $\psi_t^{(i)}$ it is assigned the dish indexed by $\theta_k$ with probability proportional to the number of tables which have previously served that dish in the franchise, in which case we set $\psi_t^{(i)} = \theta_k$, and it is assigned a new dish with probability proportional to $\alpha_0$, in which case we increment $K$, draw $\theta_K$ from $G_0$, and set $\psi_t^{(i)} = \theta_K$. Dishes are chosen with probability proportional to the number of tables which have previously served that dish in the franchise. We refer Teh et al. (2006) for additional details.

We have fitted the hierarchical Dirichlet process admixture model (7.14)–(7.15) to a set of 188 phased haplotypes from the Colombian (CLM) sample in the

1000 Genomes project. We have considered a collection of 500 bi-allelic loci from chromosomes 2. A value of $K = 3$ covers 99 % of the typed loci across individuals. This is in agreement with what is known about Colombian ancestry. Latin America has a well-documented history of extensive mixing between Native Americans and people arriving from Europe and Africa. This continental admixture, which has occurred for the past 500 years (or about 20–25 generations), gives rise to haplotype blocks. For example, in Fig. 7.2 we show posterior inference for the allocation of loci on a segment of chromosome 2 to one of the three major ancestral populations detected in the sample. The results are based on the Maximum A Posteriori clustering configuration. Notice the mosaic structure of the chromosomes.

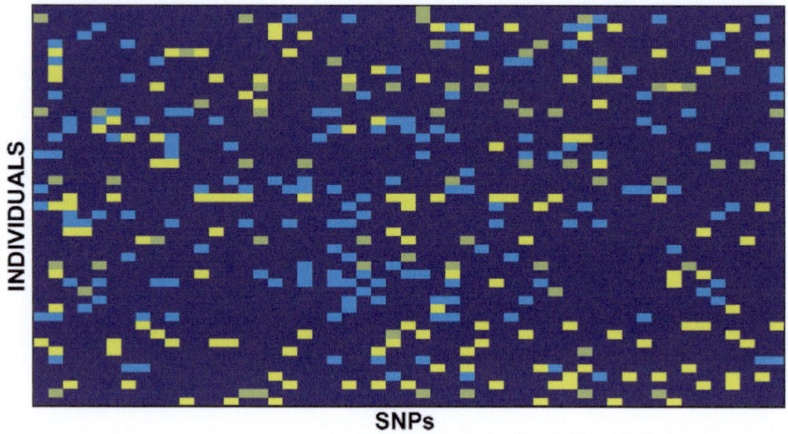


**Fig. 7.2** MAP estimates of population assignment for single nucleotide polymorphism (SNP) data. Each color corresponds to a different ancestral population

Recently, De Iorio et al. (2015) propose the Bayesian nonparametric counterpart of the linkage admixture model defined in (7.7)–(7.8), allowing for dependence in the allocation vector $z^{(i)}$. Specifically, let $d_l$ denote the genetic distance from locus $l$ to locus $l+1$, and let $s_l^{(i)}$ be a binary random variable which denotes whether locus $l$ and locus $l+1$ are on the same segment ($s_l^{(i)} = 1$) or not ($s_l^{(i)} = 0$). The Bayesian nonparametric linkage admixture model of De Iorio et al. (2015) can be specified as follows:

$$
\begin{aligned}
z_1^{(i)} \mid Q_i &\sim Q_i \\
s_l^{(i)} &\overset{\text{iid}}{\sim} \text{Bernoulli}(\mathrm{e}^{-d_l r}) \\
z_{l+1}^{(i)} \mid z_l^{(i)}, s_l^{(i)}, Q_i &\overset{\text{ind}}{\sim} s_l^{(i)} \delta_{z_l^{(i)}} + (1 - s_l^{(i)}) Q_i
\end{aligned}
$$

$$
\begin{aligned}
Q_i \,|\, Q_0 &\overset{\text{iid}}{\sim} \mathrm{DP}(\alpha, Q_0), \\
Q_0 &\sim \mathrm{DP}(\alpha_0, G_0)
\end{aligned}
\tag{7.18}
$$

and

$$
\begin{aligned}
x_l^{(i)} \,|\, Z, P &\overset{\text{ind}}{\sim} P_Z \\
P_Z &\sim \mathrm{Dirichlet}(\lambda_1, \ldots, \lambda_{J_l}),
\end{aligned}
\tag{7.19}
$$

for $i = 1, \ldots, N$ and $l = 1, \ldots, L$. $Q_i$ describes the proportion of the alleles on $x^{(i)}$ coming from each of the populations, as well as the parameters of the populations. Given $Q_i$, the sequence $x^{(i)}$ is modelled by (i) first placing segment boundaries according to a nonhomogeneous Poisson process with rate $d_l r$ (ii) and then generating alleles on each segment by picking a population according to $Q_i$ and sampling the alleles according to the population specific distribution. The model of Huelsenbeck and Andolfatto (2007) is obtained by letting $r \to +\infty$ and $\alpha \to 0$. By letting $r \to +\infty$ one obtains the standard hierarchical Dirichlet process.

We would like to conclude this section with a note of caution. In particular, within the Bayesian nonparametric settings, inference on $K$ can be sensitive to the choice of the prior on the number of populations, in particular to the prior specification on $\alpha$ and $\alpha_0$, as well as on the $\lambda_i$'s. We note that as the number of sequences and/or markers increases the model tends to generate spurious clusters, i.e. clusters with very few individuals in them. This is in agreement with recent results on the clustering properties of the Dirichlet Process. See, e.g., Miller and Harrison (2014) for details. Nevertheless, the number of clusters explaining the majority of the data, i.e. 95–99 %, is quite robust to prior specifications. In general, the biological interpretation of $K$ is difficult. See, e.g., Pritchard et al. (2000) and the references therein for a detailed discussion. See Fritsch and Ickstadt (2009) for a description of methods for summarizing posterior clustering output.

### 7.3.2 The MCMC Algorithm

We briefly present the MCMC algorithm for posterior sampling from the Bayesian nonparametric linkage admixture model (7.18)–(7.19). The conditional distributions of $(Q_i)_{0 \le i \le N}$, given $(z_l^{(i)})_{1 \le i \le N, 1 \le l \le L}$ and $(s_l^{(i)})_{1 \le i \le N, 1 \le l \le L}$, follow from standard results on the hierarchical Dirichlet process. Conditionally to $(z_l^{(i)})_{1 \le i \le N, 1 \le l \le L}$ and $(s_l^{(i)})_{1 \le i \le N, 1 \le l \le L}$, let $K^*$ be the number of populations. Then,

$$
Q_0 = \sum_{j=1}^{K^*} q_{0j} \delta_{\theta_j} + w_0 Q_0'
$$

and

$$Q_i = \sum_{j=1}^{K^*} q_{ij}\delta_{\theta_j} + w_i Q'_i,$$

for $i = 1,\dots,N$, where $Q'_0$ is independent of $(q_{01},\dots,q_{0K^*},w_0)$ and $Q'_i$ is independent of $(q_{i1},\dots,q_{iK^*},w_i)$. Let $n_{ik}$ be the number of chunks in $z_l^{(l)}$ that are assigned to population $k$, and let $m_{ik}$ be a random variable such that $m_{ik} = 0$ if $n_{ik} = 0$ and $m_{ik} \in \{1,\dots,n_{ik}\}$ if $n_{ik} > 0$. Moreover, let us define $n_{0k} = \sum_{1\le i\le N} m_{ik}$. Then, we have

$$(q_{01},\dots,q_{0K^*},w_0)\,|\,(n_{ik},m_{ik})_{0\le i\le N,1\le k\le K^*} \sim \text{Dirichlet}(n_{01},\dots,n_{0K^*},\alpha_0), \quad (7.20)$$

$$Q'_0\,|\,(n_{ik},m_{ik})_{0\le i\le N,1\le k\le K^*} \sim \text{DP}(\alpha_0,G_0), \quad (7.21)$$

$$\begin{aligned}(q_{01},\dots,q_{0K^*},w_i)\,|\,(n_{ik},m_{ik})_{0\le i\le N,1\le k\le K^*},(q_{01},\dots,q_{0K^*},w_0)\\ \sim \text{Dirichlet}(\alpha q_{01}+n_{i1},\dots,\alpha q_{0K^*}+n_{0K^*},\alpha w_0)\end{aligned} \quad (7.22)$$

and

$$Q'_i\,|\,(n_{ik},m_{ik})_{0\le i\le N,1\le k\le K^*}, \qquad G'_0 \sim \text{DP}(\alpha,G'_0). \quad (7.23)$$

Equations (7.20) and (7.22) form a hierarchy of Dirichlet distributions while Eqs. (7.21) and (7.23) form a hierarchy of Dirichlet processes. The two hierarchies are independent. The reader is referred to Teh et al. (2006) for a detailed account on Eqs. (7.20), (7.21), (7.22), and (7.23).

In order to sample from (7.21) and (7.23), De Iorio et al. (2015) adopted the slice sampling approach of Walker (2007). See also Papaspiliopoulos and Roberts (2008). The slice sampling allows to truncate the series representations of $Q'_0\,|\,(n_{ik},m_{ik})_{0\le i\le N,1\le k\le K^*}$ and $Q'_i\,|\,(n_{ik},m_{ik})_{0\le i\le N,1\le k\le K^*},G'_0$ while retaining exactness in sampling from them. The idea consists in introducing an auxiliary random variable $C_i$, the so-called slice variable, such that

$$C_i\,|\,(n_{ik},m_{ik})_{0\le i\le N,1\le k\le K^*}, \quad (Q_i)_{0\le i\le N} \sim \text{Uniform}(0,\min_l q_{iz_l^{(i)}}).$$

Conditionally on $C_i$, only the atoms with mass above the minimum threshold $\min_i C_i$ need to be simulated. This can be easily achieved by using the stick-breaking representation until the left-over mass falls below the threshold. We refer to De Iorio et al. (2015) for additional details on the implementation of the slice sampling for (7.21) and (7.23).

Finally, forward-filtering backward-sampling can be used to update the latent state sequences $(z_l^{(i)})_{1\le i\le N,1\le l\le L}$ and $(s_l^{(i)})_{1\le i\le N,1\le l\le L}$. Note that, conditionally on $C_i$, only populations with $q_{i,k} > C_i$ will have positive probability of being selected, so that the forward-filtering backward-sampling is computationally tractable. However, as the random variable $C_i$ depends on the latent state sequences $(z_l^{(i)})_{1\le i\le N,1\le l\le L}$ and $(s_l^{(i)})_{1\le i\le N,1\le l\le L}$, conditioning on $C_i$ introduces complex dependencies among the latent state variables which precludes an exact and efficient forward filtering algorithm. De Iorio et al. (2015) proposed instead to ignore the dependencies caused by the slice variable, and use the resulting efficient forward-filtering backward-sampling as a Metropolis-Hasting proposal. The forward-filtering backward-sampling has a computational scaling of the order $O(LK_i)$, linear in both the number $L$ of loci and potential populations $K_i$, and it represents the most computationally expensive part of the MCMC algorithm. MATLAB software implementing this MCMC scheme is available at http://BigBayes.github.io/HDPStructure.

## 7.4 Discussion and Concluding Remarks

The analysis of population stratification is an increasingly important component of genetic studies. Many different methods have been proposed in the literature and often software implementing such methods has been developed and made publicly available. The main goals of population structure analysis can be summarized as follows: detection of population structure in a sample of chromosomes, estimation of the number of populations present in a sample, and consequent assignment of individuals to sub-populations. In the case of genetic admixtures scientific interest focuses on inferring the number of ancestral population to a sample, estimating ancestral population proportions to admixed individuals and identifying the genetic ancestry of chromosomal segments within an individual. No single method is able to deal with the variety of research questions relating to genetic ancestry and it is helpful in applications to use a combinations of approaches.

In this chapter we have reviewed model-based clustering methods for population structure within a Bayesian framework. We have shown how the initial parametric modeling strategies have a natural counterpart in Bayesian nonparametrics, which allows for joint estimation of the number of ancestral populations and the population allocation vector for each individual. In this framework, posterior inference is usually performed through MCMC algorithms. These methods can be used for both haplotype and genotype data, although in the latter case at an extra computational cost. It is in theory straightforward to include further prior information such as geographical locations of sampled chromosomes, ethnicity and phase information. Moreover, it is possible to pre-specify the population of origin of some individuals to aid ancestry estimation for individuals of unknown origin and also to include phenotype information. The inferred clustering structure will generally be sensible and able to explain most of the variability in the data, but clusters will not necessarily correspond to "real" populations and biological interpretation of the number of clusters is often difficult, as pointed out in Pritchard et al. (2000).

**Acknowledgements** We would like to thank Kaustubh Adhikari for kindly providing the pahsed data and Lloyd Elliott for developing user-friendly MATLAB functions for the linked hierarchical Dirichlet process. Stefano Favaro is supported by the European Research Council (ERC) through StG N-BNP 306406. Yee Whye Teh is supported by the European Research Council (ERC) through the European Unions Seventh Framework Programme (FP7/2007–2013) ERC grant agreement 617411.

## References

Aldous, D. J. (1985). *Exchangeability and related topics.* Ecole d'ete de probabilites de Saint-Flour, XIII. Lecture notes in Mathematics N. 1117, Springer, Berlin.

Alexander, D.H., Novembre, J. and Lange K. (2009). Fast model-based estimation of ancestry in unrelated individuals. *Genome Research* **19**, 1655–1664.

Anderson, E.C. and Thompson, E.A. (2002). A model-based method for identifying species hybrids using multilocus genetic data. *Genetics* **160**, 1217–1229.

Balding, D.J. and Nichols, R.A. (1995). A method for quantifying differentiation between populations at multi-allelic loci and its implications for investigating identity and paternity. *Genetica* **96**, 3–12.

Blackwell, D. and MacQueen, J. B. (1973). Ferguson distributions via Pólya urn schemes. *Ann. Statist.* **1**, 353–355.

Corander, J., Waldmann, P. and Sillanpää, M.J. (2003). Bayesian analysis of genetic differentiation between populations. *Genetics* **163**, 367–374.

Corander, J., Waldmann, P., Marttinen, P. and Sillanpää, M.J. (2004). BAPS2: enhanced possibilities for the analysis of population structure. *Bioinformatics* **20**, 2363–2369.

Dawson, K.J. and Belkhir, K. (2001). A Bayesian approach to the identification of panmictic populations and the assignment of individuals. *Genet. Res.* **78**, 59–77.

De Iorio, M., Elliott, L., Favaro, S., Adhikari, K. and Teh, Y.W. (2015). Modeling population structure under hierarchical Dirichlet processes. *Preprint arXiv:1503.08278.*

Evanno, G., Regnaut, S. and Goudet, J. (2005). Detecting the number of clusters of individuals using the software Structure: a simulation study. *Mol. Ecol.* **14**, 2611–2620.

Falush, D., Stephens, M. and Pritchard, J.K. (2003). Inference of population structure from multilocus genotype data: linked loci and correlated allele frequencies. *Genetics* **164**, 1567–1587.

Falush, D., Stephens, M. and Pritchard, J.K. (2007). Inference of population structure using multi locus genotype data: dominant markers and null alleles. *Mol. Ecol. Notes* **7**, 574–578.

Ferguson, T.S. (1973). A Bayesian analysis of some nonparametric problems. *Ann. Statist.*, **1**, 209–230.

Field, D.L., Ayre, D.J., Whelan, R.J. and Young, A.G. (2011). Patterns of hybridization and asymmetrical gene flow in hybrid zones of the rare *Eucalyptus aggregata* and common *E. rubida*. *Heredity* **106**, 841–853.

Fritsch, A. and Ickstadt, K. (2009). Improved criteria for clustering based on the posterior similarity matrix. *Bayesian Analysis*, **4**, 367–392.

Hubisz, M.J., Falush, D., Stephens, M. and Pritchard, J.K. (2009). Inferring weak population structure with the assistance of sample group information. *Mol. Ecol. Resources* **9**, 1322–1332.

Huelsenbeck, J.P. and Andolfatto, P. (2007). Inference of population structure under a Dirichlet process model. *Genetics* **175**, 1787–1802.

Miller, J.W. and Harrison, M.T. (2014) Inconsistency of Pitman-Yor process mixtures for the number of components. *Journal of Machine Learning Research* **15**, 3333–3370.

Novembre, J. and Stephens, M. (2008) Interpreting principal components analyses of spatial population genetic variation. *Nature Gentics* **40**, 646–649.

Papaspiliopoulos, O. and Roberts, G.O. (2008). Retrospective Markov Chain Monte Carlo methods for Dirichlet process hierarchical models. *Biometrika* **95**, 169–186.

Parker, H.G., Kim, L.V., Sutter, N.B., Carlson, S., Lorentzen, T.D., Malek, T.B., Johnson, G.S., DeFrance, H.B., Ostrander, E.A. and Kruglya, L. (2004). Genetic structure of the purebred domestic dog. *Science* **304**, 1160–1164.

Patterson, N., Price, A.L. and Reich, D. (2006) Population structure and eigenanalysis. *PLoS Genetics* **2**, 2074–2093.

Pella, J. and Masuda, M. (2006). The Gibbs and split-merge sampler for population mixture analysis from genetic data with incomplete baselines. *Can. J. Fish. Aquat. Sci.* **63**, 576–596.

Pritchard, J.K., Stephens, M. and Donelly, P. (2000). Inference on population structure using multilocus genotype data. *Genetics* **155**, 945–959.

Ranalla, B. and Mountain, J.L. (1997). Detecting immigration by using multilocus genotypes. *Proc. Natl. Acad. Sci.* **94**, 9197–9201.

Ray, A. and Quader, S. (2014). Genetic diversity and population structure of *Lantana camara* in India indicates multiple introductions and gene flow. *Plant Biology* **16**, 651–658.

Sethuraman, J. (1994). A constructive definition of Dirichlet priors. *Statist. Sinica.* **4**, 639–650.

Teh, Y.W., Jordan, M.I., Beal, M,J. and Blei, D.M. (2006). Hierarchical Dirichlet processes. *J. Amer. Statist. Assoc.* **101**, 1566–1581.

Walker, S.G. (2007). Sampling the Dirichlet mixture model with slices. *Comm. Statist. Simulation Comput.* **36**, 45–54.

Wasser, S.K., Mailand, C., Booth, R., Mutayoba, B., Kisamo, E., Clark, B. and Stephens, M. (2007). Using DNA to track the origin of the largest ivory seizure since the 1989 trade ban. *Proceedings of the National Academy of Sciences* **104**, 4228–4233.

# Chapter 8
# Bayesian Approaches for Large Biological Networks

Yang Ni, Giovanni M. Marchetti, Veerabhadran Baladandayuthapani, and Francesco C. Stingo

**Abstract** Bayesian methods have found many successful applications in high-throughput genomics. We focus on approaches for network-based inference from gene expression data. Methods that employ sparse priors have been particularly successful, as they are properly designed to analyze large datasets in which the amount of measured variables can be greater than the number of observations. Here, we describe Bayesian approaches for both undirected and directed networks; we discuss novel approaches that are computationally efficient, do not rely on linearity assumptions, and perform comparatively better than state-of-the-art methods. We demonstrate the utility of our methods via applications to glioblastoma gene expression data.

## 8.1 Introduction

In this chapter, we review novel approaches for the analysis of large biological networks, motivated by application to cancer genomic datasets. Cancer is a set of diseases characterized by several cellular alterations, the complexity of which is

Y. Ni
Department of Statistics, Rice University, Houston, TX 77005, USA
e-mail: yn7@rice.edu

G.M. Marchetti
Dipartimento di Statistica, Informatica, Applicazioni "G. Parenti",
University of Florence, Viale Morgagni 59, 50134 Firenze, Italy
e-mail: marchetti@disia.unifi.it

V. Baladandayuthapani • F.C. Stingo (✉)
Department of Biostatistics, The University of Texas MD
Anderson Cancer Center, Houston, TX, USA
e-mail: veera@mdanderson.org; FStingo@mdanderson.org



R. Mitra, P. Müller (eds.), *Nonparametric Bayesian Inference in Biostatistics*,
Frontiers in Probability and the Statistical Sciences,
DOI 10.1007/978-3-319-19518-6_8

defined at multiple levels of cellular organization (Reimand et al. 2013; Hanahan & Weinberg 2011). Understanding the complex functions of the genes, proteins, and other aspects of the genome requires the integration of data from different sources, as realized by both the biostatistical (Hamid et al. 2009; Huang 2014; Huang et al. 2014; Ni et al. 2014; Stingo et al. 2010; Stingo et al. 2011; Wang et al. 2013; Wu et al. 2014) and medical (Chin et al. 2011; Kristensen et al. 2014; Weinstein et al. 2013; Parsons et al. 2008) communities. A primary objective of this large effort is to exploit the knowledge gained about the biological mechanisms of cancer for treatment development; this goal can be achieved if functional molecular mechanisms and novel targets/genes that can be pharmacologically modulated by targeted drugs are identified.

A key task to this end is to develop flexible and efficient quantitative models for the analysis of dependence structures of these high-throughput assays. Graphical models, which describe the conditional dependence relationships among random variables, have been widely applied in genomics and proteomics to infer various types of networks, including co-expression, gene regulatory, and protein interaction networks (Dobra et al. 2004*a*; Mukherjee & Speed 2008; Stingo et al. 2010; Telesca et al. 2012).

The chapter is organized as follows. In Sect. 8.2, we first introduce general methodological frameworks for both directed and undirected graphical models under the Bayesian paradigm. In Sect. 8.3, we focus on a recent approach for nonlinear directed graphical models, and, in Sect. 8.4, on a recent approach for undirected graphical models. In Sect. 8.5, we provide a brief discussion.

## 8.2 Introduction to Graphical Models

Graphical models are a class of statistical models for a random vector $\boldsymbol{X} = (X_1, \ldots, X_p)^{\mathrm{T}}$ that provide a graphic representation of conditional independencies among the variables, with the additional aim of describing their essential interrelations. The general theory of graphical models also plays an important role in simplifying inferences, implementing variable selection algorithms, and defining a framework for causal reasoning.

In a graphical model, variables are represented by a set of nodes, any pair of which may or may not be adjacent, i.e., joined by an edge. A missing edge represents the independence between the associated variables conditional on a set of other variables. The conditioning set depends on the chosen type of graph. Here, we focus on two types of graphs: undirected graphs and directed acyclic graphs (DAGs).

A graph is defined by a pair $G = (V, E)$ where $V = \{1, \ldots, p\}$ is a set of nodes and $E$ is a set of edges. We consider only graphs containing at most one edge between any pair of nodes. We say that $G$ is undirected if $E$ contains only undirected edges represented by $\{u, v\} = u - v$ for $(u, v) \in V \times V$. $G$ is said to be directed if all the edges are directed and represented by an ordered pair $(u, v) = u \to v$. The distinction between undirected and directed edges is important because undirected edges are typically used when variables are considered to have equal standing while directed edges, i.e., arrows such as $u \to v$, indicate that $v$ is a response and $u$ is an explanatory variable.

### 8.2.1 Undirected Graphical Models

Given an undirected graph $G = (V,E)$, with $V = \{1,\ldots,p\}$, the random vector $\boldsymbol{X} = (X_1,\ldots,X_p)^{\mathrm{T}}$ with joint distribution $P$ is said to be (globally) Markov with respect to $G$ if for any three disjoint subsets $A$, $B$, and $C$ of $V$ such that $C$ separates $A$ and $B$ (i.e., all the paths between $A$ and $B$ contain at least one node in $C$), all the variables $\boldsymbol{X}_A$ are independent of $\boldsymbol{X}_B$ given $\boldsymbol{X}_C$, written as $\boldsymbol{X}_A \perp\!\!\!\perp \boldsymbol{X}_B \mid \boldsymbol{X}_C$. This establishes a connection between the graphical concept of separation in the graph and the concept of conditional independence in the random vector $\boldsymbol{X}$.

If in addition $P > 0$, i.e., it is strictly positive, $\boldsymbol{X}$ is Markov with respect to $G$ if and only if for any pair of non-adjacent nodes $u$ and $v$, the conditional independence $\boldsymbol{X}_u \perp\!\!\!\perp \boldsymbol{X}_v \mid \boldsymbol{X}_{V\setminus\{u,v\}}$ holds. Furthermore, the Hammersley and Clifford theorem establishes that an equivalent condition is that the joint probability density function $f(\boldsymbol{x})$ factorizes according to certain components of the graph called maximal cliques. A maximal clique is a complete subgraph of $G$ that is not contained in a larger complete graph. A graph is complete if every pair of distinct edges is connected. Then the factorization is

$$f(\boldsymbol{x}) = \prod_{C\in\mathscr{C}} \psi_C(\boldsymbol{x}_C),$$

where $\mathscr{C}$ is the set of cliques of $G$ and the non-negative functions $\psi_G$ depend on $x$ only through the components $\boldsymbol{x}_C = (x_u)_{u\in C}$.

In the Gaussian case, which is the main subject of our discussion in this chapter, undirected graphical models are defined by zero constraints on the canonical parameter. The canonical parameter of a Gaussian random vector $\boldsymbol{X} \sim N_p(\boldsymbol{0},\boldsymbol{\Sigma})$ with positive definite covariance matrix $\boldsymbol{\Sigma}$ is the concentration matrix $\boldsymbol{\Omega} = \boldsymbol{\Sigma}^{-1} = (\omega_{uv})$ and the density can be written as

$$f(\boldsymbol{x}) = \exp\left(\alpha - \tfrac{1}{2}\Sigma_u\Sigma_v\omega_{u,v}x_u x_v\right),$$

so that a conditional independence $X_u \perp\!\!\!\perp X_v \mid \boldsymbol{X}_{V\setminus\{u,v\}}$ holds if and only if the concentration $\omega_{uv}$ is zero. Therefore, a Gaussian model that is Markov with respect to an undirected graph $G$ is defined by a family of multivariate normal distributions $N_p(\boldsymbol{0},\boldsymbol{\Sigma})$ with $\omega_{u,v} \doteq 0$ whenever $\{u,v\} \notin E$. The model's parameter space is the cone $M^+$ of the positive definite $p\times p$ matrices $\boldsymbol{\Omega}$ that satisfy the above zero constraints.

By the formula of the partial correlation coefficient between $X_u$ and $X_v$ given all the other variables

$$\rho_{uv.V\setminus\{u,v\}} = \frac{-\omega_{uv}}{\sqrt{\omega_{uu}\omega_{vv}}},$$

one sees that in a multivariate normal distribution, conditional independence is equivalent to zero partial correlation. This happens because the normal distribution has only linear dependencies. However, in general, this is not the case and we may have both zero partial correlation with strong nonlinear dependence and conditional independence but nonzero partial correlation (we introduce a novel approach for nonlinear dependence in Sect. 8.3).

If a random sample of $n$ iid multivariate observations $\boldsymbol{X}^{(1)},\ldots,\boldsymbol{X}^{(n)}$ from a Gaussian undirected graph model with graph $G$ is available, the log-likelihood function is

$$\log L(\boldsymbol{\Omega}|\boldsymbol{S}) \propto \log|\boldsymbol{\Omega}| - \mathrm{tr}(\boldsymbol{\Omega}\boldsymbol{S}), \quad \boldsymbol{\Omega} \in M^{+},$$

where $\boldsymbol{S} = \sum_{i=1}^{n} \boldsymbol{X}^{(i)}\boldsymbol{X}^{(i)^{\mathrm{T}}}$ is the sample sum-of-products matrix. Inference for undirected graphical models is not straightforward. For example, maximum likelihood (ML) estimation requires an equality-constrained convex optimization (Whittaker 2009).

For a subclass of graphical models, there is an explicit ML estimate of $\boldsymbol{\Omega}$. This subclass, which plays a special role in inference and computations, is that of decomposable graphs. A partition $(A,B,C)$ of the nodes $V$ is said to be a proper decomposition of the graph $G$ if $C$ separates $A$ and $B$, $C$ is complete, and $A$ and $B$ are nonempty. If no proper decomposition exists, then $G$ is said to be prime. Any graph can be recursively decomposed into its maximal prime subgraphs. Then $G$ is decomposable if all maximal prime subgraphs are cliques. It can be shown that a graph is decomposable if and only if it is chordal, meaning that every cycle of length $n \geq 4$ has a chord. The space of decomposable graphs is much smaller than the space of general undirected graphs. However, as decomposable graphical models allow for several simplifications in the computation, inference and model determination, the assumption of decomposability is often made.

#### 8.2.1.1 Bayesian Estimation of Undirected Graphical Models

The Bayesian analysis of a Gaussian undirected graphical model is based on the determination of the posterior distribution $p(G,\boldsymbol{\Omega} \mid \boldsymbol{x}) = p(G \mid \boldsymbol{x})p(\boldsymbol{\Omega} \mid \boldsymbol{x}, G)$. The first step is to define the prior distribution for the concentration matrix $\boldsymbol{\Omega}$ and the graph structure $G$: $p(G,\boldsymbol{\Omega}) = p(\boldsymbol{\Omega} \mid G)p(G)$. The standard conjugate prior for the concentration matrix $\boldsymbol{\Omega}$ is the Wishart distribution; this prior implicitly assumes that $G$ is complete. Equivalently, one can specify that the covariance matrix $\boldsymbol{\Sigma} = \boldsymbol{\Omega}^{-1}$ follows the inverse-Wishart distribution. Early work on Gaussian graphical models in the Bayesian framework (Dawid & Lauritzen 1993; Giudici & Green 1999) focused on restrictions of the inverse-Wishart to decomposable graphs, called hyper inverse-Wishart. The assumption of decomposability greatly simplifies computation, but can constitute a restrictive assumption for several real world applications.

To address this limitation, Roverato (2002) proposed the $G$-Wishart prior as the conjugate prior for arbitrary graphs. The $G$-Wishart is the Wishart distribution restricted to the space of concentration matrices with zeros as specified by a graph

$G$ that may be either decomposable or non-decomposable. The $G$-Wishart density $W_G(b,D)$ can be written as

$$f(\boldsymbol{\Omega}|G,b,D) = I_G(b,D)^{-1}|\boldsymbol{\Omega}|^{(b-2)/2}\exp\Big\{-\frac{1}{2}\mathrm{tr}(\boldsymbol{\Omega D})\Big\}, \quad \boldsymbol{\Omega} \in M^+,$$

where $b > 2$ is a scalar parameter, $\boldsymbol{D}$ is a $p \times p$ positive definite symmetric matrix, $I_G$ is the normalizing constant, and $M^+$ is the set of all $p \times p$ positive definite symmetric matrices with $\omega_{ij} = 0$ if and only if $(i,j) \notin E$.

Although this formulation is more flexible for modeling, it introduces computational difficulties. A number of approaches for sampling from the $G$-Wishart distribution have been made in recent years, including direct sampling (Wang & Carvalho 2010), Metropolis-Hastings algorithms (Mitsakakis et al. 2011; Dobra et al. 2011), and non-ordinary block Gibbs sampling (Wang & Li 2012). One particularly challenging aspect of computation for the $G$-Wishart distribution is that not only is the posterior normalizing constant intractable, but the prior normalizing constant is also intractable. Proposed approaches for estimating the $G$-Wishart normalizing constant include importance sampling (Roverato 2002), Monte Carlo sampling (Atay-Kayis & Massam 2005), and Laplace approximation (Lenkoski & Dobra 2011).

In many applications, we are interested in not just inferring the precision matrix given a graph $G$, but also in learning the graph structure itself. A joint search over the space of graphs and precision matrices poses computational challenges. Jones et al. (2005) and Lenkoski & Dobra (2011) simplified the problem by integrating out the precision matrix and performing a stochastic search to identify the most probable graphs based on their marginal likelihood. Dobra et al. (2011) proposed a reversible jump algorithm to sample over the joint space of graphs and precision matrices, which relies on perturbation of the elements of the Cholesky decomposition of the precision matrix. This method does not scale well to large graphs due to the requirement of a matrix completion step at every iteration. Wang & Li (2012) proposed a sampler that does not require a reversible jump and circumvents computation of the prior normalizing constant through the use of the exchange algorithm, improving both the accuracy and efficiency of computation.

### *8.2.2 Directed Graphical Models*

A directed acyclic graph (DAG) is a directed graph $G = (V,E)$ without cycles, i.e., with no directed paths that start from a node and return to the same node. DAG models on discrete variables are often called Bayesian networks. The name follows from the use of Bayes' rule in computations, not from the choice of the Bayesian paradigm in inference. Bayesian treatments of model selection for Bayesian networks were considered by Madigan & Raftery (1994) and Madigan et al. (1995).

We define the *parents* of a node $i$, $\mathrm{pa}(i)$, as the set of nodes $v$ distinct from $i$ such that $v \to i$. Moreover, we say that a subgraph of $G$ such that $u \to v \leftarrow w$ and with $u$ and $v$ not adjacent is a V-structure. A DAG with no V-structures is said to be *perfect*.

A distribution is said to be Markov with respect to the DAG $G$ if the joint density factorizes recursively as

$$f(\boldsymbol{x}) = \prod_{i=1}^{p} f(x_i \mid \boldsymbol{x}_{\mathrm{pa}(i)}), \tag{8.1}$$

where $\boldsymbol{x}_{\mathrm{pa}(i)} = (x_j : j \in \mathrm{pa}(i))$. This defines a directed acyclic graphical model, and it can be shown that the factorization is equivalent to a set of conditional independencies,

$$X_i \perp\!\!\!\perp \boldsymbol{X}_{\mathrm{nd}(i)\backslash \mathrm{pa}(i)} \mid \boldsymbol{X}_{\mathrm{pa}(i)} \text{ for any } i \in V.$$

This factorization decomposes the likelihood into a sequence of local likelihoods with variation independent parameters. In a Gaussian DAG model, each local likelihood is a normal linear regression model

$$f(x_i \mid \mathrm{pa}(i), \boldsymbol{\theta}_i) = N(x_i \mid \sum_{x_j \in \mathrm{pa}(i)} \beta_{ij} x_j;\, t_i), \tag{8.2}$$

where $N(x_i \mid \mu; t)$ is a normal density with mean $\mu$ and conditional variance $t$. The local parameters are $\boldsymbol{\theta}_i = (\boldsymbol{\beta}_i, t_i)$, where $\boldsymbol{\beta}_i = (\beta_{1i}, \ldots, \beta_{i-1,i})^{\mathrm{T}}$ is the vector of regression coefficients; see Geiger & Heckerman (2002).

Different DAGs may induce the same independencies, in which case they are said to be *Markov equivalent*. For instance, the graphs $u \leftarrow v \rightarrow w$ and $u \leftarrow v \leftarrow w$ induce the same independency, $u \perp\!\!\!\perp w \mid v$. We note that an undirected graph and a DAG may be Markov equivalent. In fact, any perfect DAG $G$ is Markov equivalent to a decomposable undirected graph $G$. Conversely, given an undirected decomposable graph $G$, there exists a Markov-equivalent perfect DAG $G$; see Lauritzen (1996). This result will be used in Sect. 8.4.1.

#### 8.2.2.1 Bayesian Estimation of Gaussian DAG Models

Learning the structure of DAGs, in general, is a very challenging task, in part because the dimensionality of the DAG space increases super-exponentially with the number of variables. Also, any algorithm searching in the entire DAG space would waste efficiency in evaluating Markov-equivalent DAGs. However, if there exists a known natural ordering of the variables, the problem is simplified because a DAG can be written as a system of recursive regressions, which greatly reduces the DAG space.

Specifically, suppose the joint distribution of $\boldsymbol{X} = (X_1, \ldots, X_p)^T$ is a multivariate normal distribution with mean 0. Without loss of generality, the natural ordering is defined as $\{1, 2, \ldots, p\}$. The DAG model for vector $\boldsymbol{X}$ can be written as

$$\begin{aligned} X_1 - \beta_{12}X_2 - \beta_{13}X_3 - \cdots - \beta_{1,p-1}X_{p-1} - \beta_{1p}X_p &= \varepsilon_1 \sim N(0,t_1) \\ X_2 - \beta_{23}X_3 - \cdots - \beta_{2,p-1}X_{p-1} - \beta_{2p}X_p &= \varepsilon_2 \sim N(0,t_2) \\ &\vdots \\ X_p &= \varepsilon_p \sim N(0,t_p) \end{aligned} \tag{8.3}$$

In practice, the ordering may be obtained from additional experiments, a reference network, "arrow of time" and so on. Let $\boldsymbol{E} = (\varepsilon_1, \ldots, \varepsilon_p)^T$, $\boldsymbol{T} = \text{diag}(t_1, \ldots, t_p)$ and let $\boldsymbol{A} = (-\beta_{ij})$ be a unit upper triangular matrix. Then, the system of recursive regressions can be written in matrix form:

$$\boldsymbol{AX} = \boldsymbol{E} \;\text{ with }\; \boldsymbol{E} \sim N(\boldsymbol{0}, \boldsymbol{T}). \tag{8.4}$$

The matrix $\boldsymbol{A} = (-\beta_{ij})$ encodes the network structure: there is a missing arrow from node $j$ to node $i$ if $\beta_{ij} = 0$. From (8.4) it follows that the concentration matrix of $\boldsymbol{X}$ is

$$\boldsymbol{\Omega} = \boldsymbol{A}^{\mathrm{T}}\boldsymbol{T}^{-1}\boldsymbol{A}; \tag{8.5}$$

see Wermuth (1980). This is sometimes called a *modified Cholesky decomposition* of $\boldsymbol{\Omega}$; see Pourahmadi (2007).

We assume that $\boldsymbol{A}$ is sparse, i.e., for each node, there are only a few nodes connected to it and most of the edges are missing. Learning the network structure then reduces to a variable selection problem for linear regressions, which has been extensively studied in the literature. The penalized likelihood approaches such as LASSO (Tibshirani 1996), SCAD (Fan & Li 2001), and SIS (Fan & Lv 2008) have gained popularity due to their ability to handle high-dimensional data in a time-efficient manner. Meanwhile, Bayesian approaches, including SSVS (George & McCulloch 1993) and Bayesian shrinkage (Park & Casella 2008; Carvalho et al. 2010; Griffin et al. 2010), have also been developed to tackle the high-dimensional variable selection problem in scenarios where the number of parameters greatly exceeds the number of samples. Bayesian approaches have been shown to be very effective for estimating DAGs as they naturally account for graph structure uncertainty, produce regularized estimators, and have good control of the false discovery rate, all of which are desirable features for high-dimensional complex models.

In the Bayesian paradigm, sparsity can be induced by the "spike-and-slab" prior on the regression coefficients:

$$\beta_{ij} \sim \gamma_{ij} f(\beta_{ij}) + (1-\gamma_{ij})\delta_0(\beta_{ij}),$$

where the "slab" $f(\cdot)$ is some continuous distribution and the "spike" $\delta_0(\cdot)$ is a point mass at zero. The binary indicators, $\gamma_{ij}$, are latent variables that specify the relevant variables. If $\gamma_{ij} = 1$, the $j$th variable is included in the $i$th regression and hence the edge from $j$ to $i$ is selected in the DAG; otherwise, $\gamma_{ij} = 0$. Model determination is then accomplished by evaluating the marginal likelihood and using a search algorithm to inspect the model space. The former is straightforward if we impose conjugate priors. If non-conjugate priors are used (such as non-local priors Johnson & Rossell (2010, 2012)), we can approximate the marginal likelihood via, for example, Markov chain Monte Carlo (MCMC) algorithm. The latter is, however, more complicated. Although the model space is substantially reduced through natural ordering, enumerating all possible DAGs is not feasible even for moderately large $p$. Therefore, instead of using a deterministic algorithm, a stochastic search algorithm is usually adopted.

A popular approach is MCMC, which sequentially samples all the parameters from the posterior distribution and outputs a list of visited models

$$\{\boldsymbol{\gamma},\boldsymbol{A},\boldsymbol{T}\}^{(1)},\{\boldsymbol{\gamma},\boldsymbol{A},\boldsymbol{T}\}^{(2)},\ldots,\{\boldsymbol{\gamma},\boldsymbol{A},\boldsymbol{T}\}^{(N)},$$

where $N$ is the total number of iterations. A simple approach to use in summarizing the model is to select the edges for which the (marginal) posterior inclusion probability $p(\gamma_{ij}=1|\boldsymbol{X})$ is higher than a pre-specified threshold. A common threshold is 0.5, which results in the median probability model (Barbieri & Berger 2004). Marginal posterior probabilities can be approximated by the fraction of time each $\gamma_{ij}$ is visited by the Markov chain. An alternative approach is to pick the model with the highest (joint) posterior probability based on the MCMC samples. However, when the dimension is high, the posterior probability of any model might be extremely small and consequently many models would have similar posterior probabilities.

There are recent developments in stochastic search algorithms other than MCMC, including shotgun stochastic search (SSS, Hans et al. (2007)) and feature-inclusion stochastic search (FINCS, Scott & Carvalho (2008); Altomare et al. (2013)). SSS evaluates many neighboring states (graphs) at each step in parallel and moves to a new state with a probability proportional to its marginal posterior. The advantage of SSS over MCMC is parallelization: when multi-core computing resources are available, each state can be evaluated independently on separate processors. However, when such computing power is not accessible, FINCS, a serial algorithm, provides an alternative. FINCS involves three types of moves: local moves, global moves, and resampling moves. It is motivated from the intuition that a move that has already improved some models is more likely to improve other models than a random move. While the MCMC approach aims to converge to a stationary distribution, SSS and FINCS intend to search for a list of high posterior models.

DAGs have been successfully applied in many areas such as genomics, causal inference, and expert systems. In the next section, we will introduce one specific example of using DAGs to construct biological networks (Ni et al. 2015).

## 8.3 Bayesian Nonlinear Model Selection for Gene Regulatory Networks

Gene regulatory networks (GRNs) represent the regulatory relationships between genes. We consider the problem of constructing a GRN in glioblastoma multiforme (GBM) using microarray gene expression data from The Cancer Gene Atlas (TCGA). We focus on genes that are mapped to three core pathways (Furnari et al. 2007): The (1) RTK/PI3K pathway; (2) p53 pathway; and (3) Rb pathway. Since the underlying biochemistry is known to be very complicated because of extraneous factors, some gene interactions are likely to be nonlinear. This motivates us to develop a novel semiparametric DAG model that allows for the detection of interpretable nonlinear functional relationships. Since pathway information is widely available, we incorporate such information by assuming a natural ordering of the genes obtained from reference networks. To achieve parsimonious estimation, we adopt a hierarchical two-level model selection approach. The first level, edge selection, chooses relevant gene interactions, and the second level, functional selection, which is conditional on the first level, classifies the functional relationship. Our application to gene expression data from patients with GBM found some regulatory mechanisms that are consistent with those described in the biological literature as well as a few novel nonlinear interactions that need further experimental validation.

### 8.3.1 Model

Suppose the data consist of $p$ gene expression levels for $n$ GBM patients, which can be organized into an $n \times p$ data matrix $\boldsymbol{X}$. The joint distribution of the DAG model for $\boldsymbol{X}$ can be factorized according to (8.1). Without loss of generality, the ordering is defined as $\{1,2,\ldots,p\}$. Denote $x_g^{(l)}$ to be the expression level of sample $l$ and gene $g$, for $g = 1,\ldots,p$ and $l = 1,\ldots,n$. Let $[g-]$ denote the set $\{g+1,\ldots,p\}$ and $\boldsymbol{x}_{[g-]}^{(l)}$ denote $\{x_i^{(l)} : i \in [g-]\}$. Each conditional distribution in the product term of Eq. (8.1) can be expressed by the following regressions:

$$x_g^{(l)} = f_g(\boldsymbol{x}_{[g-]}^{(l)}) + \varepsilon_g^{(l)}, \ g = 1,2,\ldots,p, \ l = 1,2,\ldots,n,$$

where the error term $\varepsilon_g^{(l)} \sim N(0,\lambda_g^{-1})$. The predictor function $f_g(x_{[g-]}^{(l)})$ is modeled semiparametrically using a set of cubic penalized spline (P-spline) basis functions:

$$f_g(x_{[g-]}^{(l)}) = \mu_g + f_{g,1}(x_1^{(l)}) + f_{g,2}(x_2^{(l)}) + \cdots + f_{g,g-1}(x_{g-1}^{(l)}), \quad g = 1,\ldots,p,$$

with intercept $\mu_g$ and $f_{g,\iota}(\,) = \sum_{k=1}^{M} \beta_{gl}^{(k)} B_{ik}(\cdot)$. Here, $B_{ik}(\cdot)$ represents the $k$th cubic B-spline basis and $\beta_{gi}^{(k)}$ are the corresponding spline coefficients, which are assumed to follow a discrete mixture ("spike-and-slab") of a second order Gaussian random walk (Lang & Brezger 2004) and a unit point mass at zero, $\boldsymbol{\beta}_{gj}|\tau_{gj},\lambda_g,\gamma_{gj} \sim \gamma_{gj}N(0,(\tau_{gj}\lambda_g\boldsymbol{K})^{-1}) + (1-\gamma_{gj})\delta_0(\boldsymbol{\beta}_{gj})$, where $\boldsymbol{K}$ is the penalty matrix constructed from the second order differences of the adjacent spline coefficients and $\gamma_{gj}$ is an indicator of edge selection. The smoothing parameter $\tau_{gj}$ controls the degree of smoothness of the fitted curve: a large value of $\tau_{gj}$ results in a smoother fit, while a small value of $\tau_{gj}$ leads to an irregular fit and essentially interpolates the data.

Note that when $\gamma_{gj} = 0$, the whole vector $\boldsymbol{\beta}_{gj}$ is set to zero and hence the corresponding edge is excluded from the DAG. Conditional on $\gamma_{gj}$, the functional form of the relationship between genes is defined through $\tau_{gj}$, as these parameters control the smoothness of the curve fitting. We enforce a discrete mixture of the inverted Pareto distribution and Gamma distribution: $\tau_{gj}|\phi_{gj} \sim \phi_{gj}Gamma(k_\tau,\theta_\tau) + (1-\phi_{gj})Ip(a_\tau,b_\tau)$, where $\phi_{gj}$ is the indicator of the mixture component. The density of the inverted Pareto distribution $Ip(a_\tau, b_\tau)$ is given by $\pi(\tau_{gj}|a_\tau, b_\tau) = \frac{a_\tau}{b_\tau}(\frac{\tau_{gj}}{b_\tau})^{a_\tau - 1}$, for $a_\tau > 0$, $0 < \tau_{gj} < b_\tau$. We set $a_\tau > 1$ so that the distribution concentrates on large values, which then encourages linear smooth fits of the data. The Gamma distribution is concentrated at small values of $\tau$, thus inducing nonlinear smoothing. Unlike a unimodal prior (such as the Gamma and inverted Pareto distributions), which is concentrated at either small values or large values (but not both), this mixture prior provides a sharper separation between "linear" and "nonlinear" relationships among genes because of its bimodal nature. Essentially, $\phi_{gj} = 1$ implies that $\tau_{gj}$ is distributed as a Gamma distribution (concentrated on small values) and hence is likely to be small, which in turn suggests a nonlinear interaction between gene $g$ and gene $j$ whereas $\phi_{gj} = 0$ implies that $\tau_{gj}$ is distributed as an inverted Pareto distribution (concentrated on large values) and therefore a linear interaction. We refer to this model that has two-level selection as the nonlinear mixture DAG (nMixDAG).

We complete our model by specifying the prior on the precision of error term $\lambda_g$, constant term $\mu_g$, network parameter $\gamma_{gj}$ and its hyperparameter $\rho$, and the mixture component indicator $\phi_{gj}$ and its hyperparameter $\omega$.

We assume conjugate priors for the error precision $\lambda_g \sim Gamma(a_\lambda, b_\lambda)$ and the constant term $\mu_g \sim N(0, (\lambda_g \kappa_\mu)^{-1})$. For the network parameter $\gamma_{gj}$, we use a Bernoulli prior with success probability $\rho$, $\gamma_{gj}|\rho \sim Bernoulli(\rho)$. The prior probability of inclusion $\rho$ follows a Beta distribution, $\rho \sim Beta(a_\rho, b_\rho)$, which yields an automatic multiplicity penalty since the posterior distribution of $\rho$ will become more concentrated at small values near 0 as the total number of variables increases (Scott & Berger 2010). Similar to the prior for $\gamma_{gj}$, a Bernoulli distribution is assumed for $\phi_{gj}$, with the success probability following a Beta hyper-prior, $\phi_{gj}|\omega \sim Bernoulli(\omega)$, $\omega \sim Beta(a_\omega, b_\omega)$. See Ni et al. (2015) for a description of the MCMC algorithm developed to fit nMixDAG.

### 8.3.2 Application to GBM Data

Our analysis involves TCGA-based microarray gene expression data for $n = 241$ GBM tumor specimens. We focus on the $p = 49$ genes that overlap with the three core pathways. Hence, in this case, $\boldsymbol{X}$ is a $241 \times 49$ matrix. We obtain the natural ordering of the 49 genes from the induced subgraph of the GBM signaling pathways (McLendon et al. 2008), which is shown in Fig. 8.1. In Fig. 8.2, we show the network reconstructed by nMixDAG. The solid lines represent linear regulations and the dotted lines represent nonlinear ones. The line width is proportional to the posterior inclusion probability, with thicker lines indicating a higher probability of the edge. The node size is proportional to its degree, i.e., the number of edges connected to the node. In total, we find 95 connections (85 are linear and 10 are nonlinear) and 5 highly connected genes (AKT1, FOXO3, SPRY2, GAB1, and PDPK1). Some of our findings are consistent with previous results reported in the biological literature.

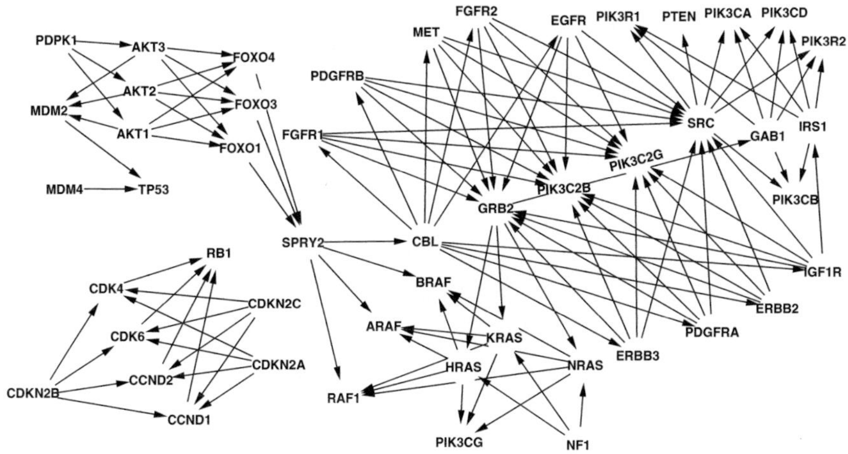


**Fig. 8.1** GBM reference signaling pathways—the RTK/PI3K, p53, and Rb signaling pathways. The prior ordering is obtained from this network

For example, the NF1 protein inhibits RAS function (Malumbres & Barbacid 2003) and RAS proteins activate PI3K complexes (Blume-Jensen & Hunter 2001). Those highly connected genes (also known as hub genes) may participate in multiple regulatory events and hence play an important role in the GRN. For instance, the AKT family is frequently amplified and the FOXO family is often mutated (McLendon et al. 2008) in GBM.

Our study also reveals some novel findings. In Fig. 8.3, we plot the nonlinear functional reconstructions of nine edges, together with their 95 % credible bands. Marginal posterior inclusion probabilities are shown at the top of each plot. We can see that the expression level of RAF1 decreases with that of ERBB3 when ERBB3 is low in expression, but starts to increase with ERBB3 after a cut-point around $-0.7$. It is even more interesting that CDKN2A manifests a sinusoidal trend with CDK4. These relationships have not been reported previously to the best of our knowledge and may deserve further validation via biological experiments. To quantify the evidence for the nonlinearity of each fitted curve, we define the *nonlinearity measure* ($\mathcal{N}$) as $\mathcal{N}_{gj} = p(\phi_{gj} = 1|\boldsymbol{Y}, \gamma_{gj} = 1)$, the probability that a given connection ($\gamma_{gj} = 1$) is nonlinear ($\phi_{gj} = 1$) *a posteriori*, which can be easily computed from MCMC samples of $\phi_{gj}, \gamma_{gj}$: $\mathcal{N}_{gj} \approx \sum_{i=1}^{N} I(\phi_{gj}^{(i)} = 1, \boldsymbol{\gamma}^{(i)} = \boldsymbol{\gamma}^{select}, \gamma_{gj}^{(i)} = 1) / \sum_{i=1}^{N} I(\boldsymbol{\gamma}^{(i)} = \boldsymbol{\gamma}^{select}, \gamma_{gj}^{(i)} = 1)$, where the superscript $(i)$ labels the $i$th MCMC sample, $N$ is the number of MCMC samples, and $\boldsymbol{\gamma}^{select}$ indicates the selected $\boldsymbol{\gamma}$ from the highest posterior model. The nonlinearity measure is also shown at the top of each plot in Fig. 8.3. For example, the evidence for nonlinearity between PIK3C2B and MDM4 is strong (0.997), while the evidence between PIK3CA and RAF1 is much weaker (0.510), which is consistent with our observations.

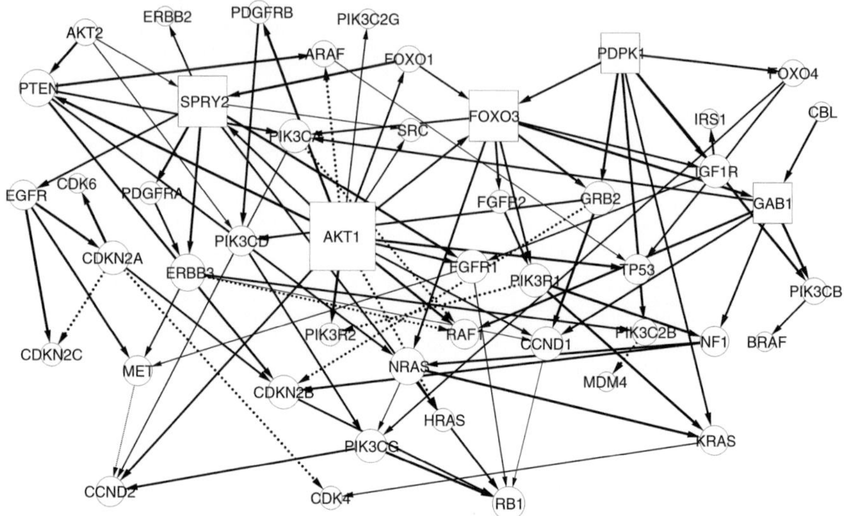


**Fig. 8.2** Recovered network from GBM data with nMixDAG. The *solid lines* indicate linear interactions; the *dotted lines* indicate nonlinear interactions; the line width is proportional to its posterior probability; and the node size is proportional to its degree, except for the hub genes (*rectangles*), whose node sizes are further enlarged for clarity

## 8.4 Efficient Approaches for Undirected Networks

Bayesian model determination for undirected graphical models is commonly performed by running a Markov chain for which the state variable has the form $(G, \boldsymbol{\Omega})$ where $G = (V, E)$ is an undirected graph, and $\boldsymbol{\Omega}$ is the concentration matrix of the random vector $\boldsymbol{X}$. The computational strategy is typically based on sequential random choices of pairs of distinct nodes: if they are currently connected, the algorithm proposes the removal of the edge; otherwise, it proposes an edge addition. If the search is limited to the space of decomposable graphs, the algorithm needs to check whether the edge perturbation is legal, i.e., yields a graph $G'$ that is still decomposable. Frydenberg & Lauritzen (1989) and Giudici & Green (1999) showed how to efficiently verify, through local computation, whether decomposability is preserved. If the proposal is accepted, then the parameter values for the new graph $G'$ are updated, and a Metropolis-Hastings acceptance ratio is computed.

Green & Thomas (2013) used recent advancements in graph theory for decomposable graphs and found that the construction of samplers is more efficient if the state variable contains a further object $J(G)$ that is associated with the graph, called the *junction tree*. The search is then based on the state variable $(G, J(G), \boldsymbol{\Omega})$ and on local updates of the junction tree. In fact, a graph is decomposable if and only if it has a *junction tree* representation (in general not unique) for the maximal cliques. A junction tree $J(G)$ is a connected graph with no cycles (thus a tree), whose nodes are the maximal cliques of $G$, and with the so-called *running intersection property*.

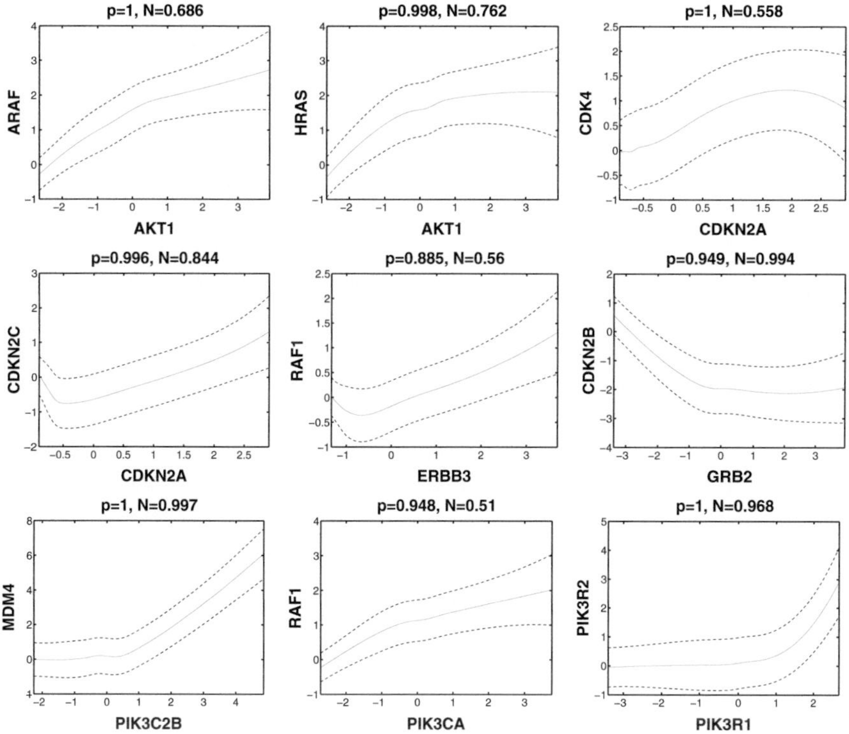


**Fig. 8.3** Nine functional reconstructions with 95 % credible bands for selected genes with nonlinear relationships. The marginal posterior inclusion probability (p) and nonlinearity measure (N) are shown at the top of each plot

This means that given any two maximal cliques, $C$ and $K$, and considering their common nodes, $C \cap K$, then all the maximal cliques along the unique path between $C$ and $K$ must contain $C \cap K$. In fact, $G$ is decomposable if and only if it admits a *perfect elimination order* of its nodes, such that for each node $v$ of the graph, the neighbors of $v$ that are previous in the ordering, plus $v$ itself form a clique (Lauritzen 1996). Whichever approach is used, within an MCMC that explores the model space of decomposable graphs by moves that add or remove one or multiple edges at a time, it is essential to have a fast method to update the junction tree and/or the perfect elimination order. At the start, the algorithm of *maximum cardinality search* (Tarjan & Yannakakis 1984) can be used to obtain an initial perfect elimination order and a junction tree. As recalculating the full junction tree at each step is highly inefficient, strategies have been developed to update it dynamically; see Green & Thomas (2013) and Stingo & Marchetti (2015). This allows for simultaneous local updates of the perfect elimination order.

### 8.4.1 *Inference on Directed Graphical Models Via Regression Modeling*

Stingo & Marchetti (2015) proposed an alternative approach that is built upon the decomposable Gaussian graphical model framework, but where the state variable is instead $(G, \sigma(G), \text{Cholesky}(\boldsymbol{\Omega}_\sigma))$ where $\sigma(G)$ is a *perfect elimination order* of the nodes of the graph, and $\text{Cholesky}(\boldsymbol{\Omega}_\sigma)$ is the *modified Cholesky decomposition* (8.5) of the concentration matrix rearranged according to $\sigma$. When the rows and columns of the concentration matrix are permuted according to a perfect elimination order, the modified Cholesky decomposition of the resulting matrix $\boldsymbol{\Omega}_\sigma$ defines a parameterization of a Gaussian DAG model which is Markov equivalent to the given decomposable model. This regression parameterization allows for greater flexibility in the prior specification.

As seen in Sect. 8.2.2.1, a Gaussian DAG model is equivalent to a recursive system of linear regression equations (8.3). The new parameterization of decomposable Gaussian graphical models as a system of linear regressions brings several advantages. The regression coefficients are variation independent, whereas the elements of the concentration matrix $\boldsymbol{\Omega}$ are not. Priors on $\Omega$ have to be defined on a subset of the cone of the positive definite matrices $P_G$ defined by $G$, such as the $G$-Wishart distributions whereas the parameter space of $\Gamma$ is $\mathbb{R}^{|E|}$. Alternative approaches to the $G$-Wishart distribution, like the Bayesian graphical lasso (Wang 2012) or similar priors marginally defined on the elements of the precision matrix, have substantial computational disadvantages since the positive definiteness of $\boldsymbol{\Omega}$ has to be checked at each step of the optimization/sampling algorithm.

From standard distribution theory, the distributions of $\beta_{ij}$ and $t_i$ are functions of the elements of the concentration matrix $\boldsymbol{\Omega}$; see Dawid & Lauritzen (1993) and Dobra et al. (2004*b*), among others. An inverse-Wishart prior on $\boldsymbol{\Sigma} \sim$ Inv-Wishart $(c, \eta^{-1}\boldsymbol{I})$, with $\boldsymbol{I}$ being an identity matrix of the same dimension of $\boldsymbol{\Sigma}$, implies that

$$\beta_{ij} \sim N(0, \eta\, t_i) \tag{8.6}$$

$$t_i \sim \text{Inv-Ga}((c + |\beta_i|)/2, 2\eta), \tag{8.7}$$

where $|\beta_i|$ is the number of covariates in equation $i$, and $\eta$ and $c$ are hyperparameters set to a fixed value.

Within this model framework, any prior distribution $p(G)$ on the graph space can be specified. Of the most commonly used among the set of decomposable graphs, we can mention (a) the uniform prior $p(G) \propto 1$, (b) the prior that assumes the edge inclusion probability to be equal to $p_1 \in (0,1)$, and (c) the prior of Armstrong et al. (2009), which gives equal probability to the size, defined as the number of selected edges, of the graph and equal probability to graphs of each size. Prior (a) gives more probability to graphs of a medium size and can be seen as a special case of prior (b), with $p_1 = 0.5$ whereas smaller values of $p_1$ favor sparse graphs. Prior (c) can be used when there is strong prior information about the expected size of the graph. All the described priors have to be normalized since every nondecomposable graph has zero probability.

A further advantage of the regression parameterization is that it suggests a probabilistic framework for learning the structure of nondecomposable Gaussian graphical models. Let $\boldsymbol{G} = (g_{ij})$ be the $p \times p$ adjacency matrix of the graph $G$ and $\boldsymbol{H} = (h_{ij})$ be a further $p \times p$ symmetric binary matrix that will be used to induce a double selection prior for the soft and hard selection of edges. An edge $\{i, j\}$ is said to be hard selected if $g_{ij} = 1$ and $h_{ij} = 1$ whereas it is soft selected if $g_{ij} = 1$ and $h_{ij} = 0$. The matrix $\boldsymbol{H}$ can assume any value on the model space of the unrestricted graphs with $p$ nodes, as long as this graph is a sub-graph of $G$. The graph space then can be defined by all pairs $(\boldsymbol{G}, \boldsymbol{H})$ such that $\boldsymbol{G}$ is an adjacency matrix of a decomposable graph and $\boldsymbol{H}$ is an adjacency matrix of an (unrestricted) sub-graph of $G$. Following the approach of George & McCulloch (1993), we introduce a soft/hard selection prior defined as a two-component mixture distribution on the regression coefficients

$$(\beta_{ij} \mid g_{ij} = 1, h_{ij}) \sim h_{ij} N(0, \eta t_i) + (1 - h_{ij}) N(0, \tau t_i), \tag{8.8}$$

with $\tau$ set to a very small value, such that the first component in the mixture puts most of its mass on values close to zero (George & McCulloch 1993), and $\eta$ is a hyperparameter to be specified. This approach encourages the *soft* selection of decomposable graphs that contain both edges supported by the data and an additional set of edges that makes the graph decomposable. Edges supported by the data are identified through the mixture prior (8.8). The model specification is completed by using conjugate inverse-Gamma priors on the $t_i$'s and by using independent Bernoulli priors on the $h_{ij}$'s, with $p_2$ being the prior probability of labeling $\{i, j\}$ as a hard edge, given $\{i, j\} \in G$; and $p_2$ can be set to a value that favors sparse graphs.

### 8.4.2 An Undirected Graphical Model Analysis of GBM Data

Our analysis involves the same TCGA-based microarray gene expression data described in Sect. 8.3.2. Here, we are interested in learning the structure of an undirected gene network (no gene ordering needs to be pre-specified) and comparing it with the findings described in Sect. 8.3.2. We used 1,000 Gibbs scans over all possible edges of the decomposable graph for the hard selection, and 200,000 Metropolis-Hastings iterations for the soft selection. We set the hyperparameters by following the guidelines given in Stingo & Marchetti (2015). We ran two Monte Carlo Markov chains with different starting points.

We select strong edges with posterior probability greater than 0.5. We find 98 connections and 8 hub genes: FOXO3 (14), SPRY2(12), NF1(11), ARAF(10), GAB1(8), IGF1R(8), PIK3CG(8), and PTEN(8). Three of these hub genes were also detected by nMixDAG (FOXO3, SPRY, GAB1). In Fig. 8.4, we show the network reconstructed by our approach for undirected networks. The line width is proportional to the posterior inclusion probability and the node size to its degree. This

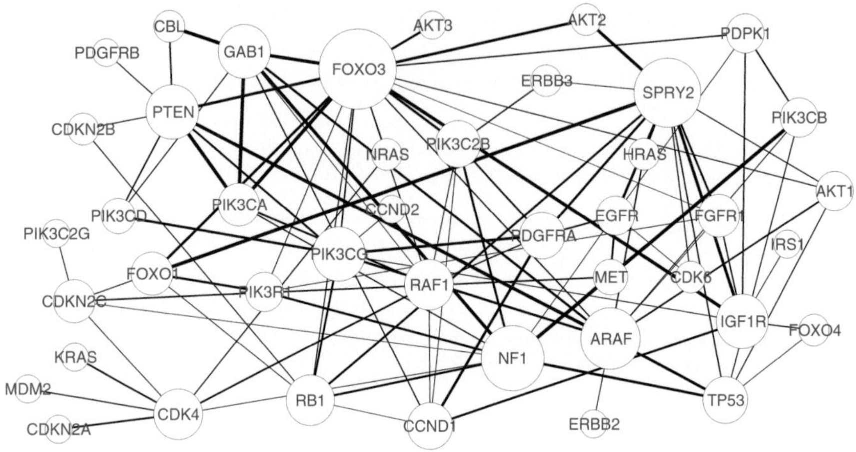


**Fig. 8.4** Recovered network from GBM data with our approach for undirected networks. The line width is proportional to its posterior probability; and the node size is proportional to its degree

graph is not decomposable, consequently some interaction patterns discovered by this analysis could not be identified by nMixDAG; on the other hand, 7 of the 10 nonlinear edges picked by nMixDAG were not selected in our undirected graph.

## 8.5 Discussion

We have reviewed Bayesian approaches for large biological networks, both directed and undirected. The Bayesian approaches we have presented offer a coherent framework in which edge selection and parameter estimation are performed simultaneously. There are several other topics related to Bayesian approaches for large biological networks that have not been discussed in this chapter. These topics include approaches for multiple graphs (Oates et al. 2014; Peterson et al. 2014), approaches for matrix-variate data (Wang & West 2009; Dobra et al. 2011), and hierarchical models that include biological networks as a component of a more complex model (Chekouo et al. 2015), among others.

Approaches for multiple graphs are most appropriate when a given population can be divided into homogeneous sub-populations, each of which can be characterized by a graphical model. Peterson et al. (2014) designed a Bayesian approach that simultaneously infers multiple undirected networks in situations where some networks may be unrelated, while others may have similar structures. Their approach infers a separate graphical model for each group but allows for shared structures when supported by the data; moreover, that approach yields a measure of relative network similarity across groups.

When the observed variables have the form of matrix-variate random variables, such as multiple platforms measured on the same subjects, matrix-variate graphical models (Wang & West 2009; Dobra et al. 2011) can be used to investigate the (conditional) dependencies along each dimension, i.e., between rows and between columns of the matrix-variate observations. This method directly considers the structural information naturally contained in the data.

We note the recent use of network approaches within regression models to guide the selection of molecular markers. We can classify these methods into two groups. Approaches that consider known biological networks as given and fixed external information (Li & Zhang 2010; Stingo & Vannucci 2011; Stingo et al. 2013) are classified in the first group. The second group is composed of methods that estimate a large biological network among the molecular predictors of a regression model (Chekouo et al. 2015). Compared to the approaches belonging to the first group, methods in the second group have the advantage of accounting for the uncertainty in the estimation of the biological network. Methods in the first group usually deal with undirected networks and can be easily extended to directed networks whereas methods belonging to the second group have been limited thus far to directed networks.

**Acknowledgements** F.C. Stingo and V. Baladandayuthapani were partially supported by the Cancer Center Support Grant (CCSG) (P30 CA016672). V. Baladandayuthapani was partially supported by NIH grant R01 CA160736. The authors are grateful to LeeAnn Chastain for editing assistance.

## References

Altomare, D., Consonni, G. & La Rocca, L. (2013), 'Objective Bayesian search of Gaussian directed acyclic graphical models for ordered variables with non-local priors', *Biometrics* **69**(2), 478–487.

Armstrong, H., Carter, C., Wong, K. & Kohn, R. (2009), 'Bayesian covariance matrix estimation using a mixture of decomposable graphical models', *Statistics and Computing* **19**, 303–316.

Atay-Kayis, A. & Massam, H. (2005), 'The marginal likelihood for decomposable and non-decomposable graphical Gaussian models', *Biometrka* **92**, 317–35.

Barbieri, M. M. & Berger, J. O. (2004), 'Optimal predictive model selection', *Annals of Statistics* pp. 870–897.

Blume-Jensen, P. & Hunter, T. (2001), 'Oncogenic kinase signalling', *Nature* **411**(6835), 355–365.

Carvalho, C. M., Polson, N. G. & Scott, J. G. (2010), 'The horseshoe estimator for sparse signals', *Biometrika* **97**, 465–480.

Chekouo, T., Stingo, F. C., Doecke, J. D. & Do, K.-A. (2015), 'miRNA-target gene regulatory networks: A Bayesian integrative approach to biomarker selection with application to kidney cancer', *Biometrics* .

Chin, L., Andersen, J. & Futreal, A. (2011), 'Cancer genomics: from discovery science to personalized medicine', *Nature Medicine* **17**, 297–303.

Dawid, A. & Lauritzen, S. (1993), 'Hyper Markov laws in the statistical analysis of decomposable graphical models', *Annals of Statistics* **3**, 1272–1317.

Dobra, A., Jones, B., Hans, C., Nevins, J. & West, M. (2004*a*), 'Sparse graphical models for exploring gene expression data', *Journal of Multivariate Analysis* **90**, 196–212.

Dobra, A., Jones, B., Hans, C., Nevins, J. & West, M. (2004*b*), 'Sparse graphical models for exploring gene expression data', *Journal of Multivariate Analysis* **90**, 196–212.

Dobra, A., Lenkoski, A. & Rodriguez, A. (2011), 'Bayesian inference for general Gaussian graphical models with application to multivariate lattice data', *Journal of the American Statistical Association* **106**(496), 1418–1433.

Fan, J. & Li, R. (2001), 'Variable selection via nonconcave penalized likelihood and its oracle properties', *Journal of the American Statistical Association* **96**(456), 1348–1360.

Fan, J. & Lv, J. (2008), 'Sure independence screening for ultrahigh dimensional feature space', *Journal of the Royal Statistical Society: Series B (Statistical Methodology)* **70**(5), 849–911.

Frydenberg, M. & Lauritzen, S. (1989), 'Decomposition of maximum likelihood in mixed graphical interaction models', *Biometrika* **76**(3), 539–555.

Furnari, F. B., Fenton, T., Bachoo, R. M., Mukasa, A., Stommel, J. M., Stegh, A., Hahn, W. C., Ligon, K. L., Louis, D. N., Brennan, C. et al. (2007), 'Malignant astrocytic glioma: genetics, biology, and paths to treatment', *Genes & Development* **21**(21), 2683–2710.

Geiger, D. & Heckerman, D. (2002), 'Parameter priors for directed acyclic graphical models and the characterization of several probability distributions', *Annals of Statistics* **5**, 1412–1440.

George, E. I. & McCulloch, R. E. (1993), 'Variable selection via Gibbs sampling', *Journal of the American Statistical Association* **88**(423), 881–889.

Giudici, P. & Green, P. (1999), 'Decomposable graphical Gaussian model determination', *Biometrika* **86**(4), 785–801.

Green, P. J. & Thomas, A. (2013), 'Sampling decomposable graphs using a Markov chain on junction trees', *Biometrika* **100**(1), 91–110.

Griffin, J. E., Brown, P. J. et al. (2010), 'Inference with normal-gamma prior distributions in regression problems', *Bayesian Analysis* **5**(1), 171–188.

Hamid, J., Hu, P., Roslin, N. M., Ling, V., Greenwood, C. & Beyene, J. (2009), 'Data integration in genetics and genomics: methods and challenges', *Human Genomics and Proteomics* **10**, 1–13.

Hanahan, D. & Weinberg, R. A. (2011), 'Hallmarks of cancer: the next generation', *Cell* **144**(5), 646–674.

Hans, C., Dobra, A. & West, M. (2007), 'Shotgun stochastic search for "large p" regression', *Journal of the American Statistical Association* **102**(478), 507–516.

Huang, Y. (2014), 'Integrative modeling of multiple genomic data from different types of genetic association studies', *Biostatistics* .

Huang, Y., VanderWeele, T. & Lin, X. (2014), 'Joint analysis of SNP and gene expression data in genetic association studies of complex diseases', *Annals of Applied Statistics* **8**, 352–376.

Johnson, V. E. & Rossell, D. (2010), 'On the use of non-local prior densities in Bayesian hypothesis tests', *Journal of the Royal Statistical Society: Series B (Statistical Methodology)* **72**(2), 143–170.

Johnson, V. E. & Rossell, D. (2012), 'Bayesian model selection in high-dimensional settings', *Journal of the American Statistical Association* **107**(498), 649–660.

Jones, B., Carvalho, C., Dobra, A., Hans, C., Carter, C. & West, M. (2005), 'Experiments in stochastic computation for high-dimensional graphical models', *Statistical Science* pp. 388–400.

Kristensen, V., Lingjaerde, O., Russnes, H., Vollan, H., Frigessi, A. & Borresen-Dale, A. (2014), 'Principles and methods of integrative genomic analyses in cancer', *Nature Reviews* **14**, 299–313.

Lang, S. & Brezger, A. (2004), 'Bayesian p-splines', *Journal of Computational and Graphical Statistics* **13**(1), 183–212.

Lauritzen, S. L. (1996), *Graphical Models*, Oxford University Press.

Lenkoski, A. & Dobra, A. (2011), 'Computational aspects related to inference in Gaussian graphical models with the G-Wishart prior', *Journal of Computational and Graphical Statistics* **20**(1), 140–157.

Li, F. & Zhang, N. R. (2010), 'Bayesian variable selection in structured high-dimensional covariate spaces with applications in genomics', *Journal of the American Statistical Association* **105**(491), 1202–1214.

Madigan, D. & Raftery, A. (1994), 'Model selection and accounting for model uncertainty in graphical models using Occam's window', *Journal of the American Statistical Association* **89**, 1535–1546.

Madigan, D., York, J. & Allard, D. (1995), 'Bayesian graphical models for discrete data', *International Statistical Review* **63**(2), 215–232.

Malumbres, M. & Barbacid, M. (2003), 'RAS oncogenes: the first 30 years', *Nature Reviews Cancer* **3**(6), 459–465.

McLendon, R., Friedman, A., Bigner, D., Van Meir, E. G., Brat, D. J., Mastrogianakis, G. M., Olson, J. J., Mikkelsen, T., Lehman, N., Aldape, K. et al. (2008), 'Comprehensive genomic characterization defines human glioblastoma genes and core pathways', *Nature* **455**(7216), 1061–1068.

Mitsakakis, M., Massam, H. & Escobar, M. D. (2011), 'A Metropolis-Hastings based method for sampling from the G-Wishart distribution in Gaussian graphical models', *Electronic Journal of Statistics* **5**, 18–30.

Mukherjee, S. & Speed, T. P. (2008), 'Network inference using informative priors', *Proceedings of the National Academy of Sciences* **105**(38), 14313–14318.

Ni, Y., Stingo, F. & Baladandayuthapani, V. (2014), 'Integrative Bayesian network analysis of genomic data', *Cancer Informatics* **2**, 39–48.

Ni, Y., Stingo, F. C. & Baladandayuthapani, V. (2015), 'Bayesian nonlinear model selection for gene regulatory networks', *Biometrics* .

Oates, C., Korkola, J., Gray, J. & Mukherjee, S. (2014), 'Joint estimation of multiple related biological networks', *The Annals of Applied Statistics* **8**(3), 1892–1919.

Park, T. & Casella, G. (2008), 'The Bayesian lasso', *Journal of the American Statistical Association* **103**(482), 681–686.

Parsons, D. W., Jones, S., Zhang, X., Lin, J. C.-H., Leary, R. J., Angenendt, P., Mankoo, P., Carter, H., Siu, I.-M., Gallia, G. L. et al. (2008), 'An integrated genomic analysis of human glioblastoma multiforme', *Science* **321**(5897), 1807–1812.

Peterson, C., Stingo, F. & Vannucci, M. (2014), 'Bayesian inference of multiple Gaussian graphical models', *Journal of the American Statistical Association* .

Pourahmadi, M. (2007), 'Cholesky decompositions and estimation of a covariance matrix: orthogonality of variance–correlation parameters', *Biometrika* **94**(4), 1006–1013.

Reimand, J., Wagih, O. & Bader, G. D. (2013), 'The mutational landscape of phosphorylation signaling in cancer', *Scientific Reports* **3**.

Roverato, A. (2002), 'Hyper inverse wishart distribution for non-decomposable graphs and its application to Bayesian inference for Gaussian graphical models', *Scandinavian Journal of Statistics* **29**(3), 391–411.

Scott, J. G. & Berger, J. O. (2010), 'Bayes and empirical-Bayes multiplicity adjustment in the variable-selection problem', *The Annals of Statistics* **38**(5), 2587–2619.

Scott, J. G. & Carvalho, C. M. (2008), 'Feature-inclusion stochastic search for Gaussian graphical models', *Journal of Computational and Graphical Statistics* **17**(4).

Stingo, F. C., Chen, Y. A., Tadesse, M. G. & Vannucci, M. (2011), 'Incorporating biological information into linear models: A Bayesian approach to the selection of pathways and genes', *The Annals of Applied Statistics* **5**(3), 1978–2002.

Stingo, F. C., Chen, Y. A., Vannucci, M., Barrier, M. & Mirkes, P. E. (2010), 'A Bayesian graphical modeling approach to microRNA regulatory network inference', *The Annals of Applied Statistics* **4**(4), 2024.

Stingo, F. C., Guindani, M., Vannucci, M. & Calhoun, V. D. (2013), 'An integrative Bayesian modeling approach to imaging genetics', *Journal of the American Statistical Association* **108**(503), 876–891.

Stingo, F. & Marchetti, G. M. (2015), 'Efficient local updates for undirected graphical models', *Statistics and Computing* **25**, 159–171.

Stingo, F. & Vannucci, M. (2011), 'Variable selection for discriminant analysis with Markov random field priors for the analysis of microarray data', *Bioinformatics* **27**(4), 495–501.

Tarjan, R. & Yannakakis, M. (1984), 'Simple linear-time algorithms to test chordality of graphs, test acyclicity of hypergraphs, and selectively reduce acyclic hypergraphs', *SIAM Journal of Computation* **13**, 566–579.

Telesca, D., Mueller, P., Kornblau, S., Suchard, M. & Ji, Y. (2012), 'Modeling protein expression and protein signaling pathways', *Journal of the American Statistical Association* **107**(500), 1372–1384.

Tibshirani, R. (1996), 'Regression shrinkage and selection via the lasso', *Journal of the Royal Statistical Society. Series B (Methodological)* pp. 267–288.

Wang, H. (2012), 'Bayesian graphical lasso models and efficient posterior computation', *Bayesian Analysis* **7**(4), 867–86.

Wang, H. & Carvalho, C. (2010), 'Simulation of hyper-inverse Wishart distributions for non-decomposable graphs', *Electronic Journal of Statistics* **4**, 1467–1470.

Wang, H. & Li, Z. (2012), 'Efficient Gaussian graphical model determination under G-Wishart prior distributions', *Electronic Journal of Statistics* **6**, 168–198.

Wang, H. & West, M. (2009), 'Bayesian analysis of matrix normal graphical models', *Biometrika* **96**(4), 821–834.

Wang, W., Baladandayuthapani, V., Morris, J. S., Broom, B. M., Manyam, G. & Do, K.-A. (2013), 'iBAG: integrative Bayesian analysis of high-dimensional multiplatform genomics data', *Bioinformatics* **29**(2), 149–159.

Weinstein, J. N., Collisson, E. A., Mills, G. B., Shaw, K. R. M., Ozenberger, B. A., Ellrott, K., Shmulevich, I., Sander, C., Stuart, J. M., Network, C. G. A. R. et al. (2013), 'The cancer genome atlas pan-cancer analysis project', *Nature Genetics* **45**(10), 1113–1120.

Wermuth, N. (1980), 'Linear recursive equations, covariance selection, and path analysis', *Journal of the American Statistical Association* **75**(372), 963–972.

Whittaker, J. (2009), *Graphical Models in Applied Multivariate Statistics*, Wiley Publishing.

Wu, J., Li, Y. & Jiang, R. (2014), 'Integrating multiple genomic data to predict disease-causing nonsynonymous single nucleotide variants in exome sequencing studies', *PLoS Genetics* **10**(3).

# Chapter 9
# Nonparametric Variable Selection, Clustering and Prediction for Large Biological Datasets

Subharup Guha, Sayantan Banerjee, Chiyu Gu,
and Veerabhadran Baladandayuthapani

**Abstract** The development of parsimonious models for reliable inference and prediction of responses in high-dimensional regression settings is often challenging due to relatively small sample sizes and the presence of complex interaction patterns between a large number of covariates. We propose an efficient, nonparametric framework for simultaneous variable selection, clustering and prediction in high-throughput regression settings with continuous outcomes. The proposed model utilizes the sparsity induced by Poisson-Dirichlet processes (PDPs) to group the covariates into lower-dimensional latent clusters consisting of covariates with similar patterns among the samples. The data are permitted to direct the choice of a suitable cluster allocation scheme, choosing between PDPs and their special case, a Dirichlet process. Subsequently, the latent clusters are used to build a nonlinear prediction model for the responses using an adaptive mixture of linear and nonlinear elements, thus achieving a balance between model parsimony and flexibility. Through analyses of gene expression microarray datasets we demonstrate the reliability of the proposed method's clustering mechanism and show that the technique compares favorably to, and often outperforms, existing methodologies in terms of the prediction accuracies of the subject specific clinical outcomes.

---

S. Guha (✉) • C. Gu
Department of Statistics, University of Missouri,
Columbia, MO, USA
e-mail: GuhaSu@missouri.edu; cgz59@mail.missouri.edu

S. Banerjee • V. Baladandayuthapani
Department of Biostatistics, The University of Texas MD
Anderson Cancer Center, Houston, TX, USA
e-mail: SBanerjee@mdanderson.org; Veera@mdanderson.org



R. Mitra, P. Müller (eds.), *Nonparametric Bayesian Inference in Biostatistics*,
Frontiers in Probability and the Statistical Sciences,
DOI 10.1007/978-3-319-19518-6_9

## 9.1 Introduction

Suppose the available data in an investigation consist of continuous responses and $p$ continuous covariates on $n$ subjects, arranged in an $n \times p$ matrix. We assume that only a subset of the covariates are statistically associated with the responses, i.e., for subjects $i = 1,\ldots,n$, the responses $w_i \in \mathscr{R}$ are assumed to be associated with an unknown subset of the covariates $x_{i1},\ldots,x_{ip}$. The goal of the analysis is two-pronged. First, we wish to infer a common, sparse set of predictor indices for all the subjects, i.e., a subset $\mathscr{S} \subset \{1,\ldots,p\}$ of dimension $q \ll p$ consisting of the indices of the covariates that are significantly associated with the responses. Second, we wish to predict the responses of $\tilde{n}$ additional subjects for whom only covariate information is available. The development of parsimonious regression models that can be used for reliable predictions is challenging. This is especially true of "small $n$, large $p$" regression problems arising in many areas such as high-throughput genomics, imaging and environmental applications.

Several innovative strategies have been developed to meet these challenges in various contexts, with reasonable degrees of success. Most (if not all) of these approaches can be classified into three broad categories based on their basic construction: (a) linear variable selection methods, (b) regression methods using low-dimensional projections of the covariate space, and (c) nonlinear prediction methods. The *linear variable selection methods* include stepwise selection (Peduzzi *et al.* 1980), penalized regression approaches such as lasso (and its variants) (Tibshirani 1997), and non-concave penalized likelihood approaches (Fan and Li 2002). Bayesian linear variable selection approaches include spike and slab mixture priors (Mitchell and Beauchamp 1988), stochastic search variable selection (George and McCulloch 1993), Gibbs-based variable selection (Dellaportas *et al.* 1982), Bayesian model averaging (Madigan and Raftery 1994; Volinsky *et al.* 1997) and indicator priors (Kuo and Mallick 1997). The stochastic search variable selection approach of George and McCulloch (1993) has been extended to multivariate settings by Brown *et al.* (1998) and to generalized linear mixed models by Cai and Dunson (2006). Effective variable selection methods have also been developed for multinomial probit models by Sha *et al.* (2004), and for microarray data with censored outcomes by Lee and Mallick (2004) and Sha *et al.* (2006). Work related to the method we present is the product partition model on covariates proposed by Müller *et al.* (2011). Methods based on *regression using low-dimensional projections of the covariate space* include partial least squares (Nguyen and Rocke 2002; Li and Gui 2004) and (supervised) principal components methods (Bair and Tibshirani 2004). *Non-linear prediction methods* include statistical and machine learning techniques such as support vector machines (Cristianini and Shawe-Taylor 2000), and ensemble methods such as random forests (Ishwaran *et al.* 2010) and Bonato *et al.* (2010).

Our motivating application arises from a high-throughput genomics setting where microarray-based expression levels of genes (usually thousands) are available for a limited number of patient samples (tens or hundreds). We wish to select important genes (variables) as well as develop efficient prediction models for continuous patient-specific clinical outcomes. To illustrate our method, we use an accelerated failure time (AFT) model (Buckley and James 1979; Cox and Oakes 1984) to analyze the motivating datasets which have the following general structure. For individuals $i = 1, \ldots, n$, the data consist of *(i)* the failure time $w_i > 0$, and *(iii)* expression levels $x_{i1}, \ldots, x_{ip}$ for $p$ genes, with $p$ being much larger than $n$. Thus, the log–failure-time $y_i$ equals $\log(w_i)$.

In a regression setting, we refer to $y_1, \ldots, y_n$ as the *regression outcomes*, and fit the model:

$$y_i \overset{indep}{\sim} N(\eta_i, \sigma^2), \tag{9.1}$$

where the regression mean $\eta_i = \beta_0 + \sum_{j \in \mathscr{S}} \beta_j x_{ij}$. George and McCulloch (1993), Kuo and Mallick (1997), and Brown, Vannucci, and Fearn (1998) have proposed the use of latent indicator variables to identify the covariate matrix columns that are associated with the regression outcomes: $\eta_i = \beta_0 + \sum_{j=1}^{p} \gamma_j \beta_j x_{ij}$, where $\gamma_j$ is an indicator that corresponds to the $j$th covariate column being a predictor. The $\gamma_j$'s are assumed to be i.i.d. Bernoulli$(\omega)$. The number of model predictors is then $|\mathscr{S}| = \sum_{j=1}^{p} \gamma_j$. With $\boldsymbol{X}_{\boldsymbol{\gamma}}$ denoting the $n$ by $(|\mathscr{S}|+1)$ predictor matrix including the intercept column consisting of all ones, a g prior (Zellner 1986) is assumed for the regression coefficients: $\boldsymbol{\beta}_{\boldsymbol{\gamma}} \sim N_{|\mathscr{S}|+1}\left(\mathbf{0}, \sigma_\beta^2 (\boldsymbol{X}_{\boldsymbol{\gamma}}' \boldsymbol{X}_{\boldsymbol{\gamma}})^{-1}\right)$ for an unknown $\sigma_\beta^2 > 0$.

For small $n$, large $p$ regression problems, we propose a nonparametric technique for clustering, variable selection, and prediction in high-dimensional regression settings (Guha and Baladandayuthapani 2014). Since the data are informative regarding the joint effects of correlated covariates rather than the individual covariates, the proposed method utilizes the sparsity-inducing (i.e., dimension reduction) property of Poisson-Dirichlet processes (PDPs) to group the $p$ columns of the covariate matrix into $q$ latent clusters, where $q \ll p$, with each cluster consisting of columns with similar patterns across the subjects. The data are allowed to direct the choice between a class of PDPs and their special case, a Dirichlet process, for a suitable allocation scheme for the covariates. The within-cluster patterns, common to all the members of the clusters, are flexibly modeled using Dirichlet processes, as opposed to linear projections such as principal components and partial least squares.

This reduces the small $n$, large $p$ problem to a "small $n$, small $q$" problem, facilitating an effective stochastic search of the indices $\mathscr{S}^* \subset \{1, \ldots, q\}$ of the *cluster predictors*, from which we may infer the indices $\mathscr{S} \subset \{1, \ldots, p\}$ of the covariate predictors associated with the responses, as opposed to the typical "black-box" nonlinear prediction methods mentioned before. In addition, the technique is capable of detecting nonlinear functional relationships through elements such as nonlinear functional kernels and basis functions such as splines or wavelets. The adaptive mixture of linear and nonlinear elements in the regression relationship aims to achieve a balance between model parsimony and flexibility. In essence, the technique specifies a random, bidirectional nested clustering of the high-dimensional covariate matrix and builds a nonlinear prediction model for the responses using the latent clusters as covariates. Together, these components define a joint model for the responses and covariates that results in an effective model-based clustering and variable selection procedure, improved posterior inferences and accurate test case predictions.

The rest of the chapter is organized as follows. We develop the proposed model in Sect. 9.2. In Sect. 9.3, we describe a posterior inference strategy based on Markov chain Monte Carlo (MCMC) techniques. In Sect. 9.4, we analyze the motivating gene expression microarray datasets in Multiple Myeloma to demonstrate the effectiveness of the proposed method and compare its prediction accuracy with those of several existing variable selection procedures for continuous outcomes.

## 9.2 Model Construction

We model the responses and covariates in a hierarchical manner. Section 9.2.1 details the models for the covariates and their allocation to the latent clusters. Section 9.2.2 describes the choice of the cluster-specific predictors and nonlinearly relates them to the subject-specific Gaussian regression outcomes. Together, these components define a coherent model that could be used for both inference and prediction.

### *9.2.1 Modeling the Covariates and Latent Clusters*

For the columns $\boldsymbol{x}_1,\ldots,\boldsymbol{x}_p$ of the (continuous) covariate matrix, suppose each column vector belongs to exactly one of $q \ll p$ clusters, where the cluster memberships and $q$ are unknown. For the covariate (column) $j = 1,\ldots,p$, the covariate-to-cluster assignment is determined by an **allocation variable** $c_j$ that equals $k$ if the $j$th covariate belongs to the $k$th cluster, where $k = 1,\ldots,q$.

Furthermore, the clusters are associated with **latent vectors** $\boldsymbol{v}_1,\ldots,\boldsymbol{v}_q$, each of length $n$. Typically, the covariates are noisy versions of the latent vector components, resulting in high correlations among covariates that belong to a cluster. However, within each cluster, the covariates of a few individuals may be highly variable. To account for this greater heterogeneity, we model the covariates of these individuals with a larger variance. Specifically, for the $j$th covariate, given that the allocation variable $c_j$ equals $k$ and given an indicator variable $z_{ik}$, we assume for $i = 1,\ldots,n$ that

$$x_{ij} \mid z_{ik}, c_j = k \overset{indep}{\sim} \begin{cases} N(v_{ik}, \tau_1^2) & \text{if } z_{ik} = 0 \\ N(v_{ik}, \tau^2) & \text{if } z_{ik} = 1 \end{cases}$$

where $\tau_1^2$ and $\tau^2$ are component-specific parameters with inverse Gamma priors such that $\tau_1^2 \gg \tau^2$. The value $z_{ik} = 0$ indicates that the covariates of subject $i$ belonging to the $k$th cluster have an unusually high variance. The indicator variables for the (individual, cluster) combinations are apriori distributed as

$$z_{ik} \overset{iid}{\sim} \text{Ber}(\xi), \qquad i = 1,\ldots,n \text{ and } k = 1,\ldots,q,$$

where $\xi \sim \text{beta}(\iota_1, \iota_0)$ with $\iota_1 \gg \iota_0$, so that $P(z_{ik} = 1)$ is high and only a small proportion of covariates have a large variance.

### Allocation Variables

To gain an intuitive understanding of an appropriate model for the covariate-to-cluster allocation, we performed an exploratory data analysis (EDA) of the multiple myeloma datasets explained in Sect. 9.4. We randomly selected $p = 500$ probes and $n = 100$ individuals, iteratively applying the k-means procedure to group the covariates into clusters.

The iterations were terminated when the following conditions were satisfied: *(i)* all within-cluster pairwise correlations of the covariates exceeded 0.3, and *(ii)* the allocation $R^2$ exceeded 0.7. Under the assumption that all the $z_{ik}$'s are equal to 1, the stopping conditions encourage within-cluster concordance and a small value of $\tau^2$. Figure 9.1 displays a barchart of the cluster sizes. The pattern we observe is uncharacteristic of a Dirichlet process, which is usually dominated by a small number of clusters with exponentially decreasing sizes. Specifically, for $p = 500$, the large number of clusters ($\hat{q} = 125$) and the predominance of relatively small clusters are strongly suggestive of a non-Dirichlet type of allocation for the covariate-cluster assignments.

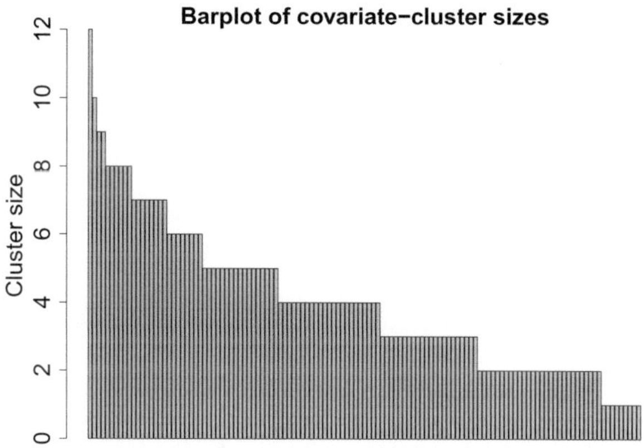


**Fig. 9.1** Barchart of cluster sizes for the exploratory data analysis

The aforementioned EDA suggests the need for a wider range of allocation patterns, such as that provided by a class of generalizations of a Dirichlet process called the *two-parameter PDP*, introduced by Perman *et al.* (1992) and further studied by Pitman (1995) and Pitman and Yor (1997). The allocation variables are apriori exchangeable for PDPs, and more generally, for product partition models (Barry and Hartigan 1993; Quintana and Iglesias 2003) and species sampling models (Ishwaran and James 2003). We assume the following prior for the allocation variables of the covariates:

$$c_1, \ldots, c_p \sim \mathbb{PDP}(\alpha_1, d) \tag{9.2}$$

where the discount parameter $0 \leq d < 1$ and mass parameter $\alpha_1 > 0$. The number of distinct clusters, $q$, is stochastically increasing in $\alpha_1$ and $d$. For a fixed $d$, all the covariates are assigned to separate clusters (i.e., $q = p$) as $\alpha_1 \to \infty$. For a fixed $\alpha_1$, setting $d = 0$ yields a Dirichlet process with mass parameter $\alpha_1$.

Conditional on the parameters $\alpha_1$ and $d$, the allocation variables of a PDP evolve as follows. We may assume without loss of generality that $c_1 = 1$. Subsequently, for $j = 2,\ldots,p$, suppose there are $q^{(j-1)}$ distinct clusters among $c_1,\ldots,c_{j-1}$, with the $k$th cluster containing $n_k^{(j-1)}$ number of covariates, where $k = 1,\ldots,q^{(j-1)}$. The predictive probability that the $j$th covariate belongs to the $k$th cluster is then

$$P\left(c_j = k \mid c_1,\ldots,c_{j-1}\right) \propto \begin{cases} n_k^{(j-1)} - d & \text{if } k = 1,\ldots,q^{(j-1)} \\ \alpha_1 + q^{(j-1)} \cdot d & \text{if } k = q^{(j-1)} + 1 \end{cases}$$

where the event $c_j = q^{(j-1)} + 1$ corresponds to the $j$th covariate opening a new cluster. When $d = 0$, we obtain the well-known Pòlya urn scheme for Dirichlet processes (Ferguson 1973). Refer to Lijoi and Prünster (2010) for a detailed discussion of Bayesian nonparametric models, including Dirichlet processes and PDPs.

The use of PDPs in this setting achieves dimension reduction for the covariate clusters because the random number of clusters, $q = q^{(p)}$, is asymptotically equivalent to

$$\begin{cases} \alpha_1 \cdot \log p & \text{if } d = 0 \quad \textit{(Dirichlet process)} \\ T_{d,\alpha_1} \cdot p^d & \text{if } 0 < d < 1 \end{cases} \tag{9.3}$$

for a random variable $T_{d,\alpha_1} > 0$. This implies that, as $p \to \infty$, the number of clusters of a Dirichlet process is of smaller order than that of a PDP with discount parameter $d > 0$. Dirichlet processes have been previously utilized for dimension reduction; for example, see Medvedovic *et al.* (2004), Kim *et al.* (2006), Dunson *et al.* (2008) and Dunson and Park (2008). In essence, this provides an effective dimension reduction clustering technique for regression settings that we exploit in our model.

The parameter $d$ in the PDP model, Eq. (9.2), is given the mixture prior $\frac{1}{2}\delta_0 + \frac{1}{2}U(0,1)$, where $\delta_0$ denotes a point mass at 0. This allows the data to flexibly choose between a Dirichlet process and a more general PDP for a suitable clustering mechanism of the covariates.

### Latent Vector Elements

The PDP prior specification is completed by a *base distribution* in $\mathscr{R}^n$ for the i.i.d. latent vectors. The $nq$ number of components of the latent vectors $\boldsymbol{v}_1,\ldots,\boldsymbol{v}_q$ are assumed to have the following distribution:

$$v_{ik} \stackrel{iid}{\sim} G \qquad i = 1,\ldots,n, \text{ and } k = 1,\ldots,q, \tag{9.4}$$

allowing the clusters to borrow strength and communicate through shared latent vector elements. Furthermore, the real-valued distribution $G$ is given a nonparametric Dirichlet process prior, which allows the latent vectors to flexibly capture the within-covariate patterns of the subjects:

$$G \sim \mathscr{D}\mathscr{P}\left(\alpha_2; N(\mu_2, \tau_2^2)\right) \tag{9.5}$$

with mass parameter $\alpha_2 > 0$ and base distribution $N(\mu_2, \tau_2^2)$. This implies that $G$ is discrete and that the number of distinct values among the $v_{ik}$'s is asymptotically equivalent to $\alpha_2 \log nq$. In Sect. 9.3, we demonstrate that this allocation scheme for the latent vector elements is validated by the real dataset.

In essence, the aforementioned probability model specifies a random, bidirectional nested clustering of the $n \times p$ covariate matrix. Unlike the model based clustering approaches of Fraley and Raftery (2002), Quintana (2006) and Freudenberg *et al.* (2010), the proposed method does not assume that it is possible to *globally* reshuffle the rows and columns of the covariate matrix to reveal a clustering pattern. Instead, somewhat similarly to the nonparametric Bayesian local clustering (NoB-LoC) approach of Lee *et al.* (2013), the proposed method clusters the covariates locally using two sets of product partition models (Hartigan 1990; Barry and Hartigan 1993; Crowley 1997). However, there are significant differences between NoB-LoC and the clustering aspect of our method, in that we are primarily motivated by high-dimensional regression problems rather than bi-clustering, which is the emphasis of NoB-LoC. In addition, NoB-LoC relies solely on Dirichlet processes for clustering whereas the proposed method permits a mixture of Dirichlet processes and PDPs.

### 9.2.2 *Modeling the Predictor Choices and Regression Outcomes*

For $k = 1, \ldots, q$, let $n_k$ be the number of covariates belonging to the $k$th cluster, so that $n_k = \sum_{j=1}^{p} \mathscr{I}(c_j = k)$ and $\sum_{k=1}^{q} n_k = p$. To gain an intuitive understanding, imagine that each cluster nominates from its covariate members a *representative* $\boldsymbol{u}_k$. In the prior, all $n_k$ covariates belonging to a cluster have an equal chance of being nominated as the representative. Let $s_k$ denote the index of the covariate belonging to the $k$th cluster that is chosen as its representative, so that $c_{s_k} = k$ and $\boldsymbol{u}_k = \boldsymbol{x}_{s_k}$. In accordance with our cluster-based strategy for dimension reduction, the responses are directly related to the cluster representatives rather than the individual covariates. The regression predictors are then chosen from the set of $q$ cluster representatives, and the indices of their clusters constitute the set of *cluster predictors*, $\mathscr{S}^* \subset \{1, \ldots, q\}$. We emphasize that the latent vectors $\boldsymbol{v}_k$ of Sect. 9.2.1 determine the allocation of the covariates to the clusters, and so indirectly but significantly inf'luence the choice of the influence of the cluster representatives. As an alternative modeling strategy, we could also choose the latent vectors themselves as the cluster representatives. The former approach is more interpretable because practitioners often think in terms of individual regressors and their corresponding effects on the outcome.

The nominated cluster representatives are featured in an additive regression model that can accommodate nonlinear functional relationships. Specifically, the log–failure-times $y_i = \log(w_i)$ are the regression outcomes and have the distribution

$$y_i \overset{indep}{\sim} N\left(\eta_i, \sigma^2\right), \quad \text{where}$$
$$\eta_i = \beta_0 + \sum_{k=1}^{q} \gamma_k^{(1)} \beta_k^{(1)} u_{ik} + \sum_{k=1}^{q} \gamma_k^{(2)} h(u_{ik}, \boldsymbol{\beta}_k^{(2)}) \tag{9.6}$$

for a nonlinear function $h$. The expression for $\eta_i$ implicitly relies on the triplet of cluster-specific indicators, $\boldsymbol{\gamma}_k = (\gamma_k^{(0)}, \gamma_k^{(1)}, \gamma_k^{(2)})$, where $\gamma_k^{(0)} + \gamma_k^{(1)} + \gamma_k^{(2)} = 1$. The value $\gamma_k^{(0)} = 1$ corresponds to the cluster representative $\boldsymbol{u}_k$ not appearing in Eq. (9.6) and none of the covariates in latent cluster $k$ being associated with the responses. The value $\gamma_k^{(1)} = 1$ corresponds to $\boldsymbol{u}_k$ appearing as a simple linear regressor in Eq. (9.6), and $\gamma_k^{(2)} = 1$ corresponds to its occurrence in a nonlinear form. This adaptive mixture of linear and nonlinear elements aims to achieve a balance between model parsimony and flexibility.

Possible options for the function $h$ in Eq. (9.6) include nonlinear function kernels such as those based on reproducible kernel Hilbert spaces (Mallick *et al.* 2005), nonlinear basis smoothing splines (Eubank 1999), and wavelets. Especially attractive due to their ease of construction and interpretability as a linear model are order-$r$ splines with $m$ number of knots (de Boor 1978; Hastie and Tibshirani 1990; Denison *et al.* 1998a):

$$h_{rm}(u_{ik}, \boldsymbol{\beta}_k^{(2)} \mid \boldsymbol{\kappa}_{s_k}) = \beta_{k,1}^{(2)} u_{ik} \cdots + \cdots \beta_{k,r}^{(2)} u_{ik}^r + \sum_{t=1}^{m} \beta_{k,r+t}^{(2)} (u_{ik} - \kappa_{s_k t})_+^r$$

where $a_+^r = (\max\{0, a\})^r$ and $\boldsymbol{\kappa}_{s_k}$ denotes the vector of $m$ knots associated with the $s_k$th covariate. This construction allows one to capture the linear dependencies, and perhaps more crucially, the nonlinear functional structures between the covariates and responses. This formulation can be viewed as a special case (without interactions) of multivariate adaptive regression splines, proposed by Friedman (1991) and extended in the Bayesian framework by Denison *et al.* (1998b) and Baladandayuthapani *et al.* (2006).

The set of covariate predictors is then $\mathscr{S} = \{s_k : \gamma_k^{(1)} + \gamma_k^{(2)} > 0, k = 1, \ldots, q\}$ and it is a subset of $\{1, \ldots, p\}$. The number of cluster predictors that appear as simple linear regressors in Eq. (9.6) is $q_1 = \sum_{j=1}^{q} \gamma_j^{(1)}$, and the number that appear as nonlinear predictors is $q_2 = \sum_{j=1}^{q} \gamma_j^{(2)}$. The number of cluster representatives that are non-predictors is $q_0 = q - q_1 - q_2$. The total number of cluster predictors is $|\mathscr{S}^*| = q_1 + q_2$, which equals the number of covariate predictors, $|\mathscr{S}|$.

For models with nonlinear functions $h$ that can be interpreted as a linear model, let $\boldsymbol{\gamma} = (\boldsymbol{\gamma}_1, \ldots, \boldsymbol{\gamma}_q)$ and $\boldsymbol{U}_{\boldsymbol{\gamma}}$ be a matrix of $n$ rows consisting of the intercept column and the independent regression variables based on the cluster representatives. Let $\text{col}(\boldsymbol{U}_{\boldsymbol{\gamma}})$ denote the number of columns of $\boldsymbol{U}_{\boldsymbol{\gamma}}$. For example, if we use order-$r$ splines with $m$ number of knots in Eq. (9.6), then $\text{col}(\boldsymbol{U}_{\boldsymbol{\gamma}}) = q_1 + (m+r) \cdot q_2 + 1$. With the symbol $[\cdot]$ representing densities, the prior for $\boldsymbol{\gamma}$ is

$$[\boldsymbol{\gamma}] \propto \omega_0^{q_0} \omega_1^{q_1} \omega_2^{q_2} \cdot \mathscr{I}\Big(\text{col}(\boldsymbol{U}_{\boldsymbol{\gamma}}) < n\Big) \tag{9.7}$$

where $\omega_0 + \omega_1 + \omega_2 = 1$, and $(\omega_0, \omega_1, \omega_2) \sim \mathscr{D}_3(1,1,1)$, a Dirichlet distribution. The restricted support of $\boldsymbol{\gamma}$ induces model sparsity, as discussed below. As before, a g prior is assumed for the regression coefficients:

$$\boldsymbol{\beta}_{\boldsymbol{\gamma}} \sim N_{|\mathscr{S}^*|+1}\Big(\mathbf{0}, \sigma_\beta^2 (\boldsymbol{U}_{\boldsymbol{\gamma}}' \boldsymbol{U}_{\boldsymbol{\gamma}})^{-1}\Big). \tag{9.8}$$

An advantage of the procedure is its ability to quantify nonlinear functional relationships between the responses and covariates. The *nonlinearity measure* $\mathscr{N} \in [0,1]$ is defined as the posterior expectation,

$$\mathscr{N} = E\big(\frac{\omega_2}{\omega_1 + \omega_2} | \boldsymbol{w}, \boldsymbol{X}\big). \tag{9.9}$$

The nonlinearity measure can be interpreted as the posterior predictive probability that a hypothetical, additional cluster appears as a predictor in Eq. (9.6) in a nonlinear form, rather than as a simple linear regressor. That is, $\mathscr{N}$ is the posterior probability that $\gamma_{q+1}^{(2)} = 1$. A value of $\mathscr{N}$ close to 0 (1) corresponds to linear (nonlinear) associations between the response and a majority of the predictors.

The schematic architecture of the model is shown in Fig. 9.2 using a directed acyclic graph.

## 9.3 Posterior Inference

Starting with an initial configuration obtained by a naïve, preliminary analysis, the model parameters are iteratively updated by MCMC methods. Section 9.3.1 describes the generation of the allocation variables. Section 9.3.2 describes the updates of the latent vector elements and their binary indicators. Section 9.3.3 describes the MCMC updates of the cluster predictors. Section 9.3.4 discusses the prediction of responses for individuals with only covariates available.

Due to the intensive nature of the posterior inference, the analysis can be done in two stages, with cluster detection followed by predictor discovery:

**Stage 1** Focusing on only the covariates and ignoring the responses:

*Stage 1a* The procedures of Sects. 9.3.1 and 9.3.2 are iteratively performed until the MCMC chain converges. Monte Carlo estimates are computed for the posterior probability of clustering for each pair of covariates.

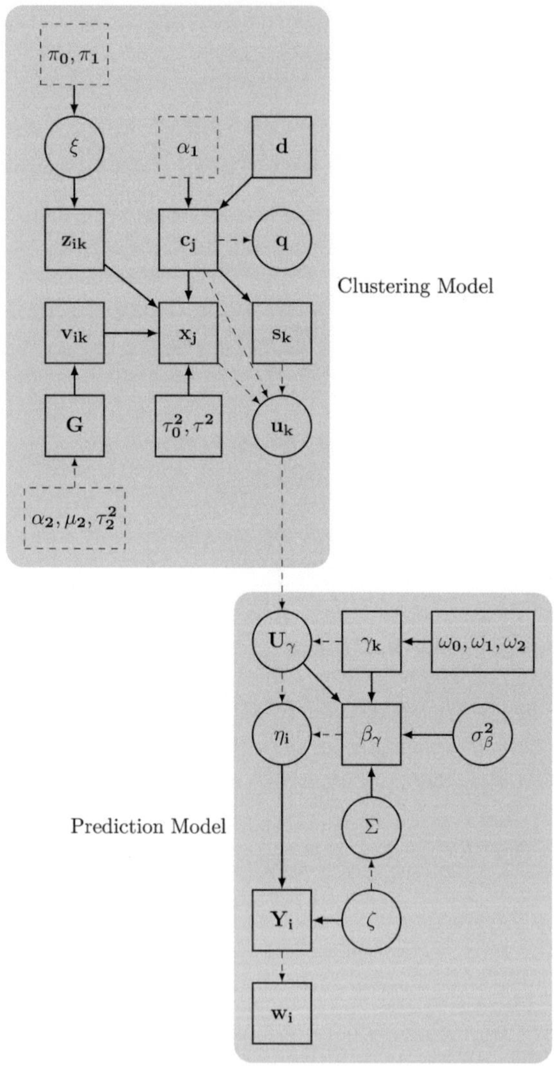


**Fig. 9.2** Directed acyclic graph of the proposed model in which the cluster representatives are chosen from the set of co-clustered covariates. *Circles* represent stochastic model parameters, *solid rectangles* represent data and deterministic variables, and *dashed rectangles* represent model constants. *Solid (dashed) arrows* represent stochastic (deterministic) relationships

Applying the technique of Dahl (2006), these pairwise probabilities are used to compute a point estimate for the allocation variables, which is called the *least-squares allocation*.

*Stage 1b* As an optional step, if the latent vector elements are parameters of interest, a second MCMC sample could be generated conditional on the least-squares allocation using the procedure described in Sect. 9.3.2.

Again applying the technique of Dahl (2006), we may compute a point estimate, called the *least-squares configuration*, for the set of latent vector elements $\{v_{ik}\}$ and indicators $\{z_{ik}\}$.

**Stage 2** Conditional on the least-squares allocation, and focussing on the responses, a third MCMC sample is generated using the strategies of Sect. 9.3.3. The sample is post-processed to obtain posterior inferences for the predictors. As described in Sect. 9.3.4, the sample can also be used to predict the outcomes of subjects with unknown responses.

As a further benefit of having a well-defined model for the covariates, as part of the MCMC procedure, we are able to perform model-based imputations of any missing covariate values.

## 9.3.1 Covariate-to-Cluster Allocation

For $j = 1,\ldots,p$, the full conditional distribution of allocation variable $c_j$ is not available in closed form. Nevertheless, we borrow ideas from sequential importance sampling (refer to Liu 2008, chap. 3) to devise a Gibbs sampler. Applying this strategy, new clusters were successfully opened 8.5 % of the time. Key to the fast mixing rate of this strategy is the assumption that the clusters borrow strength through a common distribution $G$ for their latent vector elements.

Given the full set of allocation variables, the PDP discount parameter $d \in [0,1)$ is updated by a Metropolis-Hasting algorithm. The proposal distribution for this algorithm exploits the fact that the likelihoods for a set of $d$'s can be quickly computed in closed form (Lijoi and Prünster 2010).

For the motivating data, the upper left panel of Fig. 9.3 displays the estimated posterior density of the PDP's discount parameter $d$. The estimated posterior probability of the event $[d = 0]$ is exactly zero, implying that a non-Dirichlet process clustering mechanism is strongly favored by the data, as suggested earlier by the EDA. The upper right panel of Fig. 9.3 plots the estimated posterior density of the number of clusters. The a posteriori large number of clusters (for $p = 500$ covariates) is suggestive of a PDP model with $d > 0$ (i.e., a non-Dirichlet process model). The lower left panel of Fig. 9.3 summarizes the cluster sizes of the least-squares allocation (Dahl 2006). The large number of clusters ($\hat{q} = 111$) and the multiplicity of small clusters are very unusual for a Dirichlet process, justifying the use of the more general PDP model.

## 9.3.2 Latent Vectors and Indicators

Among the allocation variables $c_1,\ldots,c_p$, suppose there are $q$ clusters, with cluster $k$ consisting of $n_k = \sum_{j=1}^{p} \mathscr{I}(c_j = k)$ covariates for $k = 1,\ldots,q$. As $i = 1,\ldots,n$

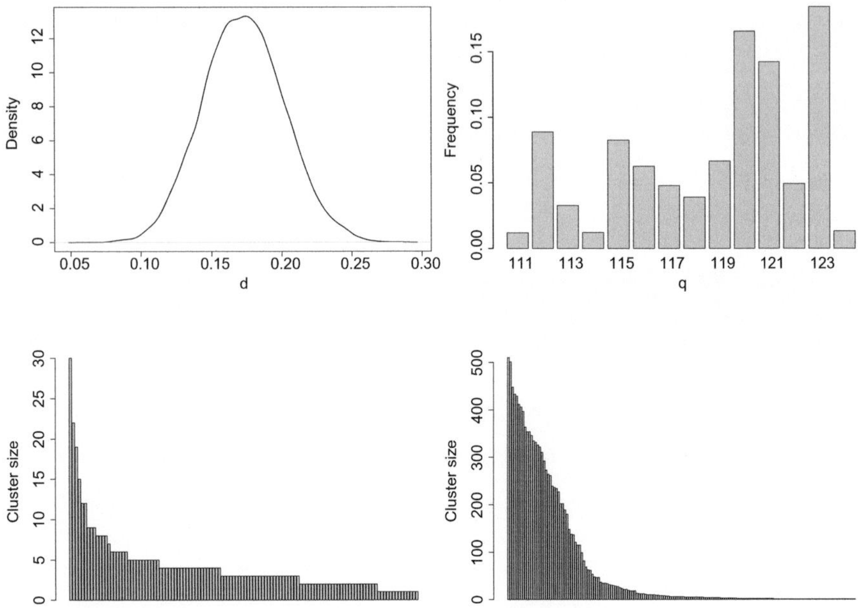


**Fig. 9.3** Posterior summaries for the motivating dataset. The top panels and the lower left panel summarize the least-squares covariate-to-cluster PDP allocation of the 500 genes. The lower right panel depicts the least-squares Dirichlet process configuration of more than 15,000 latent vector elements with binary indicators equal to 1

and $k = 1, \ldots, q$ vary, the sufficient statistics $\bar{x}_{ik} = \sum_{j=1}^{p} x_{ij} \cdot \mathscr{I}(c_j = k)/n_k$ are independently distributed as $N(0, \tau_1^2/n_k)$ if $z_{ik} = 0$, and as $N(v_{ik}, \tau^2/n_k)$ if $z_{ik} = 1$. Dirichlet process prior (9.5) is conjugate to the above distribution and to the sampling distribution of the $z_{ik}$'s. For $i = 1, \ldots, n$, and $k = 1, \ldots, q$, we can therefore update the bivariate vector $(v_{ik}, z_{ik})$ by Gibbs sampling. To accommodate the large number of latent vector elements, we apply a fast and accurate data squashing algorithm developed by Guha (2010) for high-dimensional settings.

In Stage 1b of the two-stage analysis, we computed the least-squares configuration of the latent vector elements. More than 90 % of the $n\hat{q} = 11,100$ latent vector elements have $\hat{z}_{ik} = 1$, implying that a relatively small proportion of covariate values for the motivating dataset can be regarded as random noise having no clustering structure. The lower right panel of Fig. 9.3 presents a summary of the least-squares configuration for the latent vector elements with $\hat{z}_{ik} = 1$. For more than 9,000 latent vector elements with $\hat{z}_{ik} = 1$, there are only 195 distinct values representing the estimated point masses of the distribution $G$. The configuration has mainly large clusters and closely resembles the typical configuration for a Dirichlet process model, justifying assumption (9.5).

For each of the $\hat{q} = 111$ clusters in the least-squares allocation of Stage 1a, we computed the correlations between its member covariates and the latent vector for individuals with $\hat{z}_{ik} = 1$. The cluster-wise median correlations are plotted in Fig. 9.4. The plots reveal fairly good within-cluster concordance regardless of the cluster size.

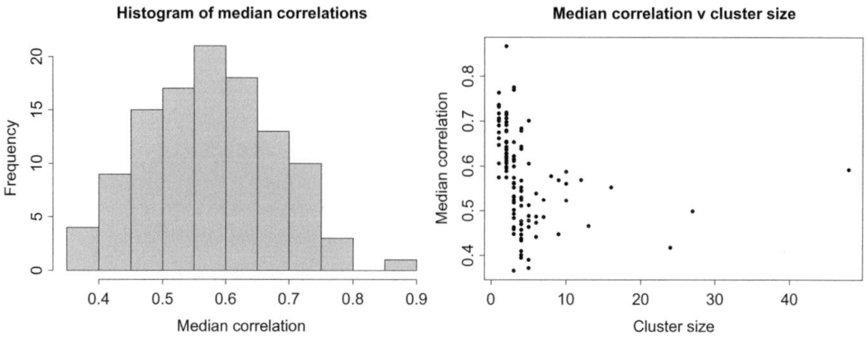


**Fig. 9.4** For the motivating dataset, median pairwise correlations for the $\hat{q} = 111$ PDP clusters in the least-squares allocation of Stage 1a

## 9.3.3 Cluster Predictors and Cluster Representatives

The choice of basis functions such as splines and wavelets for the nonlinear functionals $h$ in (9.6) results in nonlinear terms that are additive in analytic (e.g., polynomial or periodic) functions of the cluster representatives. In such cases, it is possible to integrate out the regression coefficients $\boldsymbol{\beta}_{\boldsymbol{\gamma}}$ to iteratively update the vector of indicators $\boldsymbol{\gamma}_k = (\gamma_k^{(0)}, \gamma_k^{(1)}, \gamma_k^{(2)})$, for clusters $k = 1, \ldots, q$. Given the cluster representative $\boldsymbol{u}_k$ and the set of indicators for the remaining $(q-1)$ clusters, the sub-models corresponding to $\gamma_k^{(0)} = 1$, $\gamma_k^{(1)} = 1$, and $\gamma_k^{(2)} = 1$, are then progressively nested.

Theoretical properties of Gaussian outcomes are exploited to quickly compute, up to a multiplicative constant, the likelihood functions for these three sub-models. This makes it possible to easily perform joint updates for the cluster representatives $\boldsymbol{u}_k$ and indicators $\boldsymbol{\gamma}_k$. After a cycle of updates of $q$ indicators and cluster representatives has been completed, the regression coefficients $\boldsymbol{\beta}_{\boldsymbol{\gamma}}$ may be jointly generated from the full conditional if necessary.

## 9.3.4 Predictions

Suppose there are $\tilde{n}$ additional individuals with unobserved responses but with available covariates $\tilde{x}_{i1}, \ldots, \tilde{x}_{ip}$ for $i = 1, \ldots, \tilde{n}$. As with the training set, we arrange the cluster representative elements for the test cases in an $\tilde{n} \times \text{col}(\boldsymbol{U}_{\boldsymbol{\gamma}})$ matrix. Given the set of predictors $\boldsymbol{\gamma}$, expressions for the posterior predictions of the regression outcomes, $\tilde{\mathbf{y}}$, can be obtained (Guha and Baladandayuthapani 2014).

## 9.4 Application to Gene Expression Data in Multiple Myeloma

The data we consider here comes from the Multiple Myeloma Research Consortium Reference Collection, containing a total of 304 multiple myeloma patient samples, of whom the gene expression profiles were measured using Affymetrix U133 Plus 2.0 microarrays. Robust Multichip Average (RMA) was used to normalize and quantify the expression levels of the data. We excluded samples without appropriate clinical information, which resulted in 208 patients with Multiple Myeloma (MM), and then randomly selected 100 patients for further downstream analysis. Hence, the resulting dataset consists of the results of microarray assays of gene expression levels for 500 probes/genes from the 100 patients with MM. In addition to the gene expression data, clinical information is also contained in the database, including the patient's age, gender, and measurements of clinical outcomes such as the $\beta_2$-microglublin—a serum protein which is a powerful prognostic factor and is an indicator of the severity of MM. Plots of data show the distribution of the log ratios are approximately symmetric about 0 thus justifying our gaussian assumptions.

We performed 50 independent replications of the three steps that follow. *(i)* We randomly split the data into training and test sets in a 2:1 ratio. *(ii)* We analyzed the failure times and $p = 500$ gene expression levels of the training cases using the proposed method and the techniques lasso, adaptive lasso, elastic net, and supervised principal components. *(iii)* The different techniques were used to predict the test case outcomes.

For the proposed procedure, a single covariate from each cluster was chosen to be the cluster representative. From a practical perspective, we have observed that the reliability of inferences and future predictions rapidly deteriorates as the number of cluster predictors and the number of additive nonlinear components in Eq. (9.6) increase. In spline-based models, this puts a constraint on the order of the splines, often necessitating the use of linear splines with $m = 1$ knot per cluster in Eq. (9.6). In this application, we fixed the knot for each covariate at the sample median.

Posterior inferences for some model parameters are summarized in Table 9.1. The number of clusters for the least-squares allocation of covariates, $\hat{q}$, computed in Stage 1a of the analysis, is considerably smaller for the breast cancer dataset. The relatively high estimates for the nonlinearity measure $\mathscr{N}$ indicate that the responses have nonlinear relationships with a majority of the predictors. In spite of being assigned a prior probability of 0.5, the estimated posterior probability of the Dirichlet process model (corresponding to discount parameter $d = 0$) is exactly 0 for both datasets, justifying the allocation scheme in Eq. (9.2).

**Table 9.1** Posterior inferences for selected model parameters

| Parameter | Motivating dataset |
|---|---|
| $\hat{q}$ | 111 |
| $\hat{\mathscr{N}}$ | 0.52 (0.00) |
| $\hat{P}[d = 0 \mid \text{data}]$ | 0 |

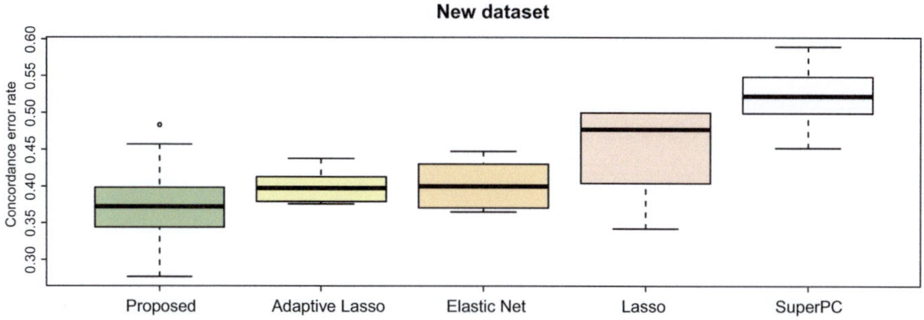


**Fig. 9.5** Side-by-side boxplots of percentage concordance error rates for the motivating dataset

Comparing the test case predictions with the actual $\beta_2$-microglublin outcomes, boxplots of numerical summaries of the concordance error rates for all the methods are presented in Fig. 9.5. The proposed method had the lowest error rate for the dataset, demonstrating its effectiveness in producing highly predictive models with small model sizes.

## 9.5 Conclusions

In summary, we offer an efficient methodology for high-dimensional clustering, variable selection, and prediction for continuous responses. The model exploits the sparsity of PDPs as dimension-reduction devices. Specifically, the covariates are grouped into lower-dimensional latent clusters consisting of covariates having similar patterns for the subjects, and are permitted to choose between PDPs and their special case, a Dirichlet process, for a suitable cluster allocation scheme. We theoretically determine how a PDP-based clustering is able to be distinguished from a Dirichlet process in terms of the number and relative sizes of their clusters.

We exploit different features of the model to develop an MCMC strategy that in cludes Metropolis-Hastings steps and a Gibbs sampler with efficient sequential importance sampling moves for cluster allocation. In predictive accuracy, the technique compares favorably with several existing methodologies for failure time datasets, consistently outperforming nonlinear techniques. These findings make a compelling case for the use of the proposed in high-dimensional regression settings such as genomics where it is critically important to detect predictive (or prognostic) models relying on a few, but important, genes that can be further biologically validated via functional experiments.

**Acknowledgements** Subharup Guha was supported by the National Science Foundation under award DMS-0906734. Veerabhadran Baladandayuthapani was partially supported by NIH grant R01 CA160736, MD Anderson Cancer Center SPORE in Multiple Myeloma (P50 CA142509) the Cancer Center Support Grant (CCSG) (P30 CA016672). The content is solely the responsibility of the authors and does not necessarily represent the official views of the National Cancer Institute or the National Institutes of Health.

## References


Bair, E. and Tibshirani, R. (2004). Semi-supervised methods to predict patient survival from gene expression data. *PLoS Biology*, **2**, 511–522.

Baladandayuthapani, V., Holmes, C. C., Mallick, B. K., and Carroll, R. J. (2006). *Modeling Nonlinear Gene Interactions using Bayesian MARS*. In Do K. A., Mueller P. and Vannucci M. (eds.) Bayesian Inference for Gene Expression and Proteomics. Cambridge University Press.

Barry, D. and Hartigan, J. A. (1993). A Bayesian analysis for change point problems. *Journal of the American Statistical Association*, **88**, 309–319.

Bonato, V., Baladandayuthapani, V., Broom, B. M., Sulman, E. P., Aldape, K. D., and Do, K.-A. (2010). Bayesian ensemble methods for survival prediction in gene expression data. *Bioinformatics*.

Brown, P. J., Vannucci, M., and Fearn, T. (1998). Multivariate Bayesian variable selection and prediction. *J. R. Stat. Soc. Series B*, **60**, 627–641.

Buckley, J. and James, I. (1979). Linear regression with censored data. *Biometrika*, **66**, 429–436.

Cai, B. and Dunson, D. (2006). Bayesian covariance selection in generalized linear mixed models. *Biometrics*, **62**, 446–457.

Cox, D. and Oakes, D. (1984). *Analysis of survival data*. London: Chapman and Hall.

Cristianini, N. and Shawe-Taylor, J. (2000). *Introduction to Support Vector Machines (and other kernel-based learning methods)*. Cambridge University Press.

Crowley, E. M. (1997). Product partition models for normal means. *Journal of the American Statistical Association*, **92**, 192–198.

Dahl, D. B. (2006). Model-based clustering for expression data via a dirichlet process mixture model. In K.-A. Do, P. Müller, and M. Vannucci, editors, *Bayesian Inference for Gene Expression and Proteomics*. Cambridge University Press.

de Boor, C. (1978). *A Practical Guide to Splines*. New York: Springer Verlag.

Dellaportas, P., Forster, J. J., and Ntzoufras, I. (1982). Bayesian variable selection using the Gibbs sampling. In D. K. Dey, S. K. Ghosh, and B. K. Mallick, editors, *Generalized linear models: a Bayesian perspective*, pages 273–286. Marcel Dekker, Inc., New York.

Denison, D. G. T., Mallick, B. K., and Smith, A. F. M. (1998a). Automatic Bayesian curve fitting. *Journal of the Royal Statistical Society, Series B*, **60**, 333–350.

Denison, D. G. T., Mallick, B. K., and Smith, A. F. M. (1998b). Bayesian mars. *Statistics and Computing*, **8**, 337–346.

Dunson, D. B. and Park, J.-H. (2008). Kernel stick-breaking processes. *Biometrika*, **95**, 307–323.

Dunson, D. B., Herring, A. H., and Engel, S. M. (2008). Bayesian selection and clustering of polymorphisms in functionally-related genes. *Journal of the American Statistical Association*, **103**, 534–546.

Eubank, R. (1999). *Nonparametric Regression and Spline Smoothing*. New York: Marcel Dekker.

Fan, J. and Li, R. (2002). Variable selection for Cox's proportional hazards model and frailty model. *Ann. Stat.*, **30**, 74–99.

Ferguson, T. S. (1973). A Bayesian analysis of some nonparametric problems. *Annals of Statistics*, **1**, 209–223.

Fraley, C. and Raftery, A. E. (2002). Model-based clustering, discriminant analysis, and density estimation. *Journal of the American Statistical Association*, **97**, 611–631.

Freudenberg, J. M., Sivaganesan, S., Wagner, M., and Medvedovic, M. (2010). A semi-parametric Bayesian model for unsupervised differential co-expression analysis. *BMC Bioinformatics*, **11**, 234.

Friedman, J. H. (1991). Multivariate adaptive regression splines. *Annals of Statistics*, **19**, 1–141.

George, E. and McCulloch, R. (1993). Variable selection via Gibbs sampling. *Journal of the American Statistical Association*, **88**, 881–889.

Guha, S. (2010). Posterior simulation in countable mixture models for large datasets. *Journal of the American Statistical Association*, **105**, 775–786.

Guha, S. and Baladandayuthapani, V. (2014). Nonparametric Variable Selection, Clustering and Prediction for High-Dimensional Regression. *ArXiv e-prints, arXiv:1407.5472.*

Hartigan, J. A. (1990). Partition models. *Communications in Statistics, Part A - Theory and Methods*, **19**, 2745–2756.

Hastie, T. J. and Tibshirani, R. J. (1990). *Generalized additive models*. London: Chapman & Hall.

Ishwaran, H. and James, L. F. (2003). Generalized weighted chinese restaurant processes for species sampling mixture models. *Statist. Sinica*, **13**, 1211–1235.

Ishwaran, H., Kogalur, U. B., *et al.* (2010). High-dimensional variable selection for survival data. *Journal of the American Statistical Association*, **105**, 205–217.

Kim, S., Tadesse, M. G., and Vannucci, M. (2006). Variable selection in clustering via Dirichlet process mixture models. *Biometrika*, **93**, 877–893.

Kuo, L. and Mallick, B. (1997). Bayesian semiparametric inference for the accelerated failure time model. *Canadian J. Stat.*, **25**, 457–472.

Lee, J., Müller, P., and Ji, Y. (2013). A nonparametric Bayesian model for local clustering. *Journal of the American Statistics Association*, **108**, 775–788.

Lee, K. and Mallick, B. (2004). Bayesian methods for variable selection in survival models with application to dna microarray data. *Sankhya*, **66**, 756–778.

Li, H. and Gui, J. (2004). Partial Cox regression analysis for high-dimensional microarray gene expression data. *Bioinformatics*, **20**, 208–215.

Lijoi, A. and Prünster, I. (2010). Models beyond the Dirichlet process. In N. Hjort, C. Holmes, P. Müller, and S. Walker, editors, *Bayesian Nonparametrics*, pages 80–136. Cambridge Series in Statistical and Probabilistic Mathematics.

Liu, J. (2008). *Monte Carlo strategies in scientific computing*. Springer Verlag.

Madigan, D. and Raftery, A. (1994). Model selection and accounting for model uncertainty in graphical models using occams window. *Journal of the American Statistical Association*, **89**, 1535–1546.

Mallick, B. K., Ghosh, D., and Ghosh, M. (2005). Bayesian classification of tumours by using gene expression data. *Journal of the Royal Statistical Society: Series B (Statistical Methodology)*, **67**, 219–234.

Medvedovic, M., Yeung, K. Y., and Bumgarner, R. E. (2004). Bayesian mixture model based clustering of replicated microarray data. *Bioinformatics*, **20**, 1222–1232.

Mitchell, T. J. and Beauchamp, J. J. (1988). Bayesian variable selection in linear regression. *Journal of the American Statistical Association*, **83**, 1023–1036.

Müller, P., Quintana, F., and Rosner, G. L. (2011). A product partition model with regression on covariates. *Journal of Computational and Graphical Statistics*, **20**, 260–278.

Nguyen, D. and Rocke, D. (2002). Partial least squares proportional hazard regression for application to dna microarray survival data. *Bioinformatics*, **18**, 1625–1632.

Peduzzi, P. N., Hardy, R. J., and Holford, T. R. (1980). A stepwise variable selection procedure for nonlinear regression models. *Biometrics*, **36**, 511–516.

Perman, M., Pitman, J., and Yor, M. (1992). Size-biased sampling of poisson point processes and excursions. *Probab. Theory Related Fields*, **92**, 21–39.

Pitman, J. (1995). Exchangeable and partially exchangeable random partitions. *Probab. Theory Related Fields*, **102**, 145–158.

Pitman, J. and Yor, M. (1997). The two-parameter poisson-dirichlet distribution derived from a stable subordinator. *Ann. Probab.*, **25**, 855–900.

Quintana, F. A. (2006). A predictive view of Bayesian clustering. *Journal of Statistical Planning and Inference*, **136**, 2407–2429.

Quintana, F. A. and Iglesias, P. L. (2003). Bayesian clustering and product partition models. *J. R. Statist. Soc. B*, **65**, 557–574.

Sha, N., Vannucci, M., Tadesse, M. G., Brown, P. J., *et al.* (2004). Bayesian variable selection in multinomial probit models to identify molecular signatures of disease stage. *Biometrics*, **60**, 812–819.

Sha, N., Tadesse, M. G., and Vannucci, M. (2006). Bayesian variable selection for the analysis of microarray data with censored outcome. *Bioinformatics*, **22**, 2262–2268.

Tibshirani, R. (1997). The lasso method for variable selection in the Cox model. *Stat. Med.*, **16**, 385–395.

Volinsky, C. *et al.* (1997). Bayesian model averaging in proportional hazard models: Assessing the risk of stroke. *App. Stat.*, **46**, 433–448.

Zellner, A. (1986). On assessing prior distributions and Bayesian regression analysis with g prior distributions. In P. Goel and A. Zellner, editors, *Bayesian Inference and Decision Techniques*, pages 233–243. New York: Elsevier.

# Part III
# Survival Analysis

# Chapter 10
# Markov Processes in Survival Analysis

Luis E. Nieto-Barajas

**Abstract** This chapter presents some discrete and continuous Markov processes that have shown to be useful in survival analysis and other biostatistics applications. Both discrete and continuous time processes are used to define Bayesian nonparametric prior distributions. The discrete time processes are constructed via latent variables in a hierarchical fashion, whereas the continuous time processes are based on Lévy increasing additive processes. To avoid discreteness of the implied random distributions, these latter processes are further used as mixing measures of the parameters in a particular kernel, which lead to the so-called Lévy-driven processes. We present the use of these models in the context of survival analysis. We include univariate and multivariate settings, regression models and cure rate models.

## 10.1 Introduction

All the chapters in this book are meant to be related to *Bayesian nonparametrics*. There is no point in me going deeper in the definition of the approach, but I will use some lines to share with you my understanding of the topic.

Bayesian nonparametric theory handles statistical inference by assuming a nonparametric sampling distribution and making decisions via the Bayesian paradigm. By nonparametric sampling distributions we mean distribution families with infinite (or large) dimensional parameter spaces. Since Bayesian decision theory requires to express prior knowledge on the unknown quantities, a Bayesian nonparametric prior is a probability measure on an infinite dimensional space. What makes a prior to be

L.E. Nieto-Barajas (✉)
ITAM, Rio Hondo 1, Progreso Tizapan, 01080 Mexico D.F., Mexico
e-mail: lnieto@itam.mx



R. Mitra, P. Müller (eds.), *Nonparametric Bayesian Inference in Biostatistics*,
Frontiers in Probability and the Statistical Sciences,
DOI 10.1007/978-3-319-19518-6_10

nonparametric was clearly stated by Ferguson (1973) "a nonparametric prior must have large support in the sense that any fixed probability measure can be arbitrarily approximated by a probability measure generated by the prior."

In notation, let $X_1,\ldots,X_n$ be a sample of random variables (r.v.) such that, conditionally on a cumulative distribution function (c.d.f.) $F$ defined on $(\mathscr{X},\mathscr{B})$, the r.v.'s are independent and identically distributed (i.i.d.), that is, $X_i|F \overset{\text{iid}}{\sim} F$, where $\mathscr{X} \subset \mathrm{IR}$ is the sample space and $\mathscr{B}$ the Borel's $\sigma$-algebra. Under a Bayesian nonparametric approach, the law $F$ that describes the behaviour of the $X_i$'s can itself be treated as unknown. To place a prior on $F$, we rely on stochastic processes whose paths are c.d.f.'s. In notation, $F \sim \mathscr{P}$, where $\mathscr{P}$, defined on $(\mathscr{F},\mathscr{A})$, is the nonparametric prior, with $\mathscr{F}$ the set of all c.d.f.'s and $\mathscr{A}$ an appropriate $\sigma$-algebra of subsets of $\mathscr{F}$. If we think on the probability measure induced by $F$, then we say that this is a random probability measure.

In this chapter we will present some stochastic processes that are used to define Bayesian nonparametric priors in survival analysis. In Sect. 10.2 we define some discrete and continuous Markov processes and present some of their properties. In Sect. 10.3 we describe survival analysis models in discrete and continuous time, we include univariate and multivariate models, regression models and cure rate models. We illustrate the behaviour of some of the models in Sect. 10.4.

Before we proceed we introduce notation: $\mathrm{Be}(a,b)$ denotes a beta density with mean $a/(a+b)$; $\mathrm{Ga}(a,b)$ denotes the gamma density with mean $a/b$; $\mathrm{Bin}(c,p)$ denotes a binomial density with number of trials $c$ and probability of success $p$; $\mathrm{Po}(c)$ is a Poisson density with mean $c$; $\mathrm{Pg}(a,b,c)$ denotes a Poisson-gamma density with mean $ca/b$; $\mathrm{N}(\mu,\sigma^2)$ is a normal density with mean $\mu$ and variance $\sigma^2$.

## 10.2 Markov Processes

A stochastic process can be thought of as a family of random variables linked via a parameter which takes values on a specific domain. According to Doob (1953), a stochastic process is the mathematical abstraction of an empirical process whose development is governed by probabilistic laws. Let $\{Z(t);\, t \in \mathscr{T}\}$ denote a stochastic process with domain or index set $\mathscr{T}$ and range or state space $\mathscr{Z}$. If $\mathscr{T}$ is a discrete (countable) set, say the natural or the integer numbers, then the process is said to be a *discrete time process*, whereas if $\mathscr{T}$ is a continuous (uncountable) set, like the real numbers or a bounded interval, then the process is said to be a *continuous time process*. To distinguish between them, we will use the notation $Z_t$ for discrete time processes and $Z(t)$ for continuous time processes.

Let $(\mathscr{Z},\mathscr{B},\mathrm{P})$ be a probability space, then the process $\{Z_t\}$ is said to be a *Markov Process* if for any $B \in \mathscr{B}$ and for every $s,t \in \mathscr{T}$ with $s<t$ the following Markovian property is satisfied: $\mathrm{P}(Z_t \in B \mid \{Z_u\}, u \le s) = \mathrm{P}(Z_t \in B \mid Z_s)$. In words, a stochastic process is Markov if the probability of the future given the present does not depend on the past.

### 10.2.1 Discrete Time Processes

There are several ways of constructing discrete time Markov processes. Here we review a generative definition that is based on latent variables with priors that belong to the class of conjugate distributions in Bayesian analysis. Therefore these processes are also called Gibbs Markov processes. Since most of these processes are used to define prior distributions for infinite dimensional parameters in a Bayesian nonparametric context, we will use Greek letters to denote them.

Let $\{\theta_t\}$ be a discrete time stochastic process such that $t \in \mathscr{T} = \{1,2,\ldots\}$. Let $\{\eta_t\}$ for $t \in \mathscr{T}$ be a latent discrete time stochastic process. Nieto-Barajas and Walker (2002) defined a Markov process with dependence of order one. Their construction can be represented graphically as in the diagram depicted in Fig. 10.1. Information between $\theta_t$ and $\theta_{t+1}$ is passed only thorough $\eta_t$.

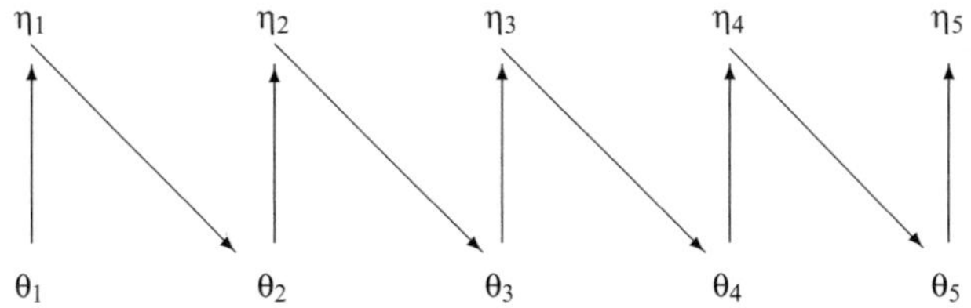


**Fig. 10.1** Graphical representation of a Markov process with dependence of order one

Specifically, Nieto-Barajas and Walker (2002) constructed beta and gamma Markov processes with the following specifications. The *beta process* relies on a latent binomial process, that is

$$\theta_1 \sim \text{Be}(a,b),\ \ \eta_t \mid \theta_t \sim \text{Bin}(c_t,\theta_t),\ \text{ and }\ \theta_{t+1} \mid \eta_t \sim \text{Be}(a+\eta_t, b+c_t-\eta_t), \quad (10.1)$$

for $t = 1,2,\ldots$. The *gamma process*, on the other hand, relies on a latent Poisson process, that is

$$\theta_1 \sim \text{Ga}(a,b),\ \ \eta_t \mid \theta_t \sim \text{Po}(c_t\theta_t),\ \text{ and }\ \theta_{t+1} \mid \eta_t \sim \text{Ga}(a+\eta_t, b+c_t), \quad (10.2)$$

for $t = 1,2,\ldots$. In any case, the joint distribution of the variables $(\theta_1,\ldots,\theta_T)$ is

$$f(\theta_1,\theta_2,\ldots) = f(\theta_1)\sum_{\eta_1}\cdots\sum_{\eta_T}\prod_{t=1}^{T} f(\eta_t \mid \theta_t) f(\theta_{t+1} \mid \eta_t)$$

from which the Markovian property can be verified. In particular, the conditional expectation of $\theta_{t+1}$ given $\theta_t$ has a linear form, $\text{E}(\theta_{t+1} \mid \theta_t)$ becomes $(a+c_t\theta_t)/(a+b+c_t)$ for the beta process, and $(a+c_t\theta_t)/(a+c_t)$ for the gamma process.

Processes defined by (10.1) and (10.2) take their name not only because they use conditional beta and gamma distributions in their constructions, but also because marginally $\theta_t$ is $\text{Be}(a,b)$ and $\text{Ga}(a,b)$, respectively. Therefore, parameters $(a,b)$

determine the form of the marginal distribution of $\theta_t$ and parameters $\{c_t\}$ determine the strength of dependence, in fact, $\text{Corr}(\theta_t+1,\theta_t)$ is $c_t/(a+b+c_t)$ for the beta process, and $c_t/(b+c_t)$ for the gamma process. Furthermore, if $c_t = c$ for all $t$, then the processes $\{\theta_t\}$ become strictly stationary.

### 10.2.2 Lévy-Driven Processes

An independent increments process, or additive process, is a continuous time process such that for $t_1 < t_2 < \cdots < t_k \in \mathscr{T}$, the increments $Z(t_i), Z(t_2)-Z(t_1), \ldots, Z(t_n)-Z(t_{n-1})$ are independent. The two principal members of this class are the Wiener and Poisson processes. According to, e.g., Gikhman and Skorokhod (1969), every stochastically continuous process with independent increments can be represented as the sum of a Wiener process and an integral of Poisson processes.

A stochastic process $Z(t)$ with independent increments is said to be homogeneous if the increments are stationary, that is, if the distribution of $Z(t+s)-Z(t)$ only depends on $s$. Lévy processes are homogeneous independent increments processes, and are usually referred to as random walks in continuous time.

Lévy processes can have non-monotonic paths. However, the Lévy processes used to construct priors in Bayesian nonparametric inference are usually nondecreasing, nonnegative and with piecewise constant paths. They are known as pure jump Lévy processes. Moreover, the homogeneity constraint that characterizes a Lévy process is relaxed to nonhomogeneous cases in the definition of nonparametric priors. Strictly speaking such processes are not Lévy any more and a more appropriate name would be increasing additive processes, but we will still refer to them as nonhomogeneous Lévy processes.

The probability law of a pure jump (homogeneous or nonhomogeneous) Lévy process is characterized by its Laplace transform and is given by

$$\mathrm{E}\left\{e^{-\phi Z(t)}\right\} = \exp\left\{-\int_{-\infty}^{t}\int_{0}^{\infty}\left(1-e^{-\phi v}\right)\nu(\mathrm{d}v,\mathrm{d}s)\right\}, \tag{10.3}$$

where $\nu(\mathrm{d}v,\mathrm{d}s) = \rho(\mathrm{d}v|s)\alpha(\mathrm{d}s)$ is called the Lévy intensity, $\rho(\cdot|s)$ is a measure on $\mathbb{R}^+$ that controls the jump sizes for every location $s$, and $\alpha(\cdot)$ is a measure on $\mathbb{R}$ that determines the jump locations. The Lévy intensity must satisfy the condition

$$\int_A\int_0^{\infty}\min\{v,1\}\nu(\mathrm{d}v,\mathrm{d}s) < \infty,$$

for any bounded $A \subset \mathscr{X}$. If the measure $\rho$ is independent of the locations $s$, i.e., $\rho(\mathrm{d}v|s) = \rho(\mathrm{d}v)$, the process $Z(t)$ is homogeneous.

There are several Lévy intensities that satisfy the previous condition, most of them can be seen as particular cases of the following two nonhomogeneous Lévy intensities:

(i) Generalized gamma: $\rho(\mathrm{d}v|s) = \Gamma(1-\varepsilon)^{-1}v^{-(1+\varepsilon)}e^{-\beta(s)v}\mathrm{d}v$, with $\varepsilon \in \{(0,1) \cup \{-1\}\}$, and
(ii) Log-beta: $\rho(\mathrm{d}v|s) = (1-e^{-v})^{-1}e^{-\beta(s)v}\mathrm{d}v$,

with a nonhomogeneous parameter function $\beta(s) \geq 0$ for all $s \in \mathbb{R}$, together with a measure $\alpha(s)$ on $\mathbb{R}$. In particular, for case (i) and with $\varepsilon = -1$, the Lévy measure becomes finite, i.e., the number of jumps in a finite interval is finite, whereas infinite Lévy measures have an infinite number of jumps in a finite interval.

A Lévy process can be generalized to include fixed jump locations $\tau_1, \tau_2, \ldots$, with independent nonnegative jump sizes $Z\{\tau_1\}, Z\{\tau_2\}, \ldots$ (also independent of the rest of the process). A general Lévy process becomes

$$Z(t) = Z_c(t) + \sum_j Z\{\tau_j\}I(\tau_j \leq t),$$

where $Z\{t\} = Z(t) - Z(t-)$ and $Z_c(t)$ is a Lévy process without fixed points of discontinuity, also known as "continuous" part, whose law is characterized by (10.3). Although $Z_c(t)$ is called "continuous", $Z_c(t)$ is almost surely discrete, so it can be represented as $Z_c(t) = \sum_j J_j^c I(\tau_j^c \leq t)$ e.g. Ferguson and Klass (1972), where $\{J_j^c\}$ are random jump sizes and $\{\tau_j^c\}$ are random locations.

Let $\mu$ denote the measure induced by the Lévy process $Z(t)$. That is, for a set $A \subset \mathscr{T}$, say $A = (a_0, a_1]$ for $a_0, a_1 \in \mathscr{T}$, define $\mu(A) := Z(a_1) - Z(a_0)$. Then, in measure theory, $\mu$ is called completely random measure. These measures are important since they can be generalized to more general complete and separable metric spaces. We refer the reader to Daley and Vere-Jones (2008) for details.

In general, a Lévy-driven process is any process defined as a function of a Lévy process. A specific form of the Lévy-driven processes used to construct Bayesian nonparametric priors are Lévy-driven mixtures of a kernel $k(x,s)$ with weights (or mixing measure) given by a Lévy process $Z(s)$. In notation, a Lévy-driven process $W(t)$ has the form

$$W(t) = \int k(t,s)Z(\mathrm{d}s) \tag{10.4}$$

Like a Lévy process, the Lévy-driven process defined in (10.4) is a Markov process. Depending on the choice of the kernel $k$, a Lévy-driven process can have piecewise constant paths, smooth increasing paths, or non-monotonic paths. Examples can be seen in Fig. 10.2, where we show random paths of Lévy-driven processes for different choices of the kernel $k$. Note that the jump sizes and locations were kept the same across the different panels in Fig. 10.2 to better appreciate the influence of the kernel. The general condition we require on a kernel $k(x,s)$ is to be nonnegative for all $x$ and $s$. Further conditions are imposed according to the specific use of the process to properly define a prior. These will be described later in this section.

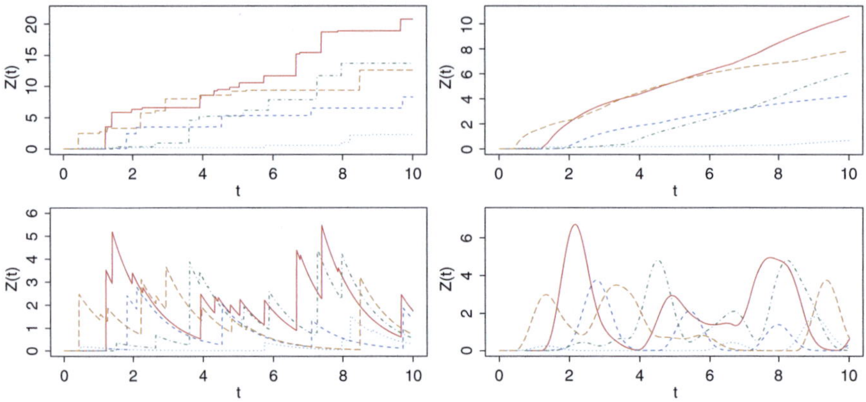


**Fig. 10.2** Random Lévy-driven paths for different kernels: $k(t,s) = I(s \leq t)$ (*top left*); $k(t,s) = \{1-(s/t)\}I(s \leq t)$ (*top right*); $k(t,s) = \exp\{-(t-s)\}I(s \leq t)$ (*bottom left*); and $k(t,s) = 3(t-s)^2 \exp\{-(t-s)^3\}I(s \leq t)$ (*bottom right*). In each panel the lines represent different random realizations of the process $W(t)$

## *10.2.3 From Discrete to Continuous Time Processes*

Nieto-Barajas and Walker (2007b) established a connection between Gibbs and autoregressive processes. Considering this connection and by taking a suitable limit, they showed that a particular autoregressive gamma process converges to a Lévy-driven process. We sketch the derivation here.

Consider the (Gibbs) gamma process (10.2) in its stationary form, that is, $c_t = c$ for all $t$, and with $b = 1$. Nieto-Barajas and Walker (2007b) derived the innovation term for an autoregressive process to be also stationary gamma. In particular they obtained that by taking

$$\theta_t = \rho\theta_{t-1} + \rho\zeta_t, \tag{10.5}$$

where $\rho = c/(1+c)$, and

$$\zeta_t|\xi_t \sim \text{Ga}(\xi_t, 1), \;\; \xi_t|\gamma_t \sim \text{Po}(\gamma_t/c) \;\text{ and }\; \gamma_t \sim \text{Ga}(a,1),$$

implies that $\{\theta_k\}$ is a strictly stationary process with $\text{Ga}(a,1)$ marginals and with conditional expected value $\text{E}(\theta_{t+1} \mid \theta_t) = (a+c\theta_t)/(1+c)$, matching the Gibbs type process (10.2).

To avoid confusion from now on we denote the discrete time index by $k$ and use $t \in \mathscr{T} = \mathbb{R}^+$ to index a process in continuous time. Nieto-Barajas and Walker (2007b) defined a partition of $\mathscr{T}$ for each $n$, via $0 = \tau_{n,0} < \tau_{n,1} < \tau_{n,2} < \cdots$, with $\tau_{n,k} = k/n$ for $k = 0,1,\ldots$, and made $c_n$ depend on $n$ via $c_n = cn$ for some $c > 0$. They thus defined a piecewise constant process $W_n(t)$ as $W_n(0) = \theta_{n,0}$ and, for $t > 0$

$$W_n(t) = \sum_{k=1}^{\infty} \theta_{n,k} I\left(\tau_{n,k-1} < t \leq \tau_{n,k}\right),$$

where $\theta_{n,0} \sim \text{Ga}(\alpha, 1)$ and $\{\theta_{n,k}\}$ is either the Gibbs process (10.2) or the autoregressive process (10.5). This continuous time process $W_n(t)$ is strictly stationary with marginal distribution $\text{Ga}(a, 1)$ for all $t \geq 0$.

Nieto-Barajas and Walker (2007b) showed that the autocorrelation function of process $W_n(t)$, regardless of the choice of $\{\theta_{n,k}\}$ to be of Gibbs type or autoregressive, is $\text{Corr}(W_n(t), W_n(0)) = \{c_n/(1+c_n)\}^{\lceil nt \rceil}$ and converges to $e^{-t/c}$ as $n \to \infty$, which is the autocorrelation function of a Lévy-driven process of the form (10.4). However, the only process that does converge to a Lévy-driven is $W_n(t)$ defined in terms of the autoregressive process. In this case, the limiting Lévy-driven process, say $W(t)$, is a shot-noise process of the form

$$W(t) = \int_0^t e^{-(t-s)/c} Z(\mathrm{d}s),$$

where $Z(s)$ is a Lévy process with a fixed jump at zero and finite Lévy measure for the continuous part $\nu(\mathrm{d}v, \mathrm{d}s) = (a/c)e^{-v}\mathrm{d}v\mathrm{d}s$. This implies that $W(t) \sim \text{Ga}(a, 1)$ marginally for all $t \geq 0$. Bottom left panel in Fig. 10.2 presents random paths from a shot-noise process. They have non-monotonic paths and present sudden increments (shots) with exponential decays.

## 10.3 Nonparametric Priors

As was mentioned in Sect. 10.1, a nonparametric prior is a probability measure $\mathscr{P}$ on the space $\mathscr{F}$ of all cumulative distribution functions. Broadly speaking, we can think of a nonparametric prior as a probability measure on the space of probability models. Therefore, we can place the prior on densities, cumulative distributions, survival functions, hazard rates, or any other function that characterizes the behaviour of the observable random variables. In survival analysis, for instance, it is customary to place the prior on the space of survival or hazard functions. For density estimation, the nonparametric prior is usually placed on the density or cumulative distribution functions.

### *10.3.1 Survival Models*

Discrete time processes (10.1) and (10.2) have been used in survival analysis to construct priors on hazard rates (Nieto-Barajas and Walker 2002). Here we review these models. In survival analysis the variable of interest is denoted by $T$ and is usually referred to as a failure time.

Let $T$ be a nonnegative discrete random variable taking values on $\{\tau_1, \tau_2, \ldots\}$ (a possible infinite set) with density function $f(t)$ such that $f(t) = \text{P}(T = \tau_j)$ if $t = \tau_j$ and $f(t) = 0$ otherwise. Let $S(t) = \text{P}(T > t)$ be the survival function associated with $f(t)$, and $h(t)$ the corresponding hazard rate function such that $h(t) = \mathrm{P}(T = t \mid T \geq t)$. This hazard rate only takes values $\theta_j = h(\tau_j)$ different than zero at times $t = \tau_j$, $j = 1,2,\ldots$. We can therefore express the density function in terms of the hazard rate $h(t)$, i.e. $\{\theta_j\}$, as

$$f(\tau_j) = \theta_j \prod_{k=1}^{j-1} (1-\theta_k) \quad j = 1,2,\ldots. \tag{10.6}$$

We can think of $h(t)$ to be defined by an infinite set of parameters $\{\theta_j\}$, then by assigning a prior on $\{\theta_j\}$ we induce a nonparametric prior on $h$ and $f$. Nieto-Barajas and Walker (2002) took $\{\theta_j\}$ to be the beta process (10.1). This beta process prior on the hazard rates allows us to borrow information across adjacent times $\tau_j$ and $\tau_{j+1}$ and thus produce a smoothed version of the Nelson-Aalen nonparametric classical estimator (Aalen 1978).

Let us now consider $T$ to be a nonnegative continuous random variable with support on the positive real line $\mathbb{R}^+$ and density function $f(t)$ for $t \geq 0$. Denote by $S(t) = \mathrm{P}(T > t)$ the survival function and by $h(t)$ the hazard (intensity) function defined as $h(t) = f(t)/S(t) = -S'(t)/S(t)$, where prime ($'$) denotes derivative with respect to $t$. From the hazard intensity function we can recover the density function via the expression

$$f(t) = h(t) \exp\left\{ \int_0^t h(s) \mathrm{d}s \right\}. \tag{10.7}$$

If we think of $\{h(t)\}$ as an infinite dimensional parameter, we can place a prior on $h$ to induce a nonparametric prior on $f$.

There are several ways of defining a prior on $h(t)$ based on the Markov processes of Sect. 10.2. One way is by defining $h(t)$ as a piecewise constant function of the form

$$h(t) = \sum_{j=1}^{J} \theta_j I(\tau_{j-1} < t \leq \tau_j), \tag{10.8}$$

with $0 = \tau_0 < \tau_1 < \cdots, \tau_J = \infty$ and $\{\tau_j\}$ forming a partition of the positive real line in intervals $(\tau_{j-1}, \tau_j]$, $j = 1,2,\ldots,J$. The value $J$ controls the flexibility of the piecewise constant hazard. $J = 1$ implies a fully parametric exponential model whereas larger values of $J$ induce a more nonparametric model. Potentially $J$ could be infinite. Nieto-Barajas and Walker (2002) took $\{\theta_j\}$ to be the gamma process (10.2) to define a nonparametric prior.

Alternatively, instead of partitioning the positive real line, we can directly put a prior on $h(t)$ based on a continuous time process. Nieto-Barajas and Walker (2004) took

$$h(t) = W(t) \tag{10.9}$$

with $W$ a Lévy-driven process of the form (10.4).

### 10.3.2 Survival Regression Models

Survival regression models with covariates are also available. Let $T_i$ be the failure time of individual $i$ with $p$-dimensional covariate vector $\mathbf{X}_i' = (X_{1i}, \ldots, X_{pi})$, for $i = 1, \ldots, n$. In its more general form, the covariates could sometimes be time dependent, that is $X_{ji} = X_{ji}(t)$, for some $j \in \{1, \ldots, p\}$.

The most widely used model that accounts for covariates is the proportional hazards model (Cox 1999). This model is specified semiparametrically by assuming that the hazard function of individual $i$ is

$$h_i(t \mid \mathbf{x}_i) = \lambda_i h_0(t), \tag{10.10}$$

with $\lambda_i = \exp\{\boldsymbol{\beta}'\mathbf{x}_i(t)\}$, $\boldsymbol{\beta}' = (\beta_1 \ldots, \beta_p)$ is a vector of regression coefficients with no intercept and $h_0(t)$ is a baseline hazard function which corresponds to the hazard function of an individual with $\mathbf{x}_i = \mathbf{0}$.

The baseline function $h_0(t)$ is unspecified. We can assign a nonparametric prior on $h_0(t)$ to induce a Bayesian semiparametric model for the survival times $T_i$'s. Nieto-Barajas (2003), for instance, used a piecewise constant function of the form (10.8) together with the gamma process prior (10.2). On the other hand, Nieto-Barajas and Walker (2005) used a continuous specification of $h_0(t)$ as in (10.9) together with a Lévy-driven process prior (10.4).

Alternative survival regression models have been proposed. Recently, based on the two groups survival model (Yang and Prentice 2005), Nieto-Barajas (2014a) proposed a regression model where the hazard rate for individual $i$ has the form

$$h_i(t \mid \mathbf{x}_i) = \frac{\lambda_i \varphi_i R'(t)}{\varphi_i + \lambda_i R(t)}, \tag{10.11}$$

where $\lambda_i = \exp(\boldsymbol{\beta}'\mathbf{x}_i)$, $\varphi_i = \exp(\boldsymbol{\gamma}'\mathbf{x}_i)$ and $R(t)$ is a baseline nonnegative monotone increasing function, with $\boldsymbol{\beta}$ and $\boldsymbol{\gamma}$ vectors of coefficients with no intercept. If we consider the "baseline individual" for which $\mathbf{x}_i - \mathbf{0}$ ($\Rightarrow \lambda_i = \varphi_i = 1$), then $R(t)$ becomes the odds function $\{1 - S_i(t)\}/S_i(t)$.

Model (10.11) can be seen as a generalization of model (10.10). To see this we consider the hazard ratio between individual $i$ and the baseline individual,

$$\frac{h(t \mid \mathbf{x}_i)}{h(t \mid \mathbf{0})} = \frac{\lambda_i \varphi_i}{\lambda_i + (\varphi_i - \lambda_i)/\{1 + R(t)\}}.$$

By taking limits as the time approaches zero or infinity we get that $\lambda_i$ corresponds to the short-term ($t \to 0$) hazard ratio whereas $\varphi_i$ corresponds to the long-term ($t \to \infty$) hazard ratio. When both parameters are equal, i.e. $\lambda_i = \varphi_i$, means that the hazard ratio remains constant for any value of $t$, which corresponds to the proportional hazards model (10.10) when the covariates do not depend on time. If $\varphi_i = 1$, the resulting model corresponds to the proportional odds model (Bennet 1983). Moreover, when $(\lambda_i - 1)(\varphi_i - 1) < 0$ the hazard functions of individual $i$ and that of the baseline individual cross.

Noting that the nonparametric function $R(t)$ behaves like a cumulative hazard function and therefore $R'(t)$ behaves like a hazard function, Nieto-Barajas (2014a) use a Lévy-driven process of the form (10.4) to model $R'(t)$.

To complete the overview of the most common survival regression models, we mention the accelerated failure time model (e.g. Klein and Moeschberger 1997). This model defines a hazard rate for individual $i$ as

$$h_i(t \mid \mathbf{x}_i) = \lambda_i h_0(\lambda_i t), \tag{10.12}$$

where again $\lambda_i = \exp\{\boldsymbol{\beta}'\mathbf{x}_i\}$ and $h_0(t)$ is a baseline hazard function. A different way of seeing model (10.12) is by writing $T_i = T_0/\lambda_i$, where $T_0$ is the failure time of the baseline individual. Therefore the covariates effect through $\lambda_i$ accelerate o decelerate the failure time, i.e., an individual with failure time $t$ under $\mathbf{x}_i = 0$ would have a failure time $t/\lambda_i$ under $\mathbf{x}_i$.

Although the Markov process defined in Sect. 10.2 could be used to define a prior on $h_0(t)$, they have not been studied. However, alternative nonparametric priors for model (10.12) have been proposed. For instance, Christensen and Johnson (1988) use Dirichlet processes, and Hanson and Johnson (2002) use Pólya trees.

### *10.3.3 Cure Rate Models*

In the study of time-to-event data for certain diseases, there exists a positive probability of not observing the failure event. In other words, there exists a fraction of the population that is cured of or insusceptible to the disease, who thus will never experience the failure. Survival models that incorporate a cure fraction are often referred to as *cure rate models*.

In a typical (proper) survival model, when the time goes to infinity, the survival function vanishes at zero which means that all individuals in the population will die or present the failure event. On the other hand, in a cure rate model, the survival function does not go to zero as time approaches infinity, but remains positive, implying an improper survival model.

Berkson and Gage (1952) represented a cure rate model as a mixture of two subpopulations, the immune people and the susceptible people. Let $g(t)$ and $G(t)$ be the density and the survival function, respectively, of the susceptible people. Then the survival function of the entire population can be written as $S(t) = \pi + (1-\pi)G(t)$, where $\pi \in (0,1)$ is the cure proportion. Note that in the previous expression the survival function for the cured sub-population is 1. In terms of the hazard function, this mixture model can be written as

$$h(t) = \frac{(1-\pi)g(t)}{\pi + (1-\pi)G(t)} \tag{10.13}$$

On the other hand, Yakovlev and Tsodikov (1996) modelled directly the (improper) survival function as $S(t) = \exp[-\lambda\{1-G(t)\}]$, where $\lambda > 0$ is a nonnegative parameter related to the cure proportion as $\pi = \exp(-\lambda)$, and $G(t)$ is a (proper) survival function, nonnecessarily associated with a susceptible group as in (10.13). In terms of the hazard function this model becomes

$$h(t) = \lambda g(t). \tag{10.14}$$

Noting that a sufficient condition to have a cure rate model is that the area under the hazard function is finite, Nieto-Barajas and Yin (2008) proposed a cure threshold model of the form

$$h(t) = h_0(t)I(t \leq \tau), \tag{10.15}$$

where $h_0(t)$ is a baseline hazard function and $\tau$ is the cure threshold time after which an individual could be considered cured or risk-free. In this model the cure proportion becomes

$$\pi = \exp\left\{-\int_0^\tau h_0(s)\mathrm{d}s\right\}.$$

The baseline hazard function $h_0$ was modelled using the piecewise function (10.8) together with the gamma process prior (10.2). Due to the piecewise nature of $h_0$, the threshold parameter becomes discrete, say $\tau_z$, with $z$ the index from where the baseline hazard becomes zero, i.e. $\theta_{z+1} = \theta_{z+2} = \cdots = 0$. The model is completed by assigning a prior distribution for $z$ of the form $z-1 \sim \mathrm{Po}(\mu)$ to ensure that $z > 0$, with $\mu > 0$ a hyperparameter.

The previous cure rate models have all been extended to include covariates in different ways. Kuk and Chen (1992) extended model (10.13) by considering a logistic regression model for the cure proportion defining $\pi_i = \exp(\boldsymbol{\beta}'\mathbf{x}_i)$. For the susceptible group only they further considered a proportional hazards model as in (10.10). However, they assumed a parametric form for the hazard $g(t)$.

Chen et al. (1999) generalized model (10.14) by taking $\lambda_i = \exp(\boldsymbol{\beta}'\mathbf{x}_i)$. This extension is equivalent to considering a proportional hazards assumption. A further generalization of model (10.14) was introduced by Yin and Nieto-Barajas (2009) to include multiplicative and additive covariates. Their model is of the form

$$h_i(t \mid \mathbf{x}_i) = \lambda_i g(t) + \varphi_i, \tag{10.16}$$

where $\lambda_i = \exp(\boldsymbol{\beta}'\mathbf{x}_i)$ and $\varphi_i = \boldsymbol{\gamma}'\mathbf{x}_i$, in which either $\boldsymbol{\beta}$ or $\boldsymbol{\gamma}$ includes and intercept but not both. Note that $\varphi_i \geq 0$ for the model to be well defined. Depending on the covariates space, this induces a constraint in the coefficients $\boldsymbol{\gamma}$.

Model (10.16) can also be used to test the existence of a cure proportion since $\boldsymbol{\gamma} = 0$ implies a positive cure proportion, whereas $\boldsymbol{\gamma} \neq 0$ induces a proper survival model where the cure proportion is zero. Yin and Nieto-Barajas (2009) took a nonparametric prior for the density $g(t)$ in (10.16) induced by a prior on the corresponding hazard function which was assumed to be piecewise as in (10.8) with the gamma process prior (10.2).

Extensions of model (10.15) to include covariates were also proposed by Nieto-Barajas and Yin (2008) in two ways. First, covariates were assumed to have a proportionality effect in the hazards, and second, covariates were also thought of determining the cure threshold time $\tau_z$. The extended model is

$$h_i(t \mid \mathbf{x}_i) = \lambda_i h_0(t) I(t \leq \tau_i), \tag{10.17}$$

where $\lambda_i = \exp(\boldsymbol{\beta}'\mathbf{x}_i)$ and in the particular case when $h_0$ is piecewise constant, as in (10.8), then $\tau_i$ becomes $\tau_{z_i}$ with $z_i - 1 \sim \text{Po}(\mu_i)$, $\mu_i = \exp(\boldsymbol{\gamma}'\mathbf{x}_i)$ and $\boldsymbol{\beta}$ and $\boldsymbol{\gamma}$ vectors of coefficients where $\boldsymbol{\beta}$ has no intercept.

### *10.3.4 Multivariate Models*

There are several settings in survival analysis where multivariate time to event observations arise. These include competing risks, recurrent events and frailty models (e.g. Klein and Moeschberger 1997). Here we describe the latter case.

Let us concentrate on the bivariate case where $(T_1, T_2)$ are two (dependent) failure times with joint survival function $S(t_1,t_2)$. Let $G_1$ and $G_2$ be two univariate survival functions. The common frailty model (Clayton 1978) is usually written in terms of the hazard function and assumes a multiplicative random effect. In terms of the bivariate survival function this model becomes a mixture of the form

$$S(t_1,t_2) = \int \{G_1(t_1)G_2(t_2)\}^{\omega} m(\omega)\mathrm{d}\omega, \tag{10.18}$$

where $\omega$ is the random effect or *frailty* with distribution $m(\omega)$.

Allowing each failure time to have its own frailty, Marshall and Olkin (1988) propose the model

$$S(t_1,t_2) = \int\int \{G_1(t_1)\}^{\omega_1} \{G_2(t_2)\}^{\omega_2} m(\omega_1,\omega_2)\mathrm{d}\omega_1\mathrm{d}\omega_2, \tag{10.19}$$

where $m(\omega_1,\omega_2)$ is a bivariate distribution of the two frailties.

In (10.18) and (10.19) the marginal survival function $S_j$ is not $G_j$. Their relation is given by $G_j(t) = \exp[-\phi_j^{-1}\{S_j(t)\}]$, where $\phi_j(\cdot)$ is the Laplace transform of $\omega_j$, $j = 1,2$. A typical additional requirement for the frailties is that $\mathrm{E}(\omega_j) = 1$ to ensure estimability when combined with proportional hazards model (10.10). Most frailty models give $G_j$ a parametric form. But a nonparametric form is certainly also possible.

Similar to the previous mixture models, Nieto-Barajas and Walker (2007a) proposed a bivariate model parametrized in terms of the marginal hazard functions. Let $h_j(t)$ be the marginal hazard function of $T_j$ with corresponding cumulative hazard function $H_j(t) = \int_0^t h_j(s)\mathrm{d}s$, $j = 1,2$. The joint density function $f(t_1,t_2)$ is expressed as

$$f(t_1,t_2) = \int\int \frac{h_1(t)}{\omega_1} I\{\omega_1 > H_1(t)\} \frac{h_2(t)}{\omega_2} I\{\omega_2 > H_2(t)\} m(\omega_1,\omega_2) \mathrm{d}\omega_1 \mathrm{d}\omega_2, \tag{10.20}$$

where $m(\omega_1,\omega_2)$ is a bivariate distribution such that marginally $m(\omega_j) = \mathrm{Ga}(\omega_j \mid 2,1)$ for $j = 1,2$. This requirement is needed so that $h_j$ is the marginal hazard function of $T_j$. Specifically, Nieto-Barajas and Walker (2007a) defined the joint frailty $m(\omega_1,\omega_2)$ by taking $\omega_j \mid \delta \sim \mathrm{Ga}(2+\delta, 1+d)$ conditionally independent for $j = 1,2$ and $\delta \sim \mathrm{Pg}(2,1,d)$ and $d > 0$ a hyper-parameter. This implies that $\omega_j \sim \mathrm{Ga}(2,1)$ marginally, for $j = 1,2$ with correlation $\mathrm{Corr}(\omega_1,\omega_2) = d/(1+d)$.

Construction (10.20) defines a copula (e.g. Nelsen 1999). Its association properties have also been studied in Nieto-Barajas and Walker (2007a). The two marginal hazard functions $h_j$, $j = 1,2$ can be assigned a nonparametric prior of the form (10.8) together with the gamma process prior (10.2), or a Lévy-driven process prior (10.4) as in (10.9). Moreover, the model can be extended to cope with covariates in a proportional hazards manner (10.10) by taking $h_{ij}(t \mid \mathbf{x}_i) = \lambda_i h_j(t)$, where as before $\lambda_i = \exp(\boldsymbol{\beta}'\mathbf{x}_i)$ for $i = 1,\ldots,n$ and $j = 1,2$.

## 10.4 Numerical Illustrations

Posterior inference in the previous models requires us to update the law that describes the underlying process. For priors that rely on discrete time processes, posterior distributions are obtained in the same way as parametric models, the only consideration is that the number of parameters to be updated can be very large. For the prior that rely on continuous time processes, and specially Lévy-driven processes, posterior distributions are slightly more difficult to obtain. The complexity depends on the specific model. But in all cases the posterior law can be characterized through the Laplace transform (10.3) conditional on the data. Specific guidelines on how to obtain the posterior characterizations of Lévy-driven processes can be found in Nieto-Barajas (2014b) and in the specific references of each model. For the regression models, posterior distributions are characterized through the full conditionals and therefore a Gibbs sampler (Smith and Roberts 1993) will often be required.

In this section we illustrate the performance of some of the survival models described above. Inference in models that rely on the beta and gamma processes (10.1) and (10.2), apart from the cure rate models and bivariate models, all are implemented in the R package `BGPhazard` that can be available from CRAN (Team 2014). The manual that comes with the package explains the use of the different routines. However, if that is not enough, further explanation and examples run with the package can be obtained from Garca-Bueno and Nieto-Barajas (2014).

### 10.4.1 Example 1

In this first example we illustrate the use of the Markov beta process (10.1) to model hazard rates as in (10.6) in a discrete survival model. We analyse the 6-MP clinical trial data (Freireich et al. 1963) which consists of remission duration times (in months) for children with acute leukemia. The study consisted in comparing drug 6-MP versus placebo. We concentrate on the 21 patients who received placebo. Observed time values range from 1 to 23 and there are no censored observations. To define the prior we took $a = b = 0.0001$ and $c_t = 50$ for all $t$. We use command *BeMRes* to fit the model and the command *BePloth* to produce graphs.

Figure 10.3 (top panel) presents the hazard rate estimates together with 95 % credible intervals. Nelson-Aalen estimates were also included for comparison. Hazard rate estimates obtained with the beta process model are available for all times, whereas the Nelson-Aalen estimates are only available for those observed times. Most of the frequentist estimates are contained in the credible intervals, but in general the beta process estimates tend to have smaller values. Although this might not look right, the implication in the survival function estimates (bottom panel) are surprising. The big steps produced by the Kaplan–Meier estimator are all smoothed out in smaller steps with the beta process model. Moreover, the uncertainty in the posterior estimates of the survival function is greatly reduced (tighter bands) as compared with the Kaplan–Meier estimator.

### 10.4.2 Example 2

For this second example we illustrate the use of the gamma process prior (10.2) to define a piecewise hazard function as in (10.8), without and with covariates, Eqs. (10.9) and (10.10), respectively. The data are survival times of 33 leukemia patients (Feigl and Zelen 1965). Times are measured in weeks from diagnosis. Reported covariates are white blood cell counts (WBC) and a binary variable AG that indicates a positive or negative test related to the white blood cell characteristics. Three of the observations were censored.

We first analyse the data without considering the covariates. The prior was defined by taking $a = b = 0.0001$ and $c_j \mid \zeta \overset{\text{iid}}{\sim} \text{Ga}(1, \zeta)$ for $j = 1, \ldots, J$ and $\zeta \sim \text{Ga}(0.01, 0.01)$. We took $J = 10$ intervals and chose the partition $\{\tau_j\}$ such that each interval contains approximately the same number of exact (not censored) observations. We fitted the model with the command *GaMRes* to fit the model and command *GaPloth* to produce graphs. Survival function estimates are included in Fig. 10.4. For comparison the Kaplan–Meier estimates are also included. As can be seen, the Bayesian nonparametric estimators follow the path of the Kaplan–Meier curve but in a lot smoother way. The uncertainty in the 95 % credible intervals is also reduced.

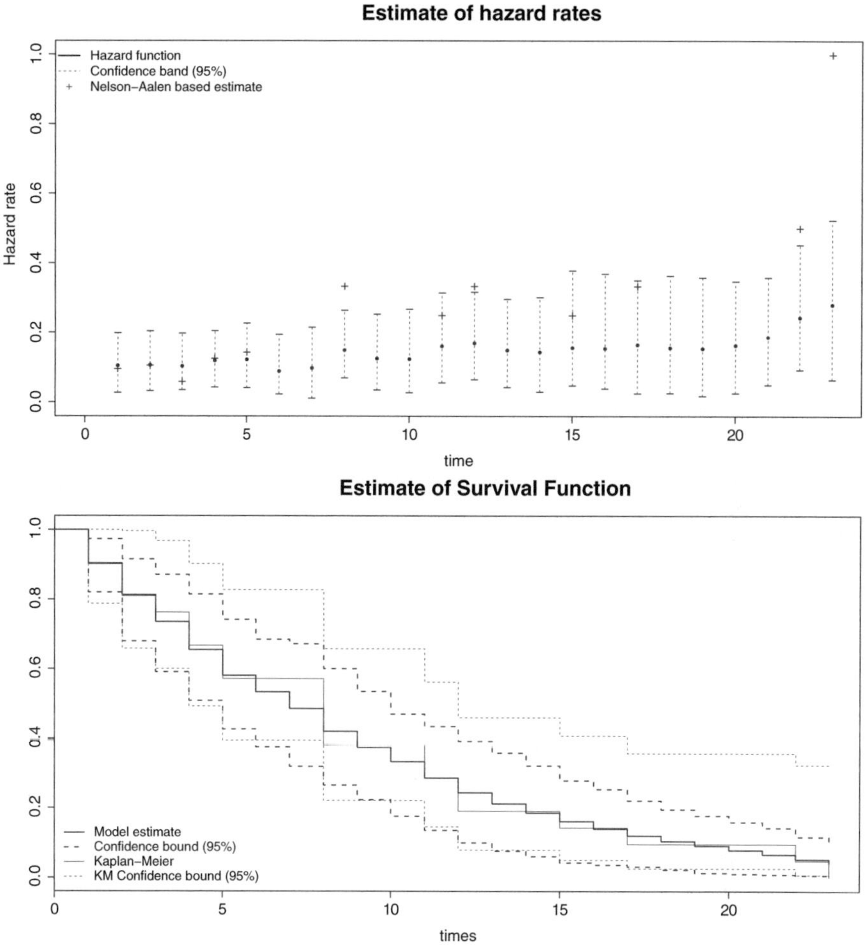


**Fig. 10.3** Hazard rate (*top*) and survival (*bottom*) estimates for the placebo group of the 6 MP dataset

We now perform a survival regression analysis using the proportional hazards model (10.10) together with the piecewise constant baseline hazard and with the gamma process prior. Both available covariates were used, with WBC on a logarithmic scale. Prior specifications are the same as in the analysis without covariates and for the covariate coefficients we took $\beta_r \sim \mathrm{N}(0,100)$, for $r = 1,2$. The model was fit with command *CGaMRes* and summaries were obtained with the command *PlotTheta*. The estimated effect of the covariates in the survival was $\hat{\beta}_1 = -1.18$ with 95 % CI $(-1.98,-0.34)$; and $\hat{\beta}_2 = 0.25$ with 95 % CI $(0.03,0.47)$. Point estimates were obtained as the posterior mean. Interpreting these values we have that an increment of one blood cell (in logarithmic scale) reduces the risk of dying in 70 %, whereas a positive AG indicator increases the risk of dying in 28 %.

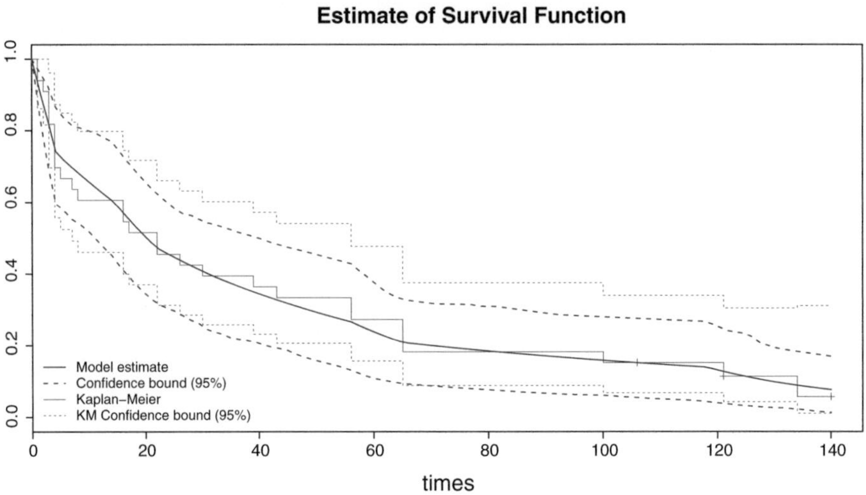


**Fig. 10.4** Survival function estimates for the leukemia dataset

### *10.4.3 Example 3*

In this example we illustrate the use of a Lévy-driven process in the survival regression model (10.10) in the presence of time dependent covariates. We consider the well-known Stanford heart transplant data. There are several versions of these data, one of them is that studied in Crowley and Hu (1977) and includes survival information of 103 patients who were accepted into the heart transplant program. Patients were accepted into the program and when a donor heart became available, medical judgement was used to select the receiving candidate. Among the 103 patients, 69 received transplants, and from them 24 were still alive at the end of the study. The data are available as the object *heart* in the R package `survival`.

The reason why this dataset has been so famous is because patients change treatment status during the course of the study, and thus defining a time dependent covariate. If we denote by $w_i$ the waiting time from acceptance to the day of transplant, for those lucky enough to have a matching donor, then $x_i(t) = I(t \geq w_i)$ is a time dependent indicator variable which takes the value of one or zero according to whether the patient has or has not received a transplant by time $t$.

To analyse these data we specify model (10.10) with a single covariate and with the baseline hazard described in terms of a Lévy-driven process of the form (10.4) with kernel

$$k(t,s) = ab(t-s)^{b-1}e^{-a(t-s)^b}I(s \leq t)$$

which corresponds to a location Weibull density. Here $b$ is a smoothing parameter and $a$ determines the rate of decay, so in particular we take $b = 2$ and a hyper prior

$a \sim \text{Ga}(1/2, 1/2)$. The Lévy intensity measure is characterized by the generalized gamma $\rho$ measure (i), with $\varepsilon = 1$ so that the measure is discrete, $\beta(s) = 1$ and $\alpha$ measure given by $\alpha(ds) = \text{Ga}(s|1, 0.001)ds$. Finally, the prior for the covariate coefficient $\beta \sim \text{N}(0, 10)$. The model was implemented in Fortran and is available upon request from the author.

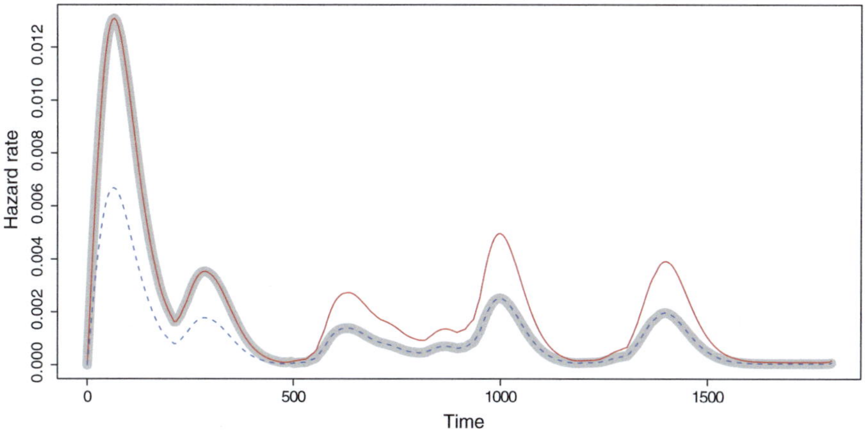


**Fig. 10.5** Hazard function estimates for stanford heart transplant data. With no transplant made (*solid thin line*), transplant made at time zero (*dashed line*), and transplant made at time 500 days (*thick grey line*)

Posterior hazard rate estimates (posterior means) are shown in Fig. 10.5. The solid thin line corresponds to the hazard rate for a patient who did not receive transplant, i.e. $w_i = \infty$. The dashed line corresponds to a patient who received a heart transplant immediately after being accepted in the program, i.e. $w_i = 0$. The effect of a heart transplant can be clearly seen by a lower hazard rate for the patient who did receive a transplant. In fact, this reduction can be quantified by the parameter $\beta$ which has a posterior mean of $-0.68$ and a 95 % credible interval $(-1.09, -0.24)$. These values imply a 50 % reduction (in average) in the risk of dying after the transplant. Moreover, the thick grey line in Fig. 10.5 corresponds to the hazard rate estimate of a patient who received a heart transplant 500 days after being accepted into the program ($w_i = 500$). In the figure, we can see that this hypothetical patient starts with the no transplant group (higher) hazard function and at time 500 it changes to the transplant group (lower) hazard function. This clearly shows the implication for a patient when changing treatment group during the course of the study.

**Acknowledgements** The author was supported by CONACYT grant 244459 and *Asociación Mexicana de Cultura, A.C.*

## References


Aalen, O. O. (1978). Nonparametric inference for a family of counting processes. *Annals of Statistics*, **6**, 701–726.

Bennet, S. (1983). Analysis of survival data by the proportional odds model. *Statistics in Medicine*, **2**, 273–277.

Berkson, J. and Gage, R. P. (1952). Survival curve for cancer patients following treatment. *Journal of the American Statistical Association*, **47**, 501–515.

Chen, M. H., Ibrahim, J. G., and Sinha, D. (1999). A new Bayesian model for survival data with a surviving fraction. *Journal of the American Statistical Association*, **94**, 909–919.

Christensen, R. and Johnson, W. (1988). Modelling accelerated failure time with a Dirichlet process. *Biometrika*, **75**, 693–704.

Clayton, D. G. (1978). A model for association in bivariate life tables and its application in epidemiological studies of familial tendency in chronic disease incidence. *Biometrika*, **65**, 141–151.

Cox, D. R. (1999). Regression models and life tables (with discussion). *Journal of the Royal Statistical Society, Series B*, **34**, 187–220.

Crowley, J. and Hu, M. (1977). Covariance analysis of heart transplant survival data. *Journal of the American Statistical Association*, **72**, 27–36.

Daley, D. J. and Vere-Jones, D. (2008). *An Introduction to the Theory of Point Processes*. Springer, New York.

Doob, J. L. (1953). *Stochastic Processes*. Wiley, New York.

Feigl, P. and Zelen, M. (1965). Estimation of exponential survival probabilities with concomitant information. *Biometrics*, **21**, 826–838.

Ferguson, T. S. (1973). A Bayesian analysis of some nonparametric problems. *Annals of Statistics*, **1**, 209–230.

Ferguson, T. S. and Klass, M. J. (1972). A representation of independent increment processes without Gaussian components. *Annals of Mathematical Statistics*, **43**, 1634–1643.

Freireich, E. J., Gehan, E., Frei, E., Schroeder, L. R., Wolman, I. J., Anbari, R., Burgert, E. O., Mills, S. D., Pinkel, D., Selawry, O. S., Moon, J. H., Gendel, B. R., Spurr, C. L., Storrs, R., Haurani, F., Hoogstraten, B., and Lee, S. (1963). The effect of 6-mercaptopurine on the duration of steroid-induced remissions in acute leukemia: A model for evaluation of other potentially useful therapy. *Blood*, **21**, 699–716.

Garca-Bueno, J. A. and Nieto-Barajas, L. E. (2014). Vignette accompanying the R package to `BGPhazard`. http://cran.itam.mx/web/packages/BGPhazard/vignettes/BGPhazard

Gikhman, I. I. and Skorokhod, A. V. (1969). *Introduction to the Theory of Random Processes*. Dover, New York.

Hanson, T. and Johnson, W. O. (2002). Modeling regression error with a mixture of Polya trees. *Journal of the American Statistical Association*, **97**, 1020–1033.

Klein, J. P. and Moeschberger, M. L. (1997). *Survival Analysis*. New York: Springer-Verlag.

Kuk, A. Y. and Chen, C. H. (1992). A mixture model combining logistic regression with proportional hazards regression. *Biometrika*, **79**, 531–541.
Marshall, A. W. and Olkin, I. (1988). Families of multivariate distributions. *Journal of the American Statistical Association*, **83**, 834–841.
Nelsen, R. B. (1999). *An Introduction to Copulas*. Springer, New York.
Nieto-Barajas, L. E. (2003). Discrete time Markov gamma processes and time dependent covariates in survival analysis. *Bulletin of the International Statistical Institute*, **54**.
Nieto-Barajas, L. E. (2014a). Bayesian semiparametric analysis of short- and long-term hazard ratios with covariates. *Computational Statistics and Data Analysis*, **71**, 477–490.
Nieto-Barajas, L. E. (2014b). Lévy-driven processes in Bayesian nonparametric inference. *Bulletin of the Mexican Mathematical Association*, **19**, 267–280.
Nieto-Barajas, L. E. and Walker, S. G. (2002). Markov beta and gamma processes for modeling hazard rates. *Scandinavian Journal of Statistics*, **29**, 413–424.
Nieto-Barajas, L. E. and Walker, S. G. (2004). Bayesian nonparametric survival analysis via Lévy driven Markov processes. *Statistica Sinica*, **14**, 1127–1146.
Nieto-Barajas, L. E. and Walker, S. G. (2005). A semiparametric Bayesian analysis of survival data based on Lévy-driven processes. *Lifetime data analysis*, **11**, 529–543.
Nieto-Barajas, L. E. and Walker, S. G. (2007a). A Bayesian semi-parametric bivariate failure time model. *Computational Statistics and Data Analysis*, **51**, 6102–6113.
Nieto-Barajas, L. E. and Walker, S. G. (2007b). Gibbs and autoregressive Markov processes. *Statistics and Probability Letters*, **77**, 1479–1485.
Nieto-Barajas, L. E. and Yin, G. (2008). Bayesian semiparametric cure rate model with an unknown threshold. *Scandinavian Journal of Statistics*, **35**, 540–556.
Smith, A. F. and Roberts, G. O. (1993). Bayesian computations via the gibbs sampler and related Markov chain monte carlo methods. *Journal of the Royal Statistical Society, Series B*, **55**, 3–23.
Team, R. C. (2014). *R: A Language and Environment for Statistical Computing*. R Foundation for Statistical Computing.
Yakovlev, A. Y. and Tsodikov, A. D. (1996). *Stochastic Models of Tumor Latency and Their Biostatistical Applications*. World Scientific, New Jersey.
Yang, S. and Prentice, R. (2005). Semiparametric analysis of short-term and long-term hazard ratios with two sample survival data. *Biometrika*, **92**, 1–17.
Yin, G. and Nieto-Barajas, L. E. (2009). Bayesian cure rate model accommodating multiplicative and additive covariates. *Statistics and Its Interface*, **2**, 513–521.

# Chapter 11
# Bayesian Spatial Survival Models

Haiming Zhou and Timothy Hanson

**Abstract** Survival analysis has received a great deal of attention as a subfield of Bayesian nonparametrics over the last 50 years. In particular, the fitting of survival models that allow for sophisticated correlation structures has become common due to computational advances in the 1990s, in particular Markov chain Monte Carlo techniques. Very large, complex spatial datasets can now be analyzed accurately including the quantification of spatiotemporal trends and risk factors. This chapter reviews four nonparametric priors on baseline survival distributions in common use, followed by a catalogue of semiparametric and nonparametric models for survival data. Generalizations of these models allowing for spatial dependence are then discussed and broadly illustrated. Throughout, practical implementation through existing software is emphasized.

## 11.1 Introduction

This chapter reviews several semiparametric Bayesian survival models, and summarizes some recent proposals to allow for spatial and covariate-adjusted dependence among the survival times. Two generalizations of the accelerated failure time model that allow crossing cumulative hazards for different covariate combinations, and hence crossing survival curves, are also discussed.

Four prior specifications in broad use are first reviewed in Sect. 11.2. A catalogue of Bayesian survival models is presented in Sect. 11.3. Section 11.4 discusses the

H. Zhou
Division of Statistics, Northern Illinois University, DeKalb, IL 60115, USA
e-mail: zhouh@stat.sc.edu

T. Hanson (✉)
Department of Statistics, University of South Carolina, Columbia, SC 29208, USA
e-mail: hansont@stat.sc.edu

 
R. Mitra, P. Müller (eds.), *Nonparametric Bayesian Inference in Biostatistics*,
Frontiers in Probability and the Statistical Sciences,
DOI 10.1007/978-3-319-19518-6_11

incorporation of dependence among survival times across the models in Sect. 11.3, focusing mostly on spatial dependence followed by several real-data illustrations in Sect. 11.5. The chapter concludes with a short discussion in Sect. 11.6. Please note at the outset that, although a review is attempted, the cited papers and approaches are biased toward what the authors are aware of and have found useful.

## 11.2 A Selection of Nonparametric Priors

A common starting point in the specification of a regression model for time-to-event data is the definition of a baseline survival function, $S_0$, that is modified (either directly or indirectly) by subject-specific covariates $\mathbf{x}$. Let $T_0$ be a random survival time from the baseline group (with all covariates equal to zero). The baseline survival function is defined by $S_0(t) = P(T_0 > t) = \exp\{-H_0(t)\}$ where $H_0(t)$ is the baseline cumulative hazard. For continuous outcomes, the baseline density and hazard functions are $f_0(t) = -\frac{d}{dt}S_0(t)$ and $h_0(t) = f_0(t)/S_0(t) = \frac{d}{dt}H_0(t)$, respectively. The cumulative distribution, survival, density, and hazard functions for a member of the population with covariates $\mathbf{x}$ will be denoted by $F_{\mathbf{x}}(t)$, $S_{\mathbf{x}}(t)$, $f_{\mathbf{x}}(t)$, and $h_{\mathbf{x}}(t)$, respectively.

A wide variety of priors have been used in Bayesian survival analysis over the last 40 years. We focus on four of these: the gamma process, B-splines, Dirichlet process mixtures, and mixtures of Polya trees. Additional reviews can be found in Sinha and Dey (1997), Ibrahim et al. (2001), Müller and Quintana (2004), Hanson et al. (2005), Nieto-Barajas (2013), and Müller et al. (2015).

### *11.2.1 Gamma Process*

Kalbfleisch (1978) proposed the gamma process (GP) to model the cumulative hazard function $H_0$ in the context of the proportional hazards (PH) model (Cox 1972). Let $H_{\boldsymbol{\theta}}(t)$ be an increasing, left-continuous function on $[0,\infty)$ indexed by $\boldsymbol{\theta}$, where $H_{\boldsymbol{\theta}}(0) = 0$; typically, $H_{\boldsymbol{\theta}}$ is parametric. Let $H_0(\cdot)$ be a stochastic process such that (i) $H_0(0) = 0$, (ii) $H_0(\cdot)$ has independent increments in disjoint intervals, and (iii) $H_0(t_2) - H_0(t_1) \sim \Gamma\{\alpha(H_{\boldsymbol{\theta}}(t_2) - H_{\boldsymbol{\theta}}(t_1)), \alpha\}$ for $t_2 > t_1$, where $\Gamma(\alpha,\beta)$ implies mean $\alpha/\beta$. Then $\{H_0(t) : t \geq 0\}$ is said to be a GP with parameter $(\alpha, H_{\boldsymbol{\theta}})$ and denoted $H_0 \sim GP(\alpha, H_{\boldsymbol{\theta}})$.

Note that $\mathrm{E}\{H_0(t)\} = H_{\boldsymbol{\theta}}(t)$ so that $H_0$ is centered at $H_{\boldsymbol{\theta}}$. Also, $\mathrm{Var}\{H_0(t)\} = H_{\boldsymbol{\theta}}(t)/\alpha$ so that, similar to the Dirichlet process and Polya trees described below, the precision parameter $\alpha$ controls how "close" $H_0$ is to $H_{\boldsymbol{\theta}}$ and provides a prior measure of how certain one is that $H_0$ is near $H_{\boldsymbol{\theta}}$. Ferguson (1973) recast the Dirichlet process (DP) as a scaled GP.

The posterior of the GP is characterized by Kalbfleisch (1978); his results for the PH model simplify when no covariates are specified. With probability one, the

GP is a monotone nondecreasing step function, implying that the corresponding survival function $S_0$ is a nonincreasing step function. Similar to the DP, matters are complicated by the presence of ties in the data with positive probability. When present in the observed data, such ties make the resulting computations awkward. Clayton (1991) described a Gibbs sampler for obtaining inferences in the PH model with a GP baseline.

Burridge (1981) and Ibrahim et al. (2001) suggest that the model as proposed by Kalbfleisch (1978) and extended by Clayton (1991) is best suited to grouped survival data. Walker and Mallick (1997) considered an approximation to the GP for continuous data. Define a partition of $(0,\infty)$ by $\{(a_{j-1},a_j]\}_{j=1}^{J} \cup (a_J,\infty)$ where $0 = a_0 < a_1 < a_2 < \cdots < a_{J+1} = \infty$. Here, $a_J$ is taken to be equal to be largest event time recorded. If $H_0 \sim GP(\alpha, H_{\boldsymbol{\theta}})$, then by definition $h_{0j} = H_0(a_j) - H_0(a_{j-1}) \stackrel{ind.}{\sim} \Gamma\{\alpha(H_{\boldsymbol{\theta}}(a_j) - H_{\boldsymbol{\theta}}(a_{j-1})), \alpha\}$. Walker and Mallick (1997) make this assumption for the given partition and further assume that $h_0(t)$ is constant and equal to $h_{0j}$ for $t \in (a_{j-1}, a_j]$, $j = 1, \ldots, J$, yielding a particular piecewise exponential model. So the piecewise exponential model, which has a long and fruitful history in both Bayesian and frequentist survival analysis, can be viewed as an approximation to the GP when gamma increments are used.

### *11.2.2 B-Splines and Bernstein Polynomials*

A flexible and popular basis expansion approach to modeling functions over a finite interval $[a,b]$ is based on B-splines (de Boor 2001). A B-spline is a piecewise-differentiable polynomial of a given degree $d$; $d = 2$ and $d = 3$ give quadratic and cubic B-splines, respectively. The B-spline is defined over the union of intervals with endpoints termed knots. The overall polynomial is continuous ($d \geq 1$) or differentiable ($d \geq 2$) over the range of the knots. Knots can be equispaced yielding a cardinal B-spline or else irregularly spaced. Computation is especially easy for equispaced knots and so we focus on that here; generalizations can be found in Kneib (2006). The B-spline includes polynomials of the same or lower degree as special cases; e.g. a quadratic B-spline includes all constant, linear, and parabolic functions over $[a,b]$.

For degree $d = 2$, the quadratic B-spline "mother" basis function is defined on $[0,3]$

$$\varphi(x) = \begin{Bmatrix} 0.5x^2 & 0 \leq x \leq 1 \\ 0.75 - (x-1.5)^2 & 1 \leq x \leq 2 \\ 0.5(3-x)^2 & 2 \leq x \leq 3 \\ 0 & \text{otherwise} \end{Bmatrix}.$$

Say the number of basis functions is $J$. The B-spline basis functions are shifted, rescaled versions of $\varphi$. Let $x_1, \ldots, x_n$ be event times of interest and $x_{(1)}, \ldots, x_{(n)}$ their order statistics. The $j$-th basis function is $B_j(x) = \varphi\left(\frac{x - x_{(1)}}{\Delta} + 3 - j\right)$, where $\Delta =$

$\frac{x_{(n)}-x_{(1)}}{J-2}$. A B-spline is typically used with a rather large number of basis functions $J$, e.g. 20–40. The B-spline model for an unknown function is

$$g(x) = \sum_{j=1}^{J} \theta_j B_j(x). \tag{11.1}$$

A global level of smoothness can be incorporated into a B-spline model by encouraging neighboring coefficients to be similar; the more regular the coefficients are, the less wiggly $g$ is. The hazard can be modeled directly as $h_0(t) = g(t)$ with the constraint $\theta_j \geq 0$ (Wang and Dunson 2011; Pan et al. 2014; Lin et al. 2015; Li et al. 2015b); typically, $\theta_1, \ldots, \theta_J$ have exponential or gamma priors. Komárek and Lesaffre (2008) consider a limiting case of the B-spline order as a model for densities and model the $\theta_j \geq 0$ via a generalized logit transformation so that $\sum_{j=1}^{J} \theta_j = 1$.

Alternatively, to avoid the positivity constraints on $\theta_i$, one can model $h_0(t) = \exp\{g(t)\}$ (Hennerfeind et al. 2006; Kneib and Fahrmeir 2007) with $\theta_j \in \mathbb{R}$. Classical spline estimation on $\{(x_i, y_i)\}_{i=1}^{n}$ proceeds by minimizing $\sum_{i=1}^{n}(y_i - g(x_i))^2$ subject to the "wiggliness" penalty $\int_a^b |g''(x)|^2 dx \leq c$ for some $c > 0$. This is equivalent to maximizing a penalized log-likelihood. Borrowing from Eilers and Marx (1996), Lang and Brezger (2004) recast and developed this idea into a Bayesian framework. Let $\mathbf{D}_2 \in \mathbb{R}^{(J-2)\times J}$ and $\mathbf{D}_1 \in \mathbb{R}^{(J-1)\times J}$ be defined as

$$\mathbf{D}_2 = \begin{bmatrix} 1 & -2 & 1 & 0 & \cdots & 0 \\ 0 & 1 & -2 & 1 & \cdots & 0 \\ \vdots & \vdots & \ddots & \ddots & \ddots & \vdots \\ 0 & 0 & \cdots & 1 & -2 & 1 \end{bmatrix} \text{ and } \mathbf{D}_1 = \begin{bmatrix} 1 & -1 & 0 & \cdots & 0 \\ 0 & 1 & -1 & \cdots & 0 \\ \vdots & \vdots & \ddots & \ddots & \vdots \\ 0 & 0 & \cdots & 1 & -1 \end{bmatrix}.$$

For equispaced, quadratic (and cubic) B-splines the penalty can be written as $\int_a^b |g''(x)|^2 dx = ||\mathbf{D}_2 \boldsymbol{\theta} \Delta||^2$, where $\boldsymbol{\theta} = (\theta_1, \ldots, \theta_J)$.

Optimization with the $\mathbf{D}_2$ penalty is equivalent to assuming a second order random-walk prior, that is, the improper prior $\mathbf{D}_2\boldsymbol{\theta} \sim N_{J-2}(\mathbf{0}, \lambda^{-1}\mathbf{I}_{J-2})$. As $\lambda$ becomes large, $g''(x)$ is forced toward zero and $g(x)$ becomes linear. Alternatively, a first order random walk prior is given by $\mathbf{D}_1\boldsymbol{\theta} \sim N_{J-1}(\mathbf{0}, \lambda^{-1}\mathbf{I}_{J-1})$. When $\lambda$ is large, adjacent basis functions are forced closer and $g'(x)$ is forced toward zero, yielding a constant $g(x)$.

The Bernstein polynomial is a special case of the B-spline with support $[0,1]$ (Petrone 1999a,b). A Bernstein polynomial prior for a function $g$ on $[0,1]$ is a discrete mixture of beta distributions with equispaced means and integer parameters; i.e., the functions $B_j(x)$ in (11.1) are

$$B_j(x) = \frac{\Gamma(J+1)}{\Gamma(j)\Gamma(J-j+1)} x^{j-1}(1-x)^{J-j}.$$

The resulting $g$ is then transformed to $[0,b)$ ($b = \infty$ for some transformations) for use in baseline survival modeling (Gelfand and Mallick 1995; Carlin and Hodges 1999; Banerjee and Dey 2005; Chang et al. 2005; Chen et al. 2014).

B-splines are now a standard tool for modeling hazard functions. Like the GP, the piecewise constant hazard is a special case, i.e. a first order B-spline with $d = 0$; piecewise exponential models have been used extensively in Bayesian survival analysis, e.g. Ibrahim et al. (2001). Existing approaches to modeling hazard functions using B-splines (Gray 1992; Hennerfeind et al. 2006; Sharef et al. 2010) choose either equispaced knots over the spread of the observed data or knots at the empirical quantiles of the observed event times. Chen et al. (2014) and Li et al. (2015b) instead choose knot locations based on an approximation of underlying parametric family, e.g. $S_{\boldsymbol{\theta}}$ indexed by $\boldsymbol{\theta}$.

### 11.2.3 Dirichlet Process Mixture Model

A random probability measure $G$ follows a DP (Ferguson 1973) with parameters $(\alpha, G_0)$, where $\alpha > 0$ and $G_0$ is an appropriate probability measure defined on $\mathbb{R}^d$, written as

$$G \mid \alpha, G_0 \sim DP(\alpha G_0), \tag{11.2}$$

if for any measurable nontrivial partition $\{B_l : 1 \le l \le k\}$ of $\mathbb{R}^d$, then the vector $(G(B_1), \ldots, G(B_k))'$ has a Dirichlet distribution with parameters $(\alpha G_0(B_1), \ldots, \alpha G_0(B_k))$. It follows that

$$G(B_l) \mid \alpha, G_0 \sim \text{Beta}(\alpha G_0(B_l), \alpha G_0(B_l^c)),$$

and therefore $E\{G(B_l) \mid \alpha, G_0\} = G_0(B_l)$ and

$$Var\{G(B_l) \mid \alpha, G_0\} = \frac{G_0(B_l)G_0(B_l^c)}{\alpha + 1}.$$

Thus $G$ is centered at $G_0$ with precision $\alpha$. The DP was used by Susarla and Van Ryzin (1976) to model and estimate the survival function for right-censored data; Müller et al. (2015) provide R code to implement this approach.

If $G|\alpha, G_0 \sim DP(\alpha G_0)$, then the process can be represented by the stick-breaking representation (Sethuraman 1994),

$$G(\cdot) = \sum_{i=1}^{\infty} w_i \delta_{\boldsymbol{\theta}}(\cdot), \tag{11.3}$$

where $\delta_{\boldsymbol{\theta}}(\cdot)$ is Dirac measure at $\boldsymbol{\theta}$, $w_i = V_i \prod_{j<i}(1 - V_j)$, with $V_i|\alpha \overset{iid}{\sim} \text{Beta}(1, \alpha)$, and $\boldsymbol{\theta}_i|G_0 \overset{iid}{\sim} G_0$. Note that $E(w_j) > E(w_{j+1})$ for all $j$, so the weights are stochastically ordered.

Convolving a DP with a parametric kernel, such as the normal, gives a DP mixture (DPM) model (Lo 1984; Escobar and West 1995). A simple DPM of Gaussian densities for continuous data $\varepsilon_1, \ldots, \varepsilon_n$ is given by

$$\varepsilon_i|G \stackrel{iid}{\sim} \int N(\mu,\sigma^2)dG(\mu,\sigma^2), \tag{11.4}$$

where $N(\mu,\sigma^2)$ denotes the normal density with mean $\mu$ and $\sigma^2$, and the mixing distribution, $G$, is a random probability measure defined on $\mathbb{R}\times\mathbb{R}^+$, following a DP. The stick-breaking representation recasts (11.4) as a countably infinite mixture of normals given by

$$\varepsilon_i|G \stackrel{iid}{\sim} \sum_{j=1}^{\infty}\left[V_j\prod_{k=1}^{j-1}(1-V_k)\right]N(\mu_j,\sigma_j^2). \tag{11.5}$$

The prior distribution on $\varepsilon_i$ is centered at the normal distribution; Griffin (2010) discusses prior specifications that control the "non-normalness" of this distribution.

### *11.2.4 Polya Tree*

A Polya tree (PT) successively partitions the reals $\mathbb{R}$ (or any other domain) into finer and finer partitions; each refinement of a partition doubles the number of partition sets by cutting the previous level's sets into two pieces; there are two sets at level 1, four sets at level 2, eight sets at 3, and so on. We focus on a PT centered at the standard normal density, that is, $N(0,1)$ is the *centering distribution* for the Polya tree. At level $j$, the Polya tree partitions the real line into $2^j$ intervals $B_{j,k}=(\Phi^{-1}((k-1)2^{-j}),\Phi^{-1}(k2^{-j}))$ of probability $2^{-j}$ under $\Phi$, $k=1,\ldots,2^j$, where $\Phi(\cdot)$ is the cumulative distribution function of $N(0,1)$. Note that $B_{j,k}=B_{j+1,2k-1}\cap B_{j+1,2k}$. Given an observation $\varepsilon$ is in set $k$ at level $j$, i.e. $\varepsilon\in B_{j,k}$, it could then be in either of the two offspring sets $B_{j+1,2k-1}$ or $B_{j+1,2k}$ at level $j+1$. The conditional probabilities associated with these sets will be denoted by $Y_{j+1,2k-1}$ and $Y_{j+1,2k}$. Clearly they must sum to one, and so a common prior for either of these probabilities is a beta distribution (Ferguson 1974; Lavine 1992, 1994; Walker and Mallick 1997, 1999; Hanson and Johnson 2002; Hanson 2006a; Zhao et al. 2009), given by

$$Y_{j,2k-1}|c \stackrel{ind.}{\sim} \text{Beta}(cj^2,cj^2),\ \ j=1,\ldots,J;\ k=1,\ldots,2^{j-1},$$

where $c>0$, which ensures that every realization of the process has a density, allowing the modeling of continuous data without the need of convolutions with continuous kernels.

The user-specified weight $c>0$ controls how closely the posterior follows $N(0,1)$ in terms of $L_1$ distance (Hanson et al. 2008), with larger values forcing the PT process $G$ closer to $N(0,1)$; often a prior is placed on $c$, e.g. $c\sim\Gamma(a,b)$. The PT is stopped at level $J$ (typically $J=5,6,7$); within the sets $\{B_{J,k}:k=1,\ldots,2^J\}$ at the level $J$, $G$ follows $N(0,1)$ (Hanson 2006a). The resulting model for data $\varepsilon_1,\ldots,\varepsilon_n$ is given by

$$\varepsilon_i|G \stackrel{iid}{\sim} G, \tag{11.6}$$

where

$$G \sim PT_J(c, N(0,1)). \tag{11.7}$$

The corresponding density is given by

$$p(\varepsilon | \{Y_{j,k}\}) = 2^J \phi(\varepsilon) \prod_{j=1}^{J} Y_{j,\lceil 2^j \phi(\varepsilon) \rceil}, \tag{11.8}$$

where $\lceil \cdot \rceil$ is the ceiling function, and so a likelihood can be formed. For the simple model, the PT is conjugate. Let $\boldsymbol{\varepsilon} = (\varepsilon_1, \ldots, \varepsilon_n)$. Then

$$Y_{j,2k-1} | \boldsymbol{\varepsilon} \overset{ind.}{\sim} \text{Beta}\left( cj^2 + \sum_{i=1}^{n} I\{\lceil 2^j \phi(\varepsilon_i) \rceil = 2k-1\}, cj^2 + \sum_{i=1}^{n} I\{\lceil 2^j \phi(\varepsilon_i) \rceil = 2k\} \right),$$

and $Y_{j,2k} = 1 - Y_{j,2k-1}$.

Location $\mu$ and spread $\sigma$ parameters are melded with expression (11.6) and the PT prior (11.7) to make a median-$\mu$ location-scale family for data $y_1, \ldots, y_n$, given by

$$y_i = \mu + \sigma \varepsilon_i,$$

where the $\varepsilon_i \mid G \overset{iid}{\sim} G$ and $G$ follows a PT prior as in expression (11.7), with the restriction $Y_{1,1} = Y_{1,2} = 0.5$. Allowing $\mu$ and $\sigma$ to be random induces a mixture of Polya trees (MPT) model for $y_1, \ldots, y_n$, smoothing out predictive inference (Lavine 1992; Hanson and Johnson 2002). Note that Jeffreys' prior under the normal model is a reasonable choice here (Berger and Guglielmi 2001), and leads to a proper posterior (Hanson 2006a).

## 11.3 Survival Models

### *11.3.1 Proportional Hazards*

A proportional hazards (PH) model (Cox 1972), for continuous data, is obtained by expressing the covariate-dependent survival function $S_{\mathbf{x}}(t)$ as

$$S_{\mathbf{x}}(t) = S_0(t)^{\exp(\mathbf{x}'\boldsymbol{\beta})}. \tag{11.9}$$

In terms of hazards, this model is

$$h_{\mathbf{x}}(t) = \exp(\mathbf{x}'\boldsymbol{\beta}) h_0(t).$$

Note then that for two individuals with covariates $\mathbf{x}_1$ and $\mathbf{x}_2$, the ratio of hazard curves is constant and proportional to $\frac{h_{\mathbf{x}_1}(t)}{h_{\mathbf{x}_2}(t)} = \exp\{(\mathbf{x}_1 - \mathbf{x}_2)'\boldsymbol{\beta}\}$, hence the name "proportional hazards." Cox (1972) is the second most cited statistical paper of all time (Ryan and Woodall 2005), and the PH model is easily the most popular semiparametric survival model in statistics, to the point where medical researchers tend to compare different populations' survival in terms of instantaneous risk (hazard) rather than mean or median survival as in common regression models. Part of the popularity of the model has to do with the incredible momentum the model has gained from how easy it is to fit the model through partial likelihood (Cox 1975) and its implementation in SAS in the procedure PHREG. The use of partial likelihood and subsequent counting process formulation (Andersen and Gill 1982) of the model has allowed ready extension to stratified analyses, proportional intensity models, frailty models, and so on (Therneau and Grambsch 2000).

The first Bayesian semiparametric approach to PH models posits a gamma process as a prior on the baseline cumulative hazard $H_0(t) = \int_0^t h_0(s)ds$ (Kalbfleisch 1978); partial likelihood emerges as a limiting case (of the marginal likelihood as the precision parameter approaches zero). The use of the gamma process prior in PH models, as well as the beta process prior (Hjort 1990), piecewise exponential priors, and correlated increments priors are covered in Ibrahim et al. (2001) (pp. 47–94) and Sinha and Dey (1997). Other approaches include what are essentially Bernstein polynomials (Gelfand and Mallick 1995; Carlin and Hodges 1999) and penalized B-splines (Hennerfeind et al. 2006; Kneib and Fahrmeir 2007). The last two models are available in the free software BayesX (Belitz et al. 2015) which can be called from R via the packages R2BayesX and BayesX (Umlauf et al. 2015). The BayesX functions allow for a general additive (including partially linear) PH model to be easily fit, including time-dependent covariates; BayesX also accommodates spatial frailties, discussed in Sect. 11.4.1. PH models with Polya tree baselines were considered by Hanson (2006a), Hanson and Yang (2007), Zhao et al. (2009), and Hanson et al. (2009) and can be fit in the SpBayesSurv package for R.

Stratified PH model posits a separate hazard function across levels of strata $s = 1,\ldots,S$,

$$h_{\mathbf{x},s}(t) = \exp(\mathbf{x}'\boldsymbol{\beta})h_{0s}(t).$$

A version of this model based on Bernstein polynomials is given by Carlin and Hodges (1999); B-splines were considered by Cai and Meyer (2011). The stratified PH model can also be fit using SAS PHREG assuming piecewise exponential priors, i.e. piecewise constant baseline hazard functions. A version of the stratified model that SAS fits, but with a "Polya tree" type prior on the hazard was considered by Dukić and Dignam (2007). Note that BayesX can also fit stratified models based on B-splines by including a time-varying regression effect for the categorical strata variable.

### 11.3.2 Accelerated Failure Time

An accelerated failure time (AFT) model is obtained by expressing the covariate-dependent survival function $S_{\mathbf{x}}(t)$ as

$$S_{\mathbf{x}}(t) = S_0\{\exp(-\mathbf{x}'\boldsymbol{\beta})t\}. \tag{11.10}$$

This is equivalent to the linear model for the log transformation of the corresponding time-to-event response variable, $T$,

$$\log T = \mathbf{x}'\boldsymbol{\beta} + \varepsilon, \tag{11.11}$$

where $\exp(\varepsilon) \sim S_0$. The mean, median, and any quantile of survival for an individual with covariates $\mathbf{x}_1$ is changed by a factor of $\exp\{(\mathbf{x}_1 - \mathbf{x}_2)'\boldsymbol{\beta}\}$ relative to those with covariates $\mathbf{x}_2$.

An early frequentist least-squares treatment of the AFT model with right-censored data is due to Buckley and James (1979); the Buckley-James estimator is implemented in Frank Harrell's `Design` library for R (Alzola and Harrell 2006). The R packages `emplik` and `bujar` have various extensions. More refined estimators followed in the 1990s (Ying et al. 1995; Yang 1999) focusing on median-regression.

From a Bayesian nonparametric perspective, the first approach, based on a Dirichlet process prior, obtained approximate marginal inferences to the AFT model (Christensen and Johnson 1988); a full Bayesian treatment using the Dirichlet process is not practically possible (Johnson and Christensen 1989). Approaches based on Dirichlet process mixture models have been considered by Kuo and Mallick (1997), Kottas and Gelfand (2001) and Hanson (2006b). Dirichlet process mixtures "fix" the discrete nature of the Dirichlet process, as do other discrete mixtures of continuous kernels. We refer the reader to Komárek and Lesaffre (2007) for an alternative approach based on finite mixtures of normal distributions, and Komárek and Lesaffre (2008) based on an approximating B-spline, both available in the R package `bayesSurv`. Polya tree priors that have continuous densities can directly model the distribution of $\varepsilon$ in expression (11.11) (Walker and Mallick 1999; Hanson and Johnson 2002; Hanson 2006a; Hanson and Yang 2007; Zhao et al. 2009). AFT models with Polya tree baseline densities can be fit in the `spBayesSurv` package for R.

Although PH is by far the most commonly used semiparametric survival model, several studies have shown vastly superior fit and interpretation from AFT models (Hanson and Yang 2007; Hanson 2006a; Kay and Kinnersley 2002; Orbe et al. 2002; Hutton and Monaghan 2002). Cox pointed out himself (Reid 1994) "*...the physical or substantive basis for ... proportional hazards models ... is one of its weaknesses ... accelerated failure time models are in many ways more appealing because of their quite direct physical interpretation ...* ". However, similar to the PH model, standard AFT models also impose constraints so that survival curves from different covariate levels are not allowed to cross, which is unrealistic in many practical applications (e.g., De Iorio et al. 2009). For these data that do not follow AFT assumptions, we next discuss two generalizations of the AFT model that allow for crossing survival and hazard curves. The two approaches are the *linear dependent Dirichlet process mixture*, which can be interpreted as a mixture of parametric AFT models, and the *linear dependent tailfree process*, which is an AFT model with very general baseline functions that are covariate-dependent. Both augmentations are examples of "density regressions," allowing the entire survival density $f_{\mathbf{x}}(t)$ to change smoothly with covariates $\mathbf{x}$.

#### 11.3.2.1 Linear Dependent Dirichlet Process

By considering a Dirichlet process mixture of normal distributions for the errors in (11.11) (Kuo and Mallick 1997), the distribution for the log survival time is the distribution of $\varepsilon_i$, given by (11.5), shifted by the linear predictor $\eta_i = \mathbf{x}_i'\boldsymbol{\beta}$. Specifically,

$$y_i|\boldsymbol{\beta}, G \overset{ind.}{\sim} \sum_{j=1}^{\infty} w_j N(\mu_j + \mathbf{x}_i'\boldsymbol{\beta}, \sigma_j^2),$$

where $G(\cdot) = \sum_{j=1}^{\infty} w_j \delta_{(\mu_j,\sigma_j^2)}(\cdot)$ is a Dirichlet process. The interpretation of the components of $\boldsymbol{\beta}$ is as usual and the model can be fit using standard algorithms for Dirichlet process mixture models (Neal 2000).

The linear dependent Dirichlet process mixture (LDDPM) (De Iorio et al. 2009; Jara et al. 2010; Jara et al. 2011; Zhou et al. 2015b) can be interpreted as a generalization of the previous model, which arises by additionally mixing over the regression coefficients, yielding a mixture of log-normal AFT models. Specifically, the LDDPM model is given by

$$y_i|G \overset{ind.}{\sim} \sum_{j=1}^{\infty} w_j N(\mathbf{x}_i'\boldsymbol{\beta}_j, \sigma_j^2), \tag{11.12}$$

where $\mathbf{x}_i$ now includes a '1' for the intercept, $w_i = V_i \prod_{j<i}(1 - V_j)$, with $V_i|\alpha \overset{iid}{\sim}$ Beta$(1,\alpha)$, and $\boldsymbol{\beta}_j \overset{iid}{\sim} N(\boldsymbol{m}_0, \mathbf{V}_0)$ and $\sigma_j^{-2} \overset{iid}{\sim} \Gamma(a_0, b_0)$.

The model trades easy interpretability offered by a single $\boldsymbol{\beta}$ for greatly increased flexibility. In particular, the LDDPM model does not stochastically order survival curves from different predictors $\mathbf{x}_{i_1}$ and $\mathbf{x}_{i_2}$, and both the survival and hazard curves can cross.

#### 11.3.2.2 Linear Dependent Tailfree Process

A Polya trees defines the conditional probabilities $Y_{j+1,2k-1}$ and $Y_{j+1,2k}$ as beta distributions. However, one can instead define a logistic regression for each of these probabilities, allowing the *entire* shape of the density to change with covariates; this is the approach considered by Jara and Hanson (2011). Given covariates $\mathbf{x}$, the linear dependent tailfree process (LDTFP) models $(Y_{j+1,2k-1}, Y_{j+1,2k})$ through logistic regressions

$$\log\{Y_{j+1,2k-1}(\mathbf{x})/Y_{j+1,2k}(\mathbf{x})\} = \mathbf{x}'\boldsymbol{\tau}_{j,k},$$

where $\mathbf{x}$ includes an intercept. There are $2^J - 1$ regression coefficient vectors $\boldsymbol{\tau} = \{\boldsymbol{\tau}_{j,k}\}$; e.g. for $J = 3$, $\{\boldsymbol{\tau}_{0,1}, \boldsymbol{\tau}_{1,1}, \boldsymbol{\tau}_{1,2}, \boldsymbol{\tau}_{2,1}, \boldsymbol{\tau}_{2,2}, \boldsymbol{\tau}_{2,3}, \boldsymbol{\tau}_{2,4}\}$. Let $\mathbf{X} = [\mathbf{x}_1 \cdots \mathbf{x}_n]'$ be the $n \times p$ design matrix. Following Jara and Hanson (2011), each is assigned an independent normal prior, $\tau_{j,k} \sim N_p\left(\mathbf{0}, \frac{2}{c(j+1)^2}\boldsymbol{\Psi}\right)$. Jara and Hanson (2011) discussed the case $\boldsymbol{\Psi} = n(\mathbf{X}'\mathbf{X})^{-1}$, generating a $g$-prior Zellner (1983) for the tailfree regression coefficients. By setting $\boldsymbol{\tau}_{0,1} \equiv \mathbf{0}$, the resulting LDTFP is almost surely a median-zero probability measure for every $\mathbf{x} \in \mathscr{X}$, important to avoid identifiability issues.

Augmenting (11.8), the random density is given by

$$g_{\mathbf{x}}(\varepsilon) = \phi(\varepsilon)2^J \prod_{j=1}^{J} Y_{j,\lceil 2^j \Phi(\varepsilon)\rceil}(\mathbf{x}).$$

Since the $\{Y_{j,k}\}$ are modeled with logistic-normal distribution instead of beta, the resulting random density is a tailfree process. The final AFT model with LDTFP baseline is given by

$$y_i = \mathbf{x}_i'\boldsymbol{\beta} + \sigma\varepsilon_i, \quad \varepsilon_i|\boldsymbol{\tau} \overset{ind.}{\sim} g_{\mathbf{x}_i}. \tag{11.13}$$

Unlike the LDDPM, the LDTFP separates survival into one distinct trend $\mathbf{x}'\boldsymbol{\beta}$ and an evolving log-baseline survival density $g_{\mathbf{x}}$. By forcing $g_{\mathbf{x}}$ to be median-zero, $e^{\beta_j}$ gives a factor by how median survival changes when $x_j$ is increased *just as in standard AFT models*. This heightened interpretability in terms of median-regression in the presence of heteroscedastic error allows a fit of the LDTFP model to easily relate covariates $\mathbf{x}$ to median survival.

The LDTFP models the probability of falling above or below quantiles of the $N(\mathbf{x}'\boldsymbol{\beta}, \sigma^2)$ distribution, but in terms of conditional probabilities. This model can be viewed as a particular kind of quantile regression model. Koenker and Hallock (2001) suggest that "*...instead of estimating linear conditional quantile models, we could instead estimate a family of binary response models for the probability that the response variable exceeded some prespecified cutoff values.*" However, Koenker and Hallock (2001) prefer the linear (in covariates) quantile specification because "*...it nests within it the iid error location shift model of classical linear regression.*" By augmenting a median-zero tailfree process with a general trend $\mathbf{x}'\boldsymbol{\beta}$ we accomplish the same objective, nesting the ubiquitous normal-errors linear model within a highly flexible median regression model, but with heteroscedastic error that changes shape with covariate levels $\mathbf{x} \in \mathscr{X}$.

Both the LDDPM and the LDTFP model the entire density at every covariate level $\mathbf{x} \in \mathscr{X}$, so full density and hazard estimates are available, accompanied by reliable interval estimates, unlike many median (and other quantile) regression models. Both models are implemented as user-friendly functions calling compiled FORTRAN in `DPpackage` or calling compiled C++ in `spBayesSurv` for `R`. These functions accommodate general interval-censored data (including current status data); the latter package also allows for spatial correlation. If only a trend function is desired one could instead use quantile regression models, such as the ones implemented in the excellent `quantreg` package in `R` (Koenker 2008).

### 11.3.3 Proportional Odds

The proportional odds (PO) model has recently gained attention as an alternative to the PH and AFT models. PO defines the survival function $S_{\mathbf{x}}(t)$ for an individual with covariate vector $\mathbf{x}$ through the relation

$$\frac{S_{\mathbf{x}}(t)}{1-S_{\mathbf{x}}(t)} = \exp\{-\mathbf{x}'\boldsymbol{\beta}\}\left(\frac{S_0(t)}{1-S_0(t)}\right). \tag{11.14}$$

The odds of dying before any time $t$ are $\exp\{(\mathbf{x}_1-\mathbf{x}_2)'\boldsymbol{\beta}\}$ times greater for those with covariates $\mathbf{x}_1$ versus $\mathbf{x}_2$.

The first semiparametric approaches to PO models involving covariates are due to Cheng et al. (1995), Murphy et al. (1997), and Yang and Prentice (1999). A semiparametric frequentist implementation of the PO model is available in the package `timereg` (Martinussen and Scheike 2006) for `R`. Bayesian nonparametric approaches for the PO model have been based on Bernstein polynomials (Banerjee and Dey 2005), B-splines (Wang and Dunson 2011; Lin and Wang 2011), and Polya trees (Hanson 2006a; Hanson and Yang 2007; Zhao et al. 2009; Hanson et al. 2011).

The PH, AFT, and PO models all make overarching assumptions about the data generating mechanism for the sake of obtaining succinct data summaries. An important aspect associated with the Bayesian nonparametric formulation of these models is that, by assuming the *same, flexible model* for the baseline survival function, they are placed on a common ground (Hanson 2006a; Hanson and Yang 2007; Zhang and Davidian 2008; Zhao et al. 2009; Hanson et al. 2011). Furthermore, parametric models are special cases of the nonparametric models. Differences in fit and/or predictive performance can therefore be attributed to the *survival* models only, rather than to additional possible differences in quite different nonparametric models or estimation methods.

Of the Bayesian approaches based on Polya trees considered by Hanson (2006a), Hanson and Yang (2007), Zhao et al. (2009) and Hanson et al. (2011), the PO model was chosen over PH and AFT according to the log-pseudo marginal likelihood (LPML) criterion (Geisser and Eddy 1979). In three of these works, the parametric log-logistic model, a special case of PO that also has the AFT property, was chosen. This may be due to the fact that the PO assumption implies that hazard ratios $\lim_{t\to\infty}\frac{h_{\mathbf{x}_1}(t)}{h_{\mathbf{x}_2}(t)}=1$, that is, eventually everyone has the same risk of dying tomorrow. These authors also found that, everything else being equal, the actual semiparametric model chosen (PO, PH or AFT) affects prediction far more than whether the baseline is modeled nonparametrically. It is worth noting that none of these papers favored the semiparametric PH model in actual applications.

### *11.3.4 Other Semiparametric Models*

PH, AFT, and PO are three of many semiparametric survival models used in practice. There are a few more hazard-based models including the additive hazards (AH) model (Aalen 1980, 1989), given by

$$h_{\mathbf{x}}(t) = h_0(t) + \mathbf{x}'\boldsymbol{\beta},$$

which is implemented in the `timereg` package for `R`. An empirical Bayes approach to this model based on the gamma process was implemented by Sinha et al. (2009). Fully Bayesian approaches require an elaborate model specification to incorporate the rather awkward constraint $h_0(t) + \mathbf{x}'\boldsymbol{\beta} \geq 0$ for $t > 0$ (Yin and Ibrahim 2005; Dunson and Herring 2005). Recently, there has been some interest in the accelerated hazards model (Chen and Wang 2000; Zhang et al. 2011; Chen et al. 2014), given by

$$h_{\mathbf{x}}(t) = h_0\{\exp(-\mathbf{x}'\boldsymbol{\beta})t\}.$$

This model allows hazard and survival curves to cross.

Finally, several interesting "super models" have been proposed in the literature, including non-proportional hazard regression models that include PH as a special case (Devarajan and Ebrahimi 2011), generalized odds-rate hazards models that include PH and PO as special cases (Dabrowska and Doksum 1988; Scharfstein et al. 1998), Box-Cox transformation regression models that include PH and AH as special cases (Yin and Ibrahim 2005; Martinussen and Scheike 2006), and extended hazard regression models that include both PH and AFT as special cases (Chen and Jewell 2001; Li et al. 2015b).

## 11.4 Spatial Dependence

When survival data are spatially correlated, it is often of scientific interest to investigate possible spatial dependence in survival outcomes after adjusting for known subject-specific covariate effects. Such spatial dependence is often due to region-specific similarities in ecological and/or social environments that are typically not measurable. We next discuss two general approaches, *frailty* and *copula*, for incorporating spatial dependence into the semiparametric models presented in Sect. 11.3, followed by some other possibilities.

### *11.4.1 Spatial Frailty Modeling*

Frailties have been frequently used to induce correlation among related survival times in models which have a linear predictor. The linear predictor is augmented

$\eta_i = \mathbf{x}_i'\boldsymbol{\beta} + v_i$, where $v_i$ is a random effect, termed "frailty," accounting for heterogeneity after adjusting for covariates. The so-called shared frailty models have one common random effect within each group, e.g. $v_i = z_{g_i}$ where $g_i \in \{1,\ldots,G\}$ is the group—e.g. county, hospital, family—to which observation $i$ belongs. Early literature considered exchangeable frailties with $z_1,\ldots,z_G \overset{iid}{\sim} H$, where $H$ was constrained to be mean or median zero to avoid confounding with the baseline function.

In the case of spatial survival data, one can extend the frailty model by including a spatial effect, e.g.,

$$\eta_i = \mathbf{x}_i'\boldsymbol{\beta} + \gamma_i,\ \gamma_i = v_i + w_i,$$

where the frailty term $\gamma_i$ incorporates the effects of both heterogeneity (via the non-spatial frailty $v_i$) and spatial dependence (through the spatial frailty $w_i$); note that, however, in applications often only spatial dependence is modeled ($\gamma_i = w_i$) or exchangeable dependence ($\gamma_i = v_i$). Spatial frailty models have been widely discussed in the literature and correspond to particular cases of hierarchical models. Such models are usually grouped into two general settings according to their underlying data structure: *point-referenced* (geostatistical) data, where the location $\mathbf{s}_i$ varies continuously throughout a fixed study region $\mathscr{D}$, and *areal* (lattice) data, where the study region is partitioned into a finite number of areal units with well-defined boundaries (Banerjee et al. 2015).

#### 11.4.1.1 Point-Referenced Data Modeling

In modeling point-referenced data, the non-spatial frailty term $v_i$ is often specified $v_i \overset{iid}{\sim} N(0,\sigma^2)$, and the spatially correlated frailties $\mathbf{w} = (w_1,\ldots,w_n)$ can be specified to have a multivariate Gaussian distribution:

$$\mathbf{w} \sim N_n(\mathbf{0}, \theta^2\mathbf{R}), \tag{11.15}$$

where $N_n$ denotes the $n$-dimensional Gaussian distribution, $\theta^2$ measures the amount of spatial variation across locations, and the $(i,j)$th element of $\mathbf{R}$, denoted by $\mathbf{R}_{ij}$, is the correlation between $w_i$ and $w_j$. An isotropic correlation function is commonly used to construct $\mathbf{R}$, where the correlation of any two subjects is a function solely of the distance $d_{ij}$ between their locations $\mathbf{s}_i$ and $\mathbf{s}_j$, i.e., $\mathbf{R}_{ij} = \rho(d_{ij})$. A flexible, frequently used correlation function is the Matérn

$$\rho(d_{ij}) = \frac{(\phi d_{ij})^{\nu} K_{\nu}(\phi d_{ij})}{2^{\nu-1}\Gamma(\nu)}, \tag{11.16}$$

where $K_\nu$ is a modified Bessel function of the third kind, $\phi > 0$ measures the spatial decay over distance, and $\nu > 0$ is a parameter controlling the smoothness of the realized random field. Interested readers are referred to Banerjee et al. (2015) for further discussion of correlation functions. Note that the Matérn reduces to the exponential $\rho(d_{ij}) = \exp(-\phi d_{ij})$ for $\nu = 0.5$ and the Gaussian $\rho(d_{ij}) = \exp(-\phi^2 d_{ij}^2)$

when $\nu \to \infty$. Under the above prior specifications of exchangeable normal $v_i$ and spatially correlated $w_i$, the resulting multivariate Gaussian distribution on frailties $\boldsymbol{\gamma} = (\gamma_1, \ldots, \gamma_n)$ is

$$\boldsymbol{\gamma} \sim N_n\left\{\mathbf{0}, \theta^2\mathbf{R} + \sigma^2\mathbf{I}\right\}. \tag{11.17}$$

With this representation, the non-spatial effect variance $\sigma^2$ is often called the *nugget*, the spatial effect variance $\theta^2$ is called the *partial sill*, and the total effect variance $\theta^2 + \sigma^2$ is called the *sill*. The rationale of including the nugget effect is that we don't expect all remaining individual heterogeneity to be accounted for by the spatial story, as other factors (e.g., measurement error, replication error, micro-scale error) may also potentially explain the heterogeneity. In Henderson et al. (2002), the term $\tau = \theta^2/(\theta^2 + \sigma^2)$ is called the *nugget effect* and interpreted as the proportion of the heterogeneity variance that is explained by spatial effects.

For posterior inference, MCMC requires computing the inverse and determinant of $n$-dimensional correlation matrix $\mathbf{R}$ in each iteration. With an increasing sample size $n$, such computation becomes very expensive and even unstable due to a large amount of numerical operations. This situation is often referred to as "the big $n$ problem." Various approaches have been developed to approximate the correlation function such as predictive process models (Banerjee et al. 2008; Finley et al. 2009), sparse approximations (Furrer et al. 2006; Kaufman et al. 2008), and the full scale approximation (FSA) method (Sang and Huang 2012). The last approximation is the summation of the former two approximations, which can capture both large- and small-scale spatial dependence. The FSA has been successfully applied to model point-referenced survival data in Zhou et al. (2015b) and implemented in the R package `spBayesSurv`.

#### 11.4.1.2 Areal Data Modeling

In the case of areal data, the whole study region $\mathscr{D}$ is often partitioned into a finite number of areas, say $B_1, \ldots, B_G$, and a common frailty is assumed for the subjects within each area, i.e.

$$\eta_i = \mathbf{x}_i'\boldsymbol{\beta} + \gamma_{g_i},\ \gamma_j = v_j + w_j, j = 1, \ldots, G.$$

Here the non-spatial frailty $v_j$ for each area is typically assigned a mean-zero normal distribution with variance $\sigma^2$. For the spatial frailty term $w_j$, there has been two general approaches. First, one can assume a fully specified mean-zero multivariate Gaussian distribution on $\mathbf{w} = (w_1, \ldots, w_G)$ with covariance matrix $\theta^2\mathbf{R}$, where $\mathbf{R}_{ij}$ is modeled using a traditional correlation function like the Matérn in (11.16) but with $d_{ij}$ representing the distance between two areal centroids. Another way is to consider an intrinsic conditionally autoregressive (ICAR) model. Let $a_{ij} = 1$ if areas $B_i$ and $B_j$ share a nontrivial border (i.e., a connected curve in $\mathbb{R}^2$ that is more than one point) and $a_{ij} = 0$ otherwise; set $a_{ii} = 0$. Then the $G \times G$ matrix $\mathbf{A} = [a_{ij}]$ is called the adjacency matrix for the region $\mathscr{D}$. The ICAR prior is defined through the set of all conditional distributions

$$w_j|\{w_i : i \neq j\} \sim N\left(\bar{w}_j,\ \theta^2/a_{j+}\right),\ j = 1,\ldots,G, \tag{11.18}$$

denoted $\mathbf{w} \sim \text{ICAR}(1/\theta^2)$, where $a_{j+}$ is the number of neighbors of area $B_j$, $\bar{w}_j = \frac{1}{a_{j+}}\sum_{i:a_{ij}=1} w_i$ is the sample mean of the $a_{j+}$ values of the neighboring areal unit frailties, and $\theta^2/a_{j+}$ is the conditional variance. Note that the ICAR model induces an improper joint density, and the constraint $\sum_{j=1}^{G} w_j = 0$ is commonly used to avoid identifiability issues. Another common fix is to assume a proper CAR model by multiplying the conditional mean $\bar{w}_j$ in (11.18) by a shrinkage scale parameter $\rho$, where $0 \leq \rho < 1$; it is generally difficult to estimate $\rho$ and $\theta^2$ simultaneously.

#### 11.4.1.3 Related Literature

Henderson et al. (2002) modeled the spatial structure of leukemia survival data using both district-level and point-referenced frailty effects in the context of the PH model. In their point-referenced analysis, a multivariate gamma distribution for $(e^{\gamma_1},\ldots,e^{\gamma_n})'$ was constructed so that each marginal has a gamma distribution with mean 1 and variance $\sigma^2 + \theta^2$, and the correlation between $e^{\gamma_i}$ and $e^{\gamma_j}$ takes the form defined in (11.17). In their district-level analysis, they considered a linear predictor with individual frailties as $\eta_i = \mathbf{x}_i'\boldsymbol{\beta} + \gamma_i$, where $e^{\gamma_i}|\mu_{g_i} \sim \Gamma(1/\xi, 1/(\xi\mu_{g_i}))$. They then assumed a multivariate Gaussian distribution on the latent effects $\boldsymbol{\mu} = (\mu_1,\ldots,\mu_G)$ with the correlation function between the $i$th and $j$th district modeled via the powered exponential and Matérn. They also considered the ICAR specification on $\boldsymbol{\mu}$ and found that the multivariate Gaussian via a Matérn correlation with $\nu = 2$ had the best fit based on the DIC goodness-of-fit criterion.

Pan et al. (2014) fitted the semiparametric PH model with ICAR frailties to interval censored data with the baseline hazard function modeled via B-splines. Lin et al. (2015) duplicated this model without the ICAR frailties. Using the same methodology, a special case of interval-censored data, current-status data, was presented in Cai et al. (2011). The aforementioned models can be fit in the `ICBayes` R package. Li and Ryan (2002) modeled the district-level frailty effect using a fully specified multivariate normal prior within the framework of PH, and applied the model to detect prognostic factors leading to childhood asthma. All of these approaches are essentially a special case of the general models previously presented in Kneib (2006) and Hennerfeind et al. (2006), which can be efficiently fit in the freely available program BayesX or the R package `R2BayesX`; the latter package uses compiled code and places the B-spline prior on the log-hazard instead of the hazard. An advantage of the models fitted in BayesX is that both areal and point-referenced data are accommodated as well as nonparametric additive effects. In addition, the R package `spatsurv` can also fit the PH model with multivariate Gaussian frailties, where the baseline hazard is modeled either parametrically or nonparametrically via B-splines.

Banerjee and Carlin (2003) developed a semiparametric PH frailty model for capturing spatio-temporal heterogeneity in survival of women diagnosed with breast cancer in Iowa, using a mixture of beta densities baseline. Banerjee et al. (2003)

applied the Weibull parametric PH frailty model to infant mortality data in Minnesota, where the county-level frailties were assumed to have either an uncorrelated zero-mean Gaussian prior, an ICAR prior, or a fully specified multivariate Gaussian prior as in (11.15). They showed that the fully specified prior provides the best model fitting in terms of DIC in the analysis of the infant data. Banerjee and Dey (2005) utilized the same frailty modeling technique for capturing spatial heterogeneity within the framework of semiparametric PO, found that the proper CAR prior yielded the best fit in the application to a subset of surveillance epidemiology and end results (SEER) breast cancer data. Zhao et al. (2009) considered either an AFT, PH, or PO model with ICAR frailties, where the baseline function was assumed to have a mixture of Polya trees prior. Zhang and Lawson (2011) and Wang et al. (2012) developed parametric and semiparametric AFT models with ICAR frailties, respectively. Chernoukhov (2013) extended the additive hazards model for allowing various spatial dependence structures in his dissertation. Zhou et al. (2015a) extended the generalized model in (11.13) by allowing frailties accommodating spatial correlation via the ICAR prior distribution. The models proposed in Zhao et al. (2009) and Zhou et al. (2015a) can be fit in the R package `spBayesSurv`. Other references focusing on spatial frailty modeling and its application include McKinley (2007), Diva et al. (2008), Darmofal (2009), Liu (2012), Ojiambo and Kang (2013), Dasgupta et al. (2014), Li et al. (2015a), and among others.

### 11.4.2 Spatial Copula Modeling

Spatial copulas are just beginning to become popular in geostatistics. The use of copulas in the spatial context was first proposed by Bárdossy (2006), where the empirical variogram is replaced by empirical copulas to investigate the spatial dependence structure. The spatial copula approach offers an appealing way to separate modeling from the spatial dependence structure for multivariate distributions. Copulas completely describe association among random variables separately from their univariate distributions and thus capture joint dependence without the influence of the marginal distribution (Li 2010). In the context of survival models, the idea of spatial copula approach is to first assume that the survival time $T_i$ at location $\mathbf{s}_i$ marginally follows a model $S_{\mathbf{x}_i}(t)$ introduced in Sect. 11.3, then model the joint distribution of $(T_1,\ldots,T_n)'$ as

$$F(t_1,\ldots,t_n) = C(F_{\mathbf{x}_1}(t_1),\ldots,F_{\mathbf{x}_n}(t_n)), \tag{11.19}$$

where $F_{\mathbf{x}_i}(t) = 1 - S_{\mathbf{x}_i}(t)$ is the cumulative distribution function and the function $C$ is an $n$-copula used to capture spatial dependence. If we let $U_i = F_{\mathbf{x}_i}(T_i)$, then the problem is reduced to constructing a copula for modeling the joint distribution of $\mathbf{U} = (U_1,\ldots,U_n)$. Hereafter we assume that $U_i$ follows a uniform distribution on $[0,1]$ for all locations $\mathbf{s}_i$; i.e., the survival model $S_{\mathbf{x}_i}(t)$ is assumed to be correctly specified. In fact, copulas are all the joint cumulative distribution functions on the unit hypercube with uniform marginal distributions. We refer interested readers to Nelsen (2006) for general introduction to copulas and to Smith (2013) for Bayesian approaches to copula modeling.

In the geostatistical framework, the multivariate spatial copula of $\mathbf{U}$ is often constructed so that for any selected two locations $\mathbf{s}_i$ and $\mathbf{s}_j$, the bivariate copula (i.e., joint distribution) of $(U_i, U_j)$ does not depend on the locations $\mathbf{s}_i$ and $\mathbf{s}_j$ but on their distance $d_{ij}$ only. However, such construction is not a trivial task. Here we introduce a spatial version of the Gaussian copula and refer readers to Li (2010) for further discussion of other theoretical spatial copulas. Define $Z_i = \Phi^{-1}\{U_i\}$, where $\Phi(\cdot)$ is the standard normal cumulative distribution function, then we have $Z_i \sim N(0,1)$ for all $i$. If we further assume that $\mathbf{Z} = (Z_1, \ldots, Z_n)'$ follows a multivariate normal distribution with mean zero and covariance $\mathbf{R}$, i.e., $\mathbf{Z} \sim N_n(\mathbf{0}, \mathbf{R})$, then the induced joint distribution of $\mathbf{U}$ is called the Gaussian copula, which is given by

$$C(u_1, \ldots, u_n) = \Phi_n\left(\Phi^{-1}\{u_1\}, \ldots, \Phi^{-1}\{u_n\}; \mathbf{R}\right), \tag{11.20}$$

where $\Phi_n(\ldots; \mathbf{R})$ denotes the distribution function of $N_n(\mathbf{0}, \mathbf{R})$. Note that all the diagonal elements of $\mathbf{R}$ are ones, so we refer to $\mathbf{R}$ as the correlation matrix thereafter. The Gaussian copula has a symmetrical density, which can be written as

$$c(u_1, \ldots, u_n) = |\mathbf{R}|^{-1/2} \exp\left\{\frac{1}{2}\mathbf{z}'(\mathbf{R}^{-1} - \mathbf{I})\mathbf{z}\right\}, \tag{11.21}$$

where $\mathbf{z} = (z_i, \ldots, z_n)'$ with $z_i = \Phi^{-1}\{u_i\}$ and $\mathbf{I}$ is the identity matrix. The spatial dependence structure of the Gaussian copula is induced by constructing the correlation matrix $\mathbf{R}$ using classical geostatistical models. For example, the $(i, j)$th element of $\mathbf{R}$ can be defined using the Matérn in (11.16) with a nugget effect $\tau$, that is, $\mathbf{R}_{ij} = \tau\rho(d_{ij})$ for $i \neq j$, where $1 < \tau < 1$. Under the spatial Gaussian copula, the joint density of $(T_1, \ldots, T_n)$ takes the form

$$f(t_1, \ldots, t_n) = |\mathbf{R}|^{-1/2} \exp\left\{-\frac{1}{2}\mathbf{z}'(\mathbf{R}^{-1} - \mathbf{I}_n)\mathbf{z}\right\} \prod_{i=1}^{n} f_{\mathbf{x}_i}(t_i), \tag{11.22}$$

where $z_i = \Phi^{-1}\{F_{\mathbf{x}_i}(t_i)\}$ and $f_{\mathbf{x}_i}(t_i)$ is the density function of $T_i$. The use of spatial copulas has not been widely applied for modeling survival data that are subject to spatial correlation. Li and Lin (2006) successfully applied the spatial Gaussian copula approach to a semiparametric PH model and proposed spatial semiparametric estimating equations that yield consistent and asymptotically normal estimators. Zhou et al. (2015b) considered the LDDPM marginal model given in (11.12) using the same Gaussian copula for capturing spatial dependence structure, where MCMC algorithms were used to obtain posterior inferences. Zhou et al. (2015b) also provided a Bayesian version of the model considered in Li and Lin (2006) using piecewise exponential baseline specifications. The R package `spBayesSurv` can fit the aforementioned copula-based Bayesian survival models.

The spatial Gaussian copula approach can also be extended for fitting lattice data, for which constructing the correlation matrix $\mathbf{R}$ of $\mathbf{Z} = (Z_1, \ldots, Z_n)$ becomes a challenging task. One may consider a random effects model for $\mathbf{Z}$ based on the partition of the domain $\mathscr{D}$ into $G$ districts, that is,

$$Z_i = \mu_{g_i} + \varepsilon_i,\ \boldsymbol{\mu} \sim N_G(\mathbf{0}, \mathbf{B}\boldsymbol{\Omega}\mathbf{B}),\ \varepsilon_i \overset{ind}{\sim} N\left(0, \frac{\sigma^2}{\omega_{g_i g_i} + \sigma^2}\right), g_i \in \{1, \ldots, G\}, \quad (11.23)$$

where $\boldsymbol{\mu} = (\mu_1, \ldots, \mu_G)'$ are the random effects, $\boldsymbol{\Omega} = [\omega_{ij}]$ is a $G \times G$ matrix introducing spatial dependence to $\boldsymbol{\mu}$, $\mathbf{B} = \text{diag}\left(1/\sqrt{\omega_{11} + \sigma^2}, \ldots, 1/\sqrt{\omega_{GG} + \sigma^2}\right)$, and $\varepsilon_i$ is the error term independent of the spatial random effects. Note that $Var(Z_i) = 1$. Popular models for $\Omega$ include multivariate Gaussian coupled with a spatial covariance function, ICAR, proper CAR, and many others. Li et al. (2015b) derived the implied correlation matrix $\mathbf{R} = cov(\mathbf{Z})$ under the ICAR model, which only involves one unknown quantity $\psi^*$. A smaller value of $\psi^*$ corresponds to stronger spatial dependence. With the specification of $\mathbf{R}$, one can model joint cumulative distribution function of $(T_1, \ldots, T_n)$ by

$$F(t_1, \ldots, t_n) = \Phi_n\left(\Phi^{-1}\{F_{\mathbf{x}_i}(t_i)\}, \ldots, \Phi^{-1}\{F_{\mathbf{x}_n}(t_n)\}; \mathbf{R}\right). \quad (11.24)$$

### *11.4.3 Other Spatial Dependence Modelings*

Zhao and Hanson (2011) considered a stratified PH model:

$$S_{\mathbf{x}_i}(t) = S_{0g_i}(t)^{\exp(\mathbf{x}_i'\boldsymbol{\beta})},\ g_i \in \{1, \ldots, G\},$$

where each region-specific baseline $S_{0j}(\cdot)$ approximately follows a mixture of Polya trees prior centered at a parametric log-logistic family. The spatial dependence among the $\{S_{01}(\cdot), \ldots, S_{0G}(\cdot)\}$ is induced through proper CAR priors on the logit transformed Polya tree conditional probabilities $\{Y_{l,k}\}$. Hanson et al. (2012) extended this idea to fit a Bayesian semiparametric temporally stratified PH model with spatial frailties. Stratified AFT models with ICAR areal frailties are considered by Zhou et al. (2015a).

In modeling areal data, spatial dependence is often due to unadjusted district-level risk factors that may potentially relate to survival outcomes. Zhao and Hanson (2011) note that spatial frailties serve as proxies to unmeasured region-level covariates, but are less-precise adjustments since region-level covariates (such as shortest distance to a clinic) are unlikely to sharply change at areal boundaries. Therefore it is natural to introduce spatial dependence by allowing frailties to depend on region-level covariates, especially when information is available on each region that may affect the survival outcome beyond the recorded covariates. For this reason, Zhou et al. (2015c) proposed a region-level covariate adjusted frailty PH model. Specifically, with the linear predictor $\eta_i = \mathbf{x}_i'\beta + \gamma_{g_i}$, they assume an LDTFP prior on the frailties, i.e., $\gamma_j | \mathbf{z}_j \sim g_{\mathbf{z}_j}(\cdot)$, where $\mathbf{z}_j$ is a vector of region-level covariates. This model can be fit in the `DPpackage` for R.

## 11.5 Illustrations

Both of the frailty and copula modeling approaches are illustrated using real-life datasets. All the analyses are implemented using the R packages `spBayesSurv`. The fitted models are compared in terms of the log pseudo marginal likelihood (LPML) developed by Geisser and Eddy (1979). Note that the frailties used in frailty models are either exchangeable $v_i$ or spatial $w_i$, but not both $v_i + w_i$.

### *11.5.1 SEER Cancer Data*

The SEER program of the National Cancer Institute (seer.cancer.gov) is an authoritative source of information on cancer incidence and survival in the US, providing county-level cancer data on an annual basis for particular states for public use. Areal-referenced SEER data have been analyzed by many authors in the context of spatial frailty models (e.g., Banerjee and Carlin 2003; Banerjee and Dey 2005; Zhao et al. 2009; Zhao and Hanson 2011; Wang et al. 2012; Zhou et al. 2015c,a).

For illustration, we analyze a subset of the Iowa SEER breast cancer survival data, which consists of a cohort of 1073 Iowan women, who were diagnosed with malignant breast cancer starting in 1995, and enrollment and follow-up continued through the end of 1998. This data set has been analyzed in Zhao et al. (2009) and Zhou et al. (2015c). The observed survival time, from 1 to 48, is defined as the number of months from diagnosis to either death or the last follow-up. Here we assume that only deaths due to metastasis of cancerous nodes in the breast are events, while the deaths from other causes are censored at the time of death. The right-censoring rate is 54.5 %. For each patient, the observed survival time and county of residence at diagnosis are recorded. The considered individual-level covariates include age at diagnosis and the stage: local, regional, or distant, where two dummy variables are created for regional and distant, respectively, and the reference group is local. Zhou et al. (2015c) point out that some county-level socioeconomic factors (e.g., median household income, poverty level, education, rurality) are also potentially associated with breast cancer and argue that rural counties present more heterogeneity in access to quality care and screening for breast cancer. Therefore, we also include a county-level covariate "Rural-Urban Continuum Codes" (RUCC) measuring degree of urbanization; see Zhou et al. (2015c) for a detailed description.

We fit each of the PH, AFT, and PO frailty models with a mixture of Polya trees prior on baseline survival $S_0(t)$ and the ICAR prior on the frailties $\boldsymbol{\gamma} \sim \text{ICAR}(\lambda)$, where the PH is centered at the Weibull $G_{\boldsymbol{\theta}}(t) = 1 - \exp\left\{-(e^{\theta_1}t)^{\exp(\theta_2)}\right\}$ and the AFT and PO are centered at the log-logistic $G_{\boldsymbol{\theta}}(t) = 1 - \{1 + (e^{\theta_1}t)^{\exp(\theta_2)}\}^{-1}$. We consider the following prior settings: $J = 4$, $c \sim \Gamma(5,1)$, $\boldsymbol{\theta} \sim N_2(\hat{\boldsymbol{\theta}}, \hat{\mathbf{V}})$, $\boldsymbol{\beta} \sim N_p(\hat{\beta}, 30\hat{\boldsymbol{\Sigma}})$ and $\lambda \sim \Gamma(1,1)$, where $\hat{\boldsymbol{\theta}}$, $\hat{\boldsymbol{\beta}}$, $\hat{\mathbf{V}}$, and $\hat{\boldsymbol{\Sigma}}$ are maximum likelihood estimates from the underlying parametric model. Using the same priors, we also fit the above models with Gaussian exchangeable frailties and without frailties. For all models considered, a burn-in of 100,000 iterations is followed by a run of 100,000 thinned down to 10,000 iterations. All these models are fitted using the `survregbayes` function available in the package `spBayesSurv`.

**Table 11.1** SEER breast cancer data: Posterior medians (95 % credible intervals) of fixed effects from various models

| Model | Centered age | Regional stage | Distant stage | RUCC |
|---|---|---|---|---|
| PH/CAR | 0.019 (0.013, 0.025) | 0.26 (0.03, 0.48) | 1.69 (1.45, 1.93) | −0.069 (−0.136, 0.002) |
| AFT/CAR | 0.018 (0.011, 0.023) | 0.22 (0.01, 0.43) | 1.51 (1.26, 1.75) | −0.045 (−0.105, 0.013) |
| PO/CAR | 0.030 (0.021, 0.038) | 0.40 (0.12, 0.69) | 2.59 (2.25, 2.95) | −0.087 (−0.174, −0.001) |
| PH/LDTFP | 0.019 (0.013, 0.025) | 0.27 (0.03, 0.49) | 1.64 (1.43, 1.88) | −0.105 (−0.185, −0.041) |

Note the AFT model is parameterized as $S_{\mathbf{x}}(t) = S_0(e^{\mathbf{x}'\beta}t)$

The LPML values under ICAR frailty PH, AFT, and PO are −2226, −2228, and −2210, respectively, while the corresponding LPMLs are −2230, −2224 and −2214 under exchangeable frailty models and are −2230, −2228, and −2214 under non-frailty models. We can observe that the ICAR frailty model has the best predictive ability within the context of either PH or PO, and the exchangeable frailty model performs best in terms of LPML under the AFT. Table 11.1 presents posterior means and equal-tailed 95 % credible intervals (CI) for covariate effects under each of above model with ICAR frailties. All individual covariate effects are significant in each model. Higher age at diagnosis increases the hazard; e.g. a 20-year increase in age is associated with an $\exp(0.019 \times 20) \approx 1.46$-fold increase in hazard. Using women with local stage of disease as the reference, the hazard rate of women of the same age who live in the same county will be $\exp(0.26) \approx 1.30$ times larger if their cancer is detected at the regional stage, and $\exp(1.69) \approx 5.42$ times larger if detected at the distant stage. Under the AFT assumption, among patients living in the same county and having same age, a woman with local stage typically survives $\exp(0.22) \approx 1.25$ times longer than a woman with regional stage, and $\exp(1.51) \approx 4.53$ times longer than a woman with distant stage. Finally, for the PO model, after adjusting for the age at diagnosis and the RUCC, the odds of dying from breast cancer before any time $t$ are $\exp(0.40) \approx 1.49$ greater for regional stage versus local stage, and are $\exp(2.59) \approx 13.33$ greater for distant stage versus local stage. These findings are confirmed in Fig. 11.1, which shows the fitted survival functions for women aged at 68.8 years and living a county with RUCC at 5 for distant and local stages under the three competing models and assuming a spatial frailty of zero. Turning to the county-level RUCC effect, only the PO model provides a significant result at the 0.05 level; living in more urban counties is associated with poorer survival after a breast cancer diagnosis on average.

Zhou et al. (2015c) fitted a PH model with LDTFP frailty terms using the package `DPpackage` and found more variability for frailties of rural counties. The resulting LPML is −2222 when RUCC is included into both the linear predictor and frailty terms. The pseudo Bayes factor for the LDTFP frailty model versus the ICAR frailty PH model is $\exp(2226 - 2222) \approx 55$, implying that allowing frailties depending on

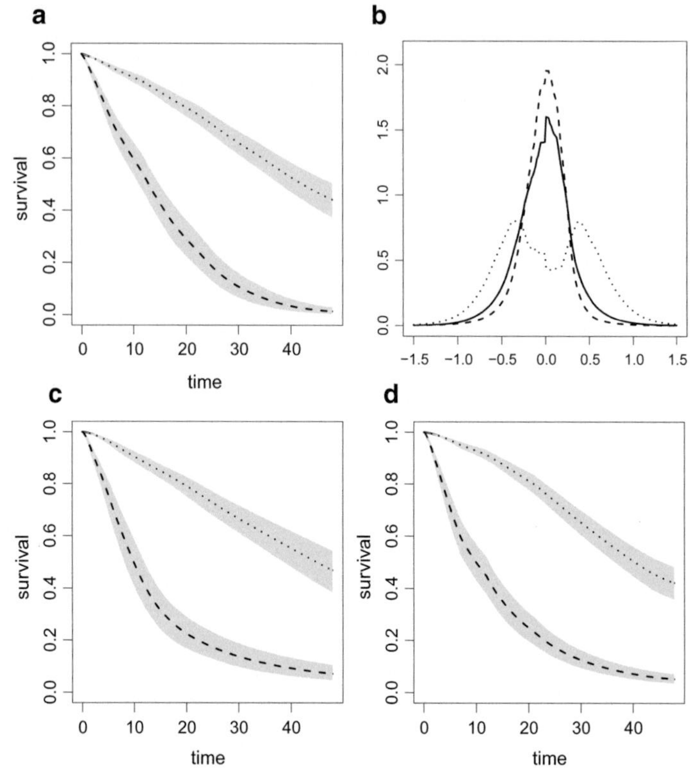


**Fig. 11.1** SEER breast cancer data. Panels (**a**), (**c**), and (**d**) show estimated survival curves for women aged at 68.8 years and living a county with RUCC at 5 for distant (*dashed lines*) and local (*dotted lines*) stages, under PH, AFT, and PO, respectively. The pointwise 95 % credible bands are also displayed as grey areas. Panel (**b**) displays frailty densities for RUCC=2, 5, and 9, which are displayed as dashed, continuous, and *dotted lines*, respectively

RUCC improves the model's predictive ability about 55 times. Table 11.1 also shows the covariate effects under the LDTFP frailty PH model. An interesting finding is that now the RUCC effect becomes significant at the 0.05 level. This may be due to the fact that frailty distributions are covariate-dependent as shown in Fig. 11.1b. After controlling for individual covariates and county, the hazard rate of women living in urban counties (with RUCC $= 2$) will be $\exp(0.105 \times 7) \approx 2$ times larger than that of women in rural counties (with RUCC $= 9$).

## *11.5.2 Leukemia Data*

We consider a dataset on the survival of acute myeloid leukemia in $n = 1{,}043$ patients, analyzed by Henderson et al. (2002) fitting a multivariate gamma frailty PH model. This dataset is available for access in Fahrmeir and Kneib (2011). It is of interest to investigate possible spatial variation in survival after accounting for known subject-specific prognostic factors, which include age, sex, white blood cell count (WBC) at diagnosis, and the Townsend score, for which higher values indicate less affluent areas. The censoring rate is 16 %. Both exact residential locations of all patients and their administrative districts (24 districts that make up the whole region) are available. Therefore, we can fit both geostatistical and lattice models.

For the geostatistical case, we fit the copula model (11.19) proposed by Zhou et al. (2015b) using the function `spCopulaDDP`, where the marginal model $F_{\mathbf{x}}(\cdot)$ is defined via the LDDPM in (11.12) and the copula function $C$ is specified through the Gaussian spatial copula in (11.20) assuming the exponential correlation function. We then use the function `spCopulaCoxph` to fit the copula model assuming a piecewise exponential PH model for $F_{\mathbf{x}}(\cdot)$, where the partition is based on $J = 20$ cut-points with each $a_k$ defined as the $\frac{k}{J}$th quantile the empirical distribution of observed survival times (see Sect. 11.2.1). For comparison, standard non-spatial LDDPM and piecewise exponential PH models are also fitted using the functions `anovaDDP` and `indeptCoxph`, respectively. The default priors are considered for above models as suggested in Zhou et al. (2015b). Regarding the lattice case, we fit each of the Polya trees PH, AFT, and PO models with ICAR frailties as in Sect. 11.5.1 using the function `survregbayes` and their corresponding non-frailty models, where the Polya trees are truncated at level $J = 5$. Finally, we fit the generalized AFT model (11.13) with and without ICAR frailties using the function `frailtyGAFT`, where $\varepsilon_i$ is allowed to depend on age and WBC. We refer readers to Zhou et al. (2015a) for discussion of prior specifications and posterior samplings. For all models, we retain $10,000$ scans thinned from $50,000$ after a burn-in period of $10,000$ iterations.

The LPML measures for the copula with LDDPM, copula with piecewise exponential PH, PH, AFT, and PO with Polya trees baselines and ICAR frailties, and generalized AFT with ICAR frailties are $-5932$, $-5939$, $-5930$, $-5953$, $-5925$, and $-5936$, respectively. Without spatial components, the above LPML values become $-5934$, $-5941$, $-5934$, $-5950$, $-5925$, and $-5942$. The PO models significantly outperform others from a predictive point of view regardless of whether spatial dependence is taken into account. Within the context of LDDPM and PH, the use of the Gaussian spatial copula slightly improves the model's predictive ability, indicating that the spatial dependence is relatively weak in this dataset. Under the framework of PH, the Polya trees prior works much better than piecewise exponential prior for modeling baseline functions. The AFT models provide the worst LPML values, while allowing the baseline varying with covariates (i.e., generalized AFT) can significantly improve the models' predictive ability; the Bayes factors for age and WBC effects on the baseline survival are 124 and 23, respectively, under the ICAR frailty model, and are 73 and 31 under the non-frailty model.

For the copula LDDPM model, the posterior median of the nugget effect parameter $\theta_1$ is 0.051 with the 95 % CI (0.000, 0.176), indicating that only 5 % of the heterogeneity variance is explained by spatial effect on average. The posterior median of $\theta_2$ is 0.831 with the 95 % CI (0.001, 3.075) indicates that the correlation decays by $1 - e^{-0.831} \approx 56\,\%$ for every kilometer increase in distance on average.

However, given such a small value $\theta_1$, the spatial decay becomes less important. Figure 11.2a shows the survival curves under the PO ICAR frailty model for female patients aged at 49 (25 %th quantile) and aged at 74 (75 %th quantile) holding other covariates at population averages, where we see that higher age is associated with lower survival probability. Figure 11.2b shows the baseline survival curves under the generalized AFT ICAR frailty model for female patients aged at 49 and aged at 74 holding WBC at its population average, where we can see that the baseline varies with age which clearly violates the AFT assumption.

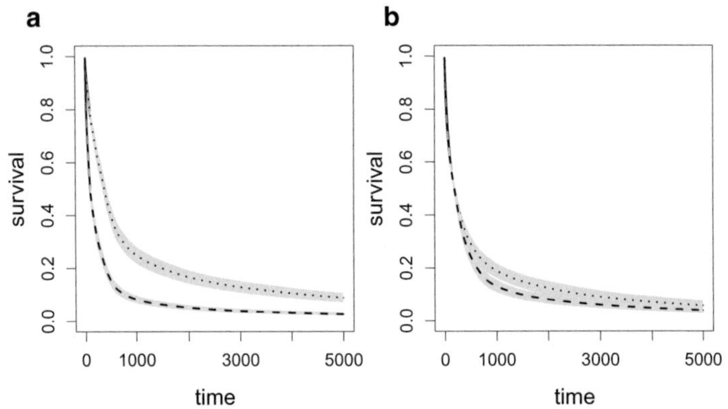


**Fig. 11.2** Leukemia data. Panel (**a**) shows estimated survival curves for women aged at 49 years (*dotted lines*) and aged at 75 years (*dashed lines*), holding other covariates at population averages and frailties at zeros, under the PO model with ICAR frailties. Panel (**b**) shows estimated baseline survival curves for women aged at 49 years (*dotted lines*) and aged at 75 years (*dashed lines*), holding WBC at its population average and frailties at zeros, under the generalized AFT with ICAR frailties. The pointwise 90 % credible bands are also displayed as grey areas

## 11.6 Concluding Remarks

We have reviewed commonly used priors on baseline functions, semiparametric and nonparametric Bayesian survival models, and recent approaches for accommodating spatial dependence, both frailty and copula. Many R packages are discussed for implementation including `DPpackage`, `spBayesSurv`, `R2BayesX`, and `spatsurv`. Two interesting data sets are illustrated, where both analyses show that PO models perform significantly better than all other models we considered including the PH, AFT, and two generalizations of AFT.

**Acknowledgements** This work was supported by federal grants 1R03CA165110 and 1R03CA176739-01A1.

## References

Aalen, O. O. (1980). A model for nonparametric regression analysis of counting processes. In *Mathematical Statistics and Probability Theory*, volume 2, pages 1–25. Springer-Verlag.

Aalen, O. O. (1989). A linear regression model for the analysis of life times. *Statistics in Medicine*, **8**(8), 907–925.

Alzola, C. and Harrell, F. (2006). *An Introduction to S and the Hmisc and Design Libraries*. Online manuscript available at http://biostat.mc.vanderbilt.edu/wiki/pub/Main/RS/sintro.pdf.

Andersen, P. K. and Gill, R. D. (1982). Cox's regression model for counting processes: A large sample study. *The Annals of Statistics*, **10**(4), 1100–1120.

Banerjee, S. and Carlin, B. P. (2003). Semiparametric spatio-temporal frailty modeling. *Environmetrics*, **14**(5), 523–535.

Banerjee, S. and Dey, D. K. (2005). Semiparametric proportional odds models for spatially correlated survival data. *Lifetime Data Analysis*, **11**(2), 175–191.

Banerjee, S., Wall, M. M., and Carlin, B. P. (2003). Frailty modeling for spatially correlated survival data, with application to infant mortality in Minnesota. *Biostatistics*, **4**(1), 123–142.

Banerjee, S., Gelfand, A. E., Finley, A. O., and Sang, H. (2008). Gaussian predictive process models for large spatial data sets. *Journal of the Royal Statistical Society: Series B (Statistical Methodology)*, **70**(4), 825–848.

Banerjee, S., Carlin, B. P., and Gelfand, A. E. (2015). *Hierarchical Modeling and Analysis for Spatial Data, Second Edition*. Chapman and Hall/CRC Press.

Bárdossy, A. (2006). Copula-based geostatistical models for groundwater quality parameters. *Water Resources Research*, **42**(11), 1–12.

Belitz, C., Brezger, A., Klein, N., Kneib, T., Lang, S., and Umlauf, N. (2015). BayesX - Software for Bayesian inference in structured additive regression models. Version 3.0. Available from http://www.bayesx.org.

Berger, J. O. and Guglielmi, A. (2001). Bayesian and conditional frequentist testing of a parametric model versus nonparametric alternatives. *Journal of the American Statistical Association*, **96**(453), 174–184.

Buckley, J. and James, I. (1979). Linear regression with censored data. *Biometrika*, **66**(3), 429–436.

Burridge, J. (1981). Empirical Bayes analysis of survival time data. *Journal of the Royal Statistical Society. Series B (Methodological)*, **43**(1), 65–75.

Cai, B. and Meyer, R. (2011). Bayesian semiparametric modeling of survival data based on mixtures of B-spline distributions. *Computational Statistics & Data Analysis*, **55**(3), 1260–1272.

Cai, B., Lin, X., and Wang, L. (2011). Bayesian proportional hazards model for current status data with monotone splines. *Computational Statistics & Data Analysis*, **55**(9), 2644–2651.

Carlin, B. P. and Hodges, J. S. (1999). Hierarchical proportional hazards regression models for highly stratified data. *Biometrics*, **55**(4), 1162–1170.

Chang, I.-S., Hsiung, C. A., Wu, Y.-J., and Yang, C.-C. (2005). Bayesian survival analysis using Bernstein polynomials. *Scandinavian Journal of Statistics*, **32**(3), 447–466.

Chen, Y., Hanson, T., and Zhang, J. (2014). Accelerated hazards model based on parametric families generalized with Bernstein polynomials. *Biometrics*, **70**(1), 192–201.

Chen, Y. Q. and Jewell, N. P. (2001). On a general class of semiparametric hazards regression models. *Biometrika*, **88**(3), 687–702.

Chen, Y. Q. and Wang, M.-C. (2000). Analysis of accelerated hazards models. *Journal of the American Statistical Association*, **95**(450), 608–618.

Cheng, S. C., Wei, L. J., and Ying, Z. (1995). Analysis of transformation models with censored data. *Biometrika*, **82**(4), 835–845.

Chernoukhov, A. (2013). Bayesian Spatial Additive Hazard Model. *Electronic Theses and Dissertations*. Paper 4965. http://scholar.uwindsor.ca/etd/4965

Christensen, R. and Johnson, W. (1988). Modeling accelerated failure time with a Dirichlet process. *Biometrika*, **75**(4), 693–704.

Clayton, D. G. (1991). A Monte Carlo method for Bayesian inference in frailty models. *Biometrics*, **47**(2), 467–485.

Cox, D. R. (1972). Regression models and life-tables (with discussion). *Journal of the Royal Statistical Society. Series B (Methodological)*, **34**(2), 187–220.

Cox, D. R. (1975). Partial likelihood. *Biometrika*, **62**(2), 269–276.

Dabrowska, D. M. and Doksum, K. A. (1988). Estimation and testing in a two-sample generalized odds-rate model. *Journal of the American Statistical Association*, **83**(403), 744–749.

Darmofal, D. (2009). Bayesian spatial survival models for political event processes. *American Journal of Political Science*, **53**(1), 241–257.

Dasgupta, P., Cramb, S. M., Aitken, J. F., Turrell, G., and Baade, P. D. (2014). Comparing multilevel and Bayesian spatial random effects survival models to assess geographical inequalities in colorectal cancer survival: a case study. *International Journal of Health Geographics*, **13**(1), 36.

de Boor, C. (2001). *A Practical Guide to Splines*. Applied Mathematical Sciences, Vol. 27. Springer-Verlag: New York.

De Iorio, M., Johnson, W. O., Müller, P., and Rosner, G. L. (2009). Bayesian nonparametric nonproportional hazards survival modeling. *Biometrics*, **65**(3), 762–771.

Devarajan, K. and Ebrahimi, N. (2011). A semi-parametric generalization of the Cox proportional hazards regression model: Inference and applications. *Computational Statistics & Data Analysis*, **55**(1), 667–676.

Diva, U., Dey, D. K., and Banerjee, S. (2008). Parametric models for spatially correlated survival data for individuals with multiple cancers. *Statistics in Medicine*, **27**(12), 2127–2144.

Dukić, V. and Dignam, J. (2007). Bayesian hierarchical multiresolution hazard model for the study of time-dependent failure patterns in early stage breast cancer. *Bayesian Analysis*, **2**, 591–610.

Dunson, D. B. and Herring, A. H. (2005). Bayesian model selection and averaging in additive and proportional hazards. *Lifetime Data Analysis*, **11**, 213–232.

Eilers, P. H. C. and Marx, B. D. (1996). Flexible smoothing with B-splines and penalties. *Statistical Science*, **11**(2), 89–102.

Escobar, M. D. and West, M. (1995). Bayesian density estimation and inference using mixtures. *Journal of the American Statistical Association*, **90**, 577–588.

Fahrmeir, L. and Kneib, T. (2011). *Bayesian Smoothing and Regression for Longitudinal, Spatial and Event History Data*. Oxford University Press.

Ferguson, T. S. (1973). A Bayesian analysis of some nonparametric problems. *The Annals of Statistics*, **1**, 209–230.

Ferguson, T. S. (1974). Prior distributions on spaces of probability measures. *The Annals of Statistics*, **2**, 615–629.

Finley, A. O., Sang, H., Banerjee, S., and Gelfand, A. E. (2009). Improving the performance of predictive process modeling for large datasets. *Computational statistics & data analysis*, **53**(8), 2873–2884.

Furrer, R., Genton, M. G., and Nychka, D. (2006). Covariance tapering for interpolation of large spatial datasets. *Journal of Computational and Graphical Statistics*, **15**(3), 502–523.

Geisser, S. and Eddy, W. F. (1979). A predictive approach to model selection. *Journal of the American Statistical Association*, **74**, 153–160.

Gelfand, A. E. and Mallick, B. K. (1995). Bayesian analysis of proportional hazards models built from monotone functions. *Biometrics*, **51**, 843–852.

Gray, R. J. (1992). Flexible methods for analyzing survival data using splines, with applications to breast cancer prognosis. *Journal of the American Statistical Association*, **87**, 942–951.

Griffin, J. (2010). Default priors for density estimation with mixture models. *Bayesian Analysis*, **5**, 45–64.

Hanson, T., Kottas, A., and Branscum, A. (2008). Modelling stochastic order in the analysis of receiver operating characteristic data: Bayesian nonparametric approaches. *Journal of the Royal Statistical Society. Series C*, **57**, 207–225.

Hanson, T., Johnson, W., and Laud, P. (2009). Semiparametric inference for survival models with step process covariates. *Canadian Journal of Statistics*, **37**(1), 60–79.

Hanson, T. E. (2006a). Inference for mixtures of finite Polya tree models. *Journal of the American Statistical Association*, **101**(476), 1548–1565.

Hanson, T. E. (2006b). Modeling censored lifetime data using a mixture of gammas baseline. *Bayesian Analysis*, **1**, 575–594.

Hanson, T. E. and Johnson, W. O. (2002). Modeling regression error with a mixture of Polya trees. *Journal of the American Statistical Association*, **97**(460), 1020–1033.

Hanson, T. E. and Yang, M. (2007). Bayesian semiparametric proportional odds models. *Biometrics*, **63**(1), 88–95.

Hanson, T. E., Branscum, A., and Johnson, W. O. (2005). Bayesian nonparametric modeling and data analysis: An introduction. In D. Dey and C. Rao, editors, *Bayesian Thinking: Modeling and Computation (Handbook of Statistics, volume 25)*, pages 245–278. Elsevier: Amsterdam.

Hanson, T. E., Branscum, A., and Johnson, W. O. (2011). Predictive comparison of joint longitudinal–survival modeling: a case study illustrating competing approaches. *Lifetime Data Analysis*, **17**, 3–28.

Hanson, T. E., Jara, A., Zhao, L., et al. (2012). A Bayesian semiparametric temporally-stratified proportional hazards model with spatial frailties. *Bayesian Analysis*, **7**(1), 147–188.

Henderson, R., Shimakura, S., and Gorst, D. (2002). Modeling spatial variation in leukemia survival data. *Journal of the American Statistical Association*, **97**(460), 965–972.

Hennerfeind, A., Brezger, A., and Fahrmeir, L. (2006). Geoadditive survival models. *Journal of the American Statistical Association*, **101**(475), 1065–1075.

Hjort, N. L. (1990). Nonparametric Bayes estimators based on beta processes in models for life history data. *The Annals of Statistics*, **18**, 1259–1294.

Hutton, J. L. and Monaghan, P. F. (2002). Choice of parametric accelerated life and proportional hazards models for survival data: Asymptotic results. *Lifetime Data Analysis*, **8**, 375–393.

Ibrahim, J. G., Chen, M. H., and Sinha, D. (2001). *Bayesian Survival Analysis*. Springer-Verlag.

Jara, A. and Hanson, T. E. (2011). A class of mixtures of dependent tailfree processes. *Biometrika*, **98**, 553–566.

Jara, A., Lesaffre, E., De Iorio, M., and Quitana, F. (2010). Bayesian semiparametric inference for multivariate doubly-interval-censored data. *The Annals of Applied Statistics*, **4**(4), 2126–2149.

Jara, A., Hanson, T. E., Quintana, F. A., Müller, P., and Rosner, G. L. (2011). DPpackage: Bayesian semi- and nonparametric modeling in R. *Journal of Statistical Software*, **40**(5), 1–30.

Johnson, W. O. and Christensen, R. (1989). Nonparametric Bayesian analysis of the accelerated failure time model. *Statistics and Probability Letters*, **8**, 179–184.

Kalbfleisch, J. D. (1978). Nonparametric Bayesian analysis of survival time data. *Journal of the Royal Statistical Society. Series B (Methodological)*, **40**, 214–221.

Kaufman, C. G., Schervish, M. J., and Nychka, D. W. (2008). Covariance tapering for likelihood-based estimation in large spatial data sets. *Journal of the American Statistical Association*, **103**(484), 1545–1555.

Kay, R. and Kinnersley, N. (2002). On the use of the accelerated failure time model as an alternative to the proportional hazards model in the treatment of time to event data: A case study in influenza. *Drug Information Journal*, **36**, 571–579.

Kneib, T. (2006). *Mixed model based inference in structured additive regression*. Ludwig-Maximilians-Universität München.

Kneib, T. and Fahrmeir, L. (2007). A mixed model approach for geoadditive hazard regression. *Scandinavian Journal of Statistics*, **34**(1), 207–228.

Koenker, R. (2008). Censored quantile regression redux. *Journal of Statistical Software*, **27**(6), 1–25.
Koenker, R. and Hallock, K. F. (2001). Quantile regression. *Journal of Economic Perspectives*, **15**, 143–156.
Komárek, A. and Lesaffre, E. (2007). Bayesian accelerated failure time model for correlated censored data with a normal mixture as an error distribution. *Statistica Sinica*, **17**, 549–569.
Komárek, A. and Lesaffre, E. (2008). Bayesian accelerated failure time model with multivariate doubly-interval-censored data and flexible distributional assumptions. *Journal of the American Statistical Association*, **103**, 523–533.
Kottas, A. and Gelfand, A. E. (2001). Bayesian semiparametric median regression modeling. *Journal of the American Statistical Association*, **95**, 1458–1468.
Kuo, L. and Mallick, B. (1997). Bayesian semiparametric inference for the accelerated failure-time model. *Canadian Journal of Statistics*, **25**, 457–472.
Lang, S. and Brezger, A. (2004). Bayesian P-splines. *Journal of Computational and Graphical Statistics*, **13**, 183–212.
Lavine, M. (1992). Some aspects of Polya tree distributions for statistical modelling. *The Annals of Statistics*, **20**, 1222–1235.
Lavine, M. (1994). More aspects of Polya tree distributions for statistical modelling. *The Annals of Statistics*, **22**, 1161–1176.
Li, J. (2010). Application of copulas as a new geostatistical tool. *Dissertation*.
Li, J., Hong, Y., Thapa, R., and Burkhart, H. E. (2015a). Survival analysis of loblolly pine trees with spatially correlated random effects. *Journal of the American Statistical Association*, **in press**.
Li, L., Hanson, T., and Zhang, J. (2015b). Spatial extended hazard model with application to prostate cancer survival. *Biometrics*, **in press**.
Li, Y. and Lin, X. (2006). Semiparametric normal transformation models for spatially correlated survival data. *Journal of the American Statistical Association*, **101**(474), 591–603.
Li, Y. and Ryan, L. (2002). Modeling spatial survival data using semiparametric frailty models. *Biometrics*, **58**(2), 287–297.
Lin, X. and Wang, L. (2011). Bayesian proportional odds models for analyzing current status data: univariate, clustered, and multivariate. *Communications in Statistics-Simulation and Computation*, **40**(8), 1171–1181.
Lin, X., Cai, B., Wang, L., and Zhang, Z. (2015). A Bayesian proportional hazards model for general interval-censored data. *Lifetime Data Analysis*, **in press**.
Liu, Y. (2012). *Bayesian analysis of spatial and survival models with applications of computation techniques*. Ph.D. thesis, University of Missouri–Columbia.
Lo, A. Y. (1984). On a class of Bayesian nonparametric estimates: I. Density estimates. *Annals of Statistics*, **12**, 351–357.
Martinussen, T. and Scheike, T. H. (2006). *Dynamic Regression Models for Survival Data*. Springer-Verlag.
McKinley, T. J. (2007). *Spatial survival analysis of infectious animal diseases*. Ph.D. thesis, University of Exeter.

Müller, P. and Quintana, F. A. (2004). Nonparametric Bayesian data analysis. *Statistical Science*, **19**(1), 95–110.
Müller, P., Quintana, F., Jara, A., and Hanson, T. (2015). *Bayesian Nonparametric Data Analysis*. Springer-Verlag: New York.
Murphy, S. A., Rossini, A. J., and van der Vaart, A. W. (1997). Maximum likelihood estimation in the proportional odds model. *Journal of the American Statistical Association*, **92**, 968–976.
Neal, R. M. (2000). Markov chain sampling methods for Dirichlet process mixture models. *Journal of Computational and Graphical Statistics*, **9**, 249–265.
Nelsen, R. B. (2006). *An Introduction to Copulas*. Springer, 2nd edition.
Nieto-Barajas, L. E. (2013). Lévy-driven processes in Bayesian nonparametric inference. *Boletín de la Sociedad Matemática Mexicana*, **19**, 267–279.
Ojiambo, P. and Kang, E. (2013). Modeling spatial frailties in survival analysis of cucurbit downy mildew epidemics. *Phytopathology*, **103**(3), 216–227.
Orbe, J., Ferreira, E., and Núñez Antón, V. (2002). Comparing proportional hazards and accelerated failure time models for survival analysis. *Statistics in Medicine*, **21**(22), 3493–3510.
Pan, C., Cai, B., Wang, L., and Lin, X. (2014). Bayesian semi-parametric model for spatial interval-censored survival data. *Computational Statistics & Data Analysis*, **74**, 198–209.
Petrone, S. (1999a). Bayesian density estimation using Bernstein polynomials. *The Canadian Journal of Statistics*, **27**, 105–126.
Petrone, S. (1999b). Random Bernstein polynomials. *Scandinavian Journal of Statistics*, **26**, 373–393.
Reid, N. (1994). A conversation with Sir David Cox. *Statistical Science*, **9**, 439–455.
Ryan, T. and Woodall, W. (2005). The most-cited statistical papers. *Journal of Applied Statistics*, **32**(5), 461–474.
Sang, H. and Huang, J. Z. (2012). A full scale approximation of covariance functions for large spatial data sets. *Journal of the Royal Statistical Society: Series B (Statistical Methodology)*, **74**(1), 111–132.
Scharfstein, D. O., Tsiatis, A. A., and Gilbert, P. B. (1998). Efficient estimation in the generalized odds-rate class of regression models for right-censored time-to-event data. *Lifetime Data Analysis*, **4**, 355–391.
Sethuraman, J. (1994). A constructive definition of Dirichlet priors. *Statistica Sinica*, **4**, 639–650.
Sharef, E., Strawderman, R. L., Ruppert, D., Cowen, M., and Halasyamani, L. (2010). Bayesian adaptive B-spline estimation in proportional hazards frailty models. *Electronic Journal of Statistics*, **4**, 606–642.
Sinha, D. and Dey, D. K. (1997). Semiparametric Bayesian analysis of survival data. *Journal of the American Statistical Association*, **92**, 1195–1212.
Sinha, D., McHenry, M. B., Lipsitz, S. R., and Ghosh, M. (2009). Empirical Bayes estimation for additive hazards regression models. *Biometrika*, **96**(3), 545–558.
Smith, M. S. (2013). Bayesian approaches to copula modelling. *Bayesian Theory and Applications*, pages 336–358.

Susarla, V. and Van Ryzin, J. (1976). Nonparametric Bayesian estimation of survival curves from incomplete observations. *Journal of the American Statistical Association*, **71**, 897–902.

Therneau, T. M. and Grambsch, P. M. (2000). *Modeling Survival Data: Extending the Cox Model*. Springer-Verlag Inc.

Umlauf, N., Adler, D., Kneib, T., Lang, S., and Zeileis, A. (2015). Structured additive regression models: An R interface to BayesX. *Journal of Statistical Software*, **63**(21), 1–46.

Walker, S. G. and Mallick, B. K. (1997). Hierarchical generalized linear models and frailty models with Bayesian nonparametric mixing. *Journal of the Royal Statistical Society, Series B: Methodological*, **59**, 845–860.

Walker, S. G. and Mallick, B. K. (1999). A Bayesian semiparametric accelerated failure time model. *Biometrics*, **55**(2), 477–483.

Wang, L. and Dunson, D. B. (2011). Semiparametric Bayes' proportional odds models for current status data with underreporting. *Biometrics*, **67**(3), 1111–1118.

Wang, S., Zhang, J., and Lawson, A. B. (2012). A Bayesian normal mixture accelerated failure time spatial model and its application to prostate cancer. *Statistical Methods in Medical Research*.

Yang, S. (1999). Censored median regression using weighted empirical survival and hazard functions. *Journal of the American Statistical Association*, **94**, 137–145.

Yang, S. and Prentice, R. L. (1999). Semiparametric inference in the proportional odds regression model. *Journal of the American Statistical Association*, **94**, 125–136.

Yin, G. and Ibrahim, J. G. (2005). A class of Bayesian shared gamma frailty models with multivariate failure time data. *Biometrics*, **61**, 208–216.

Ying, Z., Jung, S. H., and Wei, L. J. (1995). Survival analysis with median regression models. *Journal of the American Statistical Association*, **90**, 178–184.

Zellner, A. (1983). Applications of Bayesian analysis in econometrics. *The Statistician*, **32**, 23–34.

Zhang, J. and Lawson, A. B. (2011). Bayesian parametric accelerated failure time spatial model and its application to prostate cancer. *Journal of Applied Statistics*, **38**(3), 591–603.

Zhang, J., Peng, Y., and Zhao, O. (2011). A new semiparametric estimation method for accelerated hazard model. *Biometrics*, **67**, 1352–1360.

Zhang, M. and Davidian, M. (2008). "Smooth" semiparametric regression analysis for arbitrarily censored time-to-event data. *Biometrics*, **64**(2), 567–576.

Zhao, L. and Hanson, T. E. (2011). Spatially dependent Polya tree modeling for survival data. *Biometrics*, **67**(2), 391–403.

Zhao, L., Hanson, T. E., and Carlin, B. P. (2009). Mixtures of Polya trees for flexible spatial frailty survival modelling. *Biometrika*, **96**(2), 263–276.

Zhou, H., Hanson, T., and Zhang, J. (2015a). Generalized accelerated failure time spatial frailty model for arbitrarily censored data. *Lifetime Data Analysis*, **in revision**.

Zhou, H., Hanson, T., and Knapp, R. (2015b). Marginal Bayesian nonparametric model for time to disease arrival of threatened amphibian populations. *Biometrics*, **in press**.

Zhou, H., Hanson, T., Jara, A., and Zhang, J. (2015c). Modeling county level breast cancer survival data using a covariate-adjusted frailty proportional hazards model. *The Annals of Applied Statistics*, **9**(1): 43–68.

# Chapter 12
# Fully Nonparametric Regression Modelling of Misclassified Censored Time-to-Event Data

Alejandro Jara, María José García-Zattera, and Arnošt Komárek

**Abstract** We propose a fully nonparametric modelling approach for time-to-event regression data, when the response of interest can only be determined to lie in an interval obtained from a sequence of examination times and the determination of the occurrence of the event is subject to misclassification. The covariate-dependent time-to-event distributions are modelled using a linear dependent Dirichlet process mixture model. A general misclassification model is discussed, considering the possibility that different examiners were involved in the assessment of the occurrence of the events for a given subject across time. An advantage of the proposed model is that the underlying time-to-event distributions and the misclassification parameters can be estimated without any external information about the latter parameters.

## 12.1 Introduction

Considerable attention has been given to estimation of survival functions and regression coefficients from a variety of standard regression models for time-to-event data (see, e.g., Hougaard 2000; Sun 2006). Nevertheless, classical survival regression models assume that the determination of the event of interest is done without error which can be unrealistic. As a matter of fact, in many applications, ascertainment of the event of interest is based on a screening test which may not have perfect sensitivity and specificity. In this context, the use of standard survival models can lead

A. Jara (✉) • M.J. García-Zattera
Departamento de Estadística, Pontificia Universidad Católica de Chile, Casilla 306, Correo 22, Santiago, Chile
e-mail: ajara@mat.uc.cl; mjgarcia@uc.cl

A. Komárek
Charles University in Prague, Sokolovská 83, CZ-186 75 Praha 8 - Karlín, Czech Republic
e-mail: komarek@karlin.mff.cuni.cz



R. Mitra, P. Müller (eds.), *Nonparametric Bayesian Inference in Biostatistics*, Frontiers in Probability and the Statistical Sciences,
DOI 10.1007/978-3-319-19518-6_12

to wrong inferences about the distribution of the time-to-event (see, e.g., García-Zattera et al. 2014). Compared to the rich literature on methods for correcting for misclassification in regression models for categorical data (see, e.g., García-Zattera et al. 2010, 2012, and the references therein), the study of models in the context of time-to-event data has received much less attention and has been almost exclusively focussed on misclassification and measurement errors in covariates (see, e.g., Gong et al. 1990).

The effect of response misclassification on estimation and hypothesis testing has been widely investigated in the literature (see, e.g., Buonaccorsi 2010). For regression models, non-differential (covariate independent) misclassification can cause the estimates of the regression coefficients to be attenuated strongly towards the null and that, although the associated significance tests are still valid, its power may be drastically reduced. Under differential (covariate dependent) misclassification, the bias of the estimates can be in both directions, leading to an apparent effect or an apparent lack of effect of the covariate when the reverse is true (see, e.g., Buonaccorsi 2010).

In cross-sectional studies, where the data contain no information regarding the misclassification parameters, several strategies have been proposed in the literature for correcting for misclassification (see, e.g. Neuhaus 1999, 2002; Mwalili et al. 2005; Küchenhoff et al. 2006). In the context of longitudinal categorical data, generalized linear mixed models (see, e.g., Neuhaus 2002), generalized estimating equation (GEE) based approaches (see, e.g., Neuhaus 2002), and transition models (see, e.g., García-Zattera et al. 2010, 2012) have been proposed for correcting for misclassification.

Transition models for the analysis of misclassified alternating longitudinal responses have been considered in the literature by Cook et al. (2000), Rosychuk and Thompson (2001), Rosychuk and Thompson (2003), Nagelkerke et al. (1990), and Rosychuk and Islam (2009), whereas Espeland et al. (1988, 1989), Schmid et al. (1994), Singh and Rao (1995), Albert et al. (1997), García-Zattera et al. (2010), and García-Zattera et al. (2012) addressed the problem of misclassified monotone longitudinal responses. It is important to stress that in a longitudinal setting, unlike cross-sectional studies, the model parameters might be estimated without the use of external information about the misclassification parameters. García-Zattera et al. (2010, 2012) showed that under simple restrictions on the parameter space, the model parameters associated with an inhomogeneous hidden Markov model for monotone responses are identified by the available data. They also proposed univariate and multivariate models to account for predictors allowing for irregularly spaced time intervals and different classifiers.

Here we extend the misclassification modelling approach proposed by García-Zattera et al. (2010, 2012) to account for misclassification of the determination of the event in the context of continuous censored time-to-event data. To avoid specific assumptions on the relationship between the predictors and the time-to-event distribution, a dependent Bayesian nonparametric (BNP) model is considered. A general misclassification model allowing for different classifiers for each subject across examinations is discussed. The chapter is organized as follows. The commonly used regression approaches for the analysis of time-to-event data are discussed in Sect. 12.2. The dependent BNP modelling approach is introduced in Sect. 12.3. The computational implementation of the model is given in Sect. 12.4. In Sect. 12.5, the modelling approach is illustrated using simulated and real-life data. A final discussion section concludes the chapter.

## 12.2 Commonly Used Continuous Time-to-Event Regression Models

A starting point in the construction of a regression model for time-to-event data is the definition of a baseline survival function, $S_0$, that is modified (either directly or indirectly) by subject-specific covariates $\mathbf{x}$. The baseline survival function is denoted by $S_0(t)$, and corresponds to the survival function for the group with all covariates equal to zero. We assume that the time-to-event variable is continuous. Therefore, the baseline density and hazard functions are defined by $f_0(t) = -\frac{d}{dt}S_0(t)$ and $h_0(t) = f_0(t)/S_0(t)$, respectively. The survival, density, and hazard functions for a member of the population with covariates $\mathbf{x}$ will be denoted by $S_{\mathbf{x}}(t)$, $f_{\mathbf{x}}(t)$, and $h_{\mathbf{x}}(t)$, respectively.

### *12.2.1 The Proportional Hazards Model*

A proportional hazards (PH) model (Cox 1972) is obtained by expressing the covariate-dependent survival function $S_{\mathbf{x}}(t)$ as

$$S_{\mathbf{x}}(t) = S_0(t)^{\exp(\mathbf{x}'\boldsymbol{\beta})},$$

which, in terms of hazard function, reduces to

$$h_{\mathbf{x}}(t) = \exp(\mathbf{x}'\boldsymbol{\beta})h_0(t),$$

where $\boldsymbol{\beta}$ is a vector of regression coefficients. The model assumptions imply that for two individuals with covariates $\mathbf{x}_1$ and $\mathbf{x}_2$, the ratio of hazard curves is constant and proportional to

$$\frac{h_{\mathbf{x}_1}(t)}{h_{\mathbf{x}_2}(t)} = \exp\{(\mathbf{x}_1 - \mathbf{x}_2)'\boldsymbol{\beta}\},$$

hence the name "proportional hazards."

The first Bayesian semiparametric approach to PH models considered a gamma process prior for the baseline cumulative hazard function (Kalbfleisch 1978)

$$H_0(t) = \int_0^s h_0(s)ds.$$

The use of the gamma process prior in PH models, as well as the beta process prior (Hjort 1990), piecewise exponential priors, and correlated increments priors are discussed in detail in Ibrahim et al. (2001) and Sinha and Dey (1997). Other approaches include the use of Bernstein polynomials (Gelfand and Mallick 1995; Carlin and Hodges 1999).

### 12.2.2 The Accelerated Failure Time Model

An accelerated failure time (AFT) model is obtained by expressing the covariate-dependent survival function $S_{\mathbf{x}}(t)$ as

$$S_{\mathbf{x}}(t) = S_0\{\exp(-\mathbf{x}'\boldsymbol{\beta})t\}.$$

This is equivalent to the linear model for the log transformation of the corresponding time-to-event variable, $T$,

$$\log T = \mathbf{x}'\boldsymbol{\beta} + \varepsilon,$$

where $\exp(\varepsilon) \sim S_0$. The mean, median, and any quantile of survival for an individual with covariates $\mathbf{x}_1$ is changed by a factor of $\exp\{(\mathbf{x}_1 - \mathbf{x}_2)'\boldsymbol{\beta}\}$ relative to those with covariates $\mathbf{x}_2$.

The first BNP approach for AFT models was introduced in Christensen and Johnson (1988), who obtained approximate marginal inference under a Dirichlet process (DP) prior. Approaches based on DP mixture (DPM) models have been considered by Kuo and Mallick (1997), Kottas and Gelfand (2001) and Hanson (2006b). We refer the reader to Komárek and Lesaffre (2008), for an alternative approach based on mixtures of normal distributions. Tailfree priors can be used to directly model the distribution of $\varepsilon$ without using an additional mixture (Walker and Mallick 1999; Hanson and Johnson 2002; Hanson 2006a; Zhao et al. 2009).

### 12.2.3 Other Models and Extensions

PH and AFT are only two of many other time-to-event models used in practice. For instance, the proportional odds (PO) model has recently gained attention as an alternative to the PH and AFT models. PO defines the survival function $S_{\mathbf{x}}(t)$ for an individual with covariate vector $\mathbf{x}$ through the relation

$$\frac{S_{\mathbf{x}}(t)}{1 - S_{\mathbf{x}}(t)} = \exp\{-\mathbf{x}'\boldsymbol{\beta}\}\left(\frac{S_0(t)}{1 - S_0(t)}\right).$$

The odds of the occurrence of the event before any time $t$ are $\exp\{(\mathbf{x}_1 - \mathbf{x}_2)'\boldsymbol{\beta}\}$ times greater for those with covariates $\mathbf{x}_1$ versus $\mathbf{x}_2$. The PO assumption implies that hazard ratios

$$\lim_{t\to\infty} \frac{h_{\mathbf{x}_1}(t)}{h_{\mathbf{x}_2}(t)} = 1.$$

BNP approaches for the PO model have been based on Bernstein polynomials (Banerjee and Dey 2005) and Polya trees (Hanson 2006a; Hanson and Mingan 2007; Zhao et al. 2009; Hanson et al. 2011).

There are a few more hazard-based models including the additive hazards (AH) model (Aalen 1980, 1989), given by

$$h_{\mathbf{x}}(t) = h_0(t) + \mathbf{x}'\boldsymbol{\beta}.$$

An empirical Bayes approach to this model based on the gamma process was implemented by Sinha et al. (2009). Fully Bayesian approaches require an elaborated model specification to incorporate the rather awkward constraint $h_0(t) + \mathbf{x}'\boldsymbol{\beta} \geq 0$ for $t > 0$ (see, e.g., Yin and Ibrahim 2005). Recently, there has been some interest in the accelerated hazards model (Chen and Wang 2000; Zhang et al. 2011), given by

$$h_{\mathbf{x}}(t) = h_0\{\exp(-\mathbf{x}'\boldsymbol{\beta})t\}.$$

This model allows hazard and survival curves to cross. A highly interpretable model that relates covariates to the baseline residual life function is the proportional mean residual life model (Chen et al. 2005). Under this model, the residual life function for a subject with covariates **x** is given by

$$m_{\mathbf{x}}(t) = \exp(\mathbf{x}'\boldsymbol{\beta})m_0(t),$$

with $m_0(t) = E(T - t|T > t)$, where the expectation is taken with respect to the baseline distribution function. It is important to stress that there have been no Bayesian approaches to these two models to date.

Different "super models" have been proposed in the literature, including transformation models that include PH and PO as special cases (Scharfstein et al. 1998; Mallick and Walker 2003), transformation and extended regression models that include PH and AH as special cases (Yin and Ibrahim 2005; Martinussen and Scheike 2006) and hazard regression models that include both PH and AFT as special cases (Chen and Jewell 2001). While highly flexible, all these models suffer from the limitation that, once fitted, the resulting regression parameters lose any simple interpretability.

Finally, there are several generalizations of the models discussed here. For instance, a common approach for dealing with correlated data has been the introduction of frailty terms to the linear predictor (e.g., $\mathbf{x}'_{ij}\boldsymbol{\beta} + \gamma_i$ for the $j$th subject in cluster $i$). Frailty models have been widely discussed in the literature and correspond to particular cases of hierarchical models. Hazard-based models (proportional, additive, and accelerated) naturally accommodate time-dependent covariates; the linear predictor is simply augmented to be $\mathbf{x}(t)'\boldsymbol{\beta}$. Similarly, hazard-based models can also include time-dependent regression effects via $\mathbf{x}'\boldsymbol{\beta}(t)$ or even $\mathbf{x}(t)'\boldsymbol{\beta}(t)$. A traditional "quick fix" for non-proportional hazards is to introduce an interaction between a continuous covariate $x$ and time, e.g. $h_{\mathbf{x}}(t) = \exp(x\beta_1 + xt\beta_2)h_0(t)$, yielding a particular focused deviation from PH. These extensions allow one to continue using the familiar PH model in situations where the PH assumption does not hold. Other model modifications include cure rate models, joint longitudinal/survival models, recurrent events models, multi-state models, competing risks models, and multivariate models that incorporate dependence more flexibly than frailty models.

## 12.3 A Nonparametric Model with Misclassification and Censoring

Let $T_i \in \mathrm{I\!R}_+$ be the time-to-event for the $i$th subject, $i = 1,\ldots,n$. Suppose that the occurrence of the event is assessed by using a sequence of $J_i$ subject-specific evaluations. Let $0 < v_{(i,1)} < v_{(i,2)} < \cdots < v_{(i,J_i)} < +\infty$ be the ordered examination times for the $i$th subject, $i = 1,\ldots,n$. In a regular interval-censored data context, the time-to-event $T_i$ is unobserved but is exactly known to lie in an interval

$$T_i \in \left(v_{(i,l_{(i,j)}-1)}, v_{(i,l_{(i,j)})}\right],$$

obtained from the sequence of examinations, $l_i \in \left\{1,\ldots,J_i+1\right\}$, where $v_{(i,0)} \equiv 0$ and $v_{(i,J_i+1)} \equiv +\infty$. However, in our setting the determination of the event is prone to misclassification and the observed data are given by the binary variables $Y_{(i,j)}$, $j = 1,\ldots,J_i$, indicating whether the (potentially) error-corrupted evaluation indicates that the event has occurred in the time interval $(v_{i,j-1}, v_{i,j}]$ ($Y_{(i,j)} = 1$) or not ($Y_{(i,j)} = 0$).

In the following, let $\mathbf{T} = (T_i, \ldots, T_n)$ be the vector of unobserved event times, and $\mathbf{Y} = (\mathbf{Y}_1, \ldots, \mathbf{Y}_n)$, where $\mathbf{Y}_i = \left(Y_{(i,1)},\ldots,Y_{(i,J_i)}\right)$, $i = 1,\ldots,n$, is the subject-specific vector of observed binary indicators of potentially misclassified event status. Further, we assume that for each subject and unit, a $p$-dimensional design vector including exogenous covariates is recorded, $\mathbf{x}_i \in \mathscr{X} \subseteq \mathrm{I\!R}^p$, $i = 1,\ldots,n$. The aim here is to develop a fully nonparametric method to make inferences about the dependence of the event times $T_i$ on covariates $\mathbf{x}_i$, where the event times $T_i$ are observed only through sequences of possibly misclassified binary indicators $Y_{(i,j)}$ of the event status. To this end, we specify a BNP for the dependence of event times on covariates in Sect. 12.3.2. Second, link between the observable binary variables $\mathbf{Y}$ and unobservable event times $\mathbf{T}$ will be specified by the misclassification model in Sect. 12.3.1. The implied probability model for the observed data $\mathbf{Y}$ is discussed in Sect. 12.3.3.

### *12.3.1 The Misclassification Model*

Suppose that the evaluation of the event status at each visit is performed by $Q$ examiners. Denote by $\xi_{(i,j)} \in \{1,\ldots,Q\}$ the variable indicating the examiner that evaluates subject $i$ at examination time $v_{(i,j)}$, and let $\xi_i = (\xi_{(i,1)},\ldots,\xi_{(i,J_i)})$ be the vector of indicators of the examiners that score the responses of subject $i$ over time. We further assume that the scoring behavior of each examiner is the same across the study. Let $\eta_q$ and $\alpha_q$, $q = 1,\ldots,Q$, be the specificity and sensitivity parameters for the $q$th examiner, respectively. The misclassification process is characterized by the following assumptions:

(i) $\perp\!\!\!\perp_{1\le i\le n} \mathbf{Y}_i \mid T_1,\ldots,T_n,\xi_1,\ldots,\xi_n,\eta,\alpha$, i.e. the observed response vectors for each subject are independent given the true unobserved times-to-event, examiner indicators, and sensitivity and specificity parameters,

(ii) $\mathbf{Y}_i \perp\!\!\!\perp T_1,\ldots,T_{i-1},T_{i+1},\ldots,T_n \mid T_i,\xi_i,\eta,\alpha,\ \forall\ i$, i.e. the distribution of the observed response vector for a subject only depends on his true unobserved time-to-event, the examiners that score his responses, and the sensitivity and specificity parameters,

(iii) $\perp\!\!\!\perp_{1\le j\le J_i} Y_{i,j} \mid T_i,\xi_i,\eta,\alpha,\ \forall\ i$, i.e. the observed binary responses for a subject are independent across time given his unobserved time-to-event, the examiners that score his responses and the sensitivity and specificity parameters,

(iv) $Y_{(i,j)} \mid T_i,\xi_i,\eta,\alpha \sim \text{Bern}\left(\pi_{(i,j)}\right)$, where

$$\pi_{(i,j)} = P\left(Y_{(i,j)} = 1 \,\middle|\, T_i \in (0, v_{(i,j)}]\right) = \alpha_{\xi_{(i,j)}},$$

or

$$\pi_{(i,j)} = P\left(Y_{(i,j)} = 1 \,\middle|\, T_i \in (v_{(i,j)}, +\infty)\right) = 1 - \eta_{\xi_{(i,j)}},$$

if $T_i \in (0, v_{(i,j)}]$ or $T_i \in (v_{(i,j)}, +\infty)$, respectively.

Extensions of the general misclassification model are also possible, by including examiner-specific characteristics in the misclassification parameters. Following García-Zattera et al. (2010, 2012), we consider the following identifying restriction on the misclassification parameters space

$$\left\{(\eta_q,\alpha_q) \in [0,1]^2 : \eta_q + \alpha_q > 1\right\}, \qquad q = 1,\ldots,Q.$$

Finally, we assume that for every $q \in \{1,\ldots,Q\}$,

$$(\eta_q,\alpha_q) \overset{\text{ind.}}{\sim} \text{Beta}\left(a_{(q)}^{(\eta,0)}, a_{(q)}^{(\eta,1)}\right) \times \text{Beta}\left(a_{(q)}^{(\alpha,0)}, a_{(q)}^{(\alpha,1)}\right) \times \mathrm{I\!I}\left(\eta_q,\alpha_q\right)_{\left\{(\eta_q,\alpha_q):\, \eta_q+\alpha_q>1\right\}}. \tag{12.1}$$

### 12.3.2 The Underlying Time-to-Event Model

Even though the models described in Sect. 12.2 provide useful summary information in the absence of estimates of a baseline survival distribution and may be formulated in a parametric or semi-parametric fashion, they arise from specific assumptions on the relationship between the covariates and the time-to-event. All these assumptions may be considered too strong in many practical applications. This issue is particularly relevant for misclassified censored data, where the degree of available information to perform diagnostic techniques is rather reduced due to the misclassification mechanism.

We extend the recent developments on dependent nonparametric priors, initially proposed by MacEachern (1999, 2000), to provide a framework for modelling interval-censored time-to-event data. Specifically, rather than assuming a functional form for a particular functional of the survival distribution, such as the conditional hazard function, the time-to-event regression problem is cast as inference for a complete family of conditional time-to-event distributions $\mathscr{F} = \{F_{\mathbf{x}} : \mathbf{x} \in \mathscr{X} \subseteq \mathbb{R}^p\}$, where $T_i \mid \mathbf{x}_i \overset{ind.}{\sim} F_{\mathbf{x}_i}$. Here the family $\mathscr{F}$ is modelled using a dependent DP (DDP) mixture of lognormals model for its densities,

$$f_{\mathbf{x}}(t) = \int \frac{1}{t}\phi(\log t \mid \mu, \sigma^2) dG_{\mathbf{x}}(\mu, \sigma^2),$$

where $\phi(\cdot \mid \mu, \sigma^2)$ denotes the density of the Gaussian distribution with mean $\mu$ and variance $\sigma^2$, and, for every $\mathbf{x} \in \mathscr{X}$, $G_{\mathbf{x}}$ is probability measures defined on $\mathbb{R} \times \mathbb{R}_+$. The probability model for the conditional densities is induced by specifying a probability model for the collection of mixing distributions $\mathscr{G}^{\mathscr{X}} = \{G_{\mathbf{x}} : \mathbf{x} \in \mathscr{X} \subseteq \mathbb{R}^p\}$. Justified by results in Barrientos et al. (2012), on the full support of models for predictor-dependent probability measures, we focused on predictor-dependent discrete mixing distributions where only the support points are indexed by the predictor values. Specifically, we consider a "single-weights" dependent Dirichlet process prior (DDP) (MacEachern 2000) for $\mathscr{G}^{\mathscr{X}}$. A "single-weights" DDP prior defines a family of almost surely discrete random probability measures, which extends the DP stick-breaking representation (Sethuraman 1994), such that, for every $\mathbf{x} \in \mathscr{X}$,

$$G_{\mathbf{x}}(\cdot) = \sum_{l=1}^{\infty} w_l \delta_{\boldsymbol{\theta}_l(\mathbf{x})}(\cdot),$$

where $\delta_\eta(\cdot)$ denotes the Dirac measure at $\eta$, $\{\boldsymbol{\theta}_l(\mathbf{x}) : \mathbf{x} \in \mathscr{X}\}$, $l \in \mathbb{N}$, are independent stochastic processes with index set $\mathscr{X}$, and the weights arise from a stick-breaking construction: $w_1 = v_1$ and, for $k = 2, 3, \ldots$, $w_l = v_k \prod_{r=1}^{l-1}(1 - v_r)$, with

$$v_l \mid \alpha \overset{\text{i.i.d.}}{\sim} \text{Beta}(1, \alpha),$$

for $\alpha \in \mathbb{R}_+$, independent across the support point processes.

We build our proposal on a linear DDP (LDDP) prior formulation (De Iorio et al. 2004, 2009; Jara et al. 2010), which corresponds to a particular version of the "single-weights" DDP, where the component of the atoms defining the location in a DDP mixture model follows a linear regression model $\boldsymbol{\theta}_l(\mathbf{x}) = (\mathbf{x}'\boldsymbol{\beta}_l, \sigma_l^2)$. An important advantage of this model for related random probability measures is that it can be represented as the following DPM of linear (in the coefficients) regression models

$$\log T_i \mid \boldsymbol{\beta}_i, \sigma_i^2 \overset{ind.}{\sim} N\left(\mathbf{x}_i'\boldsymbol{\beta}_i, \sigma_i^2\right), \tag{12.2}$$

$$(\boldsymbol{\beta}_i, \sigma_i^2) \mid G \overset{i.i.d.}{\sim} G, \tag{12.3}$$

and

$$G \mid \alpha, G_0 \sim DP\left(\alpha G_0\right), \tag{12.4}$$

where $N(\mu, \sigma^2)$ denotes the Gaussian distribution with mean $\mu$ and variance $\sigma^2$, $G_0 \equiv N_p\left(\boldsymbol{\beta}|\mu_b, \mathbf{S}_b\right) \Gamma\left(\sigma^{-2}|\tau_1/2, \tau_2/2\right)$, with $N_p(\cdot \mid \mu, \mathbf{A})$ being the $p$-dimensional Gaussian distribution with mean $\mu$ and covariance matrix $\mathbf{A}$, and $\Gamma\left(\cdot \mid a, b\right)$ being the gamma distribution with parameter $a$ and $b$. The LDDP model specification is completed with the following hyper-priors

$$\alpha|a_0, b_0 \sim \Gamma\left(a_0, b_0\right), \tag{12.5}$$

$$\tau_2|\tau_{s_1}, \tau_{s_2} \sim \Gamma(\tau_{s_1}/2, \tau_{s_2}/2), \tag{12.6}$$

$$\mu_b|\mathbf{m}_0, \mathbf{S}_0 \sim N_p(\mathbf{m}_0, \mathbf{S}_0), \tag{12.7}$$

and

$$\mathbf{S}_b|\nu, \boldsymbol{\Psi} \sim IW_p(\nu, \boldsymbol{\Psi}), \tag{12.8}$$

where $\mathrm{IW}_p(\nu, \boldsymbol{\Psi})$ denotes a $p$-dimensional inverted—Wishart distribution with degrees of freedom $\nu$ and scale matrix $\boldsymbol{\Psi}$, parameterized such that $E(\boldsymbol{\Sigma}) = \boldsymbol{\Psi}^{-1}/(\nu - p - 1)$.

### 12.3.3 The Implied Statistical Model

The misclassification model assumptions (i)–(iv), along with the assumptions associated with the dependent mixture model for the underlying time-to-event data (12.2)–(12.8), imply that the joint probability model for the observed binary indicators and unobserved time-to-event variables for each subject is given by

$$\begin{aligned} p(\mathbf{Y}_1, \ldots \mathbf{Y}_n, T_1, \ldots, T_n \mid \alpha, \eta, G) &= \prod_{i=1}^{n} p(\mathbf{Y}_i \mid T_i, \alpha, \eta) f_{\mathbf{x}_i}(T_i \mid G), \\ &= \prod_{i=1}^{n} \left\{ \prod_{j=1}^{J_i} p\left(Y_{(i,j)} \,\middle|\, T_i, \eta_{\xi_{(i,j)}}, \alpha_{\xi_{(i,j)}}\right) \right\} f_{\mathbf{x}_i}(T_i \mid G), \end{aligned}$$

where

$$
\begin{aligned}
p\big(Y_{(i,j)} \mid T_i, \eta_{\xi_{(i,j)}}, \alpha_{\xi_{(i,j)}}\big)
&= \left\{ \alpha_{\xi_{(i,j)}}^{Y_{(i,j)}} \left(1-\alpha_{\xi_{(i,j)}}\right)^{1-Y_{(i,j)}} \right\}^{\mathbb{I}(T_i)\left\{T_i \in \left(0, \nu_{(i,j)}\right]\right\}} \times \\
&\quad \left\{ \left(1-\eta_{\xi_{(i,j)}}\right)^{Y_{(i,j)}} \eta_{\xi_{(i,j)}}^{1-Y_{(i,j)}} \right\}^{\mathbb{I}(T_i)\left\{T_i \in \left(\nu_{(i,j)}, +\infty\right)\right\}}, \\
&= \prod_{l=1}^{j} \left\{ \alpha_{\xi_{(i,j)}}^{Y_{(i,j)}} \left(1-\alpha_{\xi_{(i,j)}}\right)^{1-Y_{(i,j)}} \right\}^{\mathbb{I}(T_i)\left\{T_i \in \left(\nu_{(i,l-1)}, \nu_{(i,l)}\right]\right\}} \times \\
&\quad \prod_{l=j+1}^{J_i+1} \left\{ \left(1-\eta_{\xi_{(i,j)}}\right)^{Y_{(i,j)}} \eta_{\xi_{(i,j)}}^{1-Y_{(i,j)}} \right\}^{\mathbb{I}(T_i)\left\{T_i \in \left(\nu_{(i,l-1)}, \nu_{(i,l)}\right]\right\}}.
\end{aligned}
$$

Therefore, the induced probability model for observed data is given by

$$
\begin{aligned}
p(\mathbf{Y}_1,\ldots \mathbf{Y}_n \mid \alpha,\eta,G) &= \prod_{i=1}^{n} \int_{\mathbb{R}_+} p(\mathbf{Y}_i \mid T_i,\alpha,\eta) f_{\mathbf{x}_i}(T_i \mid G) dT_i, \\
&= \prod_{i=1}^{n} \int_{\mathbb{R}_+} \left\{ \prod_{j=1}^{J_i} p\big(Y_{(i,j)} \mid T_i, \eta_{\xi_{(i,j)}}, \alpha_{\xi_{(i,j)}}\big) \right\} \times \\
&\qquad \left\{ \int \frac{1}{T_i} \phi\left(\log T_i \mid \mathbf{x}_i' \boldsymbol{\beta}, \sigma^2\right) dG(\boldsymbol{\beta},\sigma^2), \right\} dT_i.
\end{aligned}
$$

## 12.4 The Computational Implementation

A Markov chain Monte Carlo algorithm for exploring the posterior distribution of the proposed model is implemented in the `LDDPmissurv` function of the library DPpackage (Jara 2007; Jara et al. 2011) of the R program (R Development Core Team 2015), which is available from the *Comprehensive R Archive Network (CRAN)*. This function fits a marginalized version of the model, where the random probability measure $G$ is integrated out. Full inference on the conditional density, survival, and hazard functions at a given covariate level are obtained using the $\varepsilon-$DP approximation proposed by Muliere and Tardella (1998), with $\varepsilon = 0.01$. The posterior distribution for the model parameters is explored by considering a data augmented version. Specifically, we consider the posterior distribution arising after introducing the time-to-event times $T_i$, $i = 1,\ldots,n$. The full conditionals for the characteristic parameters of the proposed models, arising from the augmented posterior are described next.

### 12.4.1 The Full Conditional for the Unobserved Time-to-Events

Assumptions (i)–(iv), along with the assumptions of the fully nonparametric regression model, imply that for $i = 1, \ldots, n$, the full conditional distribution for the corresponding time-to-event is given by

$$
\begin{aligned}
p(T_i \mid \cdots) &\propto \prod_{j=1}^{J_i} p\left(Y_{(i,j)} \mid T_i, \eta_{\xi_{(i,j)}}, \alpha_{\xi_{(i,j)}}\right) p\left(T_i \mid \boldsymbol{\beta}_i, \sigma_i^2\right), \\
&= \sum_{j=1}^{J_i+1} W_{(i,j)}(\mathbf{Y}_i, \xi_i, \alpha, \eta) \frac{1}{T_i} \phi\left(\log T_i \mid \mathbf{x}_i' \boldsymbol{\beta}_i, \sigma_i^2\right) \mathrm{I\!I}(T_i)_{\left\{T_i \in \left(v_{(i,j-1)}, v_{(i,j)}\right]\right\}},
\end{aligned}
$$

where for $j = 1, \ldots, J_i + 1$,

$$
\begin{aligned}
W_{(i,j)}(\mathbf{Y}_i, \xi_i, \alpha, \eta) = &\left\{ \prod_{l=j}^{J_i} \alpha_{\xi_{(i,l)}}^{Y_{(i,l)}} \left(1 - \alpha_{\xi_{(i,l)}}\right)^{1-Y_{(i,l)}} \right\} \\
&\times \left\{ \prod_{l=1}^{j-1} \left(1 - \eta_{\xi_{(i,l)}}\right)^{Y_{(i,l)}} \eta_{\xi_{(i,l)}}^{1-Y_{(i,l)}} \right\}. \qquad (12.9)
\end{aligned}
$$

### 12.4.2 The Full Conditional for the Misclassification Parameters

The full conditionals for the misclassification parameters are truncated beta distributions given by

$$
\eta_q \mid \cdots \sim \text{Beta}\left(a_q^{(\eta,0)} + n_q^{00}, a_q^{(\eta,1)} + n_q^{+0} - n_q^{00}\right) I(\eta_q)_{\{\eta_q : \eta_q > 1 - \alpha_q\}},
$$

and

$$
\alpha_q \mid \cdots \sim \text{Beta}\left(a_q^{(\alpha,0)} + n_q^{11}, a_q^{(\alpha,1)} + n_q^{+1} - n_q^{11}\right) I(\alpha_q)_{\{\alpha_q : \alpha_q > 1 - \eta_q\}},
$$

where

$$
n_q^{00} = \sum_{i=1}^{n} \sum_{j=1}^{J_i} I\left(Y_{(i,j)}, T_i\right)_{\left\{Y_{(i,j,k)} = 0, T_{(i,j)} \in \left(v_{(i,k)}, +\infty\right)\right\}} I\left(\xi_{(i,k)}\right)_{\left\{\xi_{(q,k)} = i\right\}},
$$

$$
n_q^{+0} = \sum_{i=1}^{N} \sum_{j=1}^{J} \sum_{k=1}^{K_i} I\left(T_{(i,j)}\right)_{\left\{T_{(i,j)} \in \left(v_{(i,k)}, +\infty\right)\right\}} I\left(\xi_{(i,k)}\right)_{\left\{\xi_{(q,k)} = i\right\}},
$$

$$
n_q^{11} = \sum_{i=1}^{N} \sum_{j=1}^{J} \sum_{k=1}^{K_i} I\left(Y_{(i,j,k)}, T_{(i,j)}\right)_{\left\{Y_{(i,j,k)} = 1, T_{(i,j)} \in \left(0, v_{(i,k)}\right]\right\}} I\left(\xi_{(i,k)}\right)_{\left\{\xi_{(q,k)} = i\right\}},
$$

and

$$n_q^{+1} = \sum_{i=1}^{N}\sum_{j=1}^{J}\sum_{k=1}^{K_i} I\left(T_{(i,j)}\right)_{\{T_{(i,j)}\in\left(0,v_{(i,k)}\right]\}} I\left(\xi_{(i,k)}\right)_{\{\xi_{(q,k)}=i\}}.$$

## 12.5 Illustrations

### 12.5.1 Simulated Data

To illustrate our approach, we conducted the analysis of two simulated data sets. In both cases, we consider the time-to-event variable for $n = 500$ subjects. We assume a binary predictor, $d_i$, and 250 subjects in each level (groups A and B). Different distributions were assumed for each level of the predictor such that

$$\log T_1,\ldots,\log T_{250} \overset{i.i.d.}{\sim} f_A \equiv 0.5 \times N(1.8, 0.005) + 0.5 \times N(2.4, 0.0025),$$

and

$$\log T_{251},\ldots,\log T_{500} \overset{i.i.d.}{\sim} f_B \equiv N(2.1, 0.0324).$$

An important characteristic of the simulation scenario is the bimodal behavior of the distribution of the time-to-event in group A. In group B, a unimodal behavior for the distribution of the time-to-event variables was assumed. The true time-to-event for each subject was interval-censored by simulating subject-specific visit times. A total number of $J_i = 8$ visit times were considered. The first visit was drawn from an $N(6, 0.0025)$ distribution. Each of the distances between the consecutive visits was drawn from an $N(0.8, 0.0125^2)$ distribution.

Two misclassification scenarios were considered. In both cases, we assume that the assessment of the occurrence of the event was performed by $Q = 10$ examiners, allocated randomly to each subject and visit. In Scenario I, an imperfect determination of the event was assumed for all examiners. In Scenario II, a perfect determination of the event was assumed for all examiners. The values for the examiner-specific sensitivity and specificity parameters under Scenario I are given in Fig. 12.2.

The proposed model was fitted to both simulated data sets by considering $\mathbf{x}_i = (1, d_i)'$ and using the following values for the hyper-parameters: $a_0 = 10$, $b_0 = 1$, $\tau_1 = \tau_{s_1} = 6.01$, $\tau_{s_2} = 2.01$, $\nu = 4$, $\boldsymbol{\Psi} = \mathbf{I}_2$, $\mathbf{m}_0 = \mathbf{0}_2$, $\mathbf{S}_0 = 10^2 \times \mathbf{I}_2$, $a_1^{(\eta,0)} = \cdots = a_Q^{(\eta,0)} = a_1^{(\eta,1)} = \cdots = a_Q^{(\eta,1)} = 1$ and $a_1^{(\alpha,0)} = \cdots = a_Q^{(\alpha,0)} = a_1^{(\alpha,1)} = \cdots = a_Q^{(\alpha,1)} = 1$. In each analysis 110,000 samples of a Markov chain cycle were completed. Because of storage limitations and dependence, the full chain was subsampled every 10 steps after a burn-in period of 10,000 samples, to give a reduced chain of length 10,000.

Figure 12.1 displays the true and estimated survival curves for the time-to-event in both groups under scenarios I and II. The predictive survival functions closely approximate the true survival ones, which were almost entirely enclosed in point-wise 95 % highest posterior density (HPD) intervals. We note that these results are for one random sample from two particular densities, and these conclusions should not be overinterpreted. Nonetheless, these examples do show that our proposal is highly flexible and is able to capture different behaviors of the time-to-event survival functions. The examples also show that when misclassification is not present, the proposed model does not overfit the data.

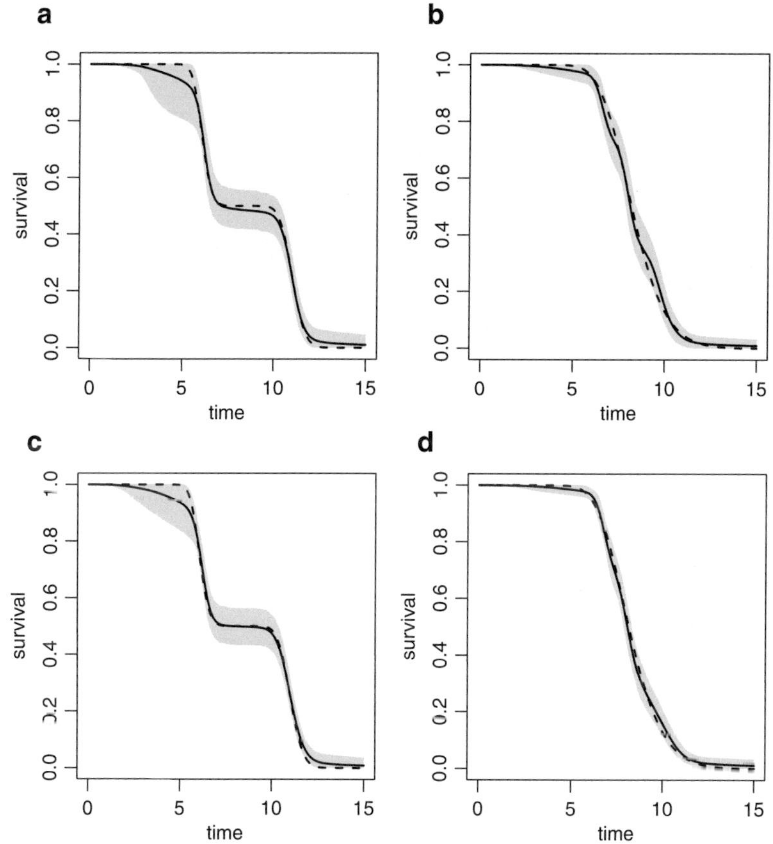


**Fig. 12.1** Simulated data: True (*dotted line*) and posterior mean (*continuous line*) of the marginal survival functions for group A (panel (**a**) and (**c**)) and B (panel (**b**) and (**d**)). A point-wise 95 % credible band is displayed as a *gray* area in each case. Panels (**a**)–(**b**) and (**c**)–(**d**) display the results for the simulation scenario I (misclassification) and II (no misclassification), respectively

Figure 12.2 shows the posterior mean and 95 % HPD credible interval for the sensitivity and specificity of each examiner. Similarly to the observed for the time-to-event parameters, the results suggest that the misclassification parameters can be estimated with only a minimal bias and with a reasonable precision. The results of both simulation scenarios strongly suggest that concentrated information on the misclassification parameters is not needed to obtain precise estimates for the model parameters. Thus, they can be estimated from the raw data without extra information on the misclassification parameters.

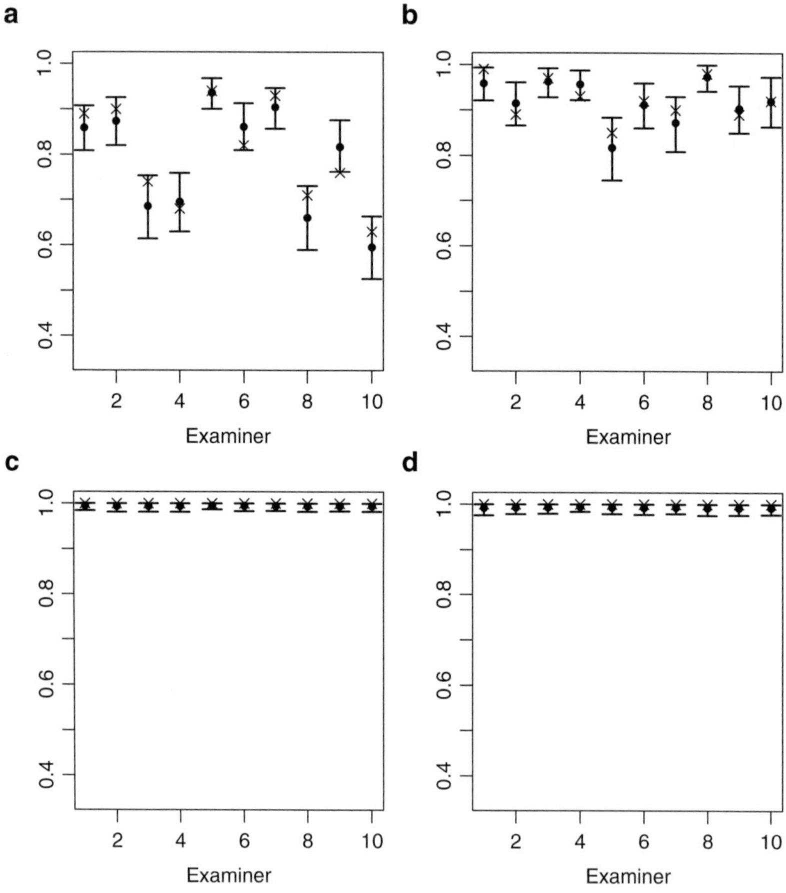


**Fig. 12.2** Simulated data: True value (*cross*), posterior mean (*filledcircle*), and 95 % HPD credible intervals for the examiner's sensitivity (panel (**a**) and (**c**)) and specificity (panel (**b**) and (**d**)). Panels (**a**)–(**b**) and (**c**)–(**d**) display the results for the simulation scenario I (misclassification) and II (no misclassification), respectively

## *12.5.2 The Signal Tandmobiel® Data*

We consider data gathered in a longitudinal oral health study conducted in Flanders (Belgium) between 1996 and 2001, the Signal-Tandmobiel® (ST) study (see, e.g., Vanobbergen et al. 2000). For this project, 4468 children were examined on a yearly basis during their primary school time (between 7 and 12 years of age) by one of sixteen dental examiners. The purpose of the investigation is to examine the effect of covariates on the distribution of the time-to-caries experience (CE), which is defined as a binary variable indicating whether a tooth is decayed at d3 level, missing or filled due to caries. This involves the analysis of a misclassified time-to-event data since: (1) due to the setup of the study, the tooth status was assessed yearly and, thus, the time to CE can only be determined to lie in a given interval of time, and (2) several examiners were involved in the study and their caries classification may not perfectly reflect the tooth's true condition and, therefore, the presence/absence of CE can be misdiagnosed.

We evaluated the effect of gender, frequency of brushing (once or more a day versus less than once a day) (freqrus), and geographical location (in terms of standardized $x$- and $y$-coordinates) of the municipality of the school to which the child belongs, on the time-to-CE for one of the permanent first molars on the maxilla, teeth 16. The inclusion of the geographical components, expressed in terms of the $x$- and $y$-coordinates, was motivated by the results of previous analyses (without correcting for misclassification) that showed a significant East–West gradient in the prevalences of CE in Flanders. A possible cause for the apparent trend in CE is a different scoring behavior of the 16 dental examiners and their non-homogeneous spatial distribution in the study area (Mwalili et al. 2005).

The proposed model was fitted to the ST data sets by considering $\mathbf{x}_i = (1, \text{gender}_i, \text{freqrus}_i, x_i, y_i)'$ and using the following values for the hyper-parameters: $a_0 = 10$, $b_0 = 1$, $\tau_1 = \tau_{s_1} = 6.01$, $\tau_{s_2} = 2.01$, $\nu = 4$, $\boldsymbol{\Psi} = \mathbf{I}_5$, $\mathbf{m}_0 = \mathbf{0}_5$, $\mathbf{S}_0 = 10^2 \times \mathbf{I}_5$, $a_1^{(\eta,0)} = \cdots = a_Q^{(\eta,0)} = a_1^{(\eta,1)} = \cdots = a_Q^{(\eta,1)} = 1$ and $a_1^{(\alpha,0)} = \cdots = a_Q^{(\alpha,0)} = a_1^{(\alpha,1)} = \cdots = a_Q^{(\alpha,1)} = 1$. Again, 110,000 samples of a Markov chain cycle were completed. Because of storage limitations and dependence, the full chain was subsampled every 10 steps after a burn-in period of 10,000 samples, to give a reduced chain of length 10,000.

The posterior inference about the survival curves for different values of the predictors suggest that boys have a higher probability of developing CE and that the higher the frequency of brushing the lower the probability of developing CE. The results also suggest a lack of effect of the geographical location of the time-to-event distributions. The lack of a significant geographical trend in the prevalences and incidences of CE would support the hypothesis that the observed geographical gradient is due to the different scoring behavior of the examiners rather than to real local geographical differences. However, this could also be explained by the loss of power associated with the presence of misclassification.These findings are illustrated in Fig. 12.3, where the posterior predictive survival curves for different covariates values are displayed.

Finally, Fig. 12.4 shows the posterior means and 95 % HPD credible intervals for the sensitivity and specificity for each examiner. The results suggest a greater variability in the sensitivity than in the specificity estimates, which can be explained by the low prevalences and incidences of CE. All examiners showed a sensitivity greater than 0.60, with relatively narrow 95 % HPD credible intervals, with one

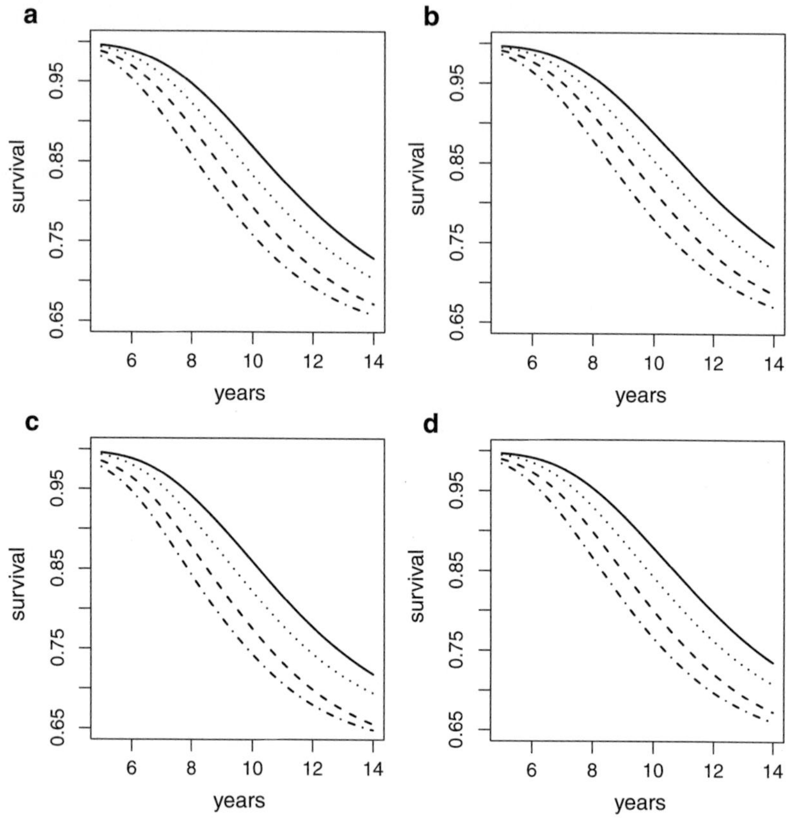


**Fig. 12.3** Signal Tandmobiel® data: Posterior mean of the survival function for different combination of the covariates. In each panel the results are display for girls that brush regularly their teeth (*solid line*), girls who do not brush regularly their teeth (*dashed line*), boys who brush regularly their teeth (*dotted line*) and boys who do not brush regularly their teeth (*dotdashed line*). Panel (**a**), (**b**), (**c**) and (**d**) display the results for $(x,y)/100 = (1.0440, 1.8120)$, $(x,y)/100 = (1.0440, 2.0400)$, $(x,y)/100 = (1.7850, 1.8120)$, and $(x,y)/100 = (1.7850, 2.0400)$, respectively

exception. The latter result is explained by the fact that this examiner was only involved in the first 2 years of the ST study, having less information for the estimation of his parameters. The posterior means for the specificity parameters were higher than 0.90 for all examiners.

## 12.6 Concluding Remarks

We have proposed a fully nonparametric regression framework for the analysis of mismeasured time-to-event response data, and where different classifiers are present. We provided empirical evidence showing that under simple restrictions on

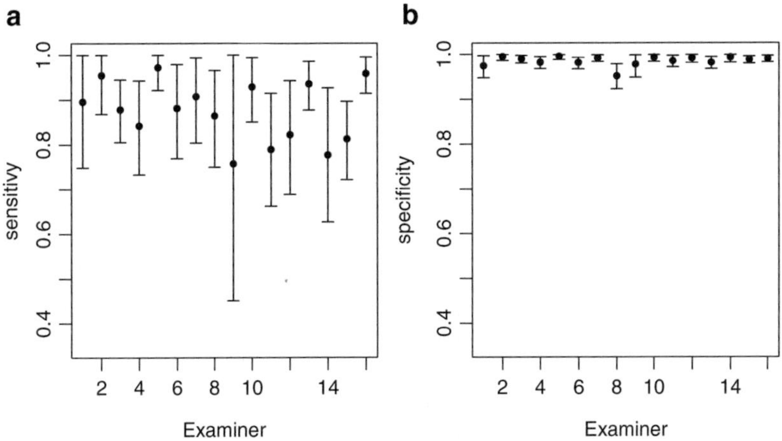


**Fig. 12.4** Signal Tandmobiel® data: posterior mean (*filledcircle*) and 95 % HPD credible intervals for the examiner's sensitivity (panel (**a**)) and specificity (panel (**b**))

the parameter space, the model parameters in the proposed model can be estimated from the raw data only, thus avoiding the need of external information on the misclassification parameters. The results suggest that even under the use of uniform priors on the misclassification parameters, correct estimates can be obtained. We noted that if external information on the misclassification parameters is available, this can be easily incorporated into the model specification.

Several extensions of this work can be done. Justified by the existence of easy/difficult to diagnose subjects, the relaxation of some of the assumptions (i)–(iv) could be of interest; for instance, a possible improvement of the scoring behavior of the examiners across the study could be considered. Finally, the extension of the proposed models for handling multivariate responses is the subject of ongoing research.

**Acknowledgements** The first author was supported by Fondecyt 1141193 grant. The second author was supported by Fondecyt 11110033 grant. The Signal-Tandmobiel® study comprises following partners: D. Declerck (Dental School, Catholic University Leuven), L. Martens (Dental School, University Ghent), J. Vanobbergen (Dental School, University Ghent), P. Bottenberg (Dental School, University Brussels), E. Lesaffre (L-BioStat, Catholic University Leuven), and K. Hoppenbrouwers (Youth Health Department, Catholic University Leuven; Flemish Association for Youth Health Care.

## References

Aalen, O. O. (1980). A model for nonparametric regression analysis of counting processes. In *Lecture Notes in Statistics, Vol. 2*, pages 1–25. Springer-Verlag.

Aalen, O. O. (1989). A linear regression model for the analysis of life times. *Statistics in Medicine*, **8**, 907–925.

Albert, P. S., Hunsberger, S. A., and Biro, F. M. (1997). Modeling repeated measures with monotonic ordinal responses and misclassification, with applications to studying maturation. *Journal of American Statistical Association*, **92**, 1304–1311.

Banerjee, S. and Dey, D. K. (2005). Semi-parametric proportional odds models for spatially correlated survival data. *Lifetime Data Analysis*, **11**, 175–191.

Barrientos, A. F., Jara, A., and Quintana, F. (2012). On the support of MacEachern's dependent Dirichlet processes and extensions. *Bayesian Analysis*, **7**, 277–310.

Buonaccorsi, J. P. (2010). *Measurement error*. Chapman & Hall/CRC.

Carlin, B. P. and Hodges, J. S. (1999). Hierarchical proportional hazards regression models for highly stratified data. *Biometrics*, **55**, 1162–1170.

Chen, Y. Q. and Jewell, N. P. (2001). On a general class of semiparametric hazards regression models. *Biometrika*, **88**, 687–702.

Chen, Y. Q. and Wang, M. C. (2000). Analysis of accelerated hazards models. *Journal of the American Statistical Association*, **95**, 608–618.

Chen, Y. Q., Jewell, N. P., Lei, X., and Cheng, S. C. (2005). Semiparametric estimation of proportional mean residual life model in presence of censoring. *Biometrics*, **61**, 170–178.

Christensen, R. and Johnson, W. O. (1988). Modeling accelerated failure time with a Dirichlet process. *Biometrika*, **75**, 693–704.

Cook, R. J., Ng, E. T. M., and Meade, M. O. (2000). Estimation of operating characteristics for dependent diagnostic tests based on latent Markov models. *Biometrics*, **56**, 1109–1117.

Cox, D. R. (1972). Regression models and life-tables (with discussion). *Journal of the Royal Statistical Society, Series B: Methodological*, **34**, 187–220.

De Iorio, M., Müller, P., Rosner, G. L., and MacEachern, S. N. (2004). An ANOVA model for dependent random measures. *Journal of the American Statistical Association*, **99**, 205–215.

De Iorio, M., Johnson, W. O., Müller, P., and Rosner, G. L. (2009). Bayesian nonparametric non-proportional hazards survival modelling. *Biometrics*, **65**, 762–771.

Espeland, M. A., Murphy, W. C., and Leverett, D. H. (1988). Assessing diagnostic reliability and estimating incidence rates associated with a strictly progressive disease: dental caries. *Statistics in Medicine*, **7**, 403–416.

Espeland, M. A., Platt, O. S., and Gallagher, D. (1989). Joint estimation of incidence and diagnostic error rates from irregular longitudinal data. *Journal of the American Statistical Association*, **84(408)**, 972–979.

García-Zattera, M. J., Mutsvari, T., Jara, A., Declerck, D., and Lesaffre, E. (2010). Correcting for misclassification for a monotone disease process with an application in dental research. *Statistics in Medicine*, **29**, 3103–3117.

García-Zattera, M. J., Jara, A., Lesaffre, E., and Marshall, G. (2012). Modeling of multivariate monotone disease processes in the presence of misclassification. *Journal of the American Statistical Association*, **107**, 976–989.

García-Zattera, M. J., Jara, A., and Komárek, A. (2014). A flexible AFT model for misclassified clustered interval-censored data. Technical report, Department of Statistics, Pontificia Universidad Católica de Chile, Chile.

Gelfand, A. E. and Mallick, B. K. (1995). Bayesian analysis of proportional hazards models built from monotone functions. *Biometrics*, **51**, 843–852.

Gong, G., Whittemore, A. S., and Grosser, S. (1990). Censored Survival Data with Misclassified Covariates: A Case Study of Breast-Cancer Mortality. *Journal of the American Statistical Association*, **85**, 20–28.

Hanson, T. E. (2006a). Inference for mixtures of finite Polya tree models. *Journal of the American Statistical Association*, **101**(476), 1548–1565.

Hanson, T. E. (2006b). Modeling censored lifetime data using a mixture of gammas baseline. *Bayesian Analysis*, **1**, 575–594.

Hanson, T. E. and Johnson, W. O. (2002). Modeling regression error with a mixture of Polya trees. *Journal of the American Statistical Association*, **97**(460), 1020–1033.

Hanson, T. E. and Mingan, Y. (2007). Bayesian semiparametric proportional odds models. *Biometrics*, **63**(1), 88–95.

Hanson, T. E., Branscum, A., and Johnson, W. O. (2011). Predictive comparison of joint longitudinal–survival modeling: a case study illustrating competing approaches. *Lifetime Data Analysis*, **17**, 3–28.

Hjort, N. L. (1990). Nonparametric Bayes estimators based on beta processes in models for life history data. *The Annals of Statistics*, **18**, 1259–1294.

Hougaard, P. (2000). *Analysis of Multivariate Survival Data*. Springer, New York, USA.

Ibrahim, J. G., Chen, M. H., and Sinha, D. (2001). *Bayesian Survival Analysis*. Springer-Verlag.

Jara, A. (2007). Applied Bayesian non- and semi-parametric inference using DPpackage. *R News*, **7**(3), 17–26.

Jara, A., Lesaffre, E., De Iorio, M., and Quitana, F. (2010). Bayesian semiparametric inference for multivariate doubly interval-censored data. *The Annals of Applied Statistics*, **4**(4), 2126–2149.

Jara, A., Hanson, T. E., Quintana, F. A., Müller, P., and Rosner, G. L. (2011). DPpackage: Bayesian semi- and nonparametric modeling in R. *Journal of Statistical Software*, **40**(5), 1–30.

Kalbfleisch, J. D. (1978). Nonparametric Bayesian analysis of survival time data. *Journal of the Royal Statistical Society, Series B: Methodological*, **40**, 214–221.

Komárek, A. and Lesaffre, E. (2008). Bayesian accelerated failure time model with multivariate doubly-interval-censored data and flexible distributional assumptions. *Journal of the American Statistical Association*, **103**, 523–533.

Kottas, A. and Gelfand, A. E. (2001). Bayesian semiparametric median regression modeling. *Journal of the American Statistical Association*, **95**, 1458–1468.

Küchenhoff, H., Mwalili, S. M., and Lesaffre, E. (2006). A general method for dealing with misclassification in regression: the misclassification SIMEX. *Biometrics*, **62**, 85–96.

Kuo, L. and Mallick, B. (1997). Bayesian semiparametric inference for the accelerated failure-time model. *Canadian Journal of Statistics*, **25**, 457–472.

MacEachern, S. N. (1999). Dependent nonparametric processes. In *ASA Proceedings of the Section on Bayesian Statistical Science, Alexandria, VA*. American Statistical Association.

MacEachern, S. N. (2000). Dependent Dirichlet processes. Technical report, Department of Statistics, The Ohio State University.

Mallick, B. K. and Walker, S. G. (2003). A bayesian semiparametric transformation model incorporating frailties. *Journal of Statistical Planning and Inference*, **112**, 159–174.

Martinussen, T. and Scheike, T. H. (2006). *Dynamic Regression Models for Survival Data*. Springer-Verlag.

Muliere, P. and Tardella, L. (1998). Approximating distributions of random functionals of Ferguson-Dirichlet priors. *The Canadian Journal of Statistics*, **26**, 283–297.

Mwalili, S. M., Lesaffre, E., and Declerck, D. (2005). A Bayesian ordinal logistic regression model to correct for interobserver measurement error in a geographical oral health study. *Journal of the Royal Statistical Society, Series C*, **54**, 77–93.

Nagelkerke, N. J. D., Chunge, R. N., and Kinot, S. N. (1990). Estimation of parasitic infection dynamics when detectability is imperfect. *Statistics in Medicine*, **9**, 1211–1219.

Neuhaus, J. M. (1999). Bias and efficiency loss due to misclassified responses in binary regression. *Biometrika*, **86(4)**, 843–855.

Neuhaus, J. M. (2002). Analysis of clustered and longitudinal binary data subject to response misclassification. *Biometrics*, **58**, 675–683.

R Development Core Team (2015). *R: A Language and Environment for Statistical Computing*. R Foundation for Statistical Computing, Vienna, Austria. ISBN 3-900051-07-0.

Rosychuk, R. J. and Islam, M. S. (2009). Parameter estimation in a model for misclassified Markov data - a Bayesian approach. *Computational Statistics and Data Analysis*, **53**, 3805–3816.

Rosychuk, R. J. and Thompson, M. E. (2001). A semi-Markov model for binary longitudinal responses subject to misclassification. *Canadian Journal of Statistics*, **19**, 394–404.

Rosychuk, R. J. and Thompson, M. E. (2003). Bias correction of two-state latent Markov process parameter estimates under misclassification. *Statistics in Medicine*, **22**, 2035–2055.

Scharfstein, D. O., Tsiatis, A. A., and Gilbert, P. B. (1998). Efficient estimation in the generalized odds-rate class of regression models for right-censored time-to-event data. *Lifetime Data Analysis*, **4**, 355–391.

Schmid, C. H., Segal, M. R., and Rosner, B. (1994). Incorporating measurement error in the estimation of autoregressive models for longitudinal data. *Journal of Statistical Planning and Inference*, **42(1–2)**, 1–18.

Sethuraman, J. (1994). A constructive definition of Dirichlet priors. *Statistica Sinica*, **4**, 639–650.

Singh, A. C. and Rao, J. N. K. (1995). On the adjustment of gross flow estimates for classification error with application to data from the Canadian labour force survey. *Journal of the American Statistical Association*, **90(430)**, 478–488.

Sinha, D. and Dey, D. K. (1997). Semiparametric Bayesian analysis of survival data. *Journal of the American Statistical Association*, **92**, 1195–1212.

Sinha, D., McHenry, M. B., Lipsitz, S. R., and Ghosh, M. (2009). Empirical bayes estimation for additive hazards regression models. *Biometrika*, **96**(3), 545–558.

Sun, J. (2006). *The Statistical Analysis of Interval-censored Failure Time Data*. Springer, New York, USA.

Vanobbergen, J., Martens, L., Lesaffre, E., and Declerck, D. (2000). The Signal-Tandmobiel project, a longitudinal intervention health promotion study in Flanders (Belgium): baseline and first year results. *European Journal of Paediatric Dentistry*, **2**, 87–96.

Walker, S. G. and Mallick, B. K. (1999). A Bayesian semiparametric accelerated failure time model. *Biometrics*, **55**(2), 477–483.

Yin, G. and Ibrahim, J. G. (2005). A class of bayesian shared gamma frailty models with multivariate failure time data. *Biometrics*, **61**, 208–216.

Zhang, J., Peng, Y., and Zhao, O. (2011). A new semiparametric estimation method for accelerated hazard model. *Biometrics*, **67**(4): 1352–1360.

Zhao, L., Hanson, T. E., and Carlin, B. P. (2009). Mixtures of Polya trees for flexible spatial frailty survival modelling. *Biometrika*, **96**(2), 263–276.

# Part IV
# Random Functions and Response Surfaces

# Chapter 13
# Neuronal Spike Train Analysis Using Gaussian Process Models

Babak Shahbaba, Sam Behseta, and Alexander Vandenberg-Rodes

**Abstract** Statistical analysis of simultaneously recorded neurons plays an important role in understanding complex behaviors, decision making process, and neurophysiological disorders. Here, we briefly review several statistical methods specifically developed for analysis of neuronal spike trains. We then focus on application of Gaussian process models for estimating time-varying firing rates of neurons and show how this approach can be extended for modeling synchrony among multiple neurons. We finish this chapter by discussing some possible future directions where more advanced nonparametric Bayesian methods can be utilized to improve existing models.

## 13.1 Introduction

A common approach in neuroscience involves recording spiking activities or action potentials of neurons using microelectrodes. Subsequently, neuronal data may be represented as the times at which spikes occur. The main objective of a considerably large number of statistical methods then is to model the temporal evolution of the firing patterns of a group of neurons (Brillinger 1988; Brown et al. 2004; Kass et al. 2005; Gerstner and Kistler 2002; Tuckwell 1988; H.C. 1989; Riccardi 1977; Holden 1976; West 2007; Rigat et al. 2006). For a comprehensive review of the topic see Kass et al. (2005).

B. Shahbaba (✉) • A. Vandenberg-Rodes
UC Irvine, CA, USA
e-mail: babaks@uci.edu; vandenbe@uci.edu

S. Behseta
CSUF, Fullerton, CA, USA
e-mail: sbehseta@exchange.fullerton.edu

 
R. Mitra, P. Müller (eds.), *Nonparametric Bayesian Inference in Biostatistics*,
Frontiers in Probability and the Statistical Sciences,
DOI 10.1007/978-3-319-19518-6_13

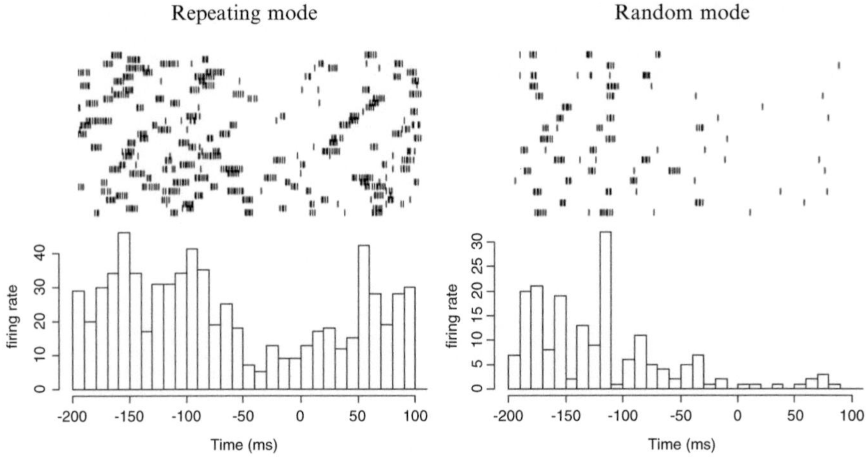


**Fig. 13.1** Raster and PSTH plots for a neuron under repeating (*left panel*) and random (*right panel*) modes

As an example, consider a study of neurons recorded from the primary motor cortex (M1) area of a Macaque monkey, performing a sequential task of reaching five targets arranged horizontally on a touch sensitive screen (Matsuzaka et al. 2007). The targets were numbered 1 to 5 from left to right and could be illuminated upon reaching them. The animal was trained to respond to the visual stimuli under two experimental conditions or modes. In the "repeating mode," a sequence of targets appeared on the screen in a repeating order. In the "random mode," targets appear in a pseudo-random order. An experimental window of 300 milliseconds (ms) was used. This time window began at 200 ms prior to the target reach and continued for 100 ms after that. The upper segment of Fig. 13.1 shows the corresponding raster plots for a neuron recorded under both modes of this task. Rows in the raster plot represent trials, and the tick-marks are spike time occurrences.

Typically, neuronal data are summarized through peri-stimulus time histograms (PSTH). For the above example, by dividing the window of 300 ms into bins of 10 ms, and pooling the spike occurrences within each bin, one can create the PSTH plots shown in the lower segment of Fig. 13.1.

Let $Y_1, \ldots, Y_n$ denote the number of spike occurrences within the bins centered at times $t_1, \ldots, t_n$. A common approach to modeling the neuronal firing rates is by discretizing an inhomogeneous Poisson point process, resulting in a hierarchical model of the form

$$\begin{aligned} Y_j &\sim p(y_j|\theta_j, \zeta) \\ \theta_j &= f(t_j), \end{aligned} \tag{13.1}$$

where the data model $p(y_j|\theta_j, \zeta)$ is usually a Poisson($\theta_j$) density. If the bins are narrow enough they can safely be assumed (or thresholded) to contain at most one spike, so that $Y_j$ can be modeled as a Bernoulli random variable. The model includes a vector of nuisance parameters $\zeta$ to allow for generality.

The (latent) firing rates $f(t_j)$ are often the key quantity of interest. In particular—ignoring the details of binning or the data model—one should think of $f(t)$ as a latent *function* over the whole time interval of interest. Thus in the Bayesian approach a key challenge is to produce an appropriately realistic and flexible prior distribution over latent functions, and to then provide computationally efficient procedures for approximating the posterior distribution of $f$ given the observed spike data $Y_1, \ldots, Y_n$.

One highly flexible approach is known as Bayesian adaptive regression splines (BARS) (Dimatteo et al. 2001). In this model the latent function $f$ is assumed to be a spline having knots at unknown locations $\xi_1, \ldots, \xi_k$. Writing $f(t)$ in terms of basis functions $b_{\xi,h}(t)$ as $f(t) = \sum_h b_{\xi,h}(t)\beta_{\xi,h}$, the function evaluations $f(t_1), \ldots, f(t_n)$ may be collected into a vector $(f(t_1), \ldots, f(t_n))^T = \mathbf{X}_\xi \beta_\xi$, where $\mathbf{X}_\xi$ is the design matrix and $\beta_\xi$ the coefficient vector. BARS then employs a reversible jump MCMC algorithm (Green 1995) to sample from a suitable approximate posterior distribution on the knot set $\xi$. Eventually, curves are fitted via model averaging.

One advantage of using BARS for modeling PSTH is the ability to develop inferential methods, suitable for comparing the patterns of spiking activities for comparative problems similar to the one depicted in Fig. 13.1 (Behseta and S. 2011; Behseta et al. 2005). Kottas et al. (2012) and Kottas and Behseta (2010) also treated the problem of comparing the spike trains resulting in the experiments similar to the ones shown in Fig. 13.1, and subsequently developed a fully-Bayesian inferential methodology for such comparative studies; however, they used a Dirichlet process mixture of Beta densities as the prior for $f$.

Although single-neuron analysis of this type has led to many interesting discoveries, it is widely perceived that complex behaviors are driven by networks of neurons instead of a single neuron (Buzsáki 2010). Therefore, investigators have been recording neuronal activity from multi-probe electrodes. From the statistical point of view, multiple channel recordings greatly facilitate assessing the temporal properties of networks of neurons in real time.

Early analysis of simultaneously recorded neurons focused on correlation of activity across pairs of neurons using cross-correlation analyses (Narayanan and Laubach 2009) and analyses of changes in correlation over time, i.e., by using a Joint Peri-Stimulus Time Histogram or PSTH (Gerstein and Perkel 1969). Similar analyses were performed in the frequency domain by using coherence analysis of neuron pairs using Fourier-transformed neural activity (Brown et al. 2004). For the Bayesian correction for attenuation of correlation in multi-trial spike see Behseta et al. (2009). There are also a number of multivariate analysis techniques for the investigation of simultaneously recorded populations of neurons (Chapin 1999; Nicolelis 1999; Grün et al. 2002; Pillow et al. 2008; Harrison et al. 2013; Brillinger 1988; Brown et al. 2004; Kass et al. 2005; West 2007; Rigat et al. 2006; Patnaik et al. 2008; Diekman et al. 2009; Sastry and Unnikrishnan 2010; Kottas et al. 2012).

Recently, Kelly and Kass (2012) proposed a new method to quantify synchrony among multiple neurons. The authors argued that separating stimulus effects from history effects would allow for a more precise estimation of the instantaneous conditional firing rate. Specifically, given the firing history $H_t$, define $\lambda^A(t|H_t^A)$, $\lambda^B(t|H_t^B)$, and $\lambda^{AB}(t|H_t^{AB})$ to be the conditional firing intensities of neuron A, neuron B, and their synchronous spikes respectively. Independence between the two point processes may be examined by testing the null hypothesis $H_0 : \zeta(t) = 1$, where $\zeta(t) = \frac{\lambda^{AB}(t|H_t^{AB})}{\lambda^A(t|H_t^A)\lambda^B(t|H_t^B)}$. where $\zeta$ represents the excess firing rate ($\zeta > 1$) or the suppression of firing rate ($\zeta < 1$) due to dependence between two neurons (Ventura et al. 2005; Kelly and Kass 2012). That is, $\zeta$ accounts for the excess joint spiking beyond what is explained by independence.

In this chapter we discuss alternative approaches that place a Gaussian process (GP) prior over the latent function in order to model the time-varying and history-dependent firing rate for each neuron. The joint distribution of spikes for multiple neurons is connected to their marginals using a parametric copula model. We first provide a brief overview of univariate GP models in Sect. 13.2. Then, in Sect. 13.3 we discuss the application of GP for single neuron analysis. The copula model for simultaneously recorded neurons is presented in Sect. 13.4. In Sect. 13.5, we discuss some future directions.

## 13.2 Gaussian Process Models

A Gaussian process (GP) on the real line is a random real-valued function $x(t)$, with statistics determined by its mean function $\mathbb{E}x(s)$ and *kernel* $\kappa(s,t) = \mathrm{Cov}(x(s),x(t))$. More precisely, all finite-dimensional distributions $(x(t_1),\ldots,x(t_n))$ are multivariate Gaussian with mean $(\mathbb{E}x(t_1),\ldots,\mathbb{E}x(t_n))$, and with covariance matrix $(\kappa(t_k,t_\ell))_{k,\ell=1}^n$. Since the latter must be positive semi-definite for every finite collection of inputs $t_1,\ldots,t_n$, only certain kernels $\kappa$ are *valid*. Thus when using Gaussian processes, a practitioner often chooses from among the few popular classes of kernels, such as the Squared Exponential (SE), Ornstein–Uhlenbeck (OU), Matérn, Polynomial, and linear combinations of these. For example, we can use the following covariance form, which combines a random constant with the SE kernel and iid observation noise (Rasmussen and Williams 2006; Neal 1998):

$$
\begin{aligned}
C_{ij} &= Cov[x(t_i),x(t_j)] \\
&= \lambda^2 + \eta^2 \exp[-\rho^2(t_i - t_j)^2] + \delta_{ij}\sigma_\varepsilon^2. \qquad (13.2)
\end{aligned}
$$

Here, $\lambda, \eta, \rho$, and $\sigma_\varepsilon$ are hyperparameters with their own hyperpriors. In general, the choice of kernel encodes our *qualitative* beliefs about the underlying signal. For instance, samples from a GP with OU kernel are always non-differentiable functions $x(t)$, and the SE kernel generates only infinitely differentiable functions. Despite such differences, both kernels have the inverse *length-scale* $\rho$ as a hyperparameter: smaller values of $\rho$ result in more slowly varying functions. In practice we only observe GPs at a finite number of points, hence local properties of GPs such as differentiability are irrelevant—in Cunningham et al. (2007), for example, it was observed that using the Matérn instead of SE kernel resulted in negligible differences when modeling spike trains.

It should be remarked that many dynamical models such as autoregressive processes with Gaussian noise are also multivariate Gaussian and hence can be situated within the GP framework, albeit with a usually less interpretable kernel.

## 13.3 Gaussian Process Model of Firing Rates

With the model (13.1), note that the latent firing rates $f(t_i)$ need to be non-negative, hence a Gaussian process cannot be directly used as a prior distribution for $f$. In the case of Poisson observations one can use an exponential link function, letting $f(t) = \exp(x(t))$, where $x(t)$ is a GP. In Cunningham et al. (2007) it was instead proposed to set the constant mean function $\mu(t) = \mu > 0$ as an additional hyperparameter for a GP, and then to let the latent rate $f$ be this GP conditioned to be non-negative.[1] In a recent work, Shahbaba et al. (2014) also use the model (13.1) to estimate the underlying firing rate of neurons, but after discretizing time so that there is at most one spike within each time interval, resulting in a binary time series $Y_1, \ldots, Y_n$ comprised of 1 s (spike) and 0 s (silence). To model the latent firing probabilities $f(t_i) = P(Y_i = 1)$, they apply the sigmoidal transformation

$$f(t_i) = \frac{1}{1 + \exp[-u(t_i)]},$$

where $u(t)$ has a GP prior. Note that as $u(t)$ increases, so does $f(t_i)$. The prior autocorrelation imposed by this model allows the firing rate to change smoothly over time. When there are $R$ trials (i.e., $R$ spike trains) for each neuron, we can model the corresponding spike trains as conditionally independent given the latent variable $u(t)$. Figure 13.2 shows the posterior expectation of firing rate (blue curve) overlaid on the PSTH plot of a single neuron with 5 ms bin intervals.

## 13.4 Detecting Synchrony Among Multiple Spike Trains

For multiple neurons, Shahbaba et al. (2014) propose to use a generalization of the method by Kelly and Kass (2012) (see Sect. 13.1) to model the joint distribution as a function of marginals. In general, models that couple the joint distribution of two (or more) variables to their individual marginal distributions are called *copula* models. See Nelsen (1998) for detailed discussion of copula models. Onken et al. (2009) and Berkes et al. (2009) also use copula models for capturing neural dependencies.

[1] Their data model is somewhat different from (13.1), as the spike times are assumed to follow a conditionally inhomogeneous gamma-interval process instead of a Poisson process.

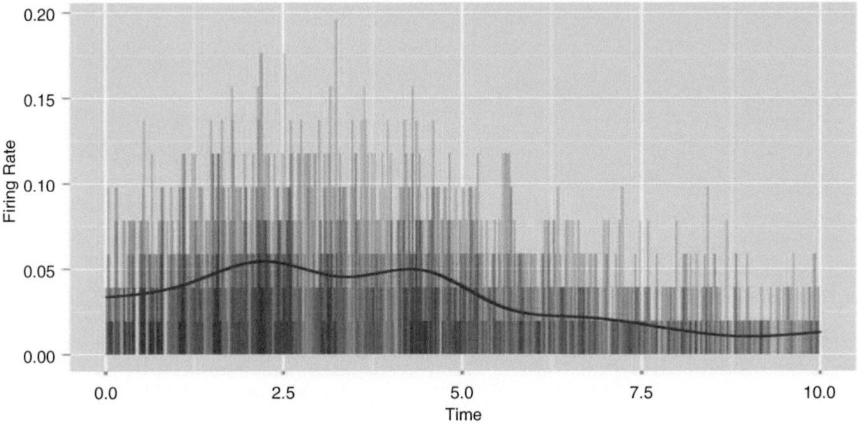


**Fig. 13.2** Using the Gaussian process model of Shahbaba et al. (2014) to capture the underlying firing rate of a single neuron from prefrontal cortical areas in rat's brain. There are 51 spike trains recorded over 10 s. The PSTH plot is generated by creating 5 ms intervals. The *curve* shows the estimated firing rate (posterior expectation)

Let $H$ be $n$-dimensional distribution functions with marginals $F_1, \ldots, F_n$. Then, an $n$-dimensional copula is a function of the following form:

$$H(y_1, \ldots, y_n) = \mathscr{H}(F_1(y_1), \ldots, F_n(y_n)), \text{ for all } y_1, \ldots, y_n.$$

Here, $\mathscr{H}$ defines the dependence structure between the marginals. For example, the Farlie–Gumbel–Morgenstern (FGM) copula family (Farlie 1960; Gumbel 1960; Morgenstern 1956; Nelsen 1998) is defined as follows:

$$\mathscr{H} = \left[1 + \sum_{k=2}^{n} \sum_{1 \le j_1 < \cdots < j_k \le n} \beta_{j_1 j_2 \ldots j_k} \prod_{l=1}^{k} (1 - F_{j_l})\right] \prod_{i=1}^{n} F_i, \tag{13.3}$$

where $F_i = F_i(y_i)$. As shown by Wilson and Ghahramani (2012), this idea can be generalized to multivariate processes. Restricting the above model to second-order interactions, we have

$$H(y_1, \ldots, y_n) = \left[1 + \sum_{1 \le j_1 < j_2 \le n} \beta_{j_1 j_2} \prod_{l=1}^{2} (1 - F_{j_l})\right] \prod_{i=1}^{n} F_i, \tag{13.4}$$

where $F_i = P(Y_i \le y_i)$. Here, we use $y_1, \ldots, y_n$ to denote the firing status of $n$ neurons at time $t$; $\beta_{j_1 j_2}$ captures the relationship between the $j_1^{\text{th}}$ and $j_2^{\text{th}}$ neurons.

For a pair of neurons with firing probabilities $p$ and $q$ respectively, we can show that $\beta = \frac{\zeta - 1}{(1-p)(1-q)}$. As discussed in Sect. 13.1, $\zeta$ represents the excess firing rate ($\zeta > 1$) or the suppression of firing rate ($\zeta < 1$) due to dependence between two neurons (Ventura et al. 2005; Kelly and Kass 2012). In our model, $\beta = 0$ indicates that the two neurons are independent; the excess firing rate and the suppression of firing rate between two dependent neurons are represented by $\beta > 0$ and $\beta < 0$ respectively.

To ensure that probability distribution functions remain within $[0,1]$, the following constraints on all $\binom{n}{2}$ parameters $\beta_{j_1 j_2}$ are imposed:

$$1+\sum_{1\leq j_1<j_2\leq n}\beta_{j_1 j_2}\prod_{l=1}^{2}\varepsilon_{j_l}\geq 0,\quad \varepsilon_1,\cdots,\varepsilon_n\in\{-1,1\}.$$

Considering all possible combinations of $\varepsilon_{j_1}$ and $\varepsilon_{j_2}$ in the above condition, there are $n(n-1)$ linear inequalities, which can be combined into the following inequality:

$$\sum_{1\leq j_1<j_2\leq n}|\beta_{j_1 j_2}|\leq 1.$$

### 13.4.1 Computation

Sampling from the posterior distribution of $\beta$'s in the above copula model is quite challenging because of the imposed constraints. Lan et al. (2014) developed a novel Markov Chain Monte Carlo algorithm for constrained target distributions of this type based on Hamiltonian Monte Carlo (HMC) (Duane et al. 1987; Neal 2011). They show that in many cases, bounded connected constrained $D$-dimensional parameter spaces can be bijectively mapped on to the $D$-dimensional unit ball. Their method then augments the original $D$-dimensional parameter $\theta$ with an extra auxiliary variable $\theta_{D+1}$ to form an extended $(D+1)$-dimensional parameter $\tilde{\theta}=(\theta,\theta_{D+1})$ such that $\|\tilde{\theta}\|_2=1$ so $\theta_{D+1}=\pm\sqrt{1-\|\theta\|_2^2}$. This way, the domain of the target distribution is changed from the unit ball to the $D$-dimensional sphere. Using the above transformation, they define the Hamiltonian dynamics on the sphere. This way, the resulting HMC sampler can move freely on the sphere, $\mathbf{S}^D$, while implicitly handling the constraints imposed on the original parameters. As illustrated in Fig. 13.3, the boundary of the constraint, i.e., $\|\theta\|_2=1$, corresponds to the equator on the sphere $\mathbf{S}^D$. Therefore, as the sampler moves on the sphere, passing across the equator from one hemisphere to the other translates to "bouncing back" off the boundary in the original parameter space.

Lan et al. (2014) show that by defining HMC on the sphere, besides handling the constraints implicitly, the computational efficiency of the sampling algorithm could be improved since the resulting dynamics has a partial analytical solution (geodesic flow on the sphere). They used this approach, called Spherical HMC, for sampling from the posterior distribution of $\beta$'s in the above copula model and showed that the resulting sampler is substantially more efficient than alternative methods.

### 13.4.2 Results for Experimental Data

We now consider an experiment designed to investigate the role of the prefrontal cortex in rats in conjunction with reward-seeking behaviors and inhibition of

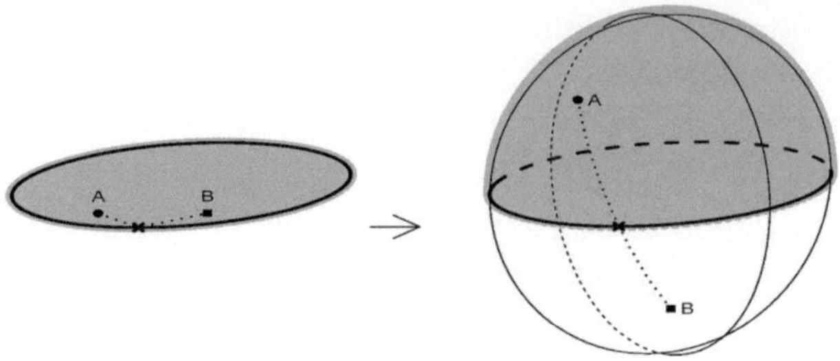


**Fig. 13.3** Transforming unit ball $\mathbf{B}_0^D(1)$ to sphere $\mathbf{S}^D$

reward-seeking in the absence of a rewarded outcome. The neural activity (spike trains) of several prefrontal neurons were recorded simultaneously. There are two conditions during the experiment: rewarded and non-rewarded. During the recording/test sessions, two different stimuli were presented: tone 1 (10 KHz) or tone 2 (5 KHz) individually and in pseudorandom order. At the same time, one of two levers was presented: an active-lever, paired with tone 1 (Rewarded-Stimulus—RS) and an inactive-lever paired with tone 2 (Non-rewarded Stimulus—NS). Pressing the active lever resulted in the offset of tone 1, retraction of the lever, and illumination of the reward receptacle. If the rat then went to the reward receptacle, 0.1 ml of 15 % sucrose solution was delivered as a reward. Pressing the inactive lever produced no effect. See Moorman and Aston-Jones (2014) for more details.

Here, we focus on five simultaneously recorded neurons. There are 51 trials per neuron under each scenario. We set the time intervals to 5 ms. Tables 13.1 and 13.2 show the estimates of $\beta_{i,j}$, which capture the association between the $i$th and $j$th neurons, under the two scenarios. Figure 13.4 shows the schematic representation of these results under the two experimental conditions. The solid line indicates significant association.

These results show that neurons recorded simultaneously in the same brain area are correlated in some conditions and not others. This strongly supports the hypothesis that population coding among neurons (here though correlated activity) is a meaningful way of signaling differences in the environment (rewarded or non-rewarded stimulus) or behavior (going to press the rewarded lever or not pressing) (Buzsáki 2010). It also shows that neurons in the same brain region are differentially involved in different tasks, an intuitive perspective but one that is neglected by much of behavioral neuroscience. Finally, these results indicate that network correlation is dynamic and that functional pairs—again, even within the same brain area—can appear and disappear depending on the environment or behavior. This suggests (but does not confirm) that correlated activity across separate populations within a single brain region can encode multiple aspects of the task. For example, the pairs that are correlated in reward and not in non-reward could be related to reward-seeking whereas pairs that are correlated in non-reward could be related to response inhibition. Characterizing neural populations within a single brain region

**Table 13.1** Estimates of $\beta$'s along with their 95 % probability intervals for the first scenario (Rewarded) based on the copula model. Statistically significant values are shown in bold

| $\beta$ | 2 | 3 | 4 | 5 |
|---|---|---|---|---|
| 1 | **0.22(0.07,0.39)** | 0.00(−0.07,0.04) | 0.03(−0.02,0.15) | 0.01(−0.04,0.08) |
| 2 | | 0.03(−0.02,0.18) | 0.06(−0.02,0.22) | **0.07(0.00,0.25)** |
| 3 | | | 0.08(−0.01,0.26) | **0.21(0.04,0.38)** |
| 4 | | | | **0.23(0.09,0.40)** |

**Table 13.2** Estimates of $\beta$'s along with their 95 % probability intervals for the second scenario (Non-rewarded) based on the copula model. Statistically significant values are shown in bold

| $\beta$ | 2 | 3 | 4 | 5 |
|---|---|---|---|---|
| 1 | 0.05(−0.02,0.25) | −0.01(−0.09,0.04) | 0.15(−0.01,0.37) | 0.05(−0.03,0.22) |
| 2 | | **0.21(0.03,0.41)** | **0.18(0.00,0.37)** | 0.03(−0.02,0.19) |
| 3 | | | **0.17(0.00,0.34)** | 0.03(−0.02,0.19) |
| 4 | | | | 0.07(-0.01,0.24) |

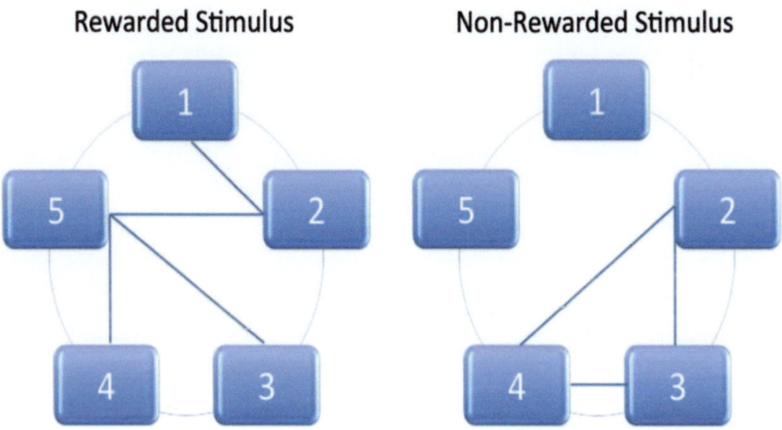


**Fig. 13.4** A schematic representation of connections between five neurons under two experimental conditions. The *solid line* indicates significant association

based on task-dependent differences in correlated firing is a less-frequently studied phenomenon compared to the frequently pursued goal of identifying the overall function of the brain region based on individual neural firing (Stokes et al. 2013).

## 13.5 Future Directions

The methods discussed here can be generalized in several ways as discussed below.

### 13.5.1 Multivariate GPs

The multivariate model presented in the previous section uses univariate Gaussian processes for the marginal distributions and a copula model for the joint distribution of multiple neurons in terms of these marginals. Alternatively, we can use a *multivariate* GP for modeling the joint distribution of multiple neurons directly. A multivariate Gaussian process can be defined in a similar way as a univariate GP, but this time the kernel function depends on two pairs of inputs. For simplicity we can assume that the mean of each process is the zero function. The kernel $\kappa$ is now defined for $i, j = 1, \ldots, p$ and $s, t \in \mathbb{R}$ as

$$\kappa([i,s],[j,t]) = \mathbb{E}x_i(s)x_j(t). \tag{13.5}$$

The initial challenge within the Gaussian process context is to produce a valid and interpretable kernel. A common technique for generating multivariate GP kernels is known as *co-kriging*, borrowed from the geostatistical literature (Cressie 1993).

One variant of co-kriging describes $(x_1(t), \ldots, x_p(t))$ as linear combinations of *latent* factors. We suppose $u_1(t), \ldots, u_q(t)$ are independent mean zero Gaussian processes, and let

$$x_i(t) = \sum_{k=1}^{q} a_{i,k} u_k(t), \qquad \text{for } i = 1, 2, \ldots p. \tag{13.6}$$

Let $\kappa_i(s,t) = \mathbb{E}u_i(s)u_i(t)$ be the kernel for the $i$th latent process. Then the observed processes $\mathbf{x}(t) = (x_1(t), \ldots, x_p(t))$ are jointly mean-zero Gaussian with covariances

$$\mathbb{E}x_i(s)x_j(t) = \sum_{k=1}^{q} a_{i,k} a_{j,k} \kappa_k(s,t). \tag{13.7}$$

This is the *semi-parametric latent factor model* of Teh et al. (2005), so-called because the linear combination of latent GPs is parameterized by the matrix of coefficients $A = (a_{i,k})$, while each Gaussian process is of course a non-parametric model. See Alvarez et al. (2011) for a survey of co-kriging and other multivariate GPs seen in the literature.

Recently, Vandenberg-Rodes and Shahbaba (2015) proposed a multivariate Gaussian processes model for multiple time series $X(t) = (x_1(t), \ldots, x_p(t))$ such that each marginal process $x_j(t)$ is a stationary mean-zero Gaussian process with Matérn kernel. Crucially, the marginal processes are not required to share the same hyperparameter values. This approach can be used to model the joint distribution of the firing rates of multiple neurons directly, and allows for significant heterogeneity among neurons while also providing a high degree of interpretability.

### *13.5.2 Dynamic Networks*

The static (stationary) model discussed here aggregates cross-neuronal spike-train interactions over time. This can lead to misleading results. Although there exist many dynamic methods developed for modeling brain functional and effective connectivity (Friston et al. 1997; Cribben et al. 2013; Ombao et al. 2005; Ombao and Van Bellegem 2008; Motta and Ombao 2012; Park et al. 2014; Lindquist et al. 2014), these approaches are primarily designed for continuous-valued signals such as functional magnetic resonance imaging (fMRI) and electroencephalogram (EEG) data. The GP-based method discussed here can be extended to model neuronal connections dynamically.

### *13.5.3 Community Detection*

Besides allowing for time-varying firing rates and interactions among neurons, the GP-based method can also be extended to cluster neurons based on their cross-dependencies in order to detect subnetworks (communities). To this end, stochastic block models could be used to identify network partitions (Holland et al. 1983). For example, Rodriguez (2012) recently proposed a stochastic block model for network analysis where interactions among factors are observed at multiple time points. This method uses a Bayesian hierarchical stochastic block model to detect possible structural changes in a network. Alternatively, one can use a method similar to the product partition model (PPM) of Müller and Quintana (2010). In general, these methods assume a prior probability on all possible partitions. The assumed prior probability could be influenced by some covariates. This approach can be used to partition neurons into subnetworks.

## References

Alvarez, M. A., Rosasco, L., and Lawrence, N. D. (2011). Kernels for Vector-Valued functions: a review.

Behseta, S. and Chenouri, S. (2011). Comparison of two population of curves with an application in neuronal data analysis. *Statistics in Medicine*, **30**, 1441–1454.

Behseta, S., Kass, R. E., and Wallstrom, G. L. (2005). Hierarchical models for assessing variability among functions. *Biometrika*, **92**(2), 419–434.

Behseta, S., Berdyyeva, T., Olson, C. R., and Kass, R. E. (2009). Bayesian correction for attenuation of correlation in Multi-Trial spike count data. *Journal of Neurophysiology*, pages 90727.2008+.

Berkes, P., Wood, F., and Pillow, J. W. (2009). Characterizing neural dependencies with copula models. In D. Koller, D. Schuurmans, Y. Bengio, and L. Bottou, editors, *Advances in Neural Information Processing Systems 21*, pages 129–136. Curran Associates, Inc.

Brillinger, D. R. (1988). Maximum likelihood analysis of spike trains of interacting nerve cells. *Biological Cybernetics*, **59**, 189–200.

Brown, E. N., Kass, R. E., and Mitra, P. P. (2004). Multiple neural spike train data analysis: state-of-the-art and future challenges. *Nature Neuroscience*, **7**(5), 456–461.

Buzsáki, G. (2010). Neural syntax: Cell assemblies, synapsembles, and readers. *Neuron*, **68**(3), 362–385.

Chapin, J. (1999). *Populatin-level analysis of multi-single neuron recording data: multivariate statistical methods.* In Methods for Neural Ensemble Recordings, M. Nicolelis Editor, CRC Press: Boca Raton (FL).

Cressie, N. (1993). *Statistics for Spatial Data.* Wiley, New York.

Cribben, I., Wager, T., and Lindquist, M. (2013). Detecting functional connectivity change points for single-subject fmri data. *Frontiers in Computational Neuroscience*, **7**(143).

Cunningham, J. P., Yu, B. M., Shenoy, K. V., and Sahani, M. (2007). Inferring neural firing rates from spike trains using gaussian processes. In J. C. Platt, D. Koller, Y. Singer, and S. T. Roweis, editors, *NIPS*.

Diekman, C. O., Sastry, P. S., and Unnikrishnan, K. P. (2009). Statistical significance of sequential firing patterns in multi-neuronal spike trains. *Journal of neuroscience methods*, **182**(2), 279–284.

Dimatteo, I., Genovese, C. R., and Kass, R. E. (2001). Bayesian curve-fitting with free-knot splines. *Biometrika*, **88**(4), 1055–1071.

Duane, S., Kennedy, A., Pendleton, B. J., and Roweth, D. (1987). Hybrid Monte Carlo. *Physics Letters B*, **195**(2), 216–222.

Farlie, D. J. G. (1960). The performance of some correlation coefficients for a general bivariate distribution. *Biometrika*, **47**(3/4).

Friston, K. J., Buechel, C., Fink, G. R., Morris, J., Rolls, E., and Dolan, R. J. (1997). Psychophysiological and modulatory interactions in neuroimaging. *NeuroImage*, **6**(3), 218–229.

Gerstein, G. L. and Perkel, D. H. (1969). Simultaneously recorded trains of action potentials: Analysis and functional interpretation. *Science*, **164**(3881), 828–830.

Gerstner, W. and Kistler, W. M. (2002). *Spiking Neuron Models: Single Neurons, Populations, Plasticity.* Cambridge University Press, 1 edition.

Green, P. J. (1995). Reversible jump Markov chain Monte Carlo computation and Bayesian model determination. *Biometrika*, **82**(4), 711–732.

Grün, S., Diesmann, M., and Aertsen, A. (2002). Unitary events in multiple single-neuron spiking activity: I. detection and significance. *Neural Computation*, **14**(1), 43–80.

Gumbel, E. J. (1960). Bivariate exponential distributions. *Journal of the American Statistical Association*, **55**, 698–707.

Harrison, M. T., Amarasingham, A., and Kass, R. E. (2013). Statistical identification of synchronous spiking. In P. D. Lorenzo and J. Victor., editors, *Spike Timing: Mechanisms and Function*. Taylor and Francis.

Tuckwell, H. C. (1989). Philadelphia, Pa. Society for Industrial and Applied Mathematics (SIAM).

Holden, A. (1976). *Models of the Stochastic Activity of Neurones*. Springer Verlag.

Holland, P. W., Laskey, K., and Leinhardt, S. (1983). Stochastic blockmodels: First steps. *Social Networks*, **5**(2), 109–137.

Kass, R. E., Ventura, V., and Brown, E. N. (2005). Statistical issues in the analysis of neuronal data. *Journal of Neurophysiology*, **94**, 8–25.

Kelly, R. C. and Kass, R. E. (2012). A framework for evaluating pairwise and multiway synchrony among Stimulus-Driven neurons. *Neural Computation*, pages 1–26.

Kottas, A. and Behseta, S. (2010). Bayesian nonparametric modeling for comparison of single-neuron firing intensities. *Biometrics*, pages 277–286.

Kottas, A., Behseta, S., Moorman, D. E., Poynor, V., and Olson, C. R. (2012). Bayesian nonparametric analysis of neuronal intensity rates. *Journal of Neuroscience Methods*, **203**(1).

Lan, S., Zhou, B., and Shahbaba, B. ((2014)). Spherical Hamiltonian Monte Carlo for constrained target distributions. In *Proceedings of the 31th International Conference on Machine Learning (ICML)*.

Lindquist, M., Xu, Y., Nebel, M., and Caffo, B. (2014). Evaluating dynamic bivariate correlations in resting-state fMRI: A comparison study and a new approach. *Neuroimage*, **101**, 531–546.

Matsuzaka, Y., Picard, N., and Strick, P. L. (2007). Skill representation in the primary motor cortex after long-term practice. *Journal of neurophysiology*, **97**(2), 1819–1832.

Moorman, D. and Aston-Jones, G. (2014). Orbitofrontal cortical neurons encode expectation-driven initiation of reward-seeking. *Journal of Neuroscience*, **34**(31), 10234–46.

Morgenstern, D. (1956). Einfache beispiele zweidimensionaler verteilungen. *Mitteilungsblatt für Mathematische Statistik*, **8**, 234–235.

Motta, G. and Ombao, H. (2012). Evolutionary factor analysis of replicated time series. *Biometrics*, **68**, 825–836.

Müller, P. and Quintana, F. (2010). Random partition models with regression on covariates. *J Stat Plan Inference*, **140**(10), 2801–2808.

Narayanan, N. S. and Laubach, M. (2009). Methods for studying functional interactions among neuronal populations. *Methods in molecular biology*, **489**, 135–165.

Neal, R. M. (1998). Regression and classification using Gaussian process priors. *Bayesian Statistics*, **6**, 471–501.

Neal, R. M. (2011). MCMC using Hamiltonian dynamics. In S. Brooks, A. Gelman, G. Jones, and X. L. Meng, editors, *Handbook of Markov Chain Monte Carlo*, pages 113–162. Chapman and Hall/CRC.

Nelsen, R. B. (1998). *An Introduction to Copulas (Lecture Notes in Statistics)*. Springer, 1 edition.

Nicolelis, M. (1999). *Methods for Neural Ensemble Recordings*. CRC Press: Boca Raton (FL).

Ombao, H. and Van Bellegem, S. (2008). Coherence analysis: A linear filtering point of view. *IEEE Transactions on Signal Processing*, **56**, 2259–2266.

Ombao, H., von Sachs, R., and Guo, W. (2005). Slex analysis of multivariate non-stationary time series. *Journal of the American Statistical Association*, **100**, 519–531.

Onken, A., Grünewälder, S., Munk, M., and Obermayer, K. (2009). Modeling short-term noise dependence of spike counts in macaque prefrontal cortex. In D. Koller, D. Schuurmans, Y. Bengio, and L. Bottou, editors, *Advances in Neural Information Processing Systems 21*, pages 1233–1240. Curran Associates, Inc.

Park, T., Eckley, I., and Ombao, H. (2014). Estimating the time-evolving partial coherence between signals via multivariate locally stationary wavelet processes. *IEEE Transactions on Signal Processing*, **accepted**.

Patnaik, D., Sastry, P., and Unnikrishnan, K. (2008). Inferring neuronal network connectivity from spike data: A temporal data mining approach. *Scientific Programming*, **16**, 49–77.

Pillow, J. W., Shlens, J., Paninski, L., Sher, A., Litke, A. M., Chichilnisky, E. J., and Simoncelli, E. P. (2008). Spatio-temporal correlations and visual signalling in a complete neuronal population. *Nature*, **454**(7207), 995–9.

Rasmussen, C. E. and Williams, C. K. I. (2006). *Gaussian Processes for Machine Learning*. MIT Press, 2nd edition.

Riccardi, L. (1977). *Diffusion Processes and Related Topics in Biology*. Springer Verlag.

Rigat, F., de Gunst, M., and van Pelt, J. (2006). Bayesian modeling and analysis of spatio-temporal neuronal networks. *Bayesian Analysis*, pages 733–764.

Rodriguez, A. (2012). Modeling the dynamics of social networks using Bayesian hierarchical blockmodels. *Statistical Analysis and Data Mining*, **5**(3), 218–234.

Sastry, P. S. and Unnikrishnan, K. P. (2010). Conditional probability-based significance tests for sequential patterns in multineuronal spike trains. *Neural Comput.*, **22**(4), 1025–1059.

Shahbaba, B., Zhou, B., Ombao, H., Moorman, D., and Behseta, S. (2014). A Semiparametric Bayesian Model for Detecting Synchrony Among Multiple Neurons. *Neural Computation*, **26**(9), 2025–51.

Stokes, M. G., Kusunoki, M., Sigala, N., Nili, H., Gaffan, D., and Duncan, J. (2013). Dynamic coding for cognitive control in prefrontal cortex. *Neuron*, **78**(2), 364–375.

Teh, Y. W., Seeger, M., and Jordan, M. I. (2005). Semiparametric latent factor models. In *Proceedings of the International Workshop on Artificial Intelligence and Statistics*, volume 10.

Tuckwell, H. (1988). *Introduction to Theoretical Neurobiology: Volume 1, Linear Cable Theory and Dendritic Structure*. Cambridge University Press.

Vandenberg-Rodes, A. and Shahbaba, B. (2015). Dependent Matérn Processes for Multivariate Time Series.

Ventura, V., Cai, C., and Kass, R. E. (2005). Statistical assessment of time-varying dependency between two neurons. *J Neurophysiol*, **94**(4), 2940–7.

West, M. (2007). Hierarchical mixture models in neurological transmission analysis. *Journal of the American Statistical Association*, **92**, 587–606.

Wilson, A. G. and Ghahramani, Z. (2012). Modelling input dependent correlations between multiple responses. In N. C. P. Flach, T. De Bie, editor, *Proceedings of the European Conference on Machine Learning and Principles and Practice of Knowledge Discovery in Databases (ECML PKDD)*, Bristol, UK. Springer.

# Chapter 14
# Bayesian Analysis of Curves Shape Variation Through Registration and Regression

Donatello Telesca

**Abstract** Misalignment of functional features in a sample of random curves leads to potentially misleading inference, when variation in timing is ignored. This chapter reviews the use of Bayesian hierarchical curve registration in Biostatistics and Bioinformatics. Several models allowing for unit-specific random time scales are discussed and applied to longitudinal data arising in biomedicine, pharmacokinetics, and time-course genomics. We consider representations of random functionals based on P-spline priors. Under this framework, straightforward posterior simulation strategies are outlined for inference. Beyond curve registration, we discuss joint regression modeling of both random effects and population level functional quantities. Finally, the use of mixture priors is discussed in the setting of differential expression analysis.

## 14.1 Introduction

Longitudinal studies in Biostatistics often aim to characterize time-dependent dynamics associated with the evolution of specific biological or bio-behavioral processes. Several examples are reported, for example, in Pinheiro and Bates (2000). A more comprehensive treatment of statistical analysis strategies for longitudinal designs has been discussed by Wakefield (2012).

In cases where observed outcomes arise as the realization of nonlinear stochastic processes, some care is needed in the characterization of its variability. In particular, it is reasonable to expect that outcomes will be observed over unit-specific random

D. Telesca (✉)
University of California Los Angeles, Los Angeles, CA, USA
e-mail: donatello.telesca@gmail.com



R. Mitra, P. Müller (eds.), *Nonparametric Bayesian Inference in Biostatistics*, Frontiers in Probability and the Statistical Sciences,
DOI 10.1007/978-3-319-19518-6_14

time scales, resulting in phase-varying random curves. Ignoring phase variability may lead to inconsistent estimates of time-dependent quantities (Kneip and Gasser 1992, 1988), as well as hard to interpret inferential summaries.

We illustrate this point by describing two simple studies and aiming to provide a simple point estimate of the mean over time. Figure 14.1a reports the growth velocity, intended as the yearly change in height, for a sample of 39 boys and 54 girls from the Berkeley growth study (Tuddenham and Snyder 1954). We observe an overall deceleration trend in growth from infancy to adulthood, with acceleration-deceleration pulses in velocity. In particular, the most prominent velocity pulse corresponds to the pubertal spurt. Even though children experience a similar sequence of hormonal events affecting growth, such events do not occur at the same rate/time in all children. Ignoring the individual timing of growth pulses, a naïve point estimate of the average growth profile is the cross-sectional mean. Clearly, this estimate appears immediately inadequate, as it misrepresents the amplitude and length of typical pubertal growth spurts.

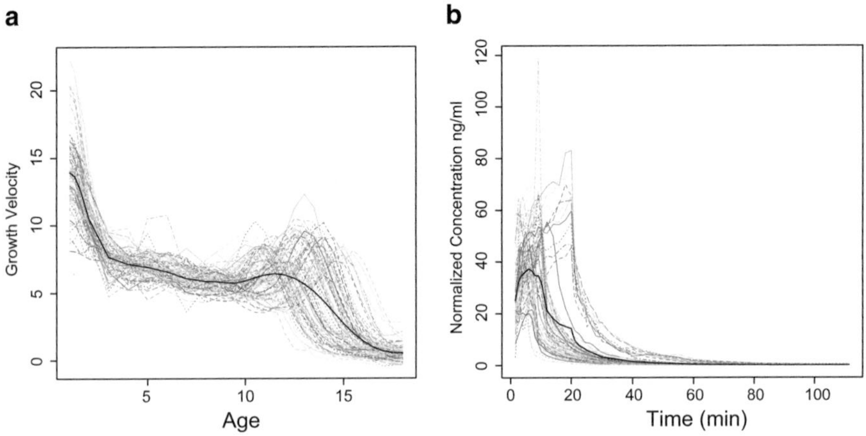


**Fig. 14.1 Phase variability of nonlinear profiles.** (**a**) Berkeley growth study: growth velocity for 93 subjects, defined as yearly changes in the subjects height. The cross-sectional mean growth velocity is superimposed. (**b**) Pharmacokinetics of Remifentanil: normalized drug blood concentration (ng/ml) monitored over phases of infusion and absorption. The cross-sectional mean PK curve is superimposed

Figure 14.1b shows the blood concentration trajectories of the drug Remifentanil for 65 post-surgical patients receiving i.v. infusion of the drug for up to 20 min until reaching a target sedation level. As the drug is infused for a length of time which varies across patients, we observe pharmacokinetic (PK) profiles over subject-specific time scales. This case study illustrates even more transparently the inadequacy of the cross-sectional mean as an estimate of average PK dynamics. In particular, timing artifacts in the study design seem to induce a two-phase excretion rate in the mean, which is not justified from a physiological perspective and clearly atypical when compared to subject-level PK profiles.

These illustrative examples highlight how different experimental and observational settings may define a differential genesis of phase variation in the observed outcome. In cases where variability in timing is related to the measurement process or the design of the study itself, technical or experimental information may be useful in devising pre-processing techniques aimed at removing timing artifacts. However, most commonly, variability in the timing of subject-level functional features relates to the very nature of the observed process and a rigorous approach to alignment strategies is needed to provide valid inference.

The structure of this chapter is as follows. We introduce the problem of curve registration in Sect. 14.2 and review Bayesian hierarchical curve registration in Sect. 14.3. In relation to this model, we introduce a simplified regression strategy in Sect. 14.4. More involved regression approaches based on varying-coefficient models are introduced in Sect. 14.5. Finally, in Sect. 14.6 we introduce and discuss applications of curve registration models to the analysis of time-course genomic data.

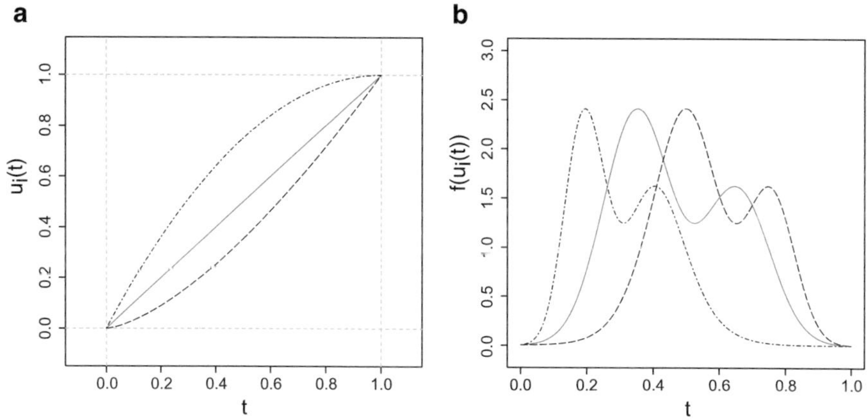


**Fig. 14.2 Time transformation.** (**a**) Time transformation functions for three random profiles. In *solid grey*, we report the identity transform. (**b**) A function of time evaluated over the random time scales in (**a**)

## 14.2 Phase Variability and Curve Registration

The case studies summarized in Fig. 14.1 illustrate how naïve estimation of a functional mean may result in inadequate inferential summaries. In order to formalize these concepts, let us define a notation for observations $y_i(t)$, as the observed outcome for subject $i$, $(i = 1, 2, \ldots, n)$ at time $t \in T$. Typically one only observes $\mathbf{y}_i = (y_i(t_{i1}), \ldots, y_i(t_{im_i}))'$ over a discrete sampling grid. However, for ease of notation, we entertain the possibility of observing $y$ continuously. A standard working assumption sees observations arising as a realization of the following compound stochastic process:

$$y_i(t) = f_i(t) + \varepsilon_i(t) = \mu\{u_i(t)\} + \varepsilon_i(t), \tag{14.1}$$

where $\mu(\cdot)$ is some shape function, $u_i(\cdot)$ is a random, unit-specific time transformation function, and $\varepsilon_i(\cdot)$ a mean-zero, stationary stochastic process. An illustration of how random time scales $u_i(t)$ act on $\mu(t)$ is reported in Fig. 14.2.

We are interested in estimating $\mu(\cdot)$ as a function quantifying a representative shape. In this setting, it may be tempting to estimate this quantity using the cross-sectional mean $\bar{y}(t) = \frac{1}{n}\sum_i y_i(t)$. However, in general, $E\{y_i(t)\} \neq \mu(t)$, therefore $\bar{y}(t)$ is not a consistent estimator of $\mu(t)$. Similar considerations extend to estimators associated with more ambitious statistical analyses, including functional PCA (Rice and Silverman 1991) and functional regression (Guo 2002; Yao et al. 2005).

This simple observation is perhaps a reflection of the deeper tension between marginal and conditional models in longitudinal data analysis. When interest centers on nonlinear dynamics, inference naturally focuses on quantities summarizing typical time-dependent evolutions (Wakefield 2012). Therefore, explicit modeling of all random components of variation is needed for meaningful inference.

There are a number of proposals that deal with the problem of phase variability. A common methodological thread aims to estimate the processes $u_i(t)$ and then compute aligned profiles $y_i^*(t) = y_i\left\{u_i^{(-1)}(t)\right\}$. The shape function $\mu(t)$ is then estimated by the structural mean $\hat{\mu}(t) = \bar{y}^*(t)$.

This procedure goes under the name of curve registration, also known as curve alignment in biology, or time warping in the engineering literature. Several time warping methods have been devised to date. In the engineering literature, Sakoe and Chiba (1978) pioneered a registration technique called dynamic time warping for pairwise alignment. Wang and Gasser (1997) introduced the technique to the statistical literature and provided large sample properties of the time transformation estimators (Wang and Gasser 1999). Gasser and Kneip (1995) proposed the landmark registration method, which consists of identifying the timing of certain features (landmarks) in the curves. Profiles are then aligned so that they occur at the same transformed times. More recently, alignment models have focused on representing time transformation functions as continuous monotone transformations (Ramsay and Li 1998; Kneip et al. 2000). Improved estimation has been reported in Gervini and Gasser (2004) and Brumback and Lindstrom (2004) as well as Liu and Müller (2004).

Bayesian approaches to the area of curve registration are more recent. Telesca and Inoue (2008) introduced a hierarchical representation of curve registration. A more recent perspective, focusing on invariance, was introduced by Cheng et al. (2013).

## 14.3 Bayesian Hierarchical Curve Registration

The problem of curve registration admits a natural probabilistic representation in terms of hierarchical models (Telesca and Inoue 2008). A Bayesian approach to the problem confers extended flexibility in modeling both the shape $\mu(t)$ and time-transformation functions $u_i(t)$. This modeling framework is well adapted to both intensive and sparse sampling time grids as information across curves is shared via partial exchangeability assumptions. As usual, extended flexibility and a formal inferential structure come at the cost of having to specify a full probability model. If the focus of analysis is more exploratory, several alternatives are available as reviewed in Sect. 14.2.

### *14.3.1 Hierarchical Model*

Let $y_i(t)$ denote the observed level of the $i$th curve at time $t$, with $i = 1,2,\ldots,n$ and $t \in T = [t_0,t_m] \subset \mathbb{R}$. In order to allow for linear shifts in the timing of functional features, we define an extended evaluation time window $\mathscr{T} \subset \mathbb{R}$, compact, with $T \subset \mathscr{T}$. The data generating mechanism in (14.1) is naturally represented via the following three-stage hierarchical model.

*Stage One*. Let $\mu(t)$ be a real valued function, s.t. $\mu(t) : \mathscr{T} \to \mathbb{R}$; let $c_i \in \mathbb{R}$ and $a_i > 0$, be two scalars. Furthermore, let $u_i(t)$ be a monotone smooth function, s.t. $u_i(t) : T \to \mathscr{T}$. The observed value of each curve $i$ at time $t$ is modeled as:

$$y_i(t) \;=\; c_i + a_i\,\mu(t) \circ u_i(t) + \varepsilon_i(t) = c_i + a_i\,\mu\{u_i(t)\} + \varepsilon_i(t); \tag{14.2}$$

where, for any $t \in T$, $\varepsilon_i(t) \sim \mathscr{N}(0,\sigma_\varepsilon^2)$. In practice we only observe $y_i(t)$ over a discrete sampling time grid, which translates into standard assumptions of iid Normal random errors. In (14.2), $\mu(t)$ denotes a common shape function generating the individual curves and $u_i(t)$ denotes a curve-specific time transformation function. Unit-specific variability in the level and amplitude of individual profiles are modeled through $c_i$ and $a_i$ respectively.

Upon registration of the curves, we identify the $i$th aligned curve at time $t$ as

$$y_i^\star(t) = y_i(t) \circ u_i^{-1}(t). \tag{14.3}$$

All functional quantities are easily represented in finite dimensional form using linear combinations of appropriate basis functions. Additional considerations about specific modeling choices are deferred to Sect. 14.3.2.

*Stage Two*. Given a common shape function $\mu(t)$, individual curves may exhibit different scales and levels of response. Assuming $c_i \sim \mathscr{N}(0;\sigma_c^2)$ and $a_i \sim \mathscr{N}(1;\sigma_a^2)I\{a_i > 0\}$, defines a straightforward mechanism for curve-specific random affine transformations. For interpretation and identifiability purposes, it is often useful to assume $\sum_i c_i = 0$ and $\sum_i a_i = n$. Normality assumptions enable conjugacy with the likelihood. Moreover, the assumption of strictly positive amplitudes can be relaxed. For example, Telesca et al. (2009) consider a mixture prior in an application to time course expression data.

Curve-specific random time transformation functions $u_i(t)$ are assumed to be smooth realizations of a functional stochastic process, monotone increasing with probability one. Additional image and identifiability constraints are usually needed for implementation. In particular, defining $0 < \delta < (t_m - t_0)/2$, it is appropriate to require $u_i(t_0) \in [-\delta, \delta]$ and $u_i(t_m) \in [t_m - \delta, t_m + \delta]$.

*Stage Three*. The hierarchical model is completed with priors over population level parameters. Common assumptions exploit conditional conjugacy to define Gamma priors[1] over precision parameters:

$$1/\sigma_a^2 \sim Ga(a_a; b_a), \;\; 1/\sigma_c^2 \sim Ga(a_c; b_c), \;\; 1/\sigma_\varepsilon^2 \sim Ga(a_\varepsilon; b_\varepsilon).$$

Additional priors are needed in the definition of random functional quantities. Specific choices are discussed in Sect. 14.3.2.

### *14.3.2 Penalized Regression Splines Representation of Random Functionals*

The shape function $\mu(t)$ is the principal object of inference in curve registration exercises. Specific parametric or semi-parametric forms may indeed be suggested by the application at hand. A general non-parametric approach is based on representing $\mu(t)$ as a random smooth function using linear combinations of B-spline basis functions (De Boor 1978).

Specifically, representation of the common shape function $\mu(t)$ may proceed selecting a set of knots $(\kappa_1, \kappa_2, \ldots, \kappa_p)$ partitioning the extended evaluation interval $\mathscr{T}$ into $p+1$ subintervals. Using piecewise polynomials of degree $r$ and given the set of interior knots, we define $\mathscr{S}_\mu(t)$ as a $K$-dimensional design vector of B-spline basis evaluated at time $t$, with $K = p + r + 1$. In this framework, letting $\boldsymbol{\beta}$ be a $K$-dimensional vector of basis coefficients, we represent the shape function as the following linear combination

$$\mu(t) = \mu(t; \boldsymbol{\beta}) = \mathscr{S}'_\mu(t)\boldsymbol{\beta}.$$

Similarly, given a set of interior knots $(\omega_1, \omega_2, \ldots, \omega_h)$, partitioning the sampling interval $T$ into $h+1$ subintervals, we may represent the individual time transformation functions $u_i(t)$ following the same strategy. In particular, let $\mathscr{S}_u(t)$ be a $Q$-dimensional vector of B-spline bases of degree $r$ evaluated at time $t$, with $Q = h + 1 + r$. Defining $\boldsymbol{\phi}_i$ as a $Q$-dimensional vector of spline coefficients, curve-specific time transformation functions may then be represented according to the following linear combination

$$u_i(t) = u_i(t; \boldsymbol{\phi}_i) = \mathscr{S}'_u(t)\boldsymbol{\phi}_i.$$

[1] In our development, $X \sim Ga(a; b)$ is parametrized so that $E[X] = a/b$.

Monotonicity and boundary conditions are insured by the following constraints on $\boldsymbol{\phi}_i$:

$$(t_1 - \delta) \leq \phi_{i1} < \cdots < \phi_{iq} < \phi_{i(q+1)} < \cdots < \phi_{iQ} \leq (t_m + \delta). \tag{14.4}$$

Similar strategies may be adopted to impose structural constraints on the form of the shape function $\mu(t;\boldsymbol{\beta})$. For an example requiring unimodality of the common shape see Telesca et al. (2012a).

The representation of functional quantities via spline bases requires choosing the degree of local spline polynomials, the number of interior knots as well as the location of the knots for both the common shape function $\mu(t;\boldsymbol{\beta})$ and the individual time transformation functions $u_i(t;\boldsymbol{\phi}_i)$. This model selection problem is often addressed with the minimization of measures of prediction error (Hastie et al. 2001) and cross-validation procedures (Gervini and Gasser 2004).

An alternative modeling strategy relies on penalized regression splines (Eilers and Marx 1996; Ruppert et al. 2003). Specifically, a relatively large number of equidistant knots is selected in order to purposely overparametrize the model. A penalty, dependent on a smoothing parameter $\lambda$, is then placed on coefficients of adjacent B-splines. In a frequentist framework the choice of $\lambda$ is usually made in the model selection stage and is based on cross-validation analysis. From a Bayesian perspective this strategy is equivalent to the definition of appropriate dependent priors for functional coefficients $\boldsymbol{\beta}$ and $\boldsymbol{\phi}_i$.

In particular, following Lang and Brezger (2004), one may consider a second-order random walk shrinkage prior on the shape coefficients $\boldsymbol{\beta}$, so that, for $k = 1,\ldots,K$:

$$\beta_k = 2\beta_{k-1} - \beta_{k-2} + e_k, \quad e_k \sim \mathcal{N}(0;\lambda_\beta). \tag{14.5}$$

Assuming $\beta_{-1} = \beta_0 = 0$, conditional on $\lambda_\beta$, $\boldsymbol{\beta}$ has a multivariate Normal distribution with null mean vector and precision matrix $\boldsymbol{\Omega}/\lambda_\beta$. Under the above second-order random walk, $\boldsymbol{\Omega}$ is a banded precision penalization matrix

$$\boldsymbol{\Omega} = \begin{pmatrix} 6 & -4 & 1 & & & & & 0 \\ -4 & 6 & -4 & 1 & & & & \\ 1 & -4 & 6 & -4 & 1 & & & \\ & \ddots & \ddots & \ddots & \ddots & \ddots & & \\ & & 1 & -4 & 6 & -4 & 1 & \\ & & & 1 & -4 & 5 & -2 & \\ 0 & & & & 1 & -2 & 1 & \end{pmatrix}. \tag{14.6}$$

Note that the random walk variance $\lambda_\beta$ can be interpreted as the smoothing parameter. In particular, small values of $\lambda_\beta$ shrink the shape function $\mu(t,\boldsymbol{\beta})$ toward a linear function of time. Following Lang and Brezger (2004) we place a relatively diffuse conjugate inverse gamma hyperprior on the variance, so that $\lambda_\beta \sim IG(a_{\lambda 1};b_{\lambda 1})$.

A similar approach may be adopted to model time transformation functions $u_i(t;\boldsymbol{\phi}_i)$. Defining identity transform coefficients $\boldsymbol{\Upsilon}' = (\Upsilon_1,\ldots,\Upsilon_Q)$, s.t. $u_i(t,\boldsymbol{\Upsilon}) = t$; for all $i = 1,2,\ldots,n$, $q = 1,\ldots,Q$:

$$(\phi_{iq} - \Upsilon_q) = (\phi_{i(q-1)} - \Upsilon_{q-1}) + \eta_q, \quad \eta_q \sim \mathcal{N}(0;\lambda_\phi). \tag{14.7}$$

Assuming that $(\phi_{i0} - \Upsilon_0) = 0$, it can be shown that $\boldsymbol{\phi}_i \sim \mathscr{N}(\boldsymbol{\Upsilon}; \boldsymbol{P}/\lambda_\phi)$, where $\boldsymbol{P}$ is a banded precision matrix and $\lambda_\phi$ is the smoothing parameter associated with the transformation functions $u_i(t;\boldsymbol{\phi}_i)$. Small values of $\lambda_\phi$ shrink $u_i(t;\boldsymbol{\phi}_i)$ toward the identity transformation. The model is completed with a prior for $\lambda_\phi$, s.t. $\lambda_\phi \sim IG(a_{\lambda 2}; b_{\lambda 2})$.

### 14.3.3 Inference for Hierarchical Curve Registration Models

In applications we observe functional data over a finite sampling grid $\mathbf{t}_i' = (t_{i1}, \ldots, t_{ij}, \ldots, t_{im_i})$. Let $\mathbf{y}_i(\mathbf{t}_i)' = (y_i(t_{i1}), \ldots, y_i(t_{im_i}))$ be an $m_i$-dimensional vector, representing the observed trajectory for unit $i$, $(i = 1, \ldots, n)$, over time. Using B-spline representations, the functional model in (14.2) simplifies into a standard hierarchical model involving random quantities of finite dimensional form. In particular, let $\mathscr{S}_\mu(\mathbf{t}_i) : m_i \times K$ and $\mathscr{S}_u(\mathbf{t}_i) : m_i \times Q$ be the shape and time transformation spline design matrices, respectively. The sampling model can be expressed as

$$\mathbf{y}_i(\mathbf{t}_i) = c_i \, 1_{m_i} + a_i \mathscr{S}_\mu(\mathbf{t}_i)\boldsymbol{\beta} \circ \mathscr{S}_u(\mathbf{t}_i)\boldsymbol{\phi}_i + \varepsilon_i(\mathbf{t}_i); \tag{14.8}$$

with $\varepsilon_i(\mathbf{t}_i) \sim \mathscr{N}_{m_i}(0, \sigma_\varepsilon^2 I_{m_i})$.

Given priors on population level quantities $\boldsymbol{\beta}$ and $\sigma_\varepsilon^2$, and unit-specific parameters $c_i$, $a_i$ and $\boldsymbol{\phi}_i$, $(i = 1, 2, \ldots, n)$; inference about all functionals of interest is directly available from their posterior distribution. In particular, MCMC simulation from the posterior is relatively straightforward. Given $\boldsymbol{\phi}_i$, for all $i$, simulation from all remaining parameters is easily implemented following any sampling strategy applicable to hierarchical linear models (Gelman et al. 2013). Some care is needed in the sampling of $\boldsymbol{\phi}_i$ as the support of these parameters is defined over random cuts insuring monotonicity of time transformation functions. However, for these quantities, relatively simple Metropolis Hastings transitions tend to work well in practice. Telesca and Inoue (2008) discuss implementation of these strategies in detail.

Let $a_i^{(j)}$, $c_i^{(j)}$, $\boldsymbol{\phi}_i^{(j)}$ and $\boldsymbol{\beta}^{(j)}$, $(j = 1, \ldots, M)$, denote $M$ draws from the marginal posterior distributions of respective parameters. To register the observed curves one may use the posterior expectation $E\{u_i(t) \mid \mathbf{y}\}$ as a point estimate of the stochastic time scale for unit $i$. That is, given posterior samples from time transformation parameters $\boldsymbol{\phi}_i^{(j)}, (j = 1, \ldots, M)$, posterior samples for the functional quantity $u_i(t)$ are easily calculated as

$$u_i^{(j)}(t) \;=\; u_i(t; \boldsymbol{\phi}_i^{(j)}) \;=\; \mathscr{S}_u(t)' \boldsymbol{\phi}_i^{(j)}. \tag{14.9}$$

Similarly, draws from the marginal posterior distribution of the shape function $\mu(t;\boldsymbol{\beta})$, for any time $t \in \mathscr{T}$, are given by:

$$\mu^{(j)}(t;\boldsymbol{\beta}) = \mathscr{S}_\mu(t)' \boldsymbol{\beta}^{(j)}. \tag{14.10}$$

Clearly, inference about several functional summaries, including extrema, differentials, etc., are obtained in the same straightforward fashion.

Simultaneous credible bands for any function of interest, say $f(.)$, are easily approximated using a fine grid of evaluation time points $t_1 < \ldots < t_\ell$ (Baladandayuthapani et al. 2005). Let $\Gamma_\alpha$ denote the $100(1-\alpha)\,\%$ sample quantile of

$$\max_{1\leq i\leq \ell} |[f(t_i) - E\{f(t_i) \mid \mathbf{y}\}]/SD\{f(t_i) \mid \mathbf{y}\}|;$$

a simultaneous $100(1-\alpha)\,\%$ credible band for $f(t)$ is estimated as

$$I(t) = E\{f(t) \mid \mathbf{y}\} \pm \Gamma_\alpha SD\{f(t) \mid \mathbf{y}\}.$$

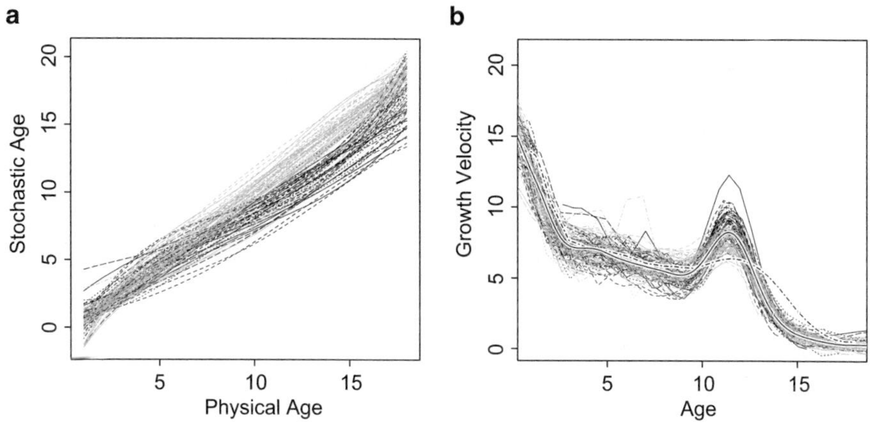


**Fig. 14.3 Berkeley Growth Study.** (**a**) Posterior expectation of time transformation functions $u_i(t)$. (**b**) Aligned growth velocity profiles, with superimposed posterior expectation of $\mu(t)$ (*solid line*). The cross-sectional mean for misaligned profiles is reported as the (*dashed line*). In both panels *grey profiles* identify girls, while *black profiles* identify boys

## *14.3.4 Case Studies in Bayesian Curve Registration*

We illustrate the application of Bayesian hierarchical registration techniques to the analysis of the two illustrative case studies reported in Fig. 14.1. In particular, Fig. 14.3 reports estimates for the posterior expected time transformation functions $u_i(t)$ (a) and aligned growth velocity profiles (b). In the same figure we report the posterior expectation for the structural average curve $\mu(t)$ (solid line). When compared to the naïve cross-sectional estimate, the structural average appears clearly as a better representation of typical growth velocity patterns.

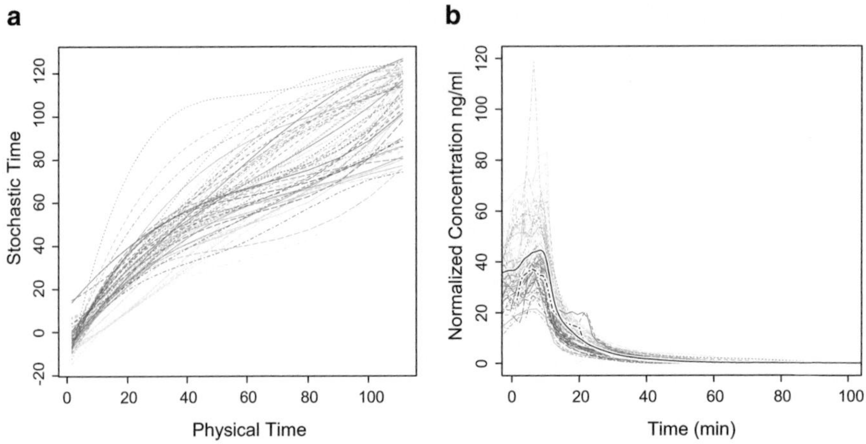


**Fig. 14.4 Pharmacokinetics of Remifentanil.** (**a**) Posterior expectation of time transformation functions $u_i(t)$. (**b**) Aligned drug concentration trajectories with superimposed posterior expectation of $\mu(t)$ (*solid line*) and cross sectional mean of misaligned concentration dynamics (*dashed line*)

Similarly Fig. 14.4, reports a hierarchical registration analysis of drug concentration dynamics associated with the drug Remifentanil. In panel (a) we plot posterior expected time transformation functions. As information about differing functional features becomes more sparse in later sampling time points, the estimated warping functions exhibit higher variance near the end of the sampling time domain. Panel (b) reports aligned drug concentration trajectories and superimposed posterior expected structural mean (solid line). As for the growth study, alignment removes artifacts in the cross-sectional average and produces estimates of average concentration kinetics, which are more representative of typical individual profiles. The application of this technique to pharmacokinetic data is indeed non-standard as one often seeks to learn about compartment model parameters in a system of differential equations. Nevertheless, we find this analysis useful and essentially informative as a primitive exploration of drug concentration dynamics.

## 14.4 Regression Models for Timing and Amplitude of Functional Features

Consider the growth study in Fig. 14.3. A more in depth look at individual profiles is indeed made easier after removing phase variability. We code individual curves in grey-levels to reflect the subject sex. Black profiles indicate boys and grey profiles indicate girls. An explorative examination of the estimated time transformation functions (a) reveals that the time scale for girls tends to lie above the identity transform, whereas boys tend to lie below it. This observation indicates that girls tend to experience pubertal growth at earlier ages, when compared to boys. Beyond variation in timing, a visual examination of the aligned profiles in (b) allows for a clearer distinction of sex-related amplitude variation patterns. In particular, pubertal spurts for girls tend to be attenuated when compared to pubertal spurts in boys.

These observations motivate a natural extension of curve registration models as formal tools to relate individual covariate information to simple and interpretable components of variation in functional data. Specifically, following Brumback and Lindstrom (2004) and Telesca et al. (2012a) we develop a class of models aimed at explaining amplitude and phase variability in a sample of curves using individual-level predictors.

### *14.4.1 Generalized Curve Registration Models*

The hierarchical model in (14.2) need not be restricted to assumptions of Gaussian sampling. In fact this assumption may be relaxed to accommodate a wider range of sampling scenarios, usually encountered in biostatistical application. In particular, we consider hidden Gaussian random fields (HGRF) models, as a suitable family amenable to straightforward adaptations of the formulation introduced in Sect. 14.3.

In HGRF observations $y_i(t_{ij})$ are equipped with mirroring latent Gaussian quantities, say $z_i(t_{ij})$. A sampling model for $y_i(t_{ij}) \mid z_i(t_{ij}), \boldsymbol{\psi}_i \sim F(z_i(t_{ij}); \boldsymbol{\psi}_i)$, is fully defined conditionally on $z_i(t_{ij})$ and a possible set of parameters $\boldsymbol{\psi}_i$. Registration is then achieved at the latent Gaussian level. More precisely, by assuming the stochastic dynamic generating $z_i(t_{ij})$ is centered around a compound process, defined as an affine transformation of a population mean trajectory $\mu(t)$, evaluated over subject-specific random time schedules $u_i(t)$, with random scales $c_i$ and amplitudes $a_i$, $(i = 1, 2, \ldots, n)$ as in (14.2).

For example, longitudinal counts arise naturally in many applications, like immunology, bioinformatics, and behavioral studies. In this case one may follow the approach outlined in Telesca et al. (2012a) and model

$$y_i(t_{ij}) \mid z_i(t_{ij}) \sim Poisson[\exp\{z_i(t_{ij})\}],$$

with $z_i(t_{ij}) \sim \mathcal{N}(g_i(t_{ij}), \sigma_\varepsilon^2)$ and $g_i(t_{ij}) = c_i + a_i \mu(t_{ij}) \circ u_i(t_{ij})$.

Similarly Erosheva et al. (2014) apply this representation to model censored Gaussian observations. In particular, given a left censoring point $\eta_0$ and right censoring $\eta_1$, common for all subjects, $z_i(t_{ij})$ is defined as an uncensored latent variable. The outcome at time $t_{ij}$, for individual $i$ is then modeled as

$$y_i(t_{ij}) = \min\left[\max\{\eta_0, z_i(t_{ij})\}, \ \eta_1\right],$$

with $z_i(t_j) \sim \mathcal{N}(g_i(t_{ij}), \sigma_\varepsilon^2)$. Other common applications of the latent Gaussian Field framework include models for binary and ordinal data (Albert and Chib 1993).

### 14.4.2 Amplitude and Phase Regression

Let $X_i$ be a $p$-dimensional vector of subject specific covariate information. The assessment of how amplitude and phase variability are explained by predictors is naturally achieved at the second stage of the hierarchical model, through covariate-dependent priors for amplitude parameters $a_i$ and time transformation coefficients $\boldsymbol{\phi}_i$.

*Amplitude regression.* Let $\mathbf{b}_a$ be a $p$-dimensional vector of amplitude regression coefficients, we explain amplitude variability by defining the hidden linear model:

$$a_i \sim N(1 + X_i'\mathbf{b}_a,\ \sigma_a^2)I(a_i > 0). \tag{14.11}$$

In the foregoing formula, regression coefficients are offset by a factor of 1, to define coefficients with respect to a reference amplitude. The coefficients $\mathbf{b}_a$ are interpreted as in common linear regression models. Additionally, it is customary to assume $\sum_i a_i = n$, for amplitude identifiability.

*Phase regression.* Let $\mathbf{b}_\phi$ be a $p$-dimensional vector of phase regression coefficients, we explain phase variability by defining the hidden autoregressive linear model:

$$\gamma_{iq} = \Upsilon_q + X_i'\mathbf{b}_\phi, \tag{14.12}$$

$$\phi_{iq} - \gamma_{iq} = \phi_{i(q-1)} - \gamma_{i(q-1)} + \eta_{iq}; \tag{14.13}$$

with $\eta_{iq} \sim \mathscr{N}(0, \sigma_\phi^2)I(\mathscr{M})$ and $\mathscr{M} = \{\phi_{iq} : \ \phi_{i(q+1)} > \phi_{iq},\ \phi_{i1} \geq (t_1 - \delta),\ \phi_{iQ} \leq (t_m + \delta),\ q = 1, 2, \ldots, Q\}$. As for the case of amplitude, regression coefficients are offset by the identity transform coefficients $\Upsilon$, in order to fix a reference time scale. Regression coefficients $\mathbf{b}_\phi$ are then interpreted as changes in the average time scale associated with changes in predictor values.

Random scales $c_i$ are most commonly treated as nuisance parameters and simply modeled as $c_i \sim \mathscr{N}(0, \sigma_c^2)$, with $\sum_i c_i = 0$ for scale identifiability.

In the setting of HGRF models, prior distributions for regression coefficients may still exploit conditional conjugacy. For example, if we denote the covariates matrix with $\mathbf{X} : n \times p$, standard Zellner priors may be considered for amplitude and phase regression as follows:

$$\mathbf{b}_a \mid \sigma_a^2 \sim \mathscr{N}\left(0, n\sigma_a^2 (X'X)^{-1}\right), \tag{14.14}$$

$$\mathbf{b}_\phi \mid \sigma_\phi^2 \sim \mathscr{N}\left(0, n\sigma_\phi^2 (X'X)^{-1}\right). \tag{14.15}$$

Variance components $\sigma_a^2$ and $\sigma_\phi^2$ are commonly assigned conditionally conjugate Inverse Gamma priors.

**Table 14.1** Amplitude and phase regression

| *Berkeley Growth Study* | | | | |
|---|---|---|---|---|
| | Amplitude | | Phase (years) | |
| Predictor | $E(b_a \mid \mathbf{y})$ | 95 % CI | $E(b_\phi \mid \mathbf{y})$ | 95 % CI |
| *Baseline* | | | | |
| Male | 0.019 | [−0.027, 0.064] | 0.44 | [−0.65, 1.49] |
| *Main Effects* | | | | |
| Female | −0.031 | [−0.093, 0.031] | −0.39 | [−1.76, 1.06] |

| *Pharmacokinetics of Remifentanil* | | | | |
|---|---|---|---|---|
| | Amplitude | | Phase (years) | |
| Predictor | $E(b_a \mid \mathbf{y})$ | 95 % CI | $E(b_\phi \mid \mathbf{y})$ | 95 % CI |
| *Baseline* | | | | |
| Female | −0.03 | [−0.19, 0.13] | −0.67 | [−13.23, 12.27] |
| *Main Effects* | | | | |
| Male | 0.05 | [−0.19, 0.282] | −0.77 | [19.22, −18.22] |
| Age | 0.01 | [−0.01, 0.022] | 0.02 | [−0.33, 0.36] |
| Weight | −0.01 | [−0.02,−0.001]* | −0.03 | [−0.63, 0.59] |

*95 % Credible Interval does not cover zero.

### *14.4.3 Growth Velocities and Drug Concentrations Revisited*

We apply the model introduced in Sect. 14.4.2 to our two case study data-sets. Regression results are reported in Table 14.1.

For the Berkeley growth study, we consider sex as a predictor of amplitude and phase variation in growth velocity. Our intuition being the plot in Fig. 14.3 is confirmed, in that girls tend to experience both attenuated amplitude ( 0.031) and accelerated timing (−0.39), when compared to boys. Our formal analysis, however, highlights that there is too much uncertainty around amplitude and phase variation in growth, therefore no significant group differences are detected.

A similar question may be asked of the PK dynamics of Remifentanil, that is, are drug concentration amplitude and phase variability explained by sex, when adjusting for potential confounding factors? In this case we perform a regression analysis involving patients sex, body weight, and age. This analysis reveals that weight plays a possible role in the concentration dynamic of Remifentanil, with heavier patients experiencing lower (−0.01) concentration amplitude per Kg.

For these analyses to produce conclusive evidence, we often require large amounts of data (large $n$), as high variability often characterizes estimates of individual level amplitude and phase. At the same time, these results warn against the meaningfulness of step-wise regression approaches, where asymptotic validity may not result in acceptable finite-sample conclusions.

## 14.5 Joint Functional Regression and Registration

From a statistical perspective our goal is to develop models that: (1) deal with registration by aligning response trajectories, so that they are defined over a standardized time scale and (2) allow for the estimation of covariate effects on a functional response, that are representative of typical response patterns. Some progress can be made using the approach outlined in Sect. 14.4. However, regression coefficients obtained using this technique average across the entire function evaluation domain and are likely to miss more nuanced, time-dependent effects. As an example, consider the growth data in Fig. 14.3. Timing differences between girls and boys are most evident between the ages of 10 and 15. However, the regression analysis reported in Table 14.1 averages across periods of homogenous timing, revealing no conclusive difference between sexes.

A more nuanced approach to joint regression and registration finds motivation in the original generative model in (14.1). In particular, one often assumes that individual trajectories are centered around common structural mean function $\mu(t)$ evaluated over unit-specific time scales $u_i(t)$. When predictors $X_i$ are available in the form of a $p$-dimensional vector of subject-level covariates, the assumption of a common mean may be relaxed and a new generative model for a sample of random trajectories is more realistically represented as:

$$y_i(t) = \mu(t, X_i) \circ u_i(t), \tag{14.16}$$

where the form of $\mu(t, X_i)$ is made dependent on the vector or predictors $X_i$.

In the following we review common approaches to the estimation of functional regression coefficients (Hastie and Tibshirani 1993; Guo 2002; Morris and Carroll 2006). Finally we discuss a natural extension of Bayesian hierarchical curve registration (Telesca and Inoue 2008) to a unified framework for functional mixed effects modeling and curve registration. We name this class of models functional mixed registration models (FMRM).

### *14.5.1 Functional Regression and Mixed Models*

Approaches extending linear models to the functional context often build on the idea of varying-coefficients (Hastie and Tibshirani 1993). Varying-coefficient models are linear in the regressors, but their coefficient are allowed to vary smoothly with the value of other variables, known as effect-modifiers. Given a set of $p$ predictors $x_1, \ldots, x_p$ and $p$ effect modifiers $r_1, \ldots, r_p$, varying coefficient models consider a general link function as $\eta = s_0 + \sum_{j=1}^{p} x_j S_j(r_j)$. Hastie and Tibshirani (1993) showed that additive and generalized additive models represent a special case of varying-coefficient models. Several authors have extended these modeling approaches to incorporate intra-curve dependence (Hart and Wehrly 1986; Wypij et al. 1993; Zeger and Diggle 1994; Wang and Gasser 1999; Verbyla et al. 1999; Silverman 1995).

From a mixed effects perspective, functional mixed models (FMMs), as proposed by Shi et al. (1996), extend the work of Laird and Ware (1982) to functional data by leaving the forms of the fixed and random effect functions unspecified. These models inherit the flexibility of mixed effects models in handling complex designs and correlation structures. We review a general and flexible view of the problem as discussed by Guo (2002) who used smoothing splines to model both fixed and random effects.

Let $y_i(t)$ denote the value of curve $i$ $(i = 1,\dots,n)$, at time $t \in \mathscr{T}$ compact. Following Guo (2002) we define:

$$y_i(t) = \mathbf{X}_i'\mathbf{B}(t) + \mathbf{Z}_i'\mathbf{U}_i(t) + \varepsilon_i(t)\,, \tag{14.17}$$

where, for any time $t \in \mathscr{T}$, $\mathbf{B}(t) = (\beta_1(t),\dots,\beta_p(t))'$ is a $p$-dimensional vector of fixed effect functions with corresponding design vector $\mathbf{X}_i = (X_{i1},\dots,X_{ip})'$ and $\mathbf{U}_i(t) = (U_{i1}(t),\dots,U_{im}(t))'$ is an $m$-dimensional vector of random effect functions with corresponding design vector $\mathbf{Z}_i = (Z_{i1},\dots,Z_{im})'$. Finally, $\varepsilon_i(t)$ represents the residual error process for curve $i$ at time $t$.

Guo (2002) discussed FMM in the context of smoothing splines proposing two estimation approaches based on restricted maximum likelihood and Kalman filters. Morris and Carroll (2006) discussed wavelet-based FMM in a Bayesian framework proposing inferences based on posterior samples of the functions of interest. FMM include several other models like linear mixed effect models, functional regression, and functional ANOVA, as special cases.

While the FMM approach is flexible enough to account for curve-specific variability, it fails to discriminate between the different sources of variation in functional data, namely, amplitude and phase variability. If phase variability is ignored, FMM tend to provide an estimate of the covariate effect which oversmoothes with respect to curve-specific functional features occurring on a stochastic time scale.

### 14.5.2 Functional Mixed Registration

We introduce our FMRM as a natural extension of the Bayesian hierarchical curve registration framework (Telesca and Inoue 2008).

Following the notation introduced in Sect. 14.5.1, we model a sample of curves $y_i(t)$ $(i = 1,\dots,n,\ t \in \mathscr{T})$ as:

$$y_i(t) = \{\mathbf{X}_i'\mathbf{B}(t) + \mathbf{Z}_i'\mathbf{U}_i(t)\} \circ u_i(t) + \varepsilon_i(t)\,, \tag{14.18}$$

so that,

$$y_i(t) = \mathbf{X}_i'\mathbf{B}\{u_i(t)\} + \mathbf{Z}_i'\mathbf{U}_i\{u_i(t)\} + \varepsilon_i(t)\,, \tag{14.19}$$

where $u_i(t)$ is a smooth monotone time transformation function as defined in Sect. 14.3.

The FMRM framework includes naturally several existing modeling strategies. In fact, given specific configurations of the time transformation functions or the covariate set, we may obtain the following models as special cases:

(a) *Functional Mixed Effect Models.* By setting the time transformation functions $\mu_i$ to the identity transformation so that $\mu_i(t) = t$, for any $t \in \mathscr{T}$, our FMRM reduces to the FMM.
(b) *Hierarchical Curve Registration Models.* By setting the random effect functions $\mathbf{U}_i(t) = c_i + a_i \mathbf{B}(t)$, with $\mathbf{X}_i = 1$ and $\mathbf{Z}_i = 1$ ($i = 1,\ldots,N$), $t \in \mathscr{T}$, our FMRM reduces to:

$$y_i(t) = \{\mathbf{B}(t) + (c_i + a_i \mathbf{B}(t)\} \circ u_i(t) + \varepsilon_i(t), \tag{14.20}$$
$$= (c_i + (1 + a_i)\mathbf{B}(t)) \circ u_i(t) + \varepsilon_i(t), \tag{14.21}$$

which is equivalent to (14.2).

The FMRM in Eq. (14.19) is, however, not identifiable. Given any fixed effect function $\mathbf{B}(t)$, a number of combinations of random effects $\mathbf{U}_i(t)$ and time transformation functions $u_i(t)$ may, in fact, lead to the same likelihood or posterior density. The identifiability issue is mainly due to the arbitrary flexibility with the random effects functions. Choosing a reference curve or considering constrained formulations may help with the identification problem. However, this is not usually a straightforward task. Here we choose to focus on random effects which are assumed to have a strictly parametric form. In particular, we will only allow for individual random scale or amplitude transformations, so that model (14.19) can be rewritten as:

$$y_i(t) = c_i + a_i \mathbf{X}_i' \mathbf{B}(t) \circ u_i(t), \quad i = 1,\ldots,N; \tag{14.22}$$

where $c_i$ is a curve-specific scale parameter and $a_i$ is a curve-specific amplitude parameter.

The finite dimensional representation of functional quantities in (14.22) may follow the penalized B-spline formulation introduced in Sect. 14.3.2. More precisely, given a $K$-dimensional set of kernels $\mathscr{S}_\beta(t)$, evaluated at time $t \in \mathscr{T}$, and a $p \times K$ matrix of regression coefficients $\boldsymbol{\beta}$, one may represent the fixed effect functions $\mathbf{B}(t)$ as $\mathbf{B}(t) = \boldsymbol{\beta} \mathscr{S}_\beta(t)$.

More generally, of course, specific choices for the kernels $\mathscr{S}_\beta(t)$ may depend on the study and should reflect reasonable assumptions about the functional form of $\mathbf{B}(t)$. For example, in the analysis of the Berkeley growth study, it is reasonable to consider functions that are smooth and continuous. Thus, one may choose $\mathscr{S}_\beta(t)$ to belong to the spline family. On the other hand, if the outcome consists of a set of long time series, characterized by highly localized features, such as in mass spectrometry data, then $\mathscr{S}_\beta(t)$ could be represented by wavelet basis functions (Morris and Carroll 2006).

Given the representation in (14.22), prior settings and MCMC simulation strategies may follow the same approach outlined in Sect. 14.3.

### 14.5.3 Functional Mixed Registration of Growth Velocities and Drug Concentrations

We apply the FMRM approach to the pharmacokinetics of Remifentanil. Our analysis replicates the regression exercise attempted in Sect. 14.4.3. The goal of our analysis is to assess differences in the pharmacokinetics of Remifentanil between males and females adjusting for age and body weight.

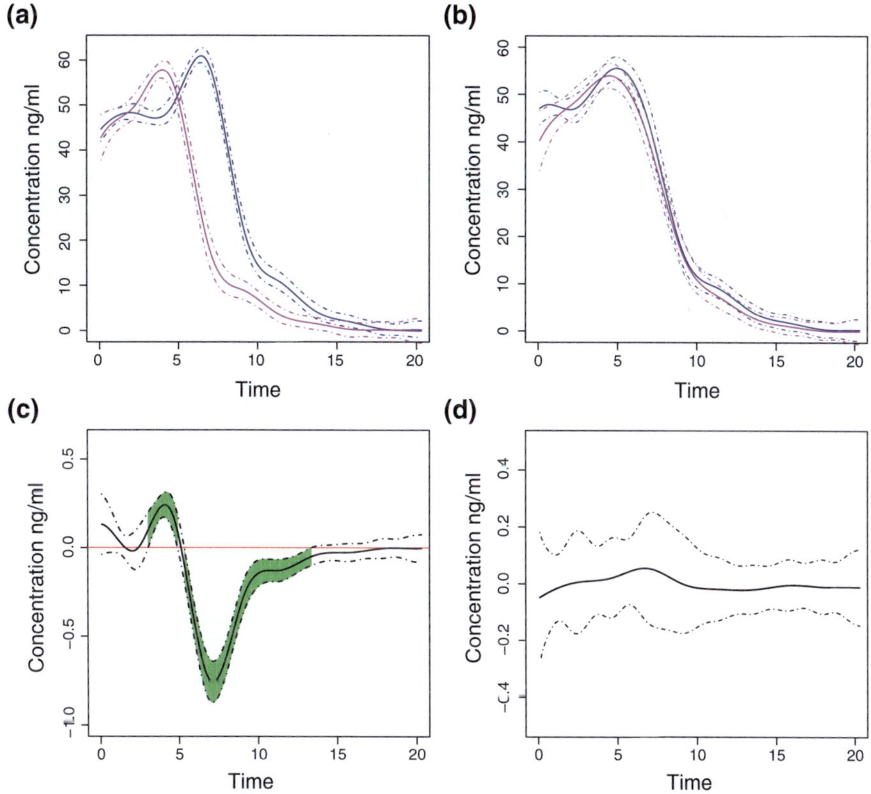


**Fig. 14.5 FMRM analysis of drug concentration**. (**a**) Unadjusted mean posterior drug concentrations for males (*blue*) and females (*magenta*). (**b**) Posterior mean drug concentration trajectories for males (*blue*) and females (*magenta*), adjusted for age and body weight. (**c**) Time varying effect of body weight. (**d**) Time varying effect of age. In all panels we report simultaneous 95 % credible bands

Figure 14.5, panels (a) through (d), shows the results from our analysis. All figures are plotted over a transformed time scale $(\log t)^2$ in order to better display differences between curves. Panel (a) shows the unadjusted posterior common scaled shape functions for the blood concentration trajectories of male (blue) and female (magenta) patients. Without adjusting for other predictors it appears that females have a faster metabolism and excretion of the drug. Panel (b) shows the mean posterior pharmacokinetic profile for males (blue) and (females) adjusted by age (years) and body weight (Kg). Panel (c) shows the time varying effect of body weight and panel (d) shows the time varying effect of age. We highlight in green areas where the effect of the predictors are significantly different from 0. We note that the differences in the metabolism of Remifentanil between males and females are now fully accounted for by differences in body weight. As one may reasonably expect, we no longer see significant sex-related effects on the pharmacokinetics of the drug.

## 14.6 Differential Expression and Gene Profile Similarities

In this section we discuss the application of registration models in Bioinformatics. Specifically, time course genomics data consist of measurements from a common set of genes collected at different time points and provide new opportunities into the understanding of gene regulation.

In particular, clues about the temporal structure of expression may be informative about co-regulation and gene–gene relationships (Qian et al. 2001; Leng and Müller 2006). In this section we discuss the approach of Telesca et al. (2009), who introduced a model-based selection of differentially expressed genes, and a probabilistic framework for the investigation of regulatory relationships between genes.

### *14.6.1 A Functional Mixture Model for Differential Expression*

Following the formulation of Sect. 14.3, we let $y_i(t)$ denote the observed expression level of gene $i$ at time $t$ where $i = 1,2,\ldots,n$ and $t \in \mathscr{T}$. We assume that gene-specific expression profiles arise following the same stochastic generative mechanism in (14.2). Given a common shape function $\mu(t)$, individual curves may exhibit different levels and amplitudes of response and different timing schedules associated with time-dependent expression features.

In this setting, the parameters $a_i$ describe the amplitude of the mRNA signal for gene $i$. A formalization of our statistical definition of differentially expressed genes may be achieved via a mixture approach. This idea follows naturally from similar formalizations introduced by Parmigiani et al. (2002) and extended by Telesca et al. (2012b).

For each gene $(i = 1,2,\ldots,n)$, we specify the following prior for the amplitude of the expression signal,

$$a_i = \pi^- N(a_0^-, \sigma^2_{a^-}) I(a_i < 0) + \pi^+ N(a_0^+, \sigma^2_{a^+}) I(a_i > 0) + \pi^0 N(0, \sigma^2_{a^0}); \quad (14.23)$$

with $(\pi^- + \pi^0 + \pi^+) = 1$. Here $\pi^0$ identifies the overall proportion of genes in their normal range of variation, while $(\pi^- + \pi^+)$ identifies the proportion of overly active genes. The mixture characterization with two truncated normals [that is, $N^-(\cdot,\cdot)I(a_i < 0)$ and $N^+(\cdot,\cdot)I(a_i > 0)$] allows us to account for genes with a synchronous expression signal of opposite sign (negative dependence).

From an inferential perspective, a decision to flag specific genes as being differentially expressed corresponds to testing the following set of hypotheses for all $i = 1,2,\ldots,n$:

$$\begin{aligned} H_{0i} &: a_i \sim N(0, \sigma^2_{a^0}) \\ H_{1i} &: a_i \sim N(a_0^+, \sigma^2_{a^+}) \text{ or } a_i \sim N(a_0^-, \sigma^2_{a^-}). \end{aligned} \tag{14.24}$$

Give posterior samples from $a_i \mid \mathbf{y}$, decision rules controlling for pre-defined error rates, like the false discovery rate (FDR) of Benjamini and Hochberg (1995), are easily derived, for example, following the approach described by Müller et al. (2006).

### 14.6.2 Posterior Measures of Profile Similarities

The underlying idea for the investigation of gene networks using time course microarray data is that genes that share similar expression profiles may share similar biological functions and thus could be related. Posterior inference about gene-specific time transformation functions may be used to derive measures of gene–gene relationships which are based on functional similarities.

In the context of the model described in Sect. 14.6.1, for differentially expressed profiles, local measures of profile similarity may be derived as follows:

*Local warping distance.* Let $\tau \subset T$, we define a *local distance* $d_{ik}(\tau)$ between genes $i$ and $k$ ($i \neq k$) as

$$d_{ik}(\tau) = \int_\tau \mid u_i(t) - u_k(t) \mid dt, \tag{14.25}$$

that is, as the absolute distance between the time transformation functions of genes $i$ and $k$ along time points $t \in \tau$. This measure may be interpreted as the average difference in the timing of expression features between the expression of two genes over a period of time $\tau$. From a global perspective one may of course consider a warping distance integrating over the entire sampling window $T$.

Relevant summaries from the marginal posterior distribution of time transformation functions may be extracted to draw inference about gene–gene relationships. In particular, one may formalize inference about profile similarities as the series of hypotheses:

$$H_{0ik} : d_{ik}(\tau) \geq \gamma, \text{ vs. } H_{1ik} : d_{ik}(\tau) < \gamma; \text{ for all } i \neq k.$$

For this series of decisions, optimal rules controlling for error rates are derived as discussed in Sect. 14.6.1.

While recognizing the importance of the timing characteristics of gene expression, the selection of an appropriate timing envelope $\gamma$ must, however, be aided by biological knowledge about the timing of gene–gene regulation in the specific process under investigation. For example, in cell cycle experiments, regulatory envelopes of interest may span only a few minutes, while in the study of androgen refractory tumors the timing of interest is of the order of days (Pound et al. 1999).

### 14.6.3 A Case Study of Time-Course Gene Expression Analysis

Here we illustrate the application of registration models with mixture priors, to the analysis of time-course expression data. In particular, we consider data on 100 gene reporters of 13 time-points mouse Affimetrix microarray gene expression from a study on primary mouse keratinocytes, with induced activation of the TRP63 transcription factor (Della Gatta et al. 2008).

The data has been processed using `rma` (affy) and the profiles are centered (zero-mean) across the time points. As the original data is composed of direct targets of TRP63, we expect all genes to be differentially expressed over time. For illustrative purposes, in order to test the performance of registration models for differential expression analysis, we augment the original data-set with 900 pseudo-genes of constant average expression.

Results are summarized in Fig. 14.6. In particular, panel (a) highlights a random sample of profiles, and panel (b) shows model fits and associated posterior predictive bands for a representative set of profiles. Even though a registration model of time-dependent expression makes what seem like restrictive assumptions about possible gene-specific time-dependent dynamics, this figure illustrates the actual flexibility of the model in its ability to recover heterogenous time-course profiles. Our analysis of differential expression is reported in panels (c) and (d), where we show the model selection, aimed at controlling the posterior expected FDR at 10 %. The model selects 96 genes as differentially expressed, all of which are in the original TRP63 target set.

A full analysis of profile similarities is beyond the scope of this chapter. For more examples we refer the reader to Telesca et al. (2009).

## 14.7 Concluding Remarks

We have reviewed the application of curve registration techniques to the analysis of functional data arising in Biostatistics and Bioinformatics. Our review is by no means exhaustive and is clearly biased toward the author's expertise.

Modeling frameworks using the idea of stochastic time scales have a strong tradition in several fields and are indeed the subject of active research efforts (Zhang and Telesca 2014; Cheng et al. 2013)

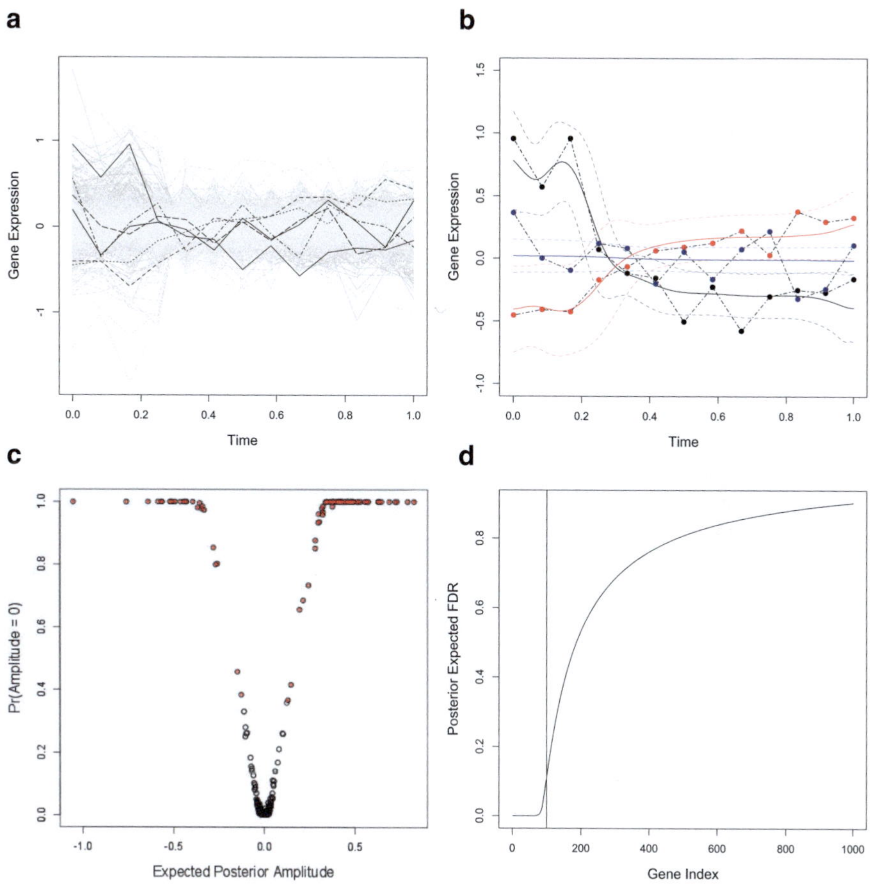


**Fig. 14.6 Time-course gene expression.** (**a**) Time course gene-expression profiles. (**b**) Individual model fit for a representative set of genes. (**c**) Volcano plot of posterior expected amplitude vs. posterior probability of no time dependent expression. (**d**) Posterior expected FDR vs. gene index

Our discussion is focused on Bayesian inference with smoothing priors. In the setting of kernel-based regression the selection of diffuse priors remains controversial and default choices are based on purely intuitive arguments. Attempts at formalization do exist, for example, Wakefield (2012), Chapter 11 discusses approaches based on effective degrees of freedom.

Finally, while inference based on posterior simulation is straightforward, the application of standard MCMC techniques may be unrealistic for cases where the analysis involves a large number of subjects. Similarly, computational feasibility is often in question for studies where technology allows for highly intensive sampling of individual profiles. In these situation, one must consider careful computational nuances and the potential development of efficient approximation techniques.

## References

Albert, J. and Chib, S. (1993). Bayesian analysis of binary and polychotomous response data. *Journal of the American Statistical Association*, **88**, 669–679.

Baladandayuthapani, V., Mallick, B. K., and Carroll, R. J. (2005). Spatially adaptive Bayesian penalized regression splines (P-splines). *Journal of Computational and Graphical Statistics*, **14**(2), 378–394.

Benjamini, Y. and Hochberg, Y. (1995). Controlling the false discovery rate: A practical and powerful approach to multiple testing. *Journal of the Royal Statistical Society, Series B*, **57**, 289–300.

Brumback, L. C. and Lindstrom, M. J. (2004). Self modeling with flexible, random time transformations. *Biometrics*, **60**(2), 461–470.

Cheng, W., Dryden, I., and Huang, X. (2013). Bayesian registrations of functions and curves. *eprint arXiv:1311.2105*.

De Boor, C. (1978). *A Practical Guide to Splines*. Berlin: Springer-Verlag.

Della Gatta, G., Bansal, M., Ambesi-Impiombato, A., Antonini, D., Missero, C., and di Bernardo, D. (2008). Direct targets of the trp63 transcription factor revealed by a combination of gene expression profiling and reverse engineering. *Genome Research*, **18**(6), 939–948.

Eilers, P. H. C. and Marx, B. D. (1996). Flexible smoothing with *B*-splines and penalties. *Statistical Science*, **11**, 89–102.

Erosheva, E. A., Matsueda, R. L., and Telesca, D. (2014). Breaking bad: Reviewing two decades of life course data analysis in criminology and beyond. *Annual Reviews of Statistics and Its Applications*, **1**, 301–332.

Gasser, T. and Kneip, A. (1995). Searching for structure in curve samples. *JASA*, **90**, 1179–1188.

Gelman, A., Carlin, J., Stern, H., Dunson, D., Vehtari, A., and Rubin, D. (2013). *Bayesian Data Analysis*. Chapman & Hall / CRC, 3rd edition.

Gervini, D. and Gasser, T. (2004). Self-modelling warping functions. *Journal of the Royal Statistical Society, Series B: Statistical Methodology*, **66**(4), 959–971.

Guo, W. (2002). Functional mixed effects models. *Biometrics*, **58**(1), 121–128.

Hart, J. D. and Wehrly, T. E. (1986). Kernel regression estimation using repeated measurements data. *Journal of the American Statistical Association*, **81**, 1080–1088.

Hastie, T. and Tibshirani, R. (1993). Varying–coefficient models. *Journal of the Royal Statistical Society*, **55**, 757–796.

Hastie, T., Tibshirani, R., and Friedman, J. H. (2001). *The Elements of Statistical Learning: Data Mining, Inference, and Prediction*. Springer–Verlag Inc.

Kneip, A. and Gasser, T. (1988). Convergence and consistency results for self-modeling nonlinear regression. *The Annals of Statistics*, **16**, 82–112.

Kneip, A. and Gasser, T. (1992). Statistical tools to analyze data representing a sample of curves. *The Annals of Statistics*, **20**, 1266–1305.

Kneip, A., Li, X., MacGibbon, K. B., and Ramsay, J. O. (2000). Curve registration by local regression. *The Canadian Journal of Statistics / La Revue Canadienne de Statistique*, **28**(1), 19–29.

Laird, N. M. and Ware, J. H. (1982). Random-effects models for longitudinal data. *Biometrics*, **38**, 963–974.

Lang, S. and Brezger, A. (2004). Bayesian P-splines. *Journal of Computational and Graphical Statistics*, **13**(1), 183–212.

Leng, X. and Müller, H. (2006). Time ordering of gene co-expression. *Biostatistics*, **7**.

Liu, X. and Müller, H. (2004). Functional averaging and synchronization for time-warped random curves. *Journal of the American Statistical Association*, **99**, 687–699.

Morris, J. S. and Carroll, R. J. (2006). Wavelet-based functional mixed models. *Journal of the Royal Statistical Society, Series B: Statistical Methodology*, **68**(2), 179–199.

Müller, P., Parmigiani, G., and Rice, K. (2006). Fdr and Bayesian multiple comparisons rules. *Proceedings of the Valencia/ISBA 8th World Meeting on Bayesian Statistics (Oxford University Press).*

Parmigiani, G., Garrett, S. E., Anbashgahn, R., and Gabrielson, E. (2002). A statistical framework for expression-based molecular classification in cancer. *Journal of The Royal Statistical Society, Series B*, **64**, 717–736.

Pinheiro, J. C. and Bates, D. M. (2000). *Mixed–Effects Models in S and S–Plus.* Springer–Verlag: New York.

Pound, C. R., Partin, A. W., Eisenberger, M. A., Chan, D. W., Pearson, J. D., and Walsh, P. C. (1999). Natural history of progression after psa elevation following radical prostatectomy. *Journal of the American Medical Association*, **281**, 1591–1597.

Qian, J., Dolled-Filhart, M., Lin, J., Yu, H., and Gerstein, M. (2001). Beyond synexpression relationships: local clustering of time-shifted and inverted gene expression profiles identifies new, biologically relevant interactions. *Journal of Molecular Biology*, **314**, 1053–1066.

Ramsay, J. O. and Li, X. (1998). Curve registration. *Journal of the Royal Statistical Society, Series B: Statistical Methodology*, **60**, 351–363.

Rice, J. A. and Silverman, B. W. (1991). Estimating the mean and covariance structure nonparametrically when the data are curves. *Journal of the Royal Statistical Society, Series B: Methodological*, **53**, 233–243.

Ruppert, D., Wand, M., and Carroll, R. J. (2003). *Semiparametric Regression*. Cambridge University Press.

Sakoe, H. and Chiba, S. (1978). Dynamic programming optimization for spoken word recognition. *IEEE Transactions of Acoustic, Speech and Signal Processing*, **ASSP–26**(1), 43–49.

Shi, M., Weiss, R. E., and Taylor, J. M. G. (1996). An analysis of paediatric CD4 counts for acquired immune deficiency syndrome using flexible random curves. *Applied Statistics*, **45**, 151–163.

Silverman, B. W. (1995). Incorporating parametric effects into functional principal components analysis. *Journal of the Royal Statistical Society, Series B: Methodological*, **57**, 673–689.

Telesca, D. and Inoue, L. Y. T. (2008). Bayesian hierarchical curve registration. *Journal of the American Statistical Association*, **103 (481)**, 328–339.

Telesca, D., Inoue, L. Y. T., Neira, M., Etzioni, R., Gleave, M., and Nelson, C. (2009). Differential expression and network inferences through functional data modeling. *Biometrics*, **65**, 793–804.

Telesca, D., Erosheva, E. A., Kreager, D. A., and Matsueda, R. L. (2012a). Modeling criminal careers as departures from a unimodal popolation age-crime curve: The case of marijuana use. *JASA*, **107**(500), 1427–1440.

Telesca, D. adn Muller, P., Parmigiani, G., and RS, F. (2012b). Modeling dependent gene expression. *Annals of Applied Statistics*, **6**(2), 542–560.

Tuddenham, R. D. and Snyder, M. M. (1954). Physical growth of Califorfornia boys and girls from birth to eighteen years. *University of California Publications in Child Development*, **1**, 183–364.

Verbyla, A. P., Arunas, P., Cullis, B. R., Kenward, M. G., and Welham, S. J. (1999). The analysis of designed experiments and longitudinal data by using smoothing splines. *Journal of the Royal Statistical Society, Series C: Applied Statistics*, **48**, 269–300.

Wakefield, J. (2012). *Bayesian and Frequentist regression methods*. Springer.

Wang, K. and Gasser, T. (1997). Alignment of curves by dynamic time warping. *The Annals of Statistics*, **25**(3), 1251–1276.

Wang, K. and Gasser, T. (1999). Synchronizing sample curves nonparametrically. *The Annals of Statistics*, **27**(2), 439–460.

Wypij, D., Pugh, M., and Ware, J. H. (1993). Modeling pulmonary function growth with regression splines. *Statistica Sinica*, **3**, 329–350.

Yao, F., Müller, H. J., and Wang, J. L. (2005). Functional data analysis of sparse longitudinal data. *JASA*, **100**(470), 577–590.

Zeger, S. L. and Diggle, P. J. (1994). Semiparametric models for longitudinal data with application to CD4 cell numbers in HIV seroconverters. *Biometrics*, **50**, 689–699.

Zhang, Y. and Telesca, D. (2014). Joint clustering and registration of functional data. Technical report, UCLA.

# Chapter 15
# Biomarker-Driven Adaptive Design

Yanxun Xu, Yuan Ji, and Peter Müller

**Abstract** We review some principles and implementations of Bayesian model-based adaptive enrichment and population finding designs that exploit biomarker information to propose adaptive treatment allocation and recommend patient subpopulations that might most benefit from the treatments under consideration.

## 15.1 Introduction

We review some recently proposed approaches to Bayesian adaptive clinical trial design based on subgroup analysis, including in particular the implementation proposed in Xu et al. (2014). In Xu et al. (2014) we use a classification and regression tree (CART) to construct a partition of the covariate space; see Fig. 15.1. Partitioning subsets are then used as candidates for identifying subgroups of patients with substantially different treatment effect, which in turn is exploited for adaptive treatment allocation. Upon conclusion of the trial final inference about the identified subgroups is reported. We refer to the proposed approach as *subgroup-based adaptive (SUBA)* design. The design was developed for a breast cancer trial that includes three candidate treatments and makes use of protein biomarkers to adaptively identify the best treatment for different patient subpopulations. Besides optimal allocation of treatment arms to patients during the trial, the main outcome is an estimated

Y. Xu (✉) • P. Müller
The University of Texas at Austin, 1, University Station, C1200 Austin, TX 78712, USA
e-mail: yxu.stat@gmail.com; pmueller@math.utexas.edu

Y. Ji
NorthShore University HealthSystem/The University of Chicago,
1001 University Place, Evanston, IL 60201, USA
e-mail: koaeraser@gmail.com



R. Mitra, P. Müller (eds.), *Nonparametric Bayesian Inference in Biostatistics*,
Frontiers in Probability and the Statistical Sciences,
DOI 10.1007/978-3-319-19518-6_15

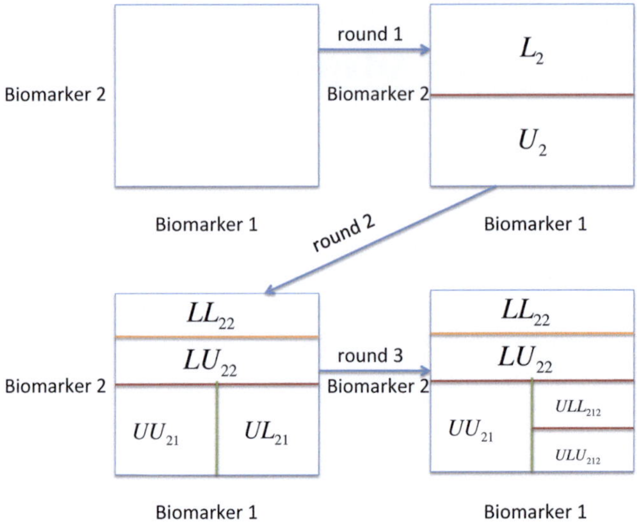


**Fig. 15.1** Partitions of the covariate space with an increasing CART tree. This illustration shows that the initial space of two biomarkers is partitioned into five sets after three rounds of split

partition of the patient population into distinct subpopulation that relate to different treatment recommendations. In this chapter we review some related literature and the proposed SUBA design.

Bayesian methods have been widely used for the identification of biomarker groups that could be truly responsive to targeted treatments. Dixon and Simon (1991) analyze subgroup effects by considering a single model that includes main effects of treatment, covariates and the first-order interactions between treatment and covariates. Simon (2002) uses a similar approach with independent priors on the interaction parameters. Sivaganesan et al. (2011) consider subgroup analysis as a model selection problem with each covariate defining a family of models. Ruberg et al. (2010) and Foster et al. (2011) develop tree-based algorithms to identify and evaluate subgroup effects by searching for regions with substantially enhanced treatment effects compared to the average effect, averaging across the covariate space. Sivaganesan et al. (2013) report subgroups within a Bayesian decision-theoretic framework. They determine rules using an extension of a 0/1/K utility function. The utility function is based on the posterior odds of subgroup models relative to the overall null and alternative models.

Several recent studies explore the use of Bayesian adaptive designs, including population finding designs. Examples include the breast cancer trial ISPY-2 (Barker et al. 2009), which uses indicators for several biomarkers and a MammaPrint risk score to define 14 subpopulations of possible practical interest. The design graduates subpopulations, that is, recommends a future phase III study; or drops subpopulations or treatment arms, that is, remove one of the 14 subpopulations or treatment arms from further consideration. A similar example is the BATTLE design of

Zhou et al. (2008) who define five subpopulations of lung cancer patients based on biomarker profiles and proceed to adaptively allocate patients to alternative treatments. Another recent discussion is Berry et al. (2013) who report an extensive comparison of Bayesian adaptive design, including a design based on a hierarchical model over different subpopulations with a comparable design using Simon's optimal Two-Stage design (Simon 2012). Some of earlier discussion appear in Simon and Maitournam (2004), Sargent et al. (2005), and Freidlin et al. (2010).

## 15.2 Bayesian CART Models

The design proposed in Xu et al. (2014) makes use of a Bayesian nonparametric (BNP) regression of the outcome on patient-specific covariates. In particular, the approach uses a variation of Bayesian CART. The Bayesian CART was introduced in Chipman et al. (1998) and Denison et al. (1998) as an attractive model for BNP regression. Consider a regression problem $y_i = f(\boldsymbol{x}_i) + \varepsilon_i$ with a multivariate covariate vector $\boldsymbol{x}_i = (x_{ik},\ k = 1,\ldots,K)$. For the moment we drop the $i$ index, considering a generic covariate vector $\boldsymbol{x}$. To start, we partition the covariate space into small enough rectangular regions $R$ described by thresholds on the covariates $x_k$, such that $E(y \mid \boldsymbol{x} \in R) \approx f_R$ is approximately constant over each rectangular region.

The partition is described by a tree $T$. The leaves of the tree correspond to the rectangular regions and are labeled with the mean response $f_R$. The tree $T$ is a recursive structure $T = (k,t,T_0,T_1)$ of splitting rules consisting of a covariate index $k$ and a threshold $t$ that identify a splitting rule $x_k < t$ and two nested trees $(T_0,T_1)$ with the tree $T_0$ defining the branch $x_k < t$ and $T_1$ defining the branch $x_k \geq t$. The recursion ends with final leaves that contain a value $f_\ell$ instead of a tree $T_\ell$. Chipman et al. (1998) and Denison et al. (1998) describe prior probability models on $T$ and posterior simulation. For a more parsimonious model the constant mean response $f_R$ in each leave can be replaced by any parametric model $p(y \mid \boldsymbol{x} \in R) = f_{\theta_R}(y)$, where $\theta_R$ are parameters specific to each leave. In general, the random partition that is indexed by the tree $T$ is interpreted as creating more homogeneous patient subpopulations. Here, homogeneity could be described in terms of sharing a common set of logistic regression parameters $\theta_R$, or any other sampling model.

The Bayesian additive regression tree (BART) of Chipman et al. (2010) creates a useful variation of the CART by constructing a random forest as a (random) sum of many small CART trees. The idea of BART is to use many small trees to approximate the desired mean function. The BART is implemented in an easy-to-use R package `BayesTree`.

The SUBA design uses a massively reduced version of the CART as a basis for model-based adaptive design. The model is restricted to a small maximum number of splits in the tree. Such restrictions on parsimony are critical for design problems that involve inference with little and no data in the early stages of the study. Upon completion of the study it is important to assure investigators and reviewers that the assumed model structure and the strength of the prior assumptions do not dominate the reported inference. This is one of the reasons for the limited use of BNP methods in clinical trial design. In fact, upon a quick review of recent texts that review Bayesian clinical trial design we find no mention of BNP methods (Berry et al. 2011; Yin 2012). The reason is the apparent conflict between the flexibility and infinite dimensional parameter space of a BNP model and the limited data and emphasis on parsimony of clinical trial design.

The SUBA design offers one way to use the flexibility of BNP methods without compromising on the restrictions of clinical trial design. The Bayesian approaches among the methods that we briefly reviewed in the introductory section all make use of parametric inference models.

## 15.3 The Model

The adaptive design proposed in Xu et al. (2014) is based on a reduced version of a Bayesian CART model. The design exploits the interpretation of the leaf nodes as characterizing a (rectangular) partition of the covariate space, that is, a partition of the patient population into subpopulations that are characterized by the corresponding thresholds on the covariates and biomarkers.

The model is a tree with up to a maximum number $M$ of splits. In the current implementation we use $M = 3$. The prior probability model on the tree is constructed by considering at each possible branching point in the tree construction a categorical choice $a$ of no further split ($a = 0$), or a split on the $k$th variable ($a = k$), $k = 1, \ldots, K$. Specifying $p(a = k) = v_k$, $k = 0, 1, \ldots, K$, defines a probability for a tree structure. In addition, if a biomarker is selected to split in previous steps, it is expected to be relevant to the response of treatment and therefore should have a higher chance being split again facilitating the identification of true subgroups. To realize this, in each possible tree $T$, we calculate $h_T$ as the number of biomarkers chosen in the three rounds of splits. Then we add an additional penalty term $\phi^{h_T - 1}$ to the prior probability of $T$, here $\phi \in (0, 1)$. Therefore, the smaller $h_T$, the larger prior probability the tree $T$ receives, which encourages repeated splits along the same biomarker. The remaining choice is the selection of the thresholds for each split. Here we again depart from the traditional CART construction and impose a deterministic threshold specification. Assume we are considering a split of a current subset $S$, and the split is decided to be on $x_k$. The threshold is the median of $x_k$ among all patients $i$ with $\boldsymbol{x}_i \in S$. The deterministic threshold selection is important to keep inference computationally efficient and allow for a parsimonious model. In summary the prior model on the tree is characterized with $(v_0, \ldots, v_K, \phi, M)$ only, that is $p(T \mid v_0, \ldots, v_K, \phi, M)$. As an example, in Fig. 15.1,

$$p(T \mid v_0, \ldots, v_K, \phi, M) \propto v_2 \times v_2 \times v_1 \times v_0 \times v_0 \times v_0 \times v_2 \times \phi^{2-1}.$$

We assume that the outcome $y_i$ for patient $i$ is a binary indicator for response, $i = 1,\ldots,N$. Conditional on the tree we assume independent Bernoulli models for each partitioning subset. Let $\{S_j,\ j = 1,\ldots,n_S\}$ denote the partition of the covariate space that is implied by the tree. Let $z_i \in \{1,\ldots,D\}$ denote the treatment assignment for patient $i$. Let $\theta_{dj}$ denote the probability of response for a patient with covariates $\boldsymbol{x}_i \in S_j$ under treatment $d$. We assume

$$p(y_i = 1 \mid \boldsymbol{x}_i \in S_j, z_i = d) = \theta_{dj}.$$

The model is completed with independent beta priors on $\theta_{dj}$ with hyperparameters $b_1$ and $b_2$, $d = 1,\ldots,D$ and $j = 1,\ldots,n_S$. The conjugate beta/Bernoulli framework allows for analytic marginalization of the joint probability model with respect to $\boldsymbol{\theta} = (\theta_{dj},\ d = 1,\ldots,D \text{ and } j = 1,\ldots,n_S)$.

One major aim of the SUBA design is to assign future patients to the treatment that is most promising for that specific patient, based on the response data from previous patients and the new patient's biomarker profile. Formally, we use the posterior predictive probability that the new patient would respond to treatment $d$ as the basis for assignment. In particular, suppose that $n$ patients have been treated in the trial and their response vector is $\boldsymbol{y}^n$. We also know their treatment allocation vector $\boldsymbol{z}^n = (z_1,\ldots,z_n)$, and their baseline biomarker profiles $\boldsymbol{x}^n$. The posterior predictive probability of response if patient $(n+1)$ is under treatment $d$ is given by

$$\begin{aligned} q_{n+1}(d) &= Pr(y_{n+1} = 1 \mid \boldsymbol{x}_{n+1}, z_{n+1} = d, \boldsymbol{y}^n, \boldsymbol{x}^n, \boldsymbol{z}^n) \\ &= \int Pr(y_{n+1} = 1 \mid \boldsymbol{x}_{n+1}, z_{n+1} = d, T)\, p(T \mid \boldsymbol{y}^n, \boldsymbol{x}^n, \boldsymbol{z}^n) dT. \end{aligned}$$

Denote $\tilde{z}_{n+1}$ the treatment decision for the $(n+1)$th patient. We allocate the $(n+1)$th patient to treatment $\tilde{z}_{n+1}$ by

$$\tilde{z}_{n+1} = \arg_d \max q_{n+1}(d). \qquad (15.1)$$

## 15.4 SUBA Design

### *15.4.1 Design*

By calculating the posterior predictive response rates of all candidate treatments, we can compare treatments and monitor the trial. For instance, if one treatment is no better than all other treatments, that treatment should be dropped. Also, we should stop the trail early if there is only one treatment left after dropping all other treatment arms as inferior treatments. We monitor the trial and update posterior distributions after each patient beyond an initial run-in phase. The trial continues until we either make an early stop decision or we reach the maximum sample size $N$.

The comparison of treatments for possible inferiority needs to account for baseline covariates. In words, we determine a treatment to be inferior if its posterior predictive probability of response is lower than all other treatments across all possible biomarker profiles. To formalize the latter condition we set up a grid in the biomarker space $R^K$. We fix $H$ grid points, $g_1,\ldots,g_H$. For example, if our biomarker space is $[-1,1]^K$, we can choose $H_0$ equally spaced points on $[-1,1]$ for each dimension. Then the total number of grid points will be $H = H_0^K$. Each grid point is treated as a possible biomarker profile. After an initial run-in phase with equal randomization, we evaluate the posterior predictive response rate under each possible biomarker profile under each treatment $d$, using $q_{g_h}(d)$ in (15.1). If we find a treatment $d^*$ satisfying

$$q_{g_h}(d^*) < q_{g_h}(d)$$

for all $d \neq d^*$ on all possible biomarker profiles $h = 1,\ldots,H$, then treatment $d^*$ is dropped from the trial. If there is only one treatment left, we stop the trial early and assign all the untreated patients to that superior treatment.

We are now ready to state the simple algorithm of the SUBA design.

**Run-in:** The first $n$ patients are equally randomized to $D$ treatment arms.

**Stopping for futility:** Drop treatment $d^\star$ if the posterior predictive response rate is uniformly inferior to all $d \neq d^\star$ based on the posterior predictive response rate. Here, uniform inferiority is across all $H$ biomarker profiles on a grid and across all other treatments.

**Adaptive treatment allocation:** Assign patient $(n+1)$ to treatment $\tilde{z}_{n+1}$ using rule (15.1).

**Posterior updating:** When the response $y_{n+1}$ is available, go back to step 2 and repeat for patients $n+2, n+3, \ldots, N$.

**Final report:** Upon conclusion of the study we report the estimated partition and the recommended treatment allocation for each subset. See Sect. 15.4.2 for details.

### *15.4.2 Posterior Inference on the Partition*

In the final report, a practical problem arises in summarizing the inference on the partition into subpopulations. It is not straightforward to summarize a posterior distribution on random partitions. There is no such thing as a posterior mean partition. To address this problem, Medvedovic et al. (2004) proposed to first compute posterior co-clustering probabilities. That is, posterior probabilities for each pair of patients $(i,j)$ we record the posterior probabilities of $i$ and $j$ sharing the same cluster. To this end we average association matrices, as follows.

For each partition $T$, an association matrix $G^T$ of dimension $N \times N$ is formed. Let $G_{ij}^T$ denote an indicator of whether patient $i$ is in the same subgroup as patient $j$

under the tree $T$. Next we evaluate $\tilde{G}$ as an element by element posterior average of these association matrices $G^T$. We refer to $\tilde{G}$ as the pairwise probability matrix. It serves as a summary of the posterior distribution $p(T \mid data)$.

Alternatively, Dahl (2006) introduced the least-squares partition for estimating a random partition. Following Dahl (2006), we propose a least-square summary

$$T^{LS} = \arg\min_{T} ||G^T - \tilde{G}||^2.$$

The minimum goes over all $T$ in a posterior Monte Carlo sample $T \sim p(T \mid data)$. That is, $T^{LS}$ is the posterior simulated partition $T$ that minimizes the sum of squared deviations of the corresponding association matrix $G^T$ from the pairwise probability matrix $\tilde{G}$. In other words, the least-square partition is the posterior simulated partition $T$ that minimizes the Frobenius distance ($L_2$ norm for matrices) between $G^T$ and $\tilde{G}$.

## 15.5 Example

### 15.5.1 Simulation Setup

We carried out simulation studies that were designed to mimic the motivating breast cancer trial. The maximum sample size was set at $N = 300$ patients in a three-arm study with three treatments labeled as 1, 2, and 3. For each patient, we measured $K = 4$ biomarkers at baseline. We generated biomarker values $x_{ik} \sim \text{Unif}(-1,1)$, $i = 1,2,\ldots,N$. In the prior specification we assumed $v_k = 1/(K+1)$, $k = 0,1,\ldots,K$, $\phi = 0.5$, $b_1 = 1$ and $b_2 = 1$. The stopping rule is implemented with $H_0 = 10$ equally spaced points on each biomarker subspace, leading to a size $H = 10{,}000$ grid of biomarker profiles. During the initial run-in phase, $n = 100$ patients were equally randomized to three candidate treatments.

We considered three scenarios and simulated 100 trials for each scenario. In scenario 1, we assumed the following simulation truth for the response rates:

$$\theta_{1i} = \Phi(1.5x_{i1} + x_{i2}),\ \theta_{2i} = \Phi(1.5x_{i1}),\ \theta_{3i} = \Phi(1.5x_{i1} - x_{i2}),$$

for treatments $1,2,3$, respectively. Here $\Phi$ is the standard normal cumulative distribution function (CDF). In this simulation truth, biomarkers 1 and 2 are relevant to response, but not biomarkers 3 and 4. Figure 15.2 plots the response rates of three treatments versus the first biomarker $x_{i1}$, given different values of the second biomarker $x_{i2}$. Treatment 3 is always the most effective arm when the second biomarker is fixed at a negative value; three treatments perform the same when the second biomarker is fixed at 0; and treatment 1 is superior when the second biomarker is fixed at a positive value. Therefore, in terms of deciding the superiority of treatment effects, the second biomarker is predictive. The response rates of three treatments increase as the measurement of first biomarker increases, but the ordering of the three treatments does not change. Therefore, the first biomarker is only predictive of response but not treatment selection.

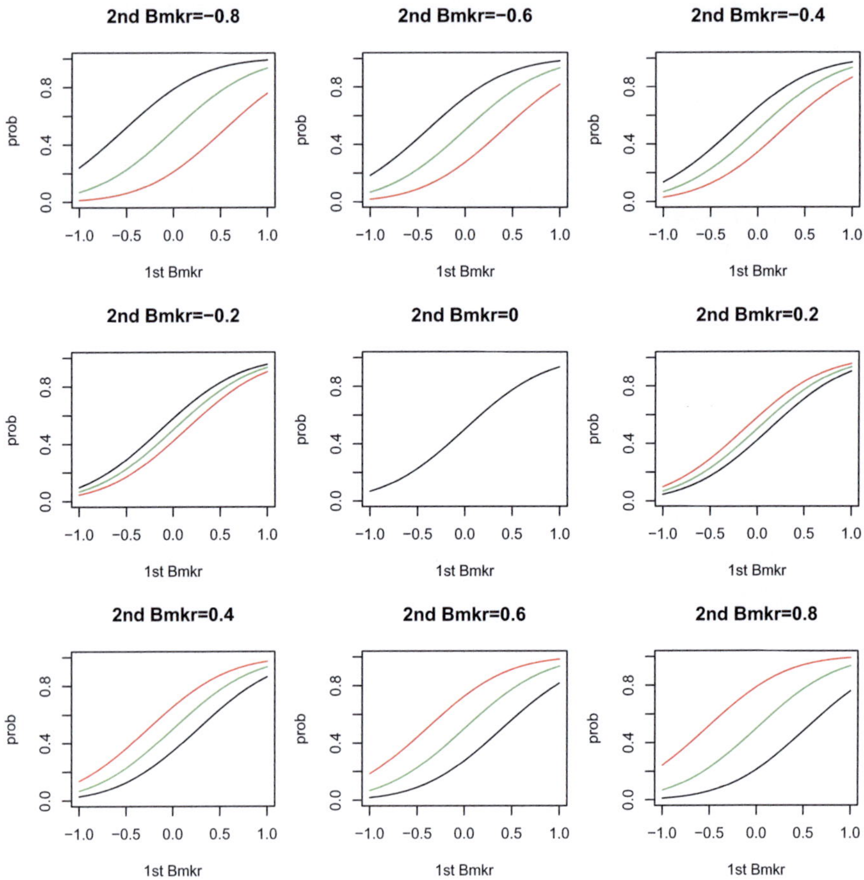


**Fig. 15.2** The probabilities of response versus the measurements of the first biomarker given fixed values of the second biomarker. *Red*, *green*, and *black lines* represent three treatments 1, 2, and 3 respectively

In scenario 2, we assumed that biomarkers 1, 2, and 3 were related to the response and we assumed an interaction between biomarkers. The response rate of patient $i$ with treatments 1, 2, and 3 were

$$\theta_{1i} = \Phi(1.5x_{i1} + 0.5x_{i2} - x_{i3} - 2x_{i1}x_{i3}),\ \theta_{2i} = \Phi(-1.5x_{i1} - 1.5x_{i3}),$$
$$\theta_{3i} = \Phi(-1.5x_{i1} - x_{i2} - 2.5x_{i1}x_{i2}),$$

for treatments 1, 2, and 3, respectively. Figure 15.3 plots a 3D illustration of response rates of three treatments versus the measurements of the first biomarker and the second biomarker given the third biomarker is fixed at 0.5 (Fig. 15.3a) or −0.5 (Fig. 15.3b). In summary, all three biomarkers together determine the optimal treatment.

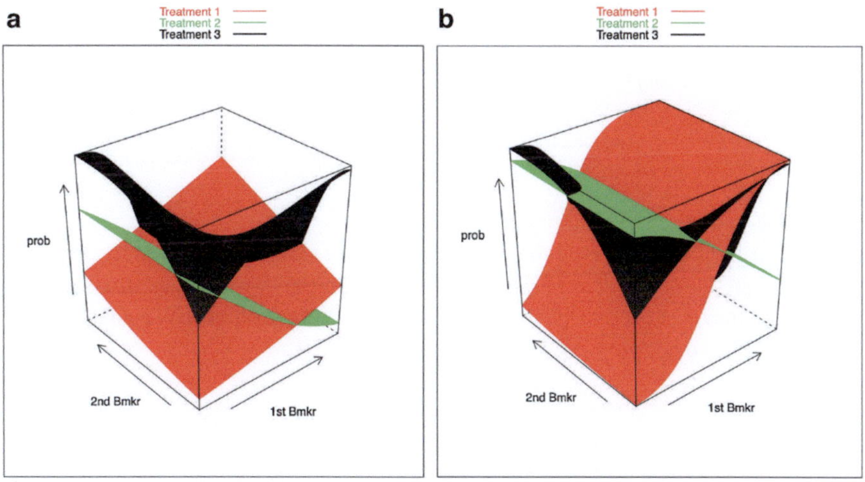


**Fig. 15.3** The probabilities of response versus the measurements of the first and the second biomarkers given the fixed values of the third biomarker at 0.5 (**a**) and −0.5 (**b**). *Red*, *green*, and *black lines* represent three treatments 1, 2, and 3, respectively

Scenario 3 was a null case with no biomarker related to response. We assumed that the response rates of three treatments were the same at 30 %, i.e., $\theta_{1i} = \theta_{2i} = \theta_{3i} = 0.3$.

### 15.5.2 Comparison

For comparison, we implemented three alternative designs. The first is a standard design with equal randomization (ER). All patients are equally randomized. The model assumed independent Bernoulli outcomes $y_i \sim \text{Bern}(\theta_{di})$ for patient $i$ under treatment $d$. The simulation truth for $\theta_{di}$ was generated as before.

The second comparison is to an outcome-adaptive randomization (AR) design. To highlight the sensitivity of AR to predefined subgroups, we defined three biomarker subgroups that were selected similar to the BATTLE trial (Kim et al. 2011) based on the first biomarker. We used the subgroups

$$\{x_{i1} < -0.5\}, \{-0.5 \leq x_{i1} \leq 0.5\} \text{ and } \{x_{i1} > 0.5\}.$$

The subgroups are deliberately selected to not match the simulation truth in all three scenarios. Let $p_{dj}$ be the response rate of treatment $d$ in subgroup $j$, and $n_{dj}$ the total number of patients receiving treatment $d$ in subgroup $j$, $d = 1,2,3$ and $j = 1,2,3$. We assumed $y_i \sim \text{Bin}(n_{dj}, p_{dj})$ and a conjugate beta prior distribution beta(1,1) on $p_{dj}$. The posterior on $p_{dj}$ can be easily computed as a $\text{Be}(n_{dj1}+1, n_{dj}-n_{dj1}+1)$ distribution, where $n_{dj1}$ is the number of patients who respond to treatment $d$ in subgroup $j$. In the AR design, we included an initial run-in with 200 patients being equally randomized to the three treatments. The next 200 patients were adaptively randomized. Let $\hat{p}_{dj} = (n_{dj1}+1)/(n_{dj}+2)$ denote the posterior mean for $p_{dj}$. The adaptive treatment allocation used

$$\pi_{jd} = \hat{p}_{dj}/(\hat{p}_{1j}+\hat{p}_{2j}+\hat{p}_{3j})$$

to assign a patient in subgroup $j$ to treatment $d$.

The third comparison is with a probit regression (Reg) design. We modeled binary outcome variables using a probit regression. In the probit model, the response rate is modeled as a probit transform of a linear combination of the biomarkers and treatment. That is,

$$Pr(y_i = 1 \mid z_i, \boldsymbol{x}_i) = \Phi(\beta_0 z_i + \boldsymbol{\beta}_1 \boldsymbol{x}_i).$$

The parameters $\beta_0$ and $\boldsymbol{\beta}_1$ are estimated using maximum likelihood method. In the Reg design, we again included an initial burn-in with 100 equally randomized patients, and then assigned the remaining patients sequentially to the treatment with the respective highest success probability under the currently estimated probit regression parameters. That is, we evaluated the probit regression with the posterior means $E(\beta_0, \beta_1 \mid data)$.

### *15.5.3 Simulation Results*

Overall Response Rate

We define the overall response rate (ORR)

$$\text{ORR} = \frac{1}{N-n} \sum_{i=n+1}^{N} I(y_i = 1),$$

as the proportion of responders among the patients who are treated after the run-in phase. We computed ORRs for the four designs under comparison, ER, AR, Reg, and SUBA for all three scenarios. Figure 15.4 plots the ORR differences between SUBA and ER, AR, Reg, respectively.

In scenario 1, SUBA and Reg are preferable to ER and AR, indicating higher efficacy. It is not surprising that Reg also performs well in this scenario since the fitted model matches the simulation truth.

In scenario 2, SUBA outperforms ER, AR, and Reg with higher ORR in all the simulated trials. ER and AR with predefined biomarker subgroups perform similarly, suggesting that no gains are obtained with AR when the biomarker subgroups are wrongly predefined.

In scenario 3, the true response rates are the same across all three treatments and not related to biomarkers, so the four designs have similar ORRs across simulated trials.

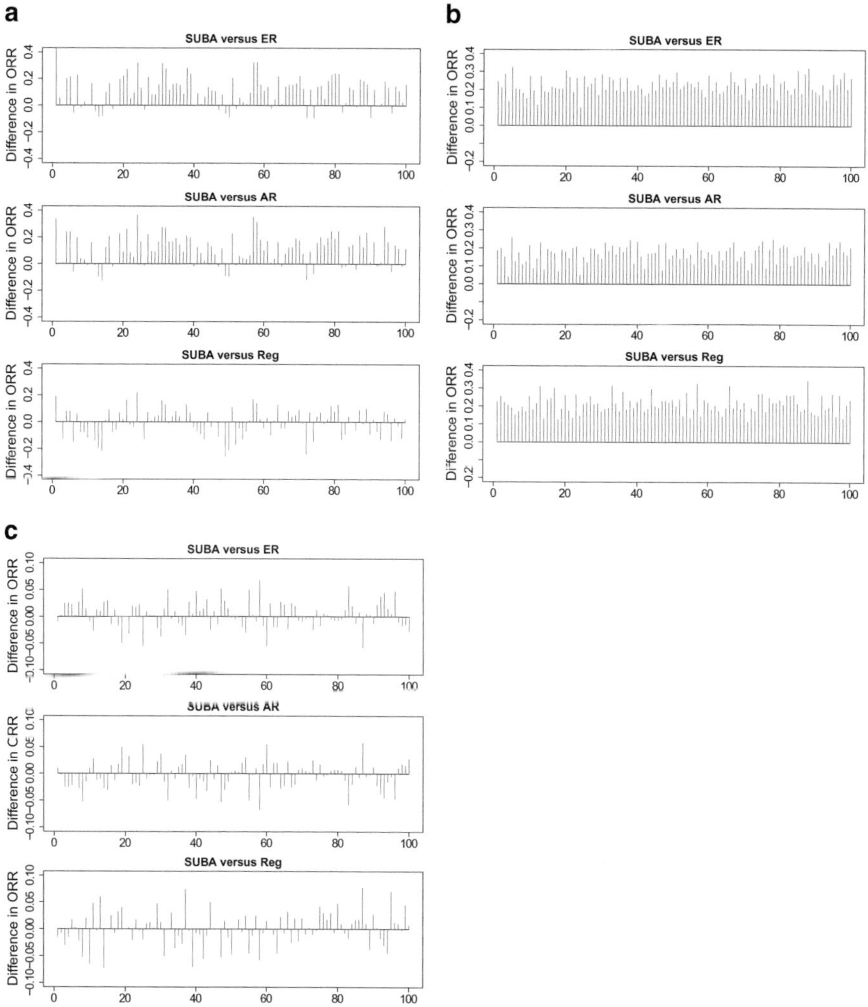


**Fig. 15.4** Overall response rates (ORR) across ER, AR, Reg, and SUBA designs in 100 simulated trials under the three scenarios. We plot the ORR differences between SUBA and ER, AR, Reg, respectively, in each scenario. *Blue color* represents that the ORR of SUBA is higher than ER, AR, or Reg; *red color* represents lower. (**a**) Scenario 1, (**b**) Scenario 2, (**c**) Scenario 3

The average numbers of patients under the SUBA design are reported in Table 15.1. In scenario 3 the early stopping rule became active and allowed for an early termination.

**Table 15.1** The average numbers of patients needed to make the decision of stopping trials early in 100 simulated trials in scenarios 1–3

| Scenario | 1 | 2 | 3 |
|---|---|---|---|
| # of patients | 298.05 | 300.00 | 202.00 |

### Average Number of Patients Assigned

Next we compare the designs on the basis of the average number of patients (ANP) assigned to treatment $d$ after the run-in phase. Denote by $\mathrm{NP}_d^w$ the number of patients assigned to treatment $d$ in the $w$th simulated trial (excluding the run-in phase). That is, $\mathrm{NP}_d^w = \sum_{i=n+1}^{N} I(z_i^w = d)$, $d = 1,2,3$ and $w = 1,\ldots,100$. We define

$$\mathrm{ANP}_d = \frac{1}{100}\sum_{w=1}^{100} \mathrm{NP}_d^w.$$

Table 15.2 shows the results. Since ER, AR, and Reg do not include early stopping rules, for a fair comparison, the summaries for SUBA are based on assigning in the case of early stopping all remaining patients until the maximum sample size $N$ to the surviving active arm.

**Table 15.2** The average numbers of patients assigned to three treatments after the run-in phase in three defined subsets by ER, AR, and SUBA in 100 simulated trials in scenarios 1–3

| Scenario | | ER | | | AR | | | Reg | | | SUBA | | |
|---|---|---|---|---|---|---|---|---|---|---|---|---|---|
| | Subset | 1 | 2 | 3 | 1 | 2 | 3 | 1 | 2 | 3 | 1 | 2 | 3 |
| 1 | $S_1$ | 33.36 | 33.19 | 33.91 | 33.86 | 33.02 | 33.58 | 74.46 | 18.67 | 7.33 | **75.78** | 16.56 | 7.48 |
| | $S_2$ | 32.28 | 33.44 | 33.82 | 33.25 | 32.83 | 33.46 | 9.65 | 17.59 | 72.30 | 8.51 | 17.70 | **73.98** |
| 2 | $S_1$ | 24.77 | 24.84 | 25.94 | 33.25 | 19.75 | 22.55 | 8.21 | 21.93 | 45.41 | **60.24** | 4.62 | 10.69 |
| | $S_2$ | 16.60 | 16.33 | 16.65 | 10.96 | 18.80 | 19.82 | 5.20 | 15.01 | 29.37 | 3.15 | **33.34** | 13.09 |
| | $S_3$ | 24.27 | 25.46 | 25.14 | 21.71 | 25.99 | 27.17 | 8.50 | 21.85 | 44.52 | 10.30 | 13.34 | **51.23** |
| 3 | / | 65.64 | 66.63 | 67.73 | 66.36 | 67.48 | 66.16 | 69.18 | 66.28 | 64.54 | 66.89 | 64.20 | 68.91 |

The results in Table 15.2 are arranged by subsets that are formed as follows. In scenario 1, we split the biomarker space to two sets $S_1^0 = \{i : x_{i2} < 0\}$ and $S_2^0 = \{i : x_{i2} > 0\}$ and separately report the average numbers of patients assigned to three treatments after the run-in phase, among those whose second biomarker is positive or negative. We separately report these two averages to demonstrate the benefits of using the SUBA design since depending on the sign of the second biomarker, different treatments should be selected as the most beneficial and effective ones for patients. For example, when the second biomarker is positive, treatment 1 is the most superior arm; when the second biomarker is negative, treatment 3 is the most effective arm according to our simulation settings. From Table 15.2, among the 200 patients after the run-in phase, about 100 patients each had positive or negative values of the second biomarker. Among the 100 patients with positive values, 76 were allocated to treatment 1, 17 to treatment 2, and 7 to treatment 3. Among the 100 with negative values, 9 were allocated to treatment 1, 18 to treatment 2, and 74 to treatment 3. According to those numbers, most of the patients were assigned to the correct superior treatments demonstrating the contribution of the SUBA design, emphasized by bold numbers. In Table 15.2, Reg assigned similar number of patients to three treatments like SUBA, while ER and AR designs assigned far fewer patients to the most effective treatments, which explained the contrast shown in Fig. 15.4a.

In scenario 2, biomarkers 1, 2, and 3 were related to the response. In a similar fashion, we report patient allocations by splitting the biomarker space to three subsets that are indicative of the true best treatment decision. Denote $\tilde{\theta}_{1i} = 1.5x_{i1} + 0.5x_{i2} - x_{i3} - 2x_{i1}x_{i3}$, $\tilde{\theta}_{2i} = -1.5x_{i1} - 1.5x_3$, and $\tilde{\theta}_{3i} = -1.5x_1 - x_2 - 2.5x_1x_2$. According to the simulation truth, we defined $S_1^0 = \{i : \tilde{\theta}_{1i} > \tilde{\theta}_{2i} \text{ and } \tilde{\theta}_{1i} > \tilde{\theta}_{3i}\}$, $S_2^0 = \{i : \tilde{\theta}_{2i} > \tilde{\theta}_{1i} \text{ and } \tilde{\theta}_{2i} > \tilde{\theta}_{3i}\}$ and $S_3^0 = \{i : \tilde{\theta}_{3i} > \tilde{\theta}_{1i} \text{ and } \tilde{\theta}_{3i} > \tilde{\theta}_{2i}\}$. Under this assumption, the best treatment for patients in set $S_d^0$ is treatment $d$ according to the simulation truth. In Table 15.2, SUBA assigned most of the patients to their correct superior treatments. In contrast, ER, AR, and Reg failed to do so.

Scenario 3 is a null case in which the biomarkers were not related to response rates that were the same across three treatments. All four designs assigned similar number of patients to three treatments.

In summary, SUBA continuously learns the biomarker subgroups to determine superior treatments with targeted patients and can substantially outperform ER and AR in terms of OOR.

We also performed sensitivity analysis with respect to the choices of $n$ and the penalty parameter $\phi$. The more patients we observe, the more accurate assignments for future patients. And for different values of $\phi$, the reported summaries vary little across the considered hyperparameter choices (results not shown), indicating robustness with respect to changes within a reasonable range of values.

### 15.5.4 Report on Partition

Figure 15.5 shows the least-square partition for scenario 1. The number in each circle represents the biomarker used to split the biomarker space. In scenario 1, treatment 1 is the best treatment when the second biomarker is positive and treatment 3 is the best one when the second biomarker is negative. The least-square partition shows that biomarker 2 is chosen to split the biomarker space in the first round of split, which agrees with the simulation truth.

## 15.6 Conclusion and Discussion

We demonstrate the importance of subgroup identification in adaptive designs, especially when such subgroups are predictive of treatment selection. The key contribution of the modeling is the construction of the random partition prior $p(T)$ which allows a flexible and simple mechanism to implement subgroup exploration. Under the Bayesian paradigm adaptive allocation based on interim outcomes is natural. Posterior inference includes formal learning about relevant subgroups. The proposed construction for $T$ is easy to interpret and, most importantly, achieves a good balance between the computational burden for posterior computation and the flexibility of the resulting prior distribution. The priors on $\theta_{dj}$ are i.i.d uniform priors in our simulation studies. If desired, this prior can be calibrated to reflect the historical response rate of the drug. The i.i.d assumption simplifies posterior inference. Alternatively, one could impose dependence across the $\boldsymbol{\theta}$'s; for example, one could assume that adjacent partition sets have similar values.

By identifying subgroups of patients who react most positively to each of the treatments, the proposed SUBA design concentrates on the treatment success for the patients. One could easily add to the SUBA design a final recommendation of a suitable patient population for a follow-up trial. Future research directions include

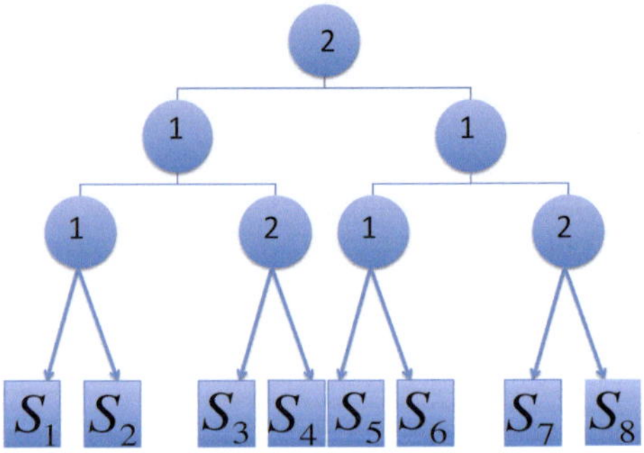


**Fig. 15.5** The tree-type least-square partition by SUBA design in scenario 1. The number in the *circle* represents the biomarker used to split the biomarker space

also an extension of SUBA to incorporate variable (biomarker) selection for trials with a large number of candidate biomarkers.

**Acknowledgements** Yuan Ji and Peter Müller's research is partially supported by NIH R01 CA132897.

## References


Barker, A., Sigman, C., Kelloff, G., Hylton, N., Berry, D., and Esserman, L. (2009). I-spy 2: an adaptive breast cancer trial design in the setting of neoadjuvant chemotherapy. *Clinical Pharmacology & Therapeutics*, **86**(1), 97–100.

Berry, S. M., Carlin, B. P., Lee, J. J., and Müller, P. (2011). *Bayesian Adaptive Methods for Clinical Trials*. CRC Press.

Berry, S. M., Broglio, K. R., Groshen, S., and Berry, D. A. (2013). Bayesian hierarchical modeling of patient subpopulations: efficient designs of Phase II oncology clinical trials. *Clin Trials*, **10**(5), 720–734.

Chipman, H. A., George, E. I., and McCulloch, R. E. (1998). Bayesian CART model search. *Journal of the American Statistical Association*, **93**(443), pp. 935–948.

Chipman, H. A., George, E. I., and McCulloch, R. E. (2010). BART: Bayesian additive regression trees. *Annals of Applied Statistics*, **4**, 266–298.

Dahl, D. (2006). Model-based clustering for expression data via a Dirichlet process mixture model. In K. Do, P. Müller, and M. Vannucci, editors, *Bayesian inference for Gene Expression and Proteomics*, pages 201–218. Cambridge: Cambridge University Press.

Denison, D. G. T., Mallick, B. K., and Smith, A. F. M. (1998). A Bayesian CART algorithm. *Biometrika*, **85**(2), pp. 363–377.

Dixon, D. O. and Simon, R. (1991). Bayesian subset analysis. *Biometrics*, **47**, 871–881.

Foster, J. C., Taylor, J. M., and Ruberg, S. J. (2011). Subgroup identification from randomized clinical trial data. *Statistics in medicine*, **30**(24), 2867–2880.

Freidlin, B., McShane, L. M., and Korn, E. L. (2010). Randomized clinical trials with biomarkers: design issues. *Journal of the National Cancer Institute*, **102**(3), 152–60.

Kim, E. S., Herbst, R. S., Wistuba, I. I., Lee, J. J., Blumenschein, G. R., Tsao, A., Stewart, D. J., Hicks, M. E., Erasmus, J., Gupta, S., *et al.* (2011). The BATTLE trial: personalizing therapy for lung cancer. *Cancer Discovery*, **1**(1), 44–53.

Medvedovic, M., Yeung, K. Y., and Bumgarner, R. E. (2004). Bayesian mixture model based clustering of replicated microarray data. *Bioinformatics*, **20**(8), 1222–1232.

Ruberg, S. J., Chen, L., and Wang, Y. (2010). The mean does not mean as much anymore: finding sub-groups for tailored therapeutics. *Clinical Trials*, **7**(5), 574–583.

Sargent, D. J., Conley, B. A., Allegra, C., and Collette, L. (2005). Clinical trial designs for predictive marker validation in cancer treatment trials. *Journal of Clinical Oncology*, **23**(9), 2020–2027.

Simon, R. (2002). Bayesian subset analysis: application to studying treatment-by-gender interactions. *Statistics in Medicine*, **21**(19), 2909–2916.

Simon, R. (2012). Clinical trials for predictive medicine. *Stat Med*, **31**(25), 3031–3040.

Simon, R. and Maitournam, A. (2004). Evaluating the efficiency of targeted designs for randomized clinical trials. *Clinical Cancer Research*, **10**(20), 6759–6763.

Sivaganesan, S., Laud, P. W., and Müller, P. (2011). A Bayesian subgroup analysis with a zero-enriched polya urn scheme. *Statistics in medicine*, **30**(4), 312–323.

Sivaganesan, S., Laud, P. W., and Müller, P. (2013). Subgroup analysis. In P. Damien, P. Dellaportas, N. Polson, and D. Stephens, editors, *Bayesian Theory and Applications*, pages 576–592. Oxford University Press.

Xu, Y., Trippa, L., Müller, P., and Ji, Y. (2014). Subgroup-Based Adaptive (SUBA) Designs for Multi-arm Biomarker Trials. *Statistics in Biosciences*, pages 1–22.

Yin, G. (2012). *Clinical Trial Design: Bayesian and Frequentist Adaptive Methods*. Wiley.

Zhou, X., Liu, S., Kim, E. S., Herbst, R. S., and Lee, J. J. (2008). Bayesian adaptive design for targeted therapy development in lung cancer–a step toward personalized medicine. *Clin Trials*, **5**(3), 181–193.

# Chapter 16
# Bayesian Nonparametric Approaches for ROC Curve Inference

Vanda Inácio de Carvalho, Alejandro Jara, and Miguel de Carvalho

**Abstract** The development of medical diagnostic tests is of great importance in clinical practice, public health, and medical research. The receiver operating characteristic (ROC) curve is a popular tool for evaluating the accuracy of such tests. We review Bayesian nonparametric methods based on Dirichlet process mixtures and the Bayesian bootstrap for ROC curve estimation and regression. The methods are illustrated by means of data concerning diagnosis of lung cancer in women.

## 16.1 Introduction

Medical diagnostic tests are designed to discriminate between alternative states of health, generally referred throughout as diseased and non-diseased/healthy states. Their ability to discriminate between these two states must be rigorously assessed trough statistical analysis before the test is approved for use in practice. In what follows, we assume the existence of a gold standard test, that is, a test that perfectly classifies the individuals as diseased and non-diseased. Compared to the truth one wants to know how well the test being evaluated performs.

The accuracy of a dichotomous test, a test that yields binary results (e.g., positive or negative), can be summarized by its sensitivity and specificity. The sensitivity (Se) is the test-specific probability of correctly detecting diseased subjects, while the specificity (Sp) is the test-specific probability of correctly detecting healthy subjects. In turn, the accuracy of a continuous scale diagnostic test is measured by the separation of test outcomes distribution in the diseased and non-diseased populations.

V. Inácio de Carvalho (✉)
Pontificia Universidad Católica de Chile, Chile
e-mail: icalhau@mat.puc.cl

A. Jara • M. de Carvalho
Departamento de Estadística, Pontifica Universidad Católica de Chile, Santiago, Chile
e-mail: ajara@mat.puc.cl; mdecarvalho@mat.puc.cl



R. Mitra, P. Müller (eds.), *Nonparametric Bayesian Inference in Biostatistics*, Frontiers in Probability and the Statistical Sciences,
DOI 10.1007/978-3-319-19518-6_16

The receiver operating characteristic (ROC) curve, which is a plot of Se against $1 - \mathrm{Sp}$ for all cutoff points that can be used to convert continuous test outcomes into dichotomous outcomes, measures exactly such amount of separation and it is probably the most widely used tool to evaluate the accuracy of continuous or ordinal tests.

A critical aspect when developing inference for ROC curves is the specification of a probability distribution for the test outcomes in the diseased and healthy groups. The main issue is that parametric models, such as the binormal model (arising when a normal distribution is assumed for both populations), are often too restrictive to capture nonstandard features of the data, such as skewness and multimodality, potentially leading to unsatisfactory inferences on the ROC curve. In these situations, we would like to relax parametric assumptions in order to gain modeling flexibility and robustness against misspecification of a parametric statistical model. Specifically, we would like to consider flexible modeling approaches that can handle nonstandard features of the data when that is needed, but that do not overfit the data when parametric assumptions are valid.

Moreover, recently, the interest on the subject has moved beyond determining the basic accuracy of a test. It has been recognized that the discriminatory power of a test is often affected by patient-specific characteristics, such as age or gender. In this situations, the parameter of interest is a collection of ROC curves associated with different covariate levels. In this context, understanding the covariate impact on the ROC curve may provide useful information regarding the test accuracy toward different populations or conditions. On the other hand, ignoring the covariate effects may lead to biased inferences about the test accuracy. As in the no-covariate case, here it is also important to consider flexible modeling approaches for assessing the effect of the covariates on test accuracy and, consequently, on the corresponding ROC curves.

In this chapter, we discuss two Bayesian nonparametric (BNP) approaches that are used to obtain data-driven inferences for a single ROC curve, based on mixtures induced by a Dirichlet process (DP) and on the Bayesian bootstrap. We also discuss an approach to model covariate-dependent ROC curves based on mixture models induced by a dependent DP (DDP), which allows for the entire distribution of the test outcomes, in each population, to smoothly change as a function of covariates. The chapter is organized as follows. In Sect. 16.2 we provide background material on ROC curves. BNP approaches for single ROC curve estimation are discussed in Sect. 16.3. A BNP ROC regression model is discussed in Sect. 16.4. In Sect. 16.5 we illustrate the methods using data concerning diagnosis of lung cancer in women. We conclude with a short discussion in Sect. 16.6.

## 16.2 ROC Curves

Let $Y_0$ and $Y_1$ be two independent random variables denoting the diagnostic test outcomes in the non-diseased and diseased populations, with cumulative distribution function (CDF) $F_0$ and $F_1$, respectively. Further, let $c$ be a cutoff value for defining a positive test result and, without loss of generality, we proceed with the assumption that a subject is classified as diseased when the test outcome is greater or equal than $c$ and as non-diseased when it is below $c$. Then, for each cutoff value $c$, the sensitivity and specificity associated with such decision criterion are

$$\text{Se}(c) = \Pr(Y_1 \geq c) = 1 - F_1(c), \qquad \text{Sp}(c) = \Pr(Y_0 < c) = F_0(c).$$

Obviously, for each value of $c$, we obtain a different sensitivity and specificity. The ROC curve summarizes the tradeoffs between Se and $1 - \text{Sp}$ (also known as false positive fraction) as the cutoff $c$ is varied and it corresponds to the set of points

$$\{(1 - F_0(c), 1 - F_1(c)) : c \in \mathbb{R}\}.$$

Alternatively, and letting $p = 1 - F_0(c)$, the ROC curve can be expressed as

$$\text{ROC}(p) = 1 - F_1\{F_0^{-1}(1-p)\}, \quad 0 \leq p \leq 1, \tag{16.1}$$

where $F_0^{-1}(1-p) = \inf\{z : F_0(z) \geq 1 - p\}$. ROC curves measure the amount of separation between the distribution of the test outcomes in the diseased and non-diseased populations. Figure 16.1 illustrates the effect of separation on the resulting ROC curve. When both distributions completely overlap, the ROC curve is the diagonal line of the unit square (that is, $\text{Se}(c) = 1 - \text{Sp}(c)$ for all $c$), thus indicating an useless test. On the other hand, the more separated the distributions the closer the ROC curve is to the point $(0,1)$ in the unit square. A curve that reaches the point $(0,1)$ has $\text{Sp}(c) = \text{Se}(c) = 1$ for some cutoff $c$, and hence corresponds to a perfect test. As it is clear from expression (16.1), estimating the ROC curve is basically a matter of estimating the distribution functions of the diseased and non-diseased populations and, hence, flexible models for estimating such distributions are in order.

Related to the ROC curve is the notion of placement value (Pepe and Cai 2004), which is simply a standardization of test outcomes with respect to a reference population. Let $U = 1 - F_0(Y_1)$ be the placement value of diseased subjects with respect to the non-diseased population. This variable quantifies the degree of separation between the two populations. Specifically, if the test outcomes in the two populations are highly separated, the placement of most diseased individuals is at the upper tail of the non-diseased distribution, so that most diseased individuals will have small placement values. In turn, if the populations overlap substantially, $U$ will have a Uniform$(0,1)$ distribution. Interestingly, the ROC curve turns out to be the CDF of $U$

$$\Pr(U \leq p) = \Pr(1 - F_0(Y_1) \leq p) = 1 - F_1\{F_0^{-1}(1-p)\} = \text{ROC}(p). \tag{16.2}$$

It is common to summarize the information of the ROC curve into a single summary index and the most widely used is the area under the ROC curve (AUC), which is defined as

$$\text{AUC} = \int_0^1 \text{ROC}(p)\text{d}p. \tag{16.3}$$

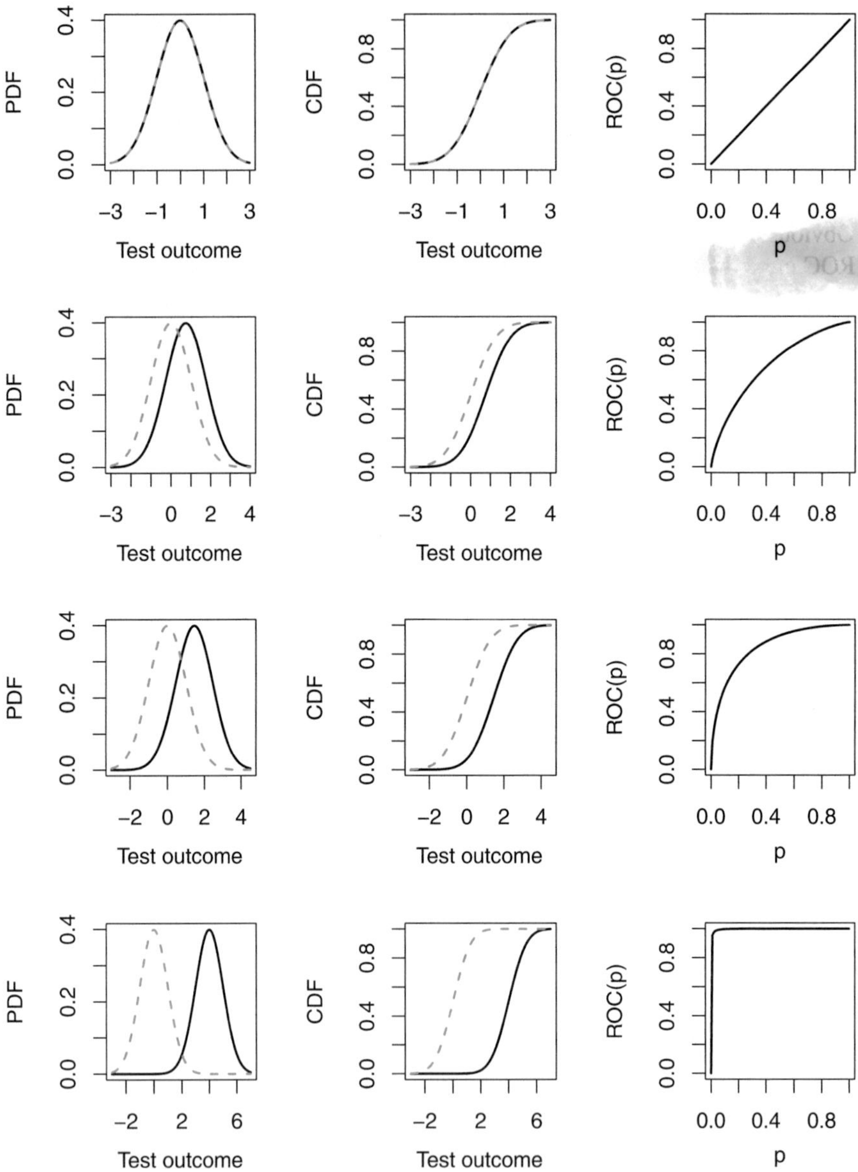


**Fig. 16.1** ROC curve illustrations: The first column displays the densities of the test outcomes for diseased (*solid black line*) and non-diseased populations (*dashed grey line*). The second column displays the corresponding distribution functions of the test outcomes for diseased (*solid black line*) and non-diseased populations (*dashed grey line*). The third column displays the corresponding ROC curves

The AUC can be interpreted as the probability that an individual chosen from the diseased population exhibits a test outcome greater than the one exhibited by a randomly selected individual from the non-diseased population, that is, AUC $= \Pr(Y_1 > Y_0)$. A test with a perfect discriminatory ability would have AUC $= 1$, while a test with no discriminatory power would have AUC $= 0.5$. Although there are some other summary indices available, such as the Youden index (Fluss et al. 2005), which has the nice feature of providing an optimal cutoff for screening subjects in practice, or the partial AUC (Dodd and Pepe 2003), which is a meaningful measure for cases where only a specific region of the ROC curve (e.g., high sensitivities or specificities) is of clinical interest, throughout this chapter we use the AUC as the preferred summary measure of diagnostic accuracy.

Now, suppose that along with $Y_0$ and $Y_1$, covariate vectors $\mathbf{x}_0$ and $\mathbf{x}_1$ are also available. Hereafter, we assume that these covariates are the same in both populations. However, this not always have to be the case. For instance, the severity of disease could play an important role on the discriminatory power of the test. As a natural extension of the ROC curve, the conditional or covariate-dependent ROC curve, for a given covariate level $\mathbf{x}$, is defined as

$$\mathrm{ROC}(p \mid \mathbf{x}) = 1 - F_1\{F_0^{-1}(1 - p \mid \mathbf{x}) \mid \mathbf{x}\}, \tag{16.4}$$

where $F_0(\cdot \mid \mathbf{x})$ and $F_1(\cdot \mid \mathbf{x})$ denote the conditional distribution function of $Y_0$ and $Y_1$ given covariate $\mathbf{x}$, respectively. For each value of $\mathbf{x}$, we possibly obtain a different ROC curve and, hence, also a possibly different AUC value, which is computed simply by replacing (16.4) in (16.3).

There is a vast literature on parametric, semiparametric, and nonparametric frequentist ROC data analysis. The books by Pepe (2003) and Zhou et al. (2011) discuss many frequentist approaches to ROC curve estimation and regression. See also the recent surveys by Gonçalves et al. (2014) and Pardo-Fernández et al. (2014). The amount of existing work in the Bayesian literature is by comparison reduced. This is particularly valid for the BNP literature, which is fairly limited. Recent work on the latter includes the DP mixture (DPM) model-based approach of Erkanli et al. (2006), the Bayesian bootstrap ROC curve estimator of Gu et al. (2008), and the stochastic ordering approach of Hanson et al. (2008a). Moreover, Branscum et al. (2008) used mixtures of finite Polya trees to analyze ROC data when the true disease status is unknown (that is, when there is no gold standard), while Hanson et al. (2008b) used bivariate mixtures of finite Polya trees to model data from two continuous tests. Additionally, Inácio et al. (2011) proposed the use of mixtures of finite Polya trees to model the ROC surface for problems where the patients have to be classified into one of three ordered classes. In what respects ROC regression, Inácio de Carvalho et al. (2013) proposed to model the conditional ROC curve using DDPs, whereas Rodríguez and Martínez (2014) used Gaussian process priors to model the mean and variance functions in each population and then computed the corresponding induced ROC curve. Finally, Branscum et al. (2014) proposed a method based on mixtures of finite Polya trees to model ROC regression data when there is not a gold standard test available.

## 16.3 Modeling Approaches for the No Covariate Case

### 16.3.1 DPM Models

When seeking for flexible modeling approaches and inferences for the distributions of the test outcomes in each population, mixture models appear as a natural option. More specifically, mixtures of normal distributions are particularly well suited for our purposes. Let $(Y_{01},\ldots,Y_{0n_0})$ and $(Y_{11},\ldots,Y_{1n_1})$ be random samples of sizes $n_0$ and $n_1$ from the non-diseased and diseased populations, respectively. It would be natural to assume that

$$Y_{01},\ldots,Y_{0n_0} \mid F_0 \overset{\text{ind.}}{\sim} F_0,$$

and

$$Y_{11},\ldots,Y_{1n_1} \mid F_1 \overset{\text{ind.}}{\sim} F_1,$$

with

$$F_h(\cdot) = \sum_{k=1}^{K_h} \omega_{hk}\Phi(\cdot \mid \mu_{hk},\sigma_{hk}^2), \quad h \in \{0,1\}, \tag{16.5}$$

where $\Phi(\cdot \mid \mu,\sigma^2)$ denotes the CDF of the normal distribution with mean $\mu$ and variance $\sigma^2$. Thus, each test outcome would arise from one of the $K_h$ mixture components, with each component having its own mean and variance. The model in (16.5) can be equivalently written as

$$F_h(\cdot) = \int \Phi(\cdot \mid \mu,\sigma^2)\mathrm{d}G_h(\mu,\sigma^2),$$

where $G_h$ is a discrete mixing distribution given by

$$G_h(\cdot) = \sum_{k=1}^{K_h} \omega_{hk}\delta_{(\mu_{hk},\sigma_{hk}^2)}(\cdot),$$

with $\delta_a(\cdot)$ denoting the Dirac measure at $a$. Usually, the weights $\{\omega_{hk}\}$ are assigned a Dirichlet distribution, while the component specific parameters $\{(\mu_{hk},\sigma_{hk}^2)\}$ arise from a prior distribution, say, $G_{0h}(\mu_h,\sigma_h^2)$, typically, a normal-inverse-gamma distribution. Hence, placing a prior on the collection

$$(\{\omega_{hk}\},\{(\mu_{hk},\sigma_{hk}^2)\}),$$

is equivalent to placing a prior on the discrete mixture distribution $G_h$. A drawback of this model specification is that we must choose the number of components $K_h$, which is not a trivial task in general. Although there are methods available that place an explicit parametric prior on $K_h$, they tend to be quite difficult to implement efficiently. An alternative is to use a DP prior (Ferguson 1973, 1974) for $G_h$, which, on one hand, offers the theoretical advantage of having full weak support on all mixing distributions and, on the other hand, the practical advantage of automatically determining the number of components that best fits a given dataset. We write $G_h \sim \text{DP}(\alpha_h, G_{0h})$ to denote that a DP prior is being assumed for $G_h$, which is defined in terms of a parametric centering distribution $G_{0h}$ (for which $\text{E}(G_h) = G_{0h}$), and a precision parameter $\alpha_h$ ($\alpha_h > 0$) which controls the uncertainty of $G_h$ about $G_{0h}$.

Undoubtedly, the most useful definition of the DP is its constructive definition (Sethuraman 1994), according to which $G_h$ has an almost sure representation of the form

$$G_h(\cdot) = \sum_{k=1}^{\infty} \omega_{hk} \delta_{(\mu_{hk}, \sigma_{hk}^2)}(\cdot), \tag{16.6}$$

where $(\mu_{hk}, \sigma_{hk}^2) \overset{\text{iid}}{\sim} G_{0h}$ and the weights arise from a stick breaking construction $\omega_{h1} = v_{h1}$, and $\omega_{hk} = v_{hk} \prod_{l<k}(1 - v_{hl})$, for $k \geq 2$, with $v_{hk} \overset{\text{iid}}{\sim} \text{Beta}(1, \alpha_h)$.

The resulting model for the test outcomes in each population is then a DPM of normals and is written as

$$F_h(\cdot) = \int \Phi(\cdot \mid \mu, \sigma^2) \text{d}G_h(\mu, \sigma^2), \quad G_h \sim \text{DP}(\alpha_h, G_{0h}), \tag{16.7}$$

where the centering distribution $G_{0h}$ is defined on $\mathbb{R} \times \mathbb{R}_+$. More specifically, we take $G_{0h}$ to be the normal-inverse-gamma distribution, that is,

$$G_{0h} \equiv \text{N}(m_h, S_h)\text{IG}(\tau_{h1}/2, \tau_{h2}/2),$$

where $\text{N}(\mu, \sigma^2)$ is the normal distribution with mean $\mu$ and variance $\sigma^2$ and $\text{IG}(a, b)$ refers to the inverse-gamma distribution with parameters $a$ and $b$. The stick-breaking representation of the DP given in (16.6) allows us to rewrite (16.7) as the following countably infinite mixture of normals

$$F_h(\cdot) - \sum_{k=1}^{\infty} \omega_{hk} \Phi( \mid \mu_{hk}, \sigma_{hk}^2).$$

The model specification is completed by assuming the following independent hyper-priors

$$\alpha_h \sim \text{G}(a_h, b_h), \quad \tau_{h2} \sim \text{G}(\tau_{sh1}/2, \tau_{sh2}/2),$$
$$m_h \sim \text{N}(\mu_{m_h}, S_{m_h}), \quad S_h \sim \text{IG}(\nu_h, \Psi_h),$$

where $\text{G}(a, b)$ refers to the gamma distribution with parameters $a$ and $b$.

Posterior inference can be conducted using two different kinds of Markov chain Monte Carlo (MCMC) strategies: (i) to employ a truncation of the stick-breaking representation (Ishwaran and James 2001) or (ii) to use a marginal Gibbs sampling where the mixing distributions are integrated out from the model (MacEachern and Müller 1998; Neal 2000). Finally, we can plug-in each MCMC realization of $F_0$ and $F_1$ in (16.1) and compute the corresponding realization of the ROC curve. Note that the computation of the ROC curve requires the evaluation of the quantile function of $F_0$, which is done numerically. A model similar to the one described here was proposed by Erkanli et al. (2006).

### 16.3.2 Bayesian Bootstrap

The Bayesian bootstrap (BB) estimator of the ROC curve was proposed by Gu et al. (2008) and it is a computationally simple, yet robust, estimator. We start by outlining how the BB works in the one-population setting. Let $(Y_1,\ldots,Y_n)$ be a random sample from an unknown distribution $F$ and suppose that $F$ itself is the parameter of interest. In Efron's frequentist bootstrap (Efron 1979), estimation and inference about $F$ are obtained by repeatedly generating bootstrap samples, where each sample is drawn with replacement from the original data. In the $b$th bootstrap replicate, $F^{(b)}$ is computed as

$$F^b(\cdot) = \sum_{i=1}^{n} \pi_i^{(b)} \delta_{Y_i}(\cdot), \tag{16.8}$$

where $\pi_i^{(b)}$ is the proportion of times $Y_i$ appears in the $b$th bootstrap sample, with $\pi_i^{(b)}$ taking values on $\{0, 1/n, \ldots, n/n\}$. By contrast, in Rubin's BB (Rubin 1981), the weights $\pi_i^{(b)}$ in expression (16.8) are assigned an Dirichlet$_n(1,\ldots,1)$ distribution and thus are smoother than those from the frequentist bootstrap. It is important to stress that in the BB the data is regarded as fixed and so we do not resample from it. The BB has connections with the DP. Specifically, it can be regarded as a non-informative version of the DP, which can be obtained by letting the precision parameter tending to zero (Gasparini 1995, Theorem 2).

The representation of the ROC curve given in (16.2) provides the rationale for the following two-step BB algorithm, which we fully describe due to its simplicity. Let us suppose, again, that $(Y_{01},\ldots,Y_{0n_0})$ and $(Y_{11},\ldots,Y_{1n_1})$ are random samples from the non-diseased and diseased populations and let $B$ be the number of BB resamples.

**Bayesian bootstrap algorithm**

For $b = 1,\ldots,B$:

**Step 1 (Compute the placement values based on the BB resampling)**

For $j = 1,\ldots,n_1$, compute the placement values

$$U_j = \sum_{i=1}^{n_0} q_i^{(b)} I(Y_{0i} \geq Y_{1j}), \quad (q_1^{(b)},\ldots,q_{n_0}^{(b)}) \overset{\text{ind.}}{\sim} \text{Dirichlet}_{n_0}(1,\ldots,1).$$

(continued)

**Step 2 (Generate a random realization of the ROC curve)**
Based on (16.2), generate a random realization of $\mathrm{ROC}(p)$, the cumulative distribution function of $(U_1,\ldots,U_{n_1})$, where

$$\mathrm{ROC}^{(b)}(p) = \sum_{j=1}^{n_1} r_j^{(b)} I(U_j \leq p), \quad (r_1^{(b)},\ldots,r_{n_1}^{(b)}) \overset{\text{ind.}}{\sim} \mathrm{Dirichlet}_{n_1}(1,\ldots,1),$$

with $0 \leq p \leq 1$. Compute the AUC associated with $\mathrm{ROC}^{(b)}(p)$, $\mathrm{AUC}^{(b)}$, using numerical integration.

The BB estimate of the ROC curve, denoted as $\widehat{\mathrm{ROC}}^{\mathrm{BB}}(p)$, is then obtained by averaging the random realizations of the ROC curve, that is,

$$\widehat{\mathrm{ROC}}^{\mathrm{BB}}(p) = \frac{1}{B}\sum_{b=1}^{B} \mathrm{ROC}^{(b)}(p), \quad 0 \leq p \leq 1.$$

Similarly,

$$\widehat{\mathrm{AUC}}^{\mathrm{BB}} = \frac{1}{B}\sum_{b=1}^{B} \mathrm{AUC}^{(b)}.$$

## 16.4 Modeling Approaches for the Covariate Case

Let $\{(\mathbf{x}_{01},Y_{01}),\ldots,(\mathbf{x}_{0n_0},Y_{0n_0})\}$ and $\{(\mathbf{x}_{11},Y_{11}),\ldots,(\mathbf{x}_{1n_1},Y_{1n_1})\}$ be regression data for the non-diseased and diseased groups, respectively, where $\mathbf{x}_{0i} \in \mathscr{X} \subseteq \mathrm{I\!R}^p$ and $\mathbf{x}_{1j} \in \mathscr{X} \subseteq \mathrm{I\!R}^p$ are $p$-dimensional covariate vectors and $Y_{0i}$ and $Y_{1j}$ are test outcomes, for $i = 1,\ldots,n_0$, $j = 1,\ldots,n_1$. It is assumed that given the covariates, the test outcomes in the diseased and non-diseased populations are independent and that

$$Y_{0i} \mid \mathbf{x}_{0i} \overset{\text{ind.}}{\sim} F_0(\cdot \mid \mathbf{x}_{0i}), \quad i = 1,\ldots,n_0,$$
$$Y_{1j} \mid \mathbf{x}_{1j} \overset{\text{ind.}}{\sim} F_1(\cdot \mid \mathbf{x}_{1j}), \quad j = 1,\ldots,n_1.$$

Here, we detail the approach proposed by Inácio de Carvalho et al. (2013) for the conditional ROC curve estimation problem, which extends the no covariate approach of Sect. 16.3.1. Specifically, these authors proposed a model for the conditional ROC curves based on the specification of a probability model for the entire collection of distributions $\mathscr{F}_h = \{F_h(\cdot \mid \mathbf{x}) : \mathbf{x} \in \mathscr{X}\}$, for $h \in \{0,1\}$, and they further modeled the conditional distributions in each population using the following covariate-dependent mixture of normal models

$$F_h(\cdot \mid \mathbf{x}) = \int \Phi(\cdot \mid \mu, \sigma^2) \mathrm{d}G_{h\mathbf{x}}(\mu, \sigma^2), \quad h \in \{0, 1\}.$$

The probability model for the conditional distributions is induced by specifying a prior for the collection of mixing distributions

$$G_{h\mathscr{X}} = \{G_{h\mathbf{x}} : \mathbf{x} \in \mathscr{X}\} \sim \mathscr{G}_h,$$

where $G_{h\mathbf{x}}$ denotes the random mixing distribution at covariate $\mathbf{x}$, which is defined on $\mathrm{IR} \times \mathrm{IR}_+$, and $\mathscr{G}_h$ is the prior for the collection $G_{h\mathscr{X}}$.

One possibility for modeling $\mathscr{G}_h$ is the DDP proposed by MacEachern (2000), which is built upon the constructive definition of the DP in (16.6), where the atoms and the components of the weights are realizations of a stochastic process over $\mathscr{X}$, and the weights arise from a stick-breaking representation. Justified by results in Barrientos et al. (2012), on the full support of MacEachern's DDPs, Inácio de Carvalho et al. (2013) considered the 'single weights' DDP (De Iorio et al. 2004, 2009; De la Cruz et al. 2007; Jara et al. 2010), where only the atoms are indexed by the covariates, thus resulting in the following specification for the conditional random mixing distribution

$$G_{h\mathbf{x}}(\cdot) = \sum_{k=1}^{\infty} \omega_{hk} \delta_{\theta_{hk}(\mathbf{x})}(\cdot), \tag{16.9}$$

where the weights $\{\omega_{hk}\}$ match those from a standard DP and the atoms are given by $\theta_{hk}(\mathbf{x}) = (m_{hk}(\mathbf{x}), \sigma^2_{hk})$, where $\{m_{hk}(\mathbf{x}) : \mathbf{x} \in \mathscr{X}\}$ are iid. Gaussian processes which are independent across $h$.

Although such formulation leads to a very flexible prior, it implies sampling realizations of the Gaussian processes at each distinct value of the covariate and, thus, inferences could take prohibitively long. This motivated Inácio de Carvalho et al. (2013) to elaborate on a linear DDP (LDDP) prior formulation (De Iorio et al. 2004, 2009; Jara et al. 2010), where the Gaussian processes are replaced by sufficiently rich linear (in the coefficients) functions, $m_{hk}(\mathbf{x}) = \mathbf{z}'\beta_{hk}$. Here $\mathbf{z}$ is a $q$-dimensional design vector possibly including nonlinear transformations of the original covariates $\mathbf{x}$. To this end, the authors considered an additive formulation based on B-splines (Eilers and Marx 1996), referred to as B-splines DDP,

$$m_{hk}(\mathbf{x}) = \beta_{hk0} + \sum_{l=1}^{p} \left( \sum_{n=1}^{K_l} \beta_{hkln} \psi(x_l, d_l) \right),$$

where $\psi_n(x, d)$ corresponds to the $n$th B-spline basis function of degree $d$ evaluated at $x$, and $\beta_{hk} = (\beta_{hk0}, \ldots, \beta_{hkpK_p})$. This formulation allows for the inclusion of discrete and continuous predictors.

Thus, under the LDDP formulation, the base stochastic processes are replaced with a group-specific distribution $G_{0h}$ that generates the component specific regression coefficients and variances. Therefore, the B-splines DDP mixture model can be equivalently formulated as a DPM of Gaussian regression models

$$F_h(\cdot \mid \mathbf{x}) = \int \Phi(\cdot \mid \mathbf{z}'\beta, \sigma^2)\mathrm{d}G_h(\beta, \sigma^2), \quad G_h \sim \mathrm{DP}(\alpha_h, G_{0h}). \tag{16.10}$$

For each group, normal-inverse-gamma distributions were used for the parametric centering distribution,

$$G_{0h} \equiv N_q(\mu_h, \mathbf{\Sigma}_h) \times \mathrm{IG}(\tau_{h1}/2, \tau_{h2}/2).$$

The model specification is completed by specifying the following hyper-priors

$$\alpha_h \sim \mathrm{G}(a_h, b_h), \quad \tau_{h2} \sim \mathrm{G}(\tau_{sh1}/2, \tau_{sh2}/2),$$
$$\mu_h \sim \mathrm{N}_q(\mathbf{m}_h, \mathbf{S}_h), \quad \mathbf{\Sigma}_h \sim \mathrm{IW}_q(\nu_h, \mathbf{\Psi}_h).$$

With regard to posterior inference, the computational strategies for Dirichlet process mixture models referred in Sect. 16.3.1 apply here in the covariate setup directly. Finally, after obtaining MCMC samples for each of the parameters, we can plug-in, for each covariate $\mathbf{x}$, each MCMC realization of $F_0(\cdot \mid \mathbf{x})$ and $F_1(\cdot \mid \mathbf{x})$ in (16.4) and compute the corresponding realization of the conditional ROC curve. The model previously described is implemented in the function `LDDProc` of the `R` library `DPpackage` (Jara et al. 2011).

## 16.5 Illustration

The accuracy of a soluble isoform of epidermal growth factor receptor (sEGFR), present in blood, as a diagnostic test for lung cancer in women is investigated. How this accuracy may vary with age is also subject of interest. The data were collected from a case–control study conducted at the Mayo clinic in Minnesota between 1998 and 2003. The dataset includes information for 140 non-diseased women and 101 lung cancer cases. This dataset was previously analyzed by Branscum et al. (2013).

Figure 16.2 shows the histogram of the $-\log(\mathrm{sEGFR})$ in both populations. The minus sign is due to the fact that the values of sEGFR tend to be lower for lung cancer cases than for controls, and so with the minus sign the usual convention that diseased individuals tend to have larger test outcomes than the non-diseased ones applies. As it can be observed, normality does not seem to apply, especially for the non-diseased population, where a bimodality is easily noticed. Figure 16.2 also displays the estimated densities, in each group of women, under the DPM of normals model, and we can see that the model captures well the bimodality in the non-diseased group, as well as, a certain skewness in the diseased group. The hyper-priors of the DPM of normals model were set to $a_h = 5$, $b_h = 1$, $\tau_{h1} = 2$, $\tau_{sh1} = 2$, $\tau_{sh2} = 10$, $\mu_{mh} = 0$, $S_{mh} = 100$, $\nu_h = 5$, and $\Psi_h = 1$, for $h \in \{0, 1\}$, while the BB estimates were obtained using 5000 resamples. With respect to the estimation of the CDFs, which are displayed in Fig. 16.3 [Panels (a) to (f)], it can be observed that the estimates provided by the DPM of normals and the BB are almost indistinguishable. When superimposing these fits (DPM and BB) with the one obtained by the binormal model, a discrepancy can be seen, especially in the non-diseased group. The resulting ROC curves, also presented in Fig. 16.3, are smooth and practically identical (except the one obtained by the binormal fit) and the corresponding posterior means (95 % credible interval) of the AUC are 0.792 (0.728, 0.848) under the DPM model and 0.792 (0.731, 0.848) under the BB method. These values reveal a quite good discriminatory ability of the sEGFR to detect lung cancer in women.

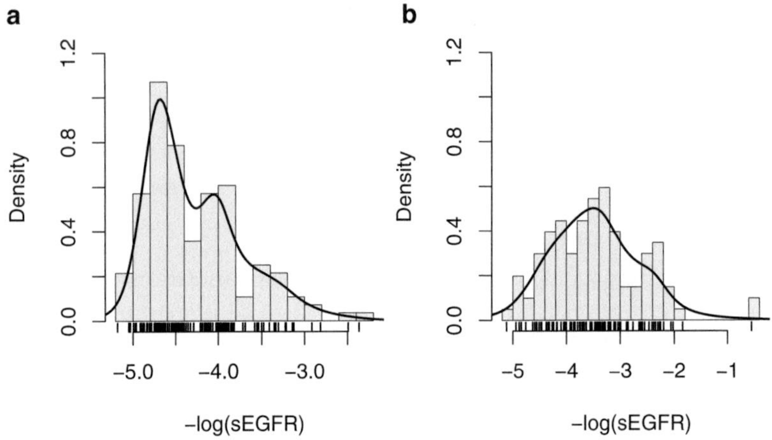


**Fig. 16.2** sEGFR data: Histogram of the $-\log(\text{sEGFR})$ in the nondiseased (Panel **a**) and diseased (Panel **b**) populations along with a rug representation of the data. The posterior mean of the density for each population under the DPM of normals models is displayed as a *solid line*

We now examine the age effect on the accuracy of the sEGFR. The B-splines dependent DPM of normals model was fit by assuming $K_1 = 3$, $a_h = 5$, $b_h = 1$, $\tau_{h1} = 2$, $\tau_{sh1} = 2$, $\tau_{sh2} = 10$, $\mathbf{m}_h = (0,0,0,0)$, $\mathbf{S}_h = 100 \times \mathbf{I}_4$, $\nu_h = 5$, and $\boldsymbol{\Psi}_h = \mathbf{I}_4$, for $h \in \{0,1\}$. Figure 16.4 shows the posterior means for the conditional mean functions, along with point-wise 95 % credible bands for $-\log(\text{sEGFR})$ levels. These estimates are overlaid on the top of the raw data. This figure suggests that the $-\log(\text{sEGFR})$ levels are more concentrated in the non-diseased than in the diseased women, across age and, further, a slightly nonlinear behavior of the conditional mean function of both groups can be observed.

Figure 16.5 presents the estimated posterior means, along with 95 % point-wise credible bands, of the conditional distribution functions in the two groups of women at three selected ages (40, 55, and 70 years old), and a change across age is clearly seen. Obviously, the same is visible in terms of the corresponding estimated ROC curves.

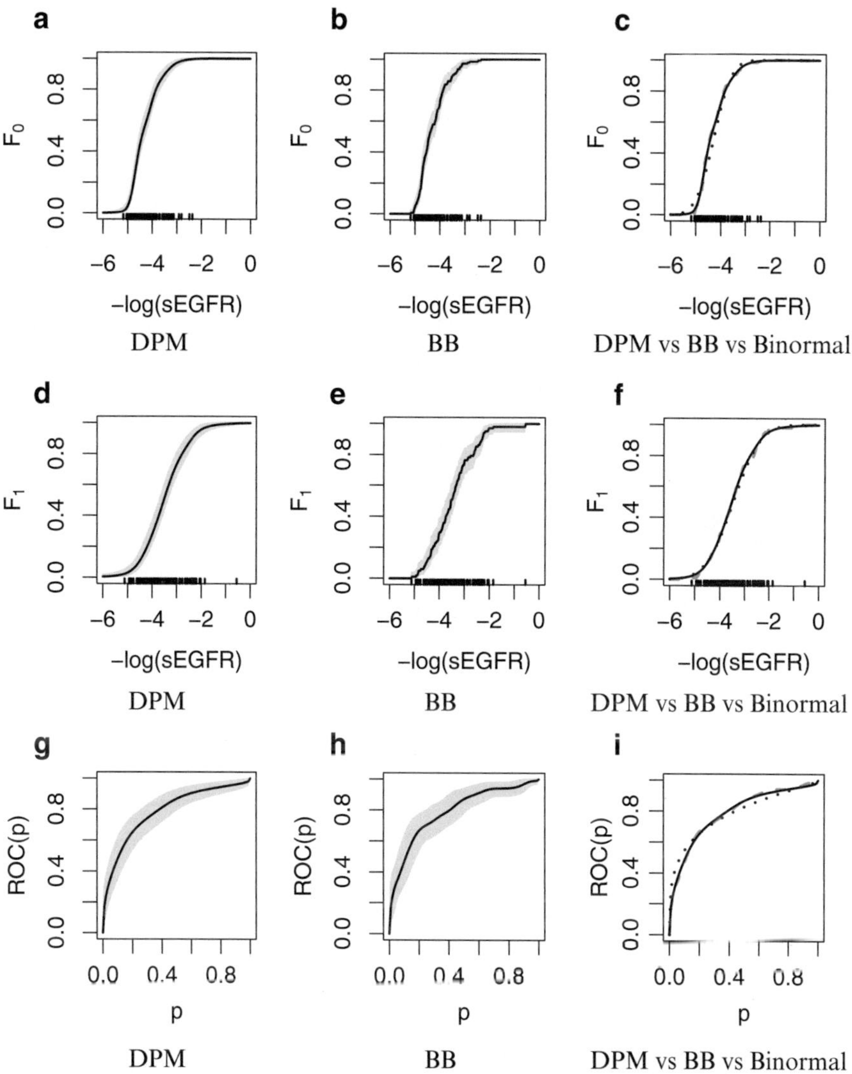


**Fig. 16.3** sEGFR data: Panels (**a**) and (**b**) show the estimated posterior mean (*solid black line*), along with the point-wise 95 % credible bands (*grey area*) of the cumulative distribution function of the non-diseased population, under the DPM of normals model and the Bayesian bootstrap, respectively. In Panel (**c**) the two estimates are superimposed along with the estimates obtained under the binormal model (*solid black line* represents the DPM estimate, *light grey dashed line* represents the BB estimate, and *dark grey dotted line* is the binormal estimate). Panels (**d**), (**e**), and (**f**) show the analogous figures but in terms of the cumulative distribution function of the diseased population, and panels (**g**), (**h**), and (**i**) in terms of the ROC curve

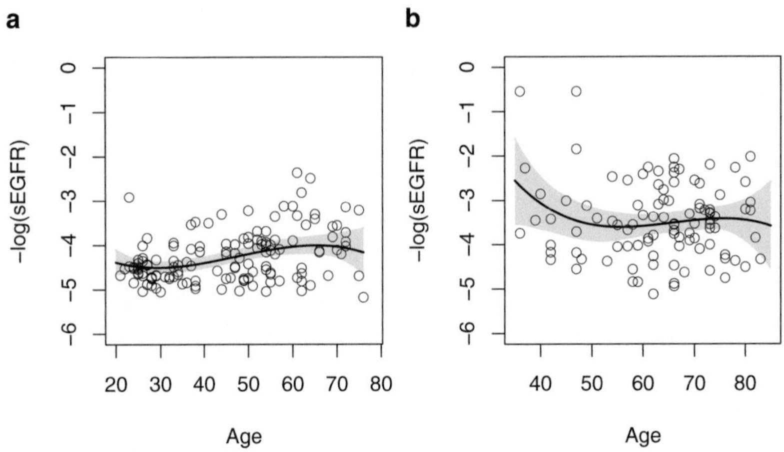


**Fig. 16.4** sEGFR data: Posterior mean (*solid line*) and 95 % point-wise credible band (*grey area*) for the conditional mean function in the group of non-diseased women (Panel (**a**)) and in the group of diseased women (Panel (**b**))

To examine the age effect further, Fig. 16.6 shows the estimated posterior mean, as well as the 95 % point-wise credible band, of the AUC as a function of age. This figure suggests a decrease in AUC until an age around 60 years old, and then a slight increase.

## 16.6 Concluding Remarks

ROC curves are a valuable tool for assessing the discriminatory power of continuous diagnostic tests. We have described and illustrated BNP approaches for ROC curve estimation and regression. Specifically, we have discussed DPM models and the BB for ROC curve estimation and an extension for the regression case based on DDP mixture models. A nice feature of the latter model is that the complete distribution of the test outcomes is allowed to smoothly change with the values of the covariates instead of just one or two characteristics (such as the mean and/or variance), as implied for most ROC regression models.

Topics of future research on BNP methods for ROC analysis include, among others, modeling diagnostic tests with mass at zero, optimal combinations of multiple tests, and time-dependent ROC curves. We end remarking that R packages for the implementation of ROC analysis tools are of great importance for practitioners.

**Acknowledgements** The authors thank Adam Branscum for sharing the lung cancer dataset with them. The first author was supported by Fondecyt grant 11130541. The second author was supported by Fondecyt grant 1141193. The third author was supported by Fondecyt grant 11121186.

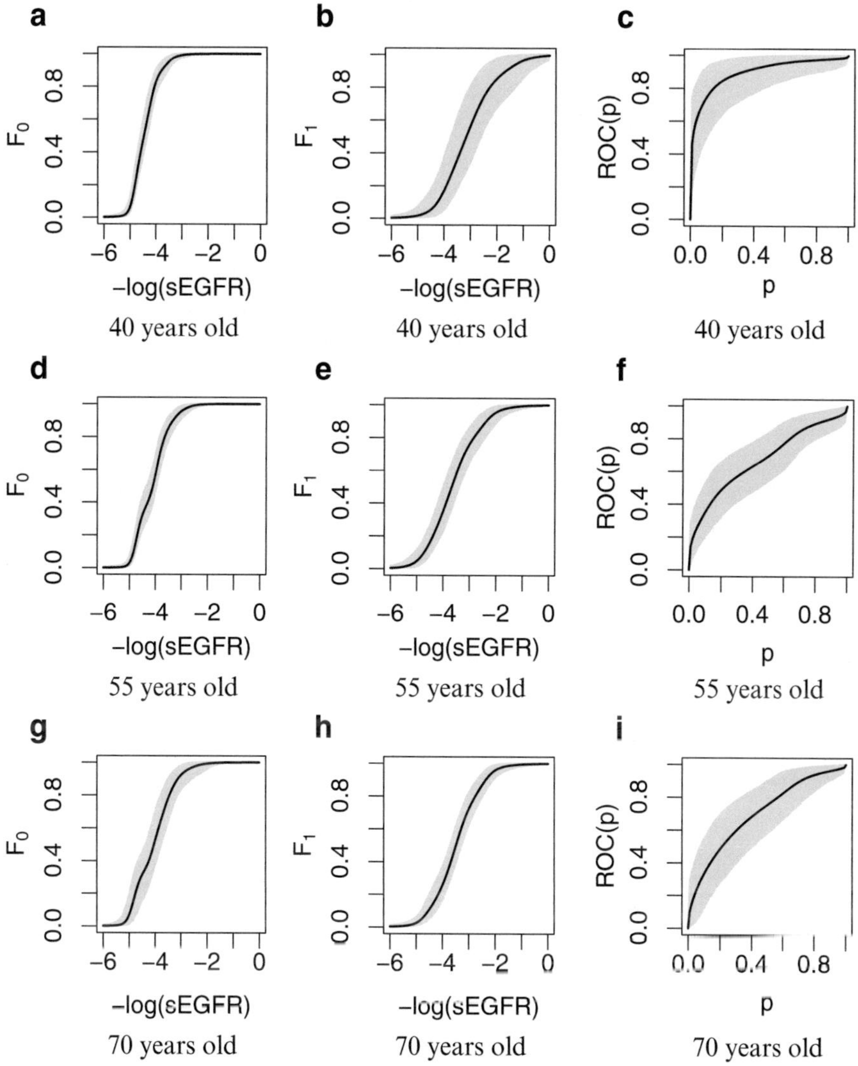


**Fig. 16.5** sEGFR data: Panels (**a**), (**d**), and (**g**) display the estimated posterior mean (*solid line*), as well as, the 95 % point-wise credible bands (*grey area*) of the conditional distribution function, in the non-diseased group, for ages of 40, 55, and 70 years old. Panels (**b**), (**e**), and (**h**) show the analogous figures for the diseased group. Panels (**c**), (**f**), and (**i**) show the corresponding ROC curves

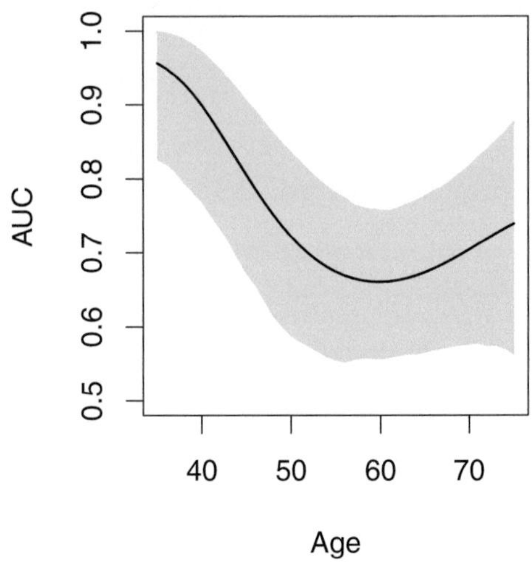


**Fig. 16.6** sEGFR data: Posterior mean (*solid line*) along with the 95 % point-wise credible band (*grey area*) for the AUC as a function of age

## References

Barrientos, A. F., Jara, A., and Quintana, F. (2012). On the support of MacEachern's dependent Dirichlet processes and extensions. *Bayesian Analysis*, **7**, 277–310.

Branscum, A. J., Johnson, W. O., Hanson, T. E., and Gardner, I. A. (2008). Bayesian semiparametric ROC curve estimation and disease diagnosis. *Statistics in Medicine*, **27**, 2474–2496.

Branscum, A. J., Johnson, W. O., and Baron, A. T. (2013). Robust medical test evaluation using flexible Bayesian semiparametric regression models. *Epidemiology Research International*, **2103**, 1–8.

Branscum, A. J., Johnson, W. O., Hanson, T. E., and Baron, A. T. (2014). Flexible regression models for ROC and risk analysis, with or without a gold standard. *Submitted*.

De Iorio, M., Müller, P., Rosner, G. L., and MacEachern, S. N. (2004). An ANOVA model for dependent random measures. *Journal of the American Statistical Association*, **99**, 205–215.

De Iorio, M., Johnson, W. O., Müller, P., and Rosner, G. L. (2009). Bayesian nonparametric non-proportional hazards survival modelling. *Biometrics*, **65**, 762–771.

De la Cruz, R., Quintana, F. A., and Müller, P. (2007). Semiparametric Bayesian classification with longitudinal markers. *Applied Statistics*, **56**, 119–137.

Dodd, L. E. and Pepe, M. S. (2003). Partial auc estimation and regression. *Biometrics*, **59**, 614–623.

Efron, B. (1979). Bootstrap methods: another look at the jackknife. *The Annals of Statistics*, **7**, 1–26.

Eilers, P. H. C. and Marx, B. D. (1996). Flexible smoothing with B-splines and penalties. *Statistical Science*, **11**, 89–121.

Erkanli, A., Sung, M., Costello, E. J., and Angold, A. (2006). Bayesian semiparametric ROC analysis. *Statistics in Medicine*, **25**, 3905–3928.

Ferguson, T. S. (1973). A Bayesian analysis of some nonparametric problems. *Annals of Statistics*, **1**, 209–230.

Ferguson, T. S. (1974). Prior distribution on the spaces of probability measures. *Annals of Statistics*, **2**, 615–629.

Fluss, R., Faraggi, D., and Reiser, B. (2005). Estimation of the Youden index and its associated cutoff point. *Biometrical Journal*, **47**, 458–472.

Gasparini, M. (1995). Exact multivariate Bayesian bootstrap distributions of moments. *The Annals of Statistics*, **23**, 762–768.

Gonçalves, L., Subtil, A., Oliveira, R., and Bermúdez, P. Z. (2014). ROC curve estimation: An overview. *REVSTAT–Statistical Journal*, **12**, 1–20.

Gu, J., Ghosal, S., and Roy, A. (2008). Bayesian bootstrap estimation of ROC curve. *Statistics in Medicine*, **27**, 5407–5420.

Hanson, T., Kottas, A., and Branscum, A. J. (2008a). Modelling stochastic order in the analysis of receiver operating characteristic data: Bayesian non-parametric approaches. *Journal of the Royal Statistical Society, Series C*, **57**, 207–225.

Hanson, T., Branscum, A., and Gardner, I. (2008b). Multivariate mixtures of Polya trees for modelling ROC data. *Statistical Modelling*, **8**, 81–96.

Inácio, V., Turkman, A. A., Nakas, C. T., and Alonzo, T. A. (2011). Nonparametric Bayesian estimation of the three-way receiver operating characteristic surface. *Biometrical Journal*, **53**, 1011–1024.

Inácio de Carvalho, V., Jara, A., Hanson, T. E., and de Carvalho, M. (2013). Bayesian nonparametric ROC regression modeling. *Bayesian Analysis*, **8**, 623–646.

Ishwaran, H. and James, L. F. (2001). Gibbs sampling methods for stick-breaking priors. *Journal of the American Statistical Association*, **96**, 161–173.

Jara, A., Lesaffre, E., De Iorio, M., and Quintana, F. A. (2010). Bayesian semiparametric inference for multivariate doubly interval-censored data. *Annals of Applied Statistics*, **4**, 2126–2149.

Jara, A., Hanson, T., Quintana, F., Müller, P., and Rosner, G. L. (2011). DPpackage: Bayesian semi- and nonparametric modeling in R. *Journal of Statistical Software*, **40**, 1–30.

MacEachern, S. N. (2000). Dependent Dirichlet processes. Technical report, Department of Statistics, The Ohio State University.

MacEachern, S. N. and Müller, P. (1998). Estimating mixture of Dirichlet process models. *Journal of Computational and Graphical Statistics*, **7**, 223–238.

Neal, R. (2000). Markov chain sampling methods for Dirichlet process mixture models. *Journal of Computational and Graphical Statistics*, **9**, 249–265.

Pardo-Fernández, J. C., Rodríguez-Álvarez, M. X., and Van Keilegom, I. (2014). A review on ROC curves in the presence of covariates. *REVSTAT–Statistical Journal*, **12**, 21–41.

Pepe, M. S. (2003). *The Statistical Evaluation of Medical Tests for Classification and Prediction*. Oxford University Press, Oxford.

Pepe, M. S. and Cai, T. (2004). The analysis of placement values for evaluating discriminatory measures. *Biometrics*, **60**, 528–535.

Rodríguez, A. and Martínez, J. C. (2014). Bayesian semiparametric estimation of covariate-dependent ROC curves. *Biostatistics*, **15**, 353–369.

Rubin, D. B. (1981). The Bayesian bootstrap. *The Annals of Statistics*, **9**, 130–134.

Sethuraman, J. (1994). A constructive definition of Dirichlet priors. *Statistica Sinica*, **2**, 639–650.

Zhou, X. H., Obuchowski, N. A., and McClish, D. K. (2011). *Statistical Methods in Diagnostic Medicine,* 2nd Ed. Wiley, New York.

# Part V
# Spatial Data

# Chapter 17
# Spatial Bayesian Nonparametric Methods

Brian James Reich and Montserrat Fuentes

**Abstract** We review nonparametric Bayesian approaches to inference for spatial data. The discussion is organized by increasing level of relaxation of traditional parametric assumptions. We start by considering nonparametric priors for covariance functions in a Gaussian process model. Next we allow for non-Gaussian marginal distributions by introducing Gaussian copulas. Finally, we go fully nonparametric and discuss Dirichlet process mixtures for the coefficients in a kernel convolution, Dirichlet process mixtures of Gaussian processes and spatial stick-breaking priors.

## 17.1 Introduction

Classical geostatistics (Cressie 1993; Banerjee et al. 2004; Gelfand et al. 2010) is based almost exclusively on Gaussian process representations of spatial data. The canonical problem is to observe the process $Y$ at $n$ spatial locations $\mathbf{s}_1,\ldots,\mathbf{s}_n$, denoted $\mathbf{Y} = [Y(\mathbf{s}_1),\ldots,Y(\mathbf{s}_n)]^T$, and to interpolate the process $Y$ to all spatial locations $\mathbf{s}$ in the measurable spatial domain $\mathscr{D} \in \mathscr{R}^2$. For example, in a study of the health impacts of air pollution, we may observe air pollution concentrations at $n$ monitoring locations and use these data to predict the concentration of the residual locations of study participants. Given that the data consist of a single partially observed realization of the process, parametric assumptions about the mean, covariance, and distribution of the process are natural. On the other hand, parametric assumptions are often questionable and difficult to verify and so more flexible

B.J. Reich • M. Fuentes (✉)
Department of Statistics, North Carolina State University, Raleigh, NC, USA
e-mail: bjreich@ncsu.edu; fuentes@ncsu.edu



R. Mitra, P. Müller (eds.), *Nonparametric Bayesian Inference in Biostatistics*,
Frontiers in Probability and the Statistical Sciences,
DOI 10.1007/978-3-319-19518-6_17

approaches are attractive. In this chapter we review the recent literature on Bayesian nonparametric (BNP) methods for geostatistical data.

Let $Y(\mathbf{s})$ be the real-valued response at spatial location $\mathbf{s}$. The Gaussian process is defined by the mean function $\mathrm{E}[Y(\mathbf{s})] = \mu(\mathbf{s})$ and covariance function $\mathrm{Cov}[Y(\mathbf{s}), Y(\mathbf{t})] = \sigma(\mathbf{s}, \mathbf{t})$. If the covariance is stationary, then we may write $\mathrm{Cov}[Y(\mathbf{s}), Y(\mathbf{t})] = C(\mathbf{h})$, where $\mathbf{h} = \mathbf{s} - \mathbf{t}$. A typical parametric analysis takes the mean function to be a linear combination of spatial covariates, $\mu(\mathbf{s}) = \mathbf{X}(\mathbf{s})^T\beta$, and the covariance to be a parametric form such as the exponential covariance $C(\mathbf{h}) = \sigma^2\exp(-||\mathbf{h}||/\psi)$. The finite dimensional distribution of the Gaussian process at the $n$ locations $\mathbf{s}_1, \ldots, \mathbf{s}_n$ is multivariate normal, $\mathbf{Y} \sim \mathrm{N}(\mu, \Sigma)$, where $\mu = [\mu(\mathbf{s}_1), \ldots, \mu(\mathbf{s}_n)]^T$ is the mean vector and the $(i, j)$ element of the covariance matrix $\Sigma$ is $\sigma(\mathbf{s}_i, \mathbf{s}_j)$. In particular, the marginal distribution at location $\mathbf{s}$ is $Y(\mathbf{s}) \sim \mathrm{N}[\mu(\mathbf{s}), \sigma(\mathbf{s}, \mathbf{s})]$.

In this chapter we discuss several extensions of this classic parametric model using BNP methods. Standard BNP methods for the mean function $\mu(\mathbf{s})$ using methods such as splines or wavelets apply here, and so we focus on flexible priors for the covariance function and BNP priors for the response distribution that relax the normality assumption. In Sect. 17.2, we discuss a BNP prior for the covariance function using spectral methods. Section 17.3 reviews methods that remove the normality assumption about the marginal distribution of $Y$, but retain the spatial dependence properties of the Gaussian process. Finally, Sect. 17.4 presents a series of BNP methods for the entire spatial process model.

## 17.2 Bayesian Non-parametric Priors for a Covariance Function

An essential step in the analysis of spatial data is the estimation of the spatial covariance function that explains the dependence between the spatial process at two different locations: The spatial covariance of Gaussian responses, or of the Gaussian random effects in a model for non-Gaussian data. The standard approach for covariance modeling is to select a parametric covariance function (Gelfand et al. 2010). Instead of restricting to a particular parametric model for the covariance function, Zheng et al. (2010) and Reich and Fuentes (2012) treat the covariance function as an unknown to be estimated from the data. This can be challenging because we need to ensure the covariance is a nonnegative definite function. A convenient way to specify a flexible model for the covariance is by modeling the corresponding spectral density, which involves fewer restrictions. In this section we follow Reich and Fuentes (2012) and specify a prior for the covariance function using spectral methods and the Dirichlet process (DP) prior.

An important result in the spectral domain is Bochner's theorem (e.g. Gelfand et al. 2010) stating that any stationary covariance function $C$ can be represented as an inverse Fourier transform

$$C(\mathbf{h}) = \int_{\mathscr{R}^2} \exp(i\mathbf{h}'\omega)G(d\omega), \tag{17.1}$$

where the function $G$ is called the spectral measure or spectrum for $Y$ and $\omega = (\omega_1, \omega_2)$ is a bivariate spectral frequency. We assume there exists a continuous differentiable spectral density $g(\omega)$ such that $G(d\omega) = \tau^2 g(\omega)d\omega$, where $\tau > 0$ and $\int g(\omega)d\omega = 1$. The spatial process $Y$ is real if and only if the spectral density is even, i.e., $g(\omega) = g(-\omega)$. The representation in (17.1) illustrates the most general strategy for constructing a stationary covariance function: we use (17.1) with an arbitrary spectral density. Conversely, any function which cannot be written in this form cannot be positive definite and hence is not the covariance of a valid stationary process.

The scaler $\tau$ controls the variance and can be given a standard prior for a variance, such as an inverse gamma prior. A prior for the spectral density $g$ is more complicated because $g$ is a density function. We model the spectral density using the DP prior (Ferguson 1973), a common nonparametric prior for unknown distributions. The DP prior can be written as the infinite mixture

$$g(\omega) = \sum_{l=1}^{\infty} \pi_l \delta(v_l), \tag{17.2}$$

where $\delta(v)$ is the point mass at $v \in \mathscr{R}$ and the mixture probabilities $\pi_l$ satisfy $\sum_{l=1}^{\infty} \pi_l = 1$ almost surely. The mixture probabilities have priors $\pi_1 = V_1$ and $\pi_l = V_l \prod_{j<l}(1 - V_j)$ for $l > 1$, where $V_l \overset{iid}{\sim} \text{Beta}(1, D)$. The frequencies $v_l$ have a parametric prior such as a bivariate student-t prior, which centers the prior for the covariance function on the Matérn covariance function.

To ensure that the spectral density is an even function, the prior can be modified as

$$g(\omega) = \frac{1}{2} \sum_{l=1}^{\infty} \pi_l \left[\delta(v_l) + \delta(-v_l)\right]. \tag{17.3}$$

The spectral representation theorem (e.g. Gelfand et al. 2010) states that the real process $Y(\mathbf{s})$ can be written as

$$Y(\mathbf{s}) = \int_{\mathscr{R}^2} \cos(\omega'\mathbf{s})dU(\omega) + \int_{\mathscr{R}^2} \sin(\omega'\mathbf{s})dW(\mathbf{s}), \tag{17.4}$$

where $U$ and $W$ are independent Gaussian processes with mean zero, orthogonal increments, and $\text{E}(|dU(\omega)|^2) + \text{E}(|dW(\omega)|^2) = G(\omega) < \infty$. Therefore, using a DP prior for $g$, (17.4) becomes a countable linear combination of sine and cosine basis functions. The spatial covariance of $Y$ is a random function of $v_j$ and $V_j$, such that given $v_j$ and $V_j$ (through $\pi_j$),

$$\text{Cov}[Y(\mathbf{s}), Y(\mathbf{t})] = \tau^2 \sum_{l=1}^{\infty} \pi_l \cos(v_l'\mathbf{h}). \tag{17.5}$$

Thus, the conditional covariance is stationary and the conditional variance of $Y$ is $\tau^2$, and although the spectral density is discrete, the corresponding covariance is a continuous function of $\mathbf{h}$.

The DP prior for the spectral density is discrete and in some situations a continuous spectral density may be desirable. Therefore, we consider a Dirichlet process mixture (DPM) model (Antoniak 1974), that substitutes the discrete point mass $\delta(\nu_l)$ with a continuous parametric distribution, $k_\gamma(\omega|\nu_l)$ with location $\nu_l$ and scale $\gamma$. To ensure the spectral density is even, we select $k$ to be a location family with location $\nu_l$ that is even in $\nu_l$,

$$g(\omega) = \frac{1}{2}\sum_{l=1}^{\infty} \pi_l \left[k_\gamma(\omega|\nu_l) + k_\gamma(\omega|-\nu_l)\right] \tag{17.6}$$

where $\pi_l$ and $\nu_l$ are modeled as before.

In this section we are able to introduce very flexible priors for spatial covariances by applying the DP prior and the DPM prior to the spectral density. This model does not require spatial locations on a regular grid. However, in practice it does require truncating the infinite mixture in (17.6) to a finite mixture of $L$ terms. After this truncation the covariance function depends on $3L$ parameters (the $V_l$ and vectors $\nu_l$). These parameters do not have conjugate full conditionals, and so Metropolis updates are required for these parameters.

## 17.3 Priors for the Marginal Distribution Using a Spatial Gaussian Copula

While the Gaussian process forms the basis of geostatistics, the normality assumption is often inappropriate. Spatial interpolation may still be reasonable using a Gaussian model even for non-Gaussian data. However, statistical inference such as coverage of prediction intervals or significance tests for covariates are more reliable when accounting for non-normality. Also, Gaussian processes are notoriously poor for estimating the probability of extreme events (Coles 2001).

The Gaussian copula (Nelsen 1999) is a tractable model for non-Gaussian spatial data. The Gaussian copula represents the response in terms of a latent Gaussian process, $Z(\mathbf{s})$. The increasing function $t$ links the latent process and the observations, $Y(\mathbf{s}) = t[Z(s)]$. To ensure identification, it is common to assume that $\mathrm{E}[Z(\mathbf{s})] = 0$ and $\mathrm{V}[Z(\mathbf{s})] = 1$, since the location and scale of the response can be included in $t$. Without loss of generality, we may express $t(x) = q[\Phi(x)]$, where $q$ is an increasing function and $\Phi$ is the standard normal distribution function. In this form, $\Phi[Z(\mathbf{s})]$ follows a Uniform(0,1) distribution, and by the probability inverse transform $q$ is the quantile (inverse distribution) function of $Y$. If $q$ is invertible, then $F(y) = q^{-1}(y)$ is the distribution function $\mathrm{Prob}[Y(\mathbf{s}) < y] = F(y)$, and $f(y) = dF(y)/dy$ is the density function.

This construction provides a general approach for non-Gaussian data, and by selecting the appropriate parametric quantile function $q$ has been used in many settings, including spatial extremes (Sang and Gelfand 2010), quantile regression (Reich 2012), and local variable selection (Boehm Vock et al. 2015). For a parametric model, the likelihood required for computation is straight-forward. If $f$ is a parametric density function with parameters $\theta$ and $\Sigma_\psi$ is the correlation matrix (with parameter $\psi$) for $\mathbf{Z} = [Z(\mathbf{s}_i),\ldots,Z(\mathbf{s}_n)]^T$, then the joint likelihood for $\mathbf{Y} = [Y(\mathbf{s}_1),\ldots,Y(\mathbf{s}_n)]^T$ is

$$p(\mathbf{Y}|\psi,\theta) = \phi(\mathbf{Z}|0,\Sigma_\psi)\prod_{i=1}^{n}\frac{f[Y(\mathbf{s}_i)|\theta]}{\phi[Z(\mathbf{s}_i)|0,1]}, \tag{17.7}$$

where $Z(\mathbf{s}) = \Phi^{-1}F[Y(\mathbf{s})|\theta]$ and $\phi$ is the multivariate normal density function.

In this chapter, our interest lies in BNP methods to estimate the marginal density, $f$. One avenue for specifying a prior for $f$ is via the corresponding quantile function, $q$. Reich (2012) proposed a semi-parametric model for $q$ as a linear combination of basis functions

$$q(x) = \alpha_0 + \sum_{l=1}^{L} B_l(x)\alpha_l, \tag{17.8}$$

where $B_l(x)$ are known basis functions and $\alpha_l$ are unknown coefficients that determine the shape of the quantile function. Restrictions are required on the basis functions and coefficients to ensure that the quantile function is increasing. Reich (2012) define the basis functions through knots $0 = \kappa_0 < \kappa_1 < \ldots < \kappa_L = 1$ and basis functions

$$B_1(x) = \begin{cases} \Phi^{-1}(x) & x \le \kappa_1 \\ \Phi^{-1}(\kappa_1) & x > \kappa_1 \end{cases} \quad \text{and} \quad B_l(x) = \begin{cases} 0 & x < \kappa_{l-1} \\ \Phi^{-1}(x) - \Phi^{-1}(\kappa_{l-1}) & \kappa_{l-1} < x \le \kappa_l \\ \Phi^{-1}(\kappa_l) - \Phi^{-1}(\kappa_{l-1}) & x > \kappa_l \end{cases} \tag{17.9}$$

for $l > 1$. Assuming these basis functions, $q(x)$ is increasing if and only if $\alpha_l > 0$ for all $l > 0$, and therefore truncated normal priors are used to satisfy this constraint. Also, if $\alpha_1 = \ldots = \alpha_L$, then $f$ is the Gaussian density with mean $\alpha_0$ and standard deviation $\alpha_1$, providing a way to center the prior on the parametric model. Finally, it can also be shown that allowing $L \to \infty$, the prior can approximate any continuous quantile function.

Several authors including Rodriguez et al. (2010) and Petrone et al. (2009) have considered the fully nonparametric approach of treating $f$ as an unknown function with Dirichlet process prior; here we outline the general approach. A flexible model for $f$ is the Dirichlet process mixture of normals (DPMN)

$$f(x) = \sum_{l=1}^{L} \pi_l \phi(x|\theta_l,\sigma_1^2), \tag{17.10}$$

where $\pi_l$ are the mixture weights with stick-breaking prior as in (17.2) and $\theta_l \overset{iid}{\sim} \mathrm{N}(0,\sigma_2^2)$ are the mixture means. In the spatial setting, the infinite-dimensional model with $L = \infty$ presents a major computational challenge because closed forms for $f$ and $F$ are required to evaluate the likelihood in (17.7). However, if the mixture is truncated using a sufficiently large $L$, MCMC can be used to implement this model, though because of the non-Gaussian responses most parameters do not have conjugate full conditional distributions.

## 17.4 Nonparametric Spatial Process Models

While the methods in Sect. 17.3 based on the Gaussian copula ease assumptions required about the marginal distribution, the form of spatial dependence remains dictated by the latent Gaussian model. This is inadequate, for example, for modeling dependence between extreme observations (Coles 2001). Other copulas are possible, such as the student-t copula (Nelsen 1999) or even the non-parametric Bayesian copula (Fuentes et al. 2013). However, to fully capture complex dependency structures, more general methods are required, such as those discussed in the remainder of this section.

### *17.4.1 Kernel Convolution of a Dirichlet Process*

Reich et al. (2013) present a BNP approach via a kernel convolution (Higdon et al. 1999) of a Dirichlet process. Spatial dependence is captured via $L$ predetermined spatial knots $\mathbf{v}_1,\ldots,\mathbf{v}_L \in \mathscr{D}$ and the kernel basis function $w_\psi(\mathbf{s},\mathbf{v})$. For example, Reich et al. (2013) consider $L = n$ knots at the $n$ data locations and the squared exponential kernel function

$$w_\psi(\mathbf{s},\mathbf{v}) = \exp\left(-\frac{||\mathbf{s}-\mathbf{v}||^2}{\psi^2}\right), \tag{17.11}$$

where $\psi > 0$ is the kernel bandwidth and determines the range of spatial dependence. The response is then modeled as

$$Y(\mathbf{s}) = \sum_{l=1}^{L} w_\psi(\mathbf{s},\mathbf{v}_l)\alpha_l + \varepsilon(\mathbf{s}), \tag{17.12}$$

where $\varepsilon(\mathbf{s}) \overset{iid}{\sim} \mathrm{N}(0,\sigma_1^2)$ and $\alpha_l \overset{iid}{\sim} f$. In this model with finite $L$, the added local variation (nugget) term $\varepsilon(\mathbf{s})$ is required to give a full-rank model.

The standard approach (Higdon et al. 1999) is to use Gaussian priors for the $\alpha_l$. In this case, letting $L \to \infty$ provides an approximation to a continuous Gaussian process. To provide modeling flexibility, the kernel coefficient density $f$ can be modeled using BNP methods. Reich et al. (2013) use a DPMN prior for $f$ as in (17.10). Unlike the Gaussian copula model, $f$ is not the marginal distribution of $Y$ because the coefficients $\alpha_l$ are convolved to obtain the response. However, Reich et al. (2013) prove that for fixed $L$ and $\psi$ and any continuous marginal distribution for $Y$, there exists a density $f$ that leads to this marginal distribution for $Y$, and therefore this construction spans the entire space of continuous marginal distributions.

With this prior, the full conditional distributions are conjugate for all parameters except the kernel bandwidth $\psi$, leading to straightforward MCMC. Computation is further improved for large datasets by using a small number of basis functions $L$ and thus reducing the dimension of the model. Additionally, allowing for non-stationarity is possible by allowing each knot to have its own bandwidth, $\psi(\mathbf{v}_l)$, and smoothing the bandwidths with a second Gaussian spatial prior for $\psi(\mathbf{v})$ (Higdon et al. 1999). Drawbacks of this approach include the requirement of the nugget term, and that the extension to an infinite dimensional process ($L \to \infty$) leads to a Gaussian process for fixed $\psi$ and $f$ because the response becomes an average of infinitely-many coefficients. However, the method is implemented with fixed $L$, and given full span of the model for any fixed $L$, further theoretical work may show that allowing $f$ to change with $L$ can produce a process that spans a wide class of marginal distributions.

### *17.4.2 Dirichlet Process Mixture of Gaussian Processes*

Gelfand et al. (2005) propose a more general BNP model for the entire joint distribution of non-Gaussian data. This approach requires replication of the process. Therefore, define $Y_t(\mathbf{s})$ as the response at location $\mathbf{s}$ and replication (e.g., time point) $t$. Let $\theta_l(\mathbf{s})$ be independent and identically distributed spatial Gaussian processes with mean zero and $\text{Cov}[\theta_l(\mathbf{s}), \theta_l(\mathbf{t})] = \sigma(\mathbf{s}, \mathbf{t})$. The marginal distribution at location $\mathbf{s}$ is given by the DPMN model

$$\sum_{l=1}^{L} \pi_l \phi[y; \theta_l(\mathbf{s}), \sigma_1^2], \tag{17.13}$$

where $\pi_l$ are the mixture probabilities and $L$ may be infinite. This extends (17.10) by allowing the mixture means $\theta_l(\mathbf{s})$ to vary spatially, and thus the marginal distribution is allowed to vary by location capturing spatial heterogeneity.

A joint distribution with this marginal distribution is given as the following clustering model. Denote $g_t \in \{1, 2, \ldots\}$ as the cluster label for replication $t$, with $\text{Prob}(g_t = j) = \pi_j$. Then (17.13) arises as the marginal (over $g_t$) distribution of the hierarchical model

$$Y_t(\mathbf{s}) | g_t = \theta_{g_t}(\mathbf{s}) + \varepsilon_t(\mathbf{s}), \tag{17.14}$$

where $\varepsilon_t(\mathbf{s})$ is a Gaussian process with mean zero and variance $\sigma_1^2$, independent across $t$. In this representation, the labels $g_t$ group the replications into clusters, with $\theta_j(\mathbf{s})$ representing the common spatial pattern in cluster $j$.

An advantage of this approach is that for any collection of $n < \infty$ spatial locations, the resulting prior for the $n$-dimensional joint distribution is an $n$-dimensional DPMN and thus this model spans the entire space of joint distributions for any collection of locations. Also, implementing this approach using MCMC is straightforward as most parameters (including $g_t$, $V_l$, and $\theta(\mathbf{s})$) have conjugate full conditional distributions, permitting Gibbs updates. A potential drawback of the global clustering model is that for a large and heterogeneous spatial domain, it may not be possible to group entire replications into meaningful clusters. In this case, the posterior may have many clusters with a single member, causing difficulty in estimating the joint distribution. The global clustering model is generalized in Duan et al. (2007), who consider local cluster allocation.

### *17.4.3 Stick-Breaking Methods*

In this section, we present several methods that introduce spatial variation into various aspects of the stick-breaking representation of a DP given in (17.2). As in Sect. 17.4.1, the response is decomposed as $Y(\mathbf{s}) = \mu(\mathbf{s}) + \varepsilon(\mathbf{s})$, where $\mu(\mathbf{s})$ is a spatial random effect and $\varepsilon(\mathbf{s}) \overset{iid}{\sim} \mathrm{N}(0, \sigma_1^2)$. Spatial extensions of the stick-breaking model for the random effect distribution lead to a broad class priors that can deal with skewness, multimodality, etc. In addition, even though the prior predictive distributions induced by these models are stationary, the posterior predictive distributions can accommodate nonstationary behavior.

Density function $f$ has stick-breaking prior if

$$f(\mu) = \sum_{l=1}^{L} \pi_l \delta(\theta_l), \tag{17.15}$$

where $\delta(\theta)$ denotes a Dirac measure at $\mu = \theta$, $\pi_l = V_l \prod_{j<l}(1 - V_l)$ where $V_l$ are independent with $V_l \sim \mathrm{Beta}(1, D)$ and $\theta_l$ are independent draws from a centering distribution $H$. The definition in (17.15) allows for either finite or infinite $L$ (with the latter corresponding to the conventional definition of nonparametrics). In order to make this nonparametric prior useful for our spatial context, we need to index it by space. This can be achieved by allowing either the masses, $\mathbf{V} = (V_1, V_2, \ldots)$, or the locations of the atoms, $\theta = (\theta_1, \theta_2, \ldots)$, to change with $\mathbf{s}$. The idea in Gelfand et al. (2005) is to introduce spatial dependence through the locations, by indexing $\theta$ with the location $\mathbf{s}$ and making $\theta(\mathbf{s})$ a realization of a random field with $H$ being a stationary Gaussian process.

An alternative approach to extending the prior in (17.15) to the spatial setting is followed by Griffin and Steel (2006), who define the ranking of the elements in the vectors $\mathbf{V}$ and $\theta$ through an ordering $o(\mathbf{s}) = [o_1(\mathbf{s}), o_2(\mathbf{s}) \ldots]$, which changes with the spatial index (or other covariates). The density function at location $\mathbf{s}$ is taken to be

$$f(\mu; \mathbf{s}) = \sum_{l=1}^{L} \pi_l(\mathbf{s}) \delta(\theta_l) \quad \text{where} \quad \pi_l(\mathbf{s}) = V_l(\mathbf{s}) \prod_{j<l} [1 - V_j(\mathbf{s})] \tag{17.16}$$

and $V_l(\mathbf{s}) = V_{o_l(\mathbf{s})}$. Since weights associated with atoms that appear earlier in the stick-breaking representation tend to be larger (i.e., $\mathrm{e}[\pi_l(\mathbf{s})] < \mathrm{e}[\pi_{l-1}(\mathbf{s})]$), this induces similarity between distributions with similar ordering. The similarity between $o(\mathbf{s}_1)$ and $o(\mathbf{s}_2)$ controls the dependence between $f(\mu;\mathbf{s}_1)$ and $f(\mu;\mathbf{s}_2)$. The induced class of models is called order-based dependent Dirichlet processes. This specification preserves the usual Dirichlet process for the marginal distribution at each location, but, in contrast with the single-$\pi$ approaches, leads to local updating, where the influence of observations decreases as they are further away. The main challenge is to define stochastic processes $o(\mathbf{s})$. Griffin and Steel (2006) use a point process and a sequence of sets which define the region in which points are relevant for determining the ordering at $\mathbf{s}$.

Reich and Fuentes (2007) introduce a spatial kernel stick-breaking (SSB) prior to extend the stick breaking construction to the spatial setting. Similar to Griffin and Steel (2006), the weights $\pi_l$ vary spatially. However, rather than random permutation of $V_l$, Reich and Fuentes (2007) introduce a series of kernel functions to allow the masses to change with space. This results in a flexible spatial model, as different kernel functions lead to different relationships between the distributions at nearby locations. This model is similar to that of Dunson and Park (2008), who use kernels to smooth the weights in the non-spatial setting.

The spatial effects $\mu(\mathbf{s})$ are assigned a random prior distribution, i.e., $\mu(\mathbf{s}) \sim f(\mu;\mathbf{s})$, that are smoothed spatially. The density $f(\mu;\mathbf{s})$ has the form of (17.16), but with

$$V_l(\mathbf{s}) = w_\psi(\mathbf{s},\mathbf{v}_l)V_l, \tag{17.17}$$

where $w_\psi(\mathbf{s},\mathbf{v}_l) \in [0,1]$ is a kernel function such as (17.11) that depends on spatial knot $\mathbf{v}_l \in \mathscr{D}$ and kernel bandwidth $\psi$. The distributions $f(\mu;\mathbf{s})$ are related through their dependence on the $V_l$ and $\theta_l$, which are given the priors $V_l \sim \text{Beta}(a,b)$ and $\theta_l \sim H$, each independent across $l$. However, the distributions vary spatially according to the kernel functions $w_\psi(\mathbf{s},\mathbf{v}_l)$.

A potential drawback of the SSB prior is that realizations of $\mu(\mathbf{s})$ are discontinuous functions of $\mathbf{s}$. To remedy this issue, Fuentes and Reich (2013) allow both the probabilities $\pi_l$ and the locations $\theta_l$ to depend on space. The prior for $f(\mu;\mathbf{s})$ is the potentially infinite mixture

$$f(\mu;\mathbf{s}) = \sum_{l=1}^{L} \pi_l(\mathbf{s})\delta[X(\mathbf{v}_l)] \text{ where } \pi_l(\mathbf{s}) = V_l(\mathbf{s})\prod_{j<l}[1-V_j(\mathbf{s})], \tag{17.18}$$

$X$ is a Gaussian process, and $V_l(\mathbf{s}) = w_\psi(\mathbf{s},\mathbf{v}_l)V_l$ as in the previous SSB model. In this representation, $X$ is a spatial process on the *knot* space. However, by construction, as the kernel bandwidth $\psi \to 0$, $\mu(\mathbf{s}) \to X(\mathbf{s})$. Therefore, by introducing the latent Gaussian process $X$, this formulation includes the parametric continuous Gaussian process as a special case.

These kernel stick-breaking models are computationally convenient in the sense that they avoid inversion of large covariance matrices which often hinder spatial data analysis. Many of the parameters have conjugate full conditionals, including the latent process $X(\mathbf{s})$. The primary challenge is updating the kernel knots $\mathbf{v}_l$, which require Metropolis updates.

## References

Antoniak, C. E. (1974). Mixtures of Dirichlet processes with applications to Bayesian nonparametric problems. *Annals of Statistics*, **2**, 1152–1174.

Banerjee, S., Carlin, B. P., and Gelfand, A. E. (2004). *Hierarchical modeling and analysis for spatial data*. Chapman & Hall/CRC.

Boehm Vock, L. F., Reich, B. J., Fuentes, M., and Dominici, F. (2015). Spatial variable selection methods for investigating acute health effects of fine particulate matter components. *Biometrics*, **71**, 167–177.

Coles, S. (2001). *An Introduction to Statistical Modeling of Extreme Values*. Springer, London.

Cressie, N. (1993). *Statistics for spatial data*. Wiley-Interscience.

Duan, J. A., Guindani, M., and Gelfand, A. E. (2007). Generalized spatial Dirichlet process models. *Biometrika*, **94**, 809–825.

Dunson, D. and Park, J. (2008). Kernel stick-breaking processes. *Biometrika*, **96**, 307–323.

Ferguson, T. (1973). A Bayesian analysis of some nonparametric problems. *The Annals of Statistics*, **1**, 209–230.

Fuentes, M. and Reich, B. J. (2013). Multivariate spatial nonparametric modeling via kernel processes mixing. *Statistica Sinica*, **23**, 75–97.

Fuentes, M., Henry, J. B., and Reich, B. J. (2013). Nonparametric spatial models for extremes: Application to extreme temperature data. *Extremes*, **16**, 75–101.

Gelfand, A., Diggle, P., Guttorp, P., and Fuentes, M. (2010). *Handbook of spatial statistics*. Chapman & Hall/CRC.

Gelfand, A. E., Kottas, A., and MacEachern, S. N. (2005). Bayesian nonparametric spatial modeling with Dirichlet process mixing. *Journal of the American Statistical Association*, **100**, 1021–1035.

Griffin, J. E. and Steel, M. F. J. (2006). Order-based dependent Dirichlet processes. *Journal of the American Statistical Association*, **101**, 179–194.

Higdon, D., Swall, J., and Kern, J. (1999). Non-stationary spatial modeling. In *Bayesian Statistics 6 - Proceedings of the Sixth Valencia Meeting*, pages 761–768. J.M. Bernardo, J.O. Berger, A.P. Dawid, and A.F.M. Smith, (editors). Clarendon Press - Oxford.

Nelsen, R. B. (1999). *An Introduction to Copulas*. Springer-Verlag, New York.

Petrone, S., Guindani, M., and Gelfand, A. E. (2009). Hybrid Dirichlet mixture models for functional data. *Journal of the Royal Statistical Society: Series B*, **71**, 755–782.

Reich, B. J. (2012). Spatiotemporal quantile regression for detecting distributional changes in environmental processes. *Journal of the Royal Statistical Society: Series C*, **64**, 535–553.

Reich, B. J. and Fuentes, M. (2007). A multivariate semiparametric Bayesian spatial modeling framework for hurricane surface wind fields. *Annals of Applied Statistics*, **1**, 249–264.

Reich, B. J. and Fuentes, M. (2012). Nonparametric Bayesian models for a spatial covariance. *Statistical Methodology*, **9**, 265–274.

Reich, B. J., Bandyopadhyay, D., and Bondell, H. D. (2013). A nonparametric spatial model for periodontal data with non-random missingness. *Journal of the American Statistical Association*, **108**, 820–831.

Rodriguez, A., Dunson, D. B., and Gelfand, A. E. (2010). Latent stick-breaking processes. *Journal of the American Statistical Association*, **105**, 647–659.

Sang, H. and Gelfand, A. E. (2010). Continuous spatial process models for spatial extreme values. *Journal of Agricultural, Biological, and Environmental Statistics*, **15**, 49–65.

Zheng, Y., Zhu, J., and Roy, A. (2010). Nonparametric Bayesian inference for the spectral density function of a random field. *Biometrika*, **97**, 238–245.

# Chapter 18
# Spatial Species Sampling and Product Partition Models

Seongil Jo, Jaeyong Lee, Garritt Page, Fernando Quintana, Lorenzo Trippa, and Peter Müller

**Abstract** Inference for spatial data arises, for example in medical imaging, epidemiology, ecology, and other areas, and gives rise to specific challenges for nonparametric Bayesian modeling. In this chapter we briefly review the fast growing related literature and discuss two specific models in more detail. The two models are the CAR SSM (species sampling with conditional autoregression) prior of Jo et al. (Dependent species sampling models for spatial density estimation. Technical report, Department of Statistics, Seoul National University, 2015) and the spatial PPM (product partition model) of Page and Quintana (Spatial product partition models. Technical report, Pontificia Universidad Católica de Chile, 2015).

S. Jo
Department of Statistics, Korea University, Seoul, South Korea
e-mail: joseongil@gmail.com

J. Lee (✉)
Department of Statistics, Seoul National University, Seoul, South Korea
e-mail: leejyc@gmail.com

F.A. Quintana • G.L. Page
Departamento de Estadística, Pontificia Universidad Católica de Chile, Santiago, Chile
e-mail: quintana@mat.uc.cl; page@mat.uc.cl

P. Müller
The University of Texas at Austin, 1, University Station, C1200, Austin, TX 78712, USA
e-mail: pmueller@math.utexas.edu

L. Trippa
Department of Biostatistics, Harvard University, Cambridge, MA, USA
e-mail: ltrippa@jimmy.harvard.edu



R. Mitra, P. Müller (eds.), *Nonparametric Bayesian Inference in Biostatistics*,
Frontiers in Probability and the Statistical Sciences,
DOI 10.1007/978-3-319-19518-6_18

## 18.1 Introduction

An important class of statistical inference problems arise with the analysis of spatial data. Spatial data is broadly understood as measurements $\{Y(\boldsymbol{s}_i) : i = 1, \ldots, n\}$ made at coordinates $\boldsymbol{s}_1, \ldots, \boldsymbol{s}_n \in \mathscr{D}$ for some set $\mathscr{D} \subset \mathbb{R}^d$. In *point-referenced data*, the coordinates may vary continuously over the fixed set $\mathscr{D}$. The case of *areal data* arises when $\mathscr{D}$ is partitioned into a finite number of smaller (areal) units. Finally, the case of *point-pattern data* follows when $\mathscr{D}$ is itself random. See further details in Banerjee et al. (2015). *Spatio-temporal data* includes additional indexing by time.

Much of the effort in spatial data analysis revolves around ways to introduce spatial dependence. This makes Gaussian process models an attractive modeling choice and a common element of many nonparametric Bayesian spatio-temporal models.

### 18.1.1 Models Based on the Dirichlet Process

Many approaches are based on variations of the Dirichlet process (DP) model. One of the earliest attempts at a nonparametric model for spatial data appears in Gelfand et al. (2005). They considered replicated point-referenced data. Let $\boldsymbol{Y} = (Y(\boldsymbol{s}_1), \ldots, Y(\boldsymbol{s}_n))$ generically denote the complete responses for a given replicate, and let $\boldsymbol{Y}_1, \ldots, \boldsymbol{Y}_T$ denote the entire dataset. Assuming replicate-specific covariate vectors $\boldsymbol{x}_t$, they assume

$$\boldsymbol{Y}_t \mid \boldsymbol{\mu}_t, \boldsymbol{\beta}, \tau^2 \overset{\text{ind}}{\sim} N(\boldsymbol{x}_t'\boldsymbol{\beta} + \boldsymbol{\mu}_t, \tau^2 \boldsymbol{I}), \quad \boldsymbol{\mu}_t \mid G \overset{\text{iid}}{\sim} G, \quad G \sim \text{DP}(M, G_0), \tag{18.1}$$

where $G_0$ is a zero-mean Gaussian process (GP) with covariance function $\sigma^2 \boldsymbol{H}_\phi$ $(\cdot,\cdot)$, and $G \sim \text{DP}(M, G_0)$ denotes a Dirichlet process prior. A random distribution

$$G = \sum p_h \delta_{\boldsymbol{\theta}_h} \tag{18.2}$$

is said to follow a Dirichlet process with parameter $M \cdot G_0$ if $p_h = V_h \prod_{l<h}(1 - V_l)$ with $V_h \sim Beta(1, M)$, i.i.d. and $\boldsymbol{\theta}_h \sim G_0$, i.i.d. Here $M > 0$ and $G_0$ is a distribution. This representation is called Sethuraman representation (Sethuraman 1994) of the Dirichlet process and the construction of $p_h$ from the i.i.d. sequence $(V_h)$ is known as the stick-breaking representation. The DP in (18.1) generates a random distribution $G$. It is a discrete distribution with point masses $p_h$ at surfaces $\boldsymbol{\theta}_h$. The surfaces $\boldsymbol{\theta}_h$ are generated from a GP prior.

Model (18.1) is completed with conjugate hyper priors for $\boldsymbol{\beta}$, $\tau^2$, $\sigma^2$, and a gamma prior for $M$. Here $\boldsymbol{H}(\phi)$ is a suitable covariance function with parameter $\phi$, for instance, an exponential function of the form $\boldsymbol{H}(\phi)_{i,j} = \exp(-\phi \|\boldsymbol{s}_j - \boldsymbol{s}_i\|)$, in which case $\phi$ is assigned a uniform prior on $(0, b_\phi]$.

Model (18.1) introduces spatial correlation through the random effects vector $\boldsymbol{\mu}_t$, which are in turn assumed to originate from a DP model. This describes the sampling model as a countable location mixture of normals. Thus, given the collection of surfaces, any realization $\mu_t \sim G$ of the process selects a single sampled surface $\boldsymbol{\theta}_h$. A modification of the spatial DP by Duan et al. (2007) yields the generalized spatial DP, where the random effects distributions allow for different surface selection at different sites. Their model involves a multivariate stick-breaking construction, where weights may be allowed to depend on spatial coordinates. They provide details for a specific case where weights vary smoothly with $\boldsymbol{s}$, and show that the spatial DP is a particular case of the proposed construction.

A spatial stick-breaking prior was introduced by Reich and Fuentes (2007) in the context of the analysis of hurricane surface wind fields. Their model can be described as

$$Y(\boldsymbol{s}) = \mu(\boldsymbol{s}) + x(\boldsymbol{s})'\boldsymbol{\beta} + \varepsilon(\boldsymbol{s}),$$

where $\boldsymbol{s} = (s_1, s_2)$, $\mu(\boldsymbol{s}) \sim F_{\boldsymbol{s}}$, and $F_{\boldsymbol{s}}$ is a (potentially infinite) mixture of point masses with weights carrying the dependence on spatial coordinates $\boldsymbol{s}$. Specifically, they considered

$$F_{\boldsymbol{s}}(\cdot) = \sum_{h=1}^{m} p_h(\boldsymbol{s}) \delta_{\boldsymbol{\theta}_h}(\cdot) \quad \text{with} \quad p_h(\boldsymbol{s}) = V_h(\boldsymbol{s}) \prod_{\ell<h} (1 - V_\ell(\boldsymbol{s})), \quad h > 1,$$

including $p_1(\boldsymbol{s}) = V_1(\boldsymbol{s})$ and $\boldsymbol{\theta}_h \overset{\text{iid}}{\sim} N(0, \tau^2)$, similar to the stick-breaking construction. The difference lies in the definition of the location-dependent fractions $V_h(\boldsymbol{s}) = V_h(\boldsymbol{s}) = w_h(\boldsymbol{s}) V_h$, which introduces spatial dependence in the weights via the kernel functions $w_i(\boldsymbol{s})$. Here, $w_i(\boldsymbol{s})$ is centered at a knot $\boldsymbol{\psi}_i = (\psi_{i1}, \psi_{i2})$ and the spread is controlled by a bandwidth parameter $\boldsymbol{\varepsilon}_i = (\varepsilon_{i1}, \varepsilon_{i2})$. Reich and Fuentes (2007) considered several examples of kernel functions, including $w_i(\boldsymbol{s}) = \prod_{j=1}^{2} I(|s_j - \psi_{ji}| < \varepsilon_{ji}/2)$, with suitable hyper priors on $\boldsymbol{\psi}_i$ and $\boldsymbol{\varepsilon}_i$. Another option is a square-exponential kernel of the form $w_i(\boldsymbol{s}) = \prod_{j=1}^{2} \exp\{-(s_j - \psi_{ji})^2 / \varepsilon_{ji}^2\}$. Reich and Fuentes (2007) also discuss how to ensure posterior propriety and details of posterior simulation.

### *18.1.2 Approaches Based on the Product Partition Model*

Some alternative nonparametric Bayesian models for spatial data are based on the product partition model (PPM) introduced in Hartigan (1990) and Barry and Hartigan (1992). We briefly review the PPM construction. Assume we have observations (either scalar or vector-valued) $y_1, \ldots, y_n$ for a set of $n$ individuals and suppose we adopt a sampling model that involves a corresponding set of parameters $\theta_1, \ldots, \theta_n$. The PPM starts by assuming conditional independence in the sampling model, i.e., $p(y^n \mid \theta^n) = \prod_{i=1}^{n} p(y_i \mid \theta_i)$, where $y^n = (y_1, \ldots, y_n)$ and $\theta^n = (\theta_1, \ldots, \theta_n)$ are the complete sets of observations and parameters. A key feature of the PPM is a partition $\{1, \ldots, n\} = \bigcup_{j=1}^{k_n} S_j$ so that all observations in a cluster share a common parameter value, $\theta_i = \theta_j^\star$ for all $i \in S_j$. That is, $\{\theta_1^\star, \ldots, \theta_{k_n}^\star\}$ are the $k_n \leq n$ unique values of $\theta_i$ and observations are clustered by the arrangement of ties among the $\theta_i$. Observe that a partition $\rho_n = (S_1, \ldots, S_{k_n})$ can be equivalently described by a set of cluster membership indicators $c_1, \ldots, c_n$ with $c_i = j$ if $i \in S_j$, i.e., $\theta_i = \theta_{c_i}^\star$. The sampling model then becomes

$$p(y^n \mid \theta^n, \rho_n) = \prod_{j=1}^{k_n} \prod_{i \in S_j} p(y_i \mid \theta_j^\star), \tag{18.3}$$

i.e., with conditional independence within clusters. To complete the model, the PPM considers a distribution for $\rho_n$, that is, a random partition model $p(\rho_n)$. The PPM assumes

$$P(\rho_n = (S_1, \ldots, S_{k_n})) \propto \prod_{j=1}^{k_n} c(S_j), \tag{18.4}$$

where $c(S)$ is the *cohesion* of $S \subset [n]$, which states how tightly grouped the elements of $S$ are thought to be a priori, and with normalization constant $\sum_{\rho_n \in \mathscr{R}_n} \prod_{j=1}^{k_n} c(S_j)$. A common definition of cohesion function is $c(S) = M \times (|S| - 1)!$, which agrees with the marginal distribution on $\mathscr{R}_n$ that arises from the well-known clustering property of the DP. See, e.g., Quintana and Iglesias (2003).

Hegarty and Barry (2008) adapt the PPM (18.4) for spatial inference, in an application to disease mapping. Their goal is to identify areas of unusually high or low risk. To this effect, they propose a model that involves a random partition of a set of areas $A_i$, $i = \{1, \ldots, n\}$. They define the desired partition using a PPM with cohesion functions for sets of areas, in what they refer to as *short boundary model*. Specifically, they define the cohesion of any subset $S \subset \{1, \ldots, n\}$ of areas, as $c(S) = \beta^{\ell(S)}$, where $\ell(S) = \sum_{A_i \in S} \ell(A_i)$ and $\ell(A_i)$ is the number of neighbors of $A_i$ not in $S$. With this definition, they discourage maps with a large number of fragmented components. The model includes a Poisson-style likelihood function. See further details such as posterior simulation and applications in Hegarty and Barry (2008).

In the rest of this chapter we discuss in more detail two nonparametric Bayesian models for spatial data proposed in Jo et al. (2015) and Page and Quintana (2015). In Sect. 18.2 we discuss a spatial variation of species sampling models (SSM), which are introduced in Jo et al. (2015) as a generalization of DP priors. In Sect. 18.3 we discuss the approach proposed in Page and Quintana (2015) who construct a model based on the computationally attractive form of PPM's.

## 18.2 A Dependent Species Sampling Models for Spatial Density Estimation

### *18.2.1 Species Sampling Models*

Jo et al. (2015) introduce a nonparametric Bayesian model for spatial data based on a generalization of the DP prior known as SSM. In words, the proposed model introduces spatial dependence for a family of random probability measures $\{G_i = \sum_{h \geq 1} p_{hi} \delta_{\theta_h},\ i = 1, \ldots, n\}$, indexed by spatial location $s_i$, $i = 1, \ldots, n$. The desired dependence is introduced by defining a conditionally autoregressive (CAR) prior for the weights $p_{hi}$. Below we first introduce the specific SSM prior. Then we define the CAR prior, and finally combine the two to construct the proposed nonparametric Bayesian CAR SSM.

A SSM is a discrete random probability distribution represented as $G = \sum_{h=1}^{\infty} p_h \delta_{\theta_h}$, where $(p_h, h \geq 1)$ is a sequence of positive random variables with $\sum_{h \geq 1} p_h = 1$ a.s., $(\theta_h, h \geq 1)$ is a sequence of i.i.d. (independent and identically distributed) random variables sampled from a nonatomic distribution $G_0$, and $(p_h)$ and $(\theta_h)$ are independent. The class of SSM is a large class of random discrete distributions which includes the DP (Ferguson 1973) and the Pitman-Yor (PY) process (Pitman and Yor 1997) as special cases. The PY process is a variation of the DP prior, which is defined by replacing the prior for $V_h$ in (18.2) by $V_h \sim Beta(1-a, b+ha)$, for $h \geq 1$ and with $0 \leq a < 1$, $b > -a$. We say $G \sim PY(a, b, G_0)$.

The family of SSMs includes any random discrete probability measure with independent prior on the location of the point masses. For example, Lee et al. (2013) construct SSMs from an arbitrary sequence of positive random variables. Let $(w_h,\ h \geq 1)$ be a sequence of independent positive random variables and define weights as

$$p_h = \frac{w_h}{\sum_{l=1}^{\infty} w_l}, \qquad h = 1, 2, \ldots. \tag{18.5}$$

For the normalization to be well defined, the infinite sum of $(w_h)$ needs to be finite a.s., i.e., $\sum_h w_h < \infty, a.s.$ A simple set of sufficient conditions exists for the finiteness of the sum. If

$$\sum_h E(u_h) < \infty \quad \text{and} \quad \sum_h Var(u_h) < \infty,$$

then $\sum_h w_h < \infty$ with probability 1. We will use (18.5) as a basic building block for the upcoming construction of the spatial SSM.

### 18.2.2 Gaussian CAR Model

We will use a CAR model to introduce the desired spatial dependence for $p_h$. We therefore briefly review the definition of CAR models. The CAR model is a popular model for spatial dependence, that is, for dependence between random variables indexed by locations. Let $\mathscr{D} = \{1, 2, \ldots, n\}$ be the set of locations and $y = (y_1, \ldots, y_n)$ be the collection of random variables where $y_i = Y(s_i)$ corresponds to location $s_i$, $i = 1, 2, \ldots, n$. The CAR is a method of modeling the joint distribution by specifying the full conditional distributions

$$y_i \mid y_{-i}, \quad i = 1, 2, \ldots, n,$$

where $y_{-i}$ is the collection of $y_i$'s omitting $y_i$. The CAR is intuitive to understand, because the full conditionals are a natural way to describe the dependency between variables. But not every set of full conditionals defines a joint distribution, and the conditions that ensure a valid joint distribution need to be checked. For the normal distribution, a set of conditions that ensure a joint distribution is known.

Suppose

$$y_i \mid y_{-i} \sim N(\mu_i - \sum_{j:j\neq i} \xi_{ij}(y_j - \mu_j), \tau_i^2), \quad i = 1,2,\ldots,n, \tag{18.6}$$

where $\tau_i^2 > 0$ and $\xi_{ij}$, $\mu_i \in \mathbb{R}$. Define an $n \times n$ matrix

$$\Lambda = (\lambda_{ij},\ i,j = 1,2,\ldots,n),$$

where $\lambda_{ii} = 1/\tau_i^2$, $i = 1,2,\ldots,n$, $\lambda_{ij} = \xi_{ij}/\tau_i^2$, $i,j = 1,2,\ldots,n$. If $\Lambda$ is symmetric and positive definite, then

$$y \sim N(\mu, \Lambda^{-1}),$$

where $\mu = (\mu_1, \mu_2, \ldots, \mu_n)$. See Rue and Held (2005), Besag (1974) or Cressie (1993) for details.

We are now ready to define a spatial SSM using a CAR model on the weights. Jo et al. (2015) consider two specifications of Gaussian CAR models for this construction.

1. *Mercer CAR model.* A Mercer kernel $K(s,t)$ is a kernel function defined on $\mathscr{D}$ such that for all integer $n$ and $s_1,\ldots,s_n \in \mathscr{D}$, the $n\times n$ matrix $H = (K(s_i,s_j),\ i,j = 1,2,\ldots,n)$ is positive definite. For example, $K(\cdot,\cdot)$ could be a negative squared exponential kernel, $K(s,t) = \exp(-\rho||s-t||^2)$. The matrix $H$ is called a Gram matrix. Let $K(s,t)$ be a Mercer kernel defined on $\mathscr{D}$ with $K(s,s) = 1$ for all $s \in \mathscr{D}$. Define $\xi_{ij} = K(s_i,s_j)$ and $\tau_i^2 = \tau^2 > 0$. Then, the joint distribution of $y$ with full conditionals (18.6) exists. We call the distribution of $y$ the Mercer CAR model.
2. *Clayton- Kaldor CAR model* (Clayton and Kaldor 1987; Sun et al. 1999). Let $N(s_i)$ be the set of neighbors of location $s_i$ and

$$\xi_{il} = \begin{cases} -\rho, & \text{if } l \in N(s_i); \\ 0, & \text{if } l \notin N(s_i), \end{cases} \tag{18.7}$$

where $\rho \in (1/\psi_1, 1/\psi_n)$ and $\psi_1$ and $\psi_n$ are the minimum and maximum of the adjacency matrix $C = (c_{ij},\ i,j = 1,2,\ldots,n)$ with $c_{ii} = 0$ and $c_{ij} = I(i \in N(s_j))$, $i,j = 1,2,\ldots,n$, and $I(a)$ is the indicator function whose value is 1 if $a$ is true and 0, otherwise. This also allows the joint distribution of $y$. In our application, we defined the neighbor of $s_i$ by $N(s_i) = \{s_j : ||s_i - s_j|| < B\}$ with $B > 0$.

### *18.2.3 CAR SSM and CAR SSM Mixtures*

Let $G_i$ be a random distribution corresponding to location $s_i$. We define a collection of random distributions $\{G_i,\ i = 1,2,\ldots,n\}$. Jo et al. (2015) propose the following construction:

$$G_i = \sum_{h=1}^{\infty} p_{i,h}\delta_{\theta_h}, \quad i = 1,2,\ldots,n, \tag{18.8}$$

where $(p_{ih})$ is the normalization of positive random variables $(w_{i,h})$ and $(\theta_h)$ is a sequence of i.i.d. random variables following an non-atomic distribution $G_0$. The positive random variables $w_{i,h}$ are defined by

$$w_{i,h} = e^{u_{i,h}}, \quad i = 1,2,\ldots,n,\ h = 1,2,\ldots$$

and $(u_{i,h},\ i = 1,2,\ldots,n)$ is generated by either the Mercer or the Clayton- Kaldor CAR model.

In particular,

$$u_{i,h} \mid u_{l,h}, l \neq i \sim N(m_{i,h} - \sum_{l:l\neq i} \xi_{il}(u_{l,h} - m_{l,h}), \tau^2), \quad i = 1,2,\ldots,n,$$

where $\tau^2 > 0$, $m_{ih} = \log(1-(1+e^{b-ah})^{-1})$, $i = 1,2,\ldots,n$ with $a,b > 0$ and the coefficients $\xi_{il}$ are defined by either the Mercer kernel

$$K(s,t) = e^{-\rho\|s-t\|^2}, \quad s,t \in \mathscr{D},$$

or by the Clayton- Kaldor CAR model (18.7). We refer to the two models as Mercer CAR SSM (MCS) and Clayton- Kaldor CAR SSM (CKCS) and write

$$\{G_s, s \in \mathscr{D}\} \sim MCS(G_0,a,b,\tau^2,\rho) \text{ and } \{G_s, s \in \mathscr{D}\} \sim CKCS(G_0,a,b,\tau^2,\rho,B).$$

Here $a,b,\tau^2$ and $\rho$ are fixed hyperparameters. For example, a typical choice is $a = 1$, $b = 10$, $\tau = 0.1$, $B = 1.1$, $\rho = 0.1$. See also the upcoming example. The CAR SSMs are flexible enough to cover all possible collections of distributions indexed by locations, i.e., it has full weak supports. See Jo et al. (2015). One remaining limitation for many applications is the discrete nature of SSMs. Similar to the popular DP mixture models, the discrete nature of the CAR SSM is easily avoided by introducing an additional convolution with a continuous kernel.

We define the CAR SSM mixture model for a collection of densities. Let

$$y_{i,j} \overset{i.i.d.}{\sim} \int f(y \mid \theta) dG_i(\theta), \quad i = 1,2,\ldots,n,\ j = 1,2,\ldots,N_i, \tag{18.9}$$

where $(G_i,\ i = 1,2,\ldots,n) \sim MCS$ or $CKCS$. Here $f(y \mid \theta)$ is a kernel density with parameter $\theta$. The integral in (18.9) is then a sum over all point masses in (18.8).

### *18.2.4 Posterior Computation*

For the posterior computation in a CAR SSM mixture model, the following equivalent formulation of mixtures of CAR SSM is useful. Let $y_{ij}$, $j = 1,\ldots,N_i$, denote the data at spatial locations $s_i$, $i = 1,\ldots,n$. We replace the mixture in (18.9) by an equivalent hierarchical model with latent variables $z_{ij}$. Also, we approximate the full infinite mixture by a finite truncation after $K$ terms. The construction is similar to the finite DP prior (Ishwaran and James 2001a):

$$y_{ij} \mid \phi, z_{ij} \sim f(\cdot \mid \phi_{z_{ij}}), \quad j = 1, 2, \ldots, N_i,$$

$$z_{ij} \mid u \sim \sum_{h=1}^{K} \frac{e^{u_{i,h}}}{\sum_{k=1}^{K} e^{u_{i,k}}} \delta_h,$$

with the CAR prior on the transformed weights $\boldsymbol{u}_h = (u_{1h}, \ldots, u_{n,h})$,

$$\boldsymbol{u}_h \mid \boldsymbol{s} \sim p(\mathbf{u}_h \mid \eta_2), \quad h = 1, 2, \ldots, K,$$

where $p(\boldsymbol{u}_h \mid \eta_2)$ is the density under the MCS or CKCS with parameter $\eta_2$. For example, in the case of the MCS, $\eta_2 = (a, b, \tau^2, \rho)$. And, finally, the independent prior on the atoms of the SSM random measure,

$$\theta_h \overset{\text{iid}}{\sim} G_0(\cdot \mid \eta_1),$$

with hyperprior $\eta_1 \sim \pi_1(\eta_1)$.

Posterior simulation is based on the blocked Gibbs algorithm (Ishwaran and James 2001b) and a data augmentation method for multinomial logit model (Scott 2011) and the Albert and Chib (1993) algorithm. The details of the algorithm can be found in Jo et al. (2015).

### *18.2.5 Data Analysis*

We analyze monthly average apartment market price in Seoul, South Korea. A data set was obtained from KB apartment market price database of Kookmin bank. These data cover the 3-year period between July 2003 and July 2006, for a total of 37 months. Data and additional details are available at http://www.nland.kbstar.com. Figure 18.1 shows the apartment market price (in South Korean won, KRW) in each period.

After scaling observations by 10,000,000, we applied the mixtures of CAR SSMs using an $MCS(G_0, a, b, \tau^2, \rho)$ and an $CKCS(G_0, a, b, \tau^2, \rho, B)$ model. We applied the CAR SSMs with hyperparameters: $\mu_{00} = 0$, $\kappa^2 = 1000$, $\nu_1 = 1$, $\nu_2 = 100$, $\nu = 4$, and $\psi = 1/2$. For the spatial parameters of the CKCS and the MCS we used $\rho = 0.1$ and $\rho = 4$, respectively. Inference is based on 5000 samples of the Markov chain Monte Carlo (MCMC) posterior simulation which are thinned by a factor 10, after a burn-in period of 10,000 samples.

Figures 18.2 and 18.3 show the posterior mean estimates and a 95 % credible interval of the $MCS(G_0, a, b, \tau^2, \rho)$ and $CKCS(G_0, a, b, \tau^2, \rho, B)$, respectively. The apartment price rose rapidly in the period of data collection. Interestingly, the density estimates show two modes clearly indicating that apartments are clustered into

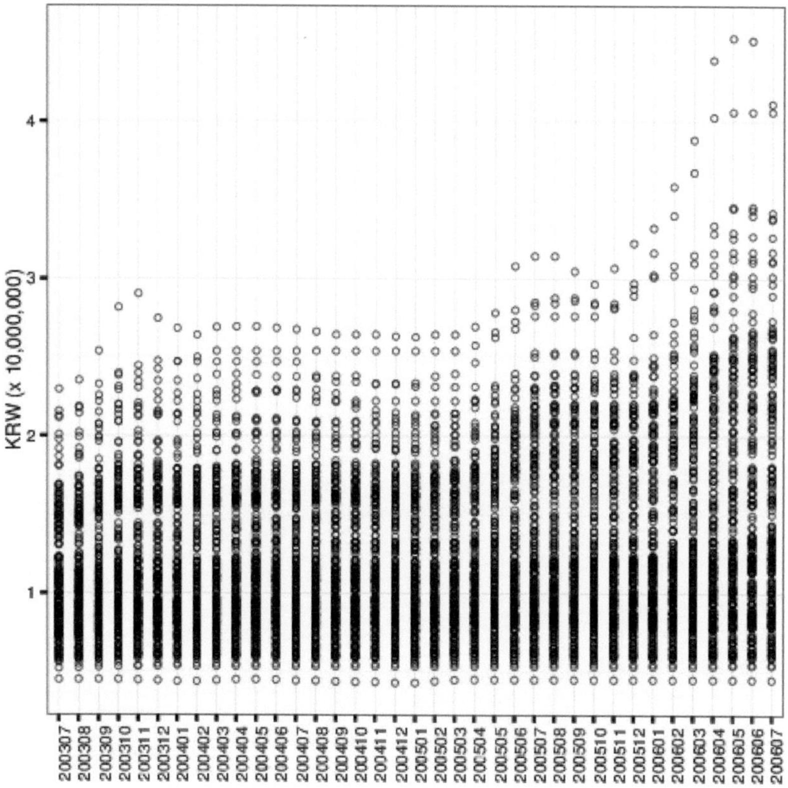


**Fig. 18.1** Apartment market price per 3.3 $m^2$ for a total of 37 months

two groups, high-priced apartments and moderate-priced apartments. The time series of density estimates show that in the period of inflation the prices of the high-priced apartments tend to rise faster than those of the moderate-priced apartments.

## 18.3 Spatial Product Partition Models

### *18.3.1 The Model*

Page and Quintana (2015) propose an interesting class of spatial models based on the PPM. They build on the covariate-dependent product partition models (PPMx) developed by Müller et al. (2011). To introduce covariate dependence in the prior product distribution, Müller et al. (2011) add an extra factor in the prior (18.4). Let $\boldsymbol{x}_i$ be a $p$-dimensional covariate vector of possibly mixed types for the $i$th subject, $x^n = (\boldsymbol{x}_1, \ldots, \boldsymbol{x}_n)$, and let $x_j^\star = \{\boldsymbol{x}_i : i \in S_j\}$ and similarly $y_j^\star = \{y_i : i \in S_j\}$ denote the covariates and responses arranged by cluster $j = 1, \ldots, k_n$. Let $g(x_j^\star)$ denote any

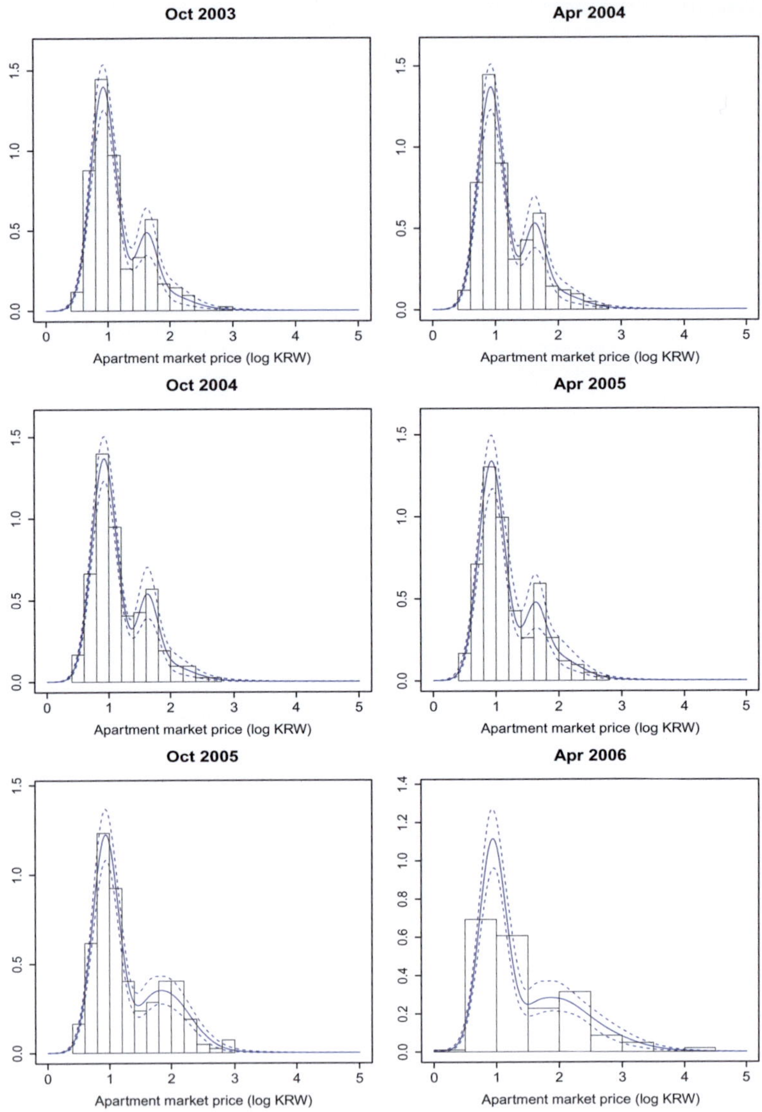


**Fig. 18.2** Results under the $MCS(G_0, a, b, \tau^2, \rho)$ model for the apartment market price data. The figure shows the posterior mean estimates of the location-specific densities (*blue solid*), 95 % credible intervals (*blue dashed*)

function of $x_j^\star$ that assigns high values for homogeneous elements in $x_j^\star$, and low values otherwise (this function will be referred to as the *similarity function*). The PPMx model changes (18.4) by

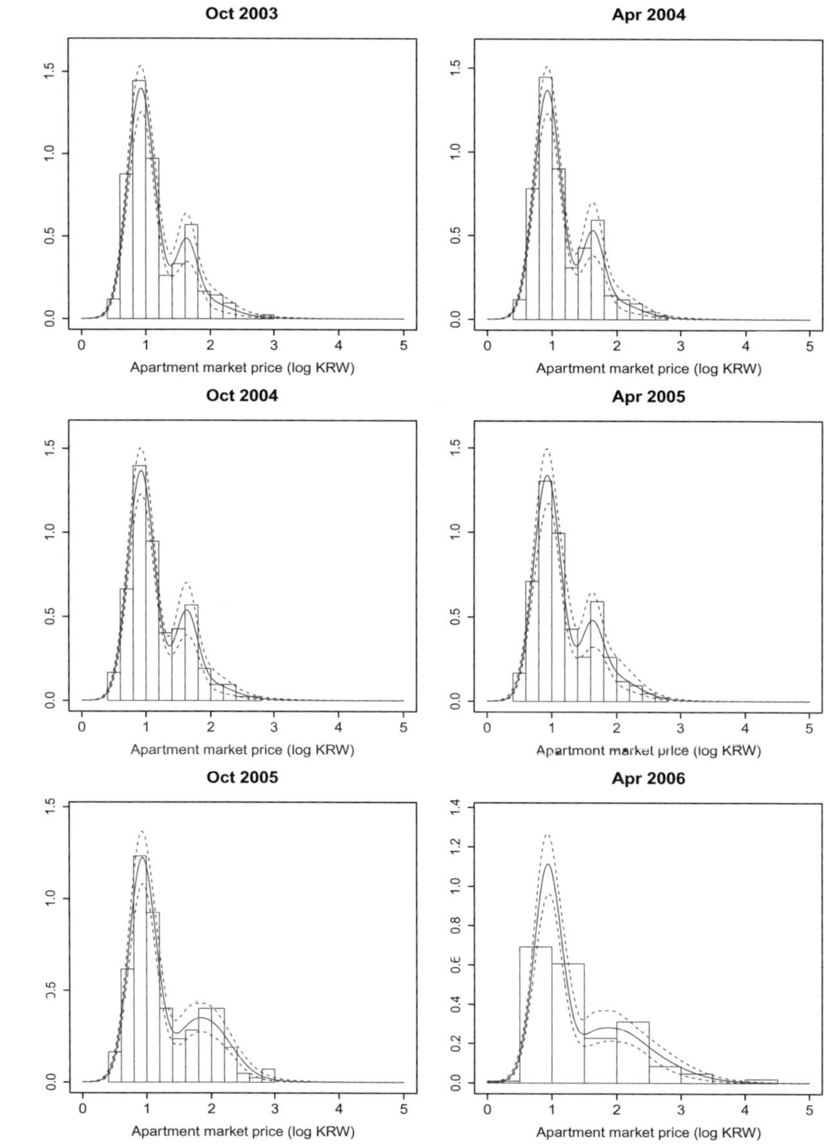


**Fig. 18.3** Results under the $CKCS(G_0,a,b,\tau^2,\rho,B)$ model. The figure shows the posterior mean estimates of the location-specific densities (*blue solid*), 95 % credible intervals (*blue dashed*)

$$P(\rho_n = (S_1,\ldots,S_{k_n})) \propto \prod_{j=1}^{k_n} c(S_j)g(x_j^\star), \tag{18.10}$$

that is, the cohesion of $S_j$ is replaced by $c(S_j)g(x_j^\star)$, which then, a priori, encourages clusters that group individuals with similar covariate values. This is seen to be useful for the purpose of predictive inference. See further details in Müller and Quintana (2010) and Müller et al. (2011), particularly, on default choices for the similarity function considering various covariate types. In particular, writing $\boldsymbol{x}_i = (x_{i1}, \ldots, x_{ip})$, they suggest to consider $g(x_j^\star) = \prod_{\ell=1}^{p} g_\ell(x_{j\ell}^\star)$, where $g_\ell(x_{j\ell}^\star)$ is a cohesion function defined solely in terms of $x_{j\ell}^\star = \{x_{i\ell} : i \in S_j\}$, i.e., the values of the $\ell$-th covariate for the $j$th cluster.

From a spatial modeling perspective, the covariate vector $x^n$ could be comprised of a set of spatial coordinates $s^n = (s_1, \ldots, s_n)$. Here, $s_i$ may be a two- or three-dimensional vector with a continuous range. To focus the discussion on spatial aspects, we restrict the attention to the case of PPMx models for which $x^n = s^n$. Likewise, write $s_j^\star = \{s_i : i \in S_j\}$. In what follows we assume that the $s_i$ vectors have been standardized to zero mean and unit standard deviation.

A simple method to define spatially oriented clusters within the PPMx framework is to adapt the proposal in Müller et al. (2011) and consider a similarity function of the form

$$g^1(s_j^\star) = \int \prod_{i \in S_j} N(s_i; 0, \Sigma)\, d\, IW(\Sigma; \nu, V), \tag{18.11}$$

where $N(s_i; 0, \Sigma)$ is the multivariate normal evaluated at $s_i$ with zero-mean and covariance matrix $\Sigma$, and $IW(\Sigma; \nu, V)$ is the inverse-Wishart density evaluated at $\Sigma$ with $\nu$ degrees of freedom and scale matrix $V$.

Invoking the PPMx to incorporate spatial information in modeling is certainly appealing as Müller et al. (2011) show that employing the similarity function found in (18.11) results in the PPMx retaining some nice properties (e.g., coherent across sample sizes). However, $g^1(\cdot)$ was not necessarily constructed with spatial structure in mind. From a spatial modeling perspective, it would be more natural to consider $\rho$'s influence on the correlation between two observations as a function of distance. Now if spatial structure exists among the realizations of a response variable measured at various locations, then the values measured at locations near each other would tend to be more similar than those that are far apart. However, this does not exclude the possibility that two locations far apart produce similar response values and clustering in the absence of spatial information would group these two locations together (as would be the case in a non-spatial PPM). As a result, the marginal correlation between observations far apart could possibly be stronger than that of observations near each other. Since this runs counter to correlation structures often desired in spatial modeling it would be appealing for $\rho$ to contain clusters that are in some sense "local." One way to encourage these types of clusters is to construct the cohesion/similarity functions so that with high probability a priori $\rho$ is made up of local clusters.

This idea is carried out by constructing an alternative similarity function that employs tessellation ideas found in Denison and Holmes (2001). Let $\bar{s}_j$ denote the centroid of cluster $S_j$ and $\mathscr{D}_j = \sum_{i \in S_j} d(s_i, \bar{s}_j)$ the sum of all distances from the centroid (typically euclidean norm $||\cdot||$ will be employed). Penalizing partitions with large $\mathscr{D}_j$ would certainly produce partitions with small local clusters, but would also encourage the creation of many singleton clusters. To counteract this, the regular

DP linked cohesion $c(S_j) = M \times \Gamma(|S_j|)$ is employed which favors a small number of large clusters. Now since $\Gamma(|S_j|)$ would overwhelm $\mathscr{D}_j$ as cluster membership grows, $\Gamma(\mathscr{D}_j)$ is considered instead. Finally, to provide more flexibility regarding penalization of distances, we introduce a tuning parameter ($\alpha$) resulting in the following cohesion function

$$g^2(s_j^\star) = \begin{cases} (\Gamma(\alpha\mathscr{D}_j)\mathbb{I}[\mathscr{D}_j \geq 1] + (\mathscr{D}_j)\mathbb{I}[\mathscr{D}_j < 1])^{-1} & \text{if } |S_j| > 1 \\ 1 & \text{if } |S_j| = 1. \end{cases} \tag{18.12}$$

where the partitioning of $\mathscr{D}_j$ was motivated by the fact that the gamma function is not monotone on $[0,1]$ and does not tend to zero as $\mathscr{D}_j$ tends to zero. Also, we set $g^2(s_j^\star) = 1$ for $|S_j| = 1$ to avoid issues associated with $\mathscr{D}_j = 0$. Notice that since all $s_1, \ldots, s_n$ are distinct $\mathscr{D}_j = 0 \iff |S_j| = 1$. See further details in Page and Quintana (2015).

## *18.3.2 Example*

To illustrate clustering and predictions available from both similarity functions, consider the scallops data found in Banerjee et al. (2015). In this data set the total scallop catch was measured at 148 locations in the New York/New Jersey Bight. Figure 18.4 displays the raw data with the circle circumference proportional to total catch amount. Letting $z(s_i)$ be the total scallop catch at location $i$, we follow suggestion of Banerjee et al. (2015) and model the log transformed total scallop catch $y(s_i) = \log(z(s_i) + 1)$.

In what follows we refer to a PPM that incorporates spatial information as a spatial PPM (sPPM). Now after introducing cluster labels $c_1, \ldots, c_n$, the hierarchical model we employ to model the scallops data is

$$\begin{aligned} y(s_i) \mid c_i, \theta^*_{c_i}(s_i), \sigma^2 &\overset{ind}{\sim} N(\theta^*_{c_i}(s_i), \sigma^2) \text{ and } \sigma \sim UN(0,10)) \\ \theta^*_j(s_i) &\overset{iid}{\sim} N(\theta_0, \sigma_0^2) \text{ for } j = 1, \ldots, k_n \text{ and } \theta_0 \sim N(0, 10^2),\ \sigma_0 \sim UN(0,10) \\ \{c_i\}_{i=1}^n &\sim sPPM. \end{aligned} \tag{18.13}$$

To provide a bit of context regarding model fit, we also fit the spatial stick breaking (SSB) process of (Reich and Fuentes 2007) to the data. This procedure is operationally similar to the sPPM. More precisely, given cluster labels $\{c_i\}_{i=1}^n$, the SSB can be expressed as the following hierarchical model

$$\begin{aligned} y(s_i) \mid c_i, \theta^*_{c_i}(s_i), \sigma^2 &\overset{ind}{\sim} N(\theta^*_{c_i}(s_i), \sigma^2) \text{ with } \sigma \sim UN(0,10) \\ \theta^*_j(s_i) &\overset{iid}{\sim} N(\theta_0, \sigma_0^2) \text{ for } h = 1, \ldots, k_n \text{ and } \theta_0 \sim N(0, 10^2),\ \sigma_0 \sim UN(0,10) \\ c_i &\sim Categorical(p_1(s_i), \ldots, p_m(s_i)), \end{aligned}$$

**Fig. 18.4** Locations at which scallop catch amounts of New York/New Jersey coast were recorded. Total catch is proportional to circle circumference

where $p_j(s) = w_j(s)V_j\prod_{k<j}[1 - w_k(s)V_k]$ for $V_j \overset{iid}{\sim} beta(1,M)$. The $w_j(s)$ are location weighted Gaussian kernels which introduce spatial dependence in the model.

Following suggestions found in Page and Quintana (2015), we set $M = 1$ for $g^1(\cdot)$ and $M = 0.0001$ for $g^2(\cdot)$. The tuning parameters associated with $g^1(\cdot)$ and $g^2(\cdot)$ are set to $\nu = 2$ and $V = \text{diag}(1,1)$ and $\alpha = 1$ respectively. For the SSB $M = 1$.

Three model fit metrics are used to assess model fit. The log pseudo marginal likelihood (LPML) which is a goodness-of-fit metric (see Christensen et al. 2011) that takes into account model complexity, $MSE = \frac{1}{n}\sum_{i=1}(y(s_i) - \hat{y}(s_i))^2$, and the Watanabe-Akaike information criterion (WAIC) which is a fairly new hierarchical model selection metric advocated in Gelman et al. (2014). Table 18.1 contains the results for the three procedures. It appears that the sPPM with $g^1(\cdot)$ fits the data better than the other two procedures. We also note that the sPPM with $g^2(\cdot)$ fits better than the SSB.

**Table 18.1** Model fit comparisons between sPPM and SSB models for the scallops data

| Procedure | LPML | MSE | WAIC |
|---|---|---|---|
| sPPM $g^1$ | −138.66 | 0.09 | 196.98 |
| sPPM $g^2$ | −140.01 | 0.06 | 182.31 |
| SSB | −167.37 | 0.11 | 243.22 |

The average number of clusters *a posteriori* for the two procedures were very similar (10.9 for $g^1(\cdot)$ and 10.1 for $g^2(\cdot)$). Using the least squares method of Dahl (2006) a point estimate of $\rho$ for each similarity function was obtained and is provided in Fig. 18.5. Both partitions appear to contain clusters that are visually "spatially pleasing" as all cluster members are located in the same general area. In Fig. 18.5 predictive maps associated with a regular grid of locations that belonged to the convex hull created by the observed locations are provided. These two maps are somewhat similar but with the "hot spot" of larger total catch associated with $g^1(\cdot)$ stretching further southwest compared to that of $g^2(\cdot)$.

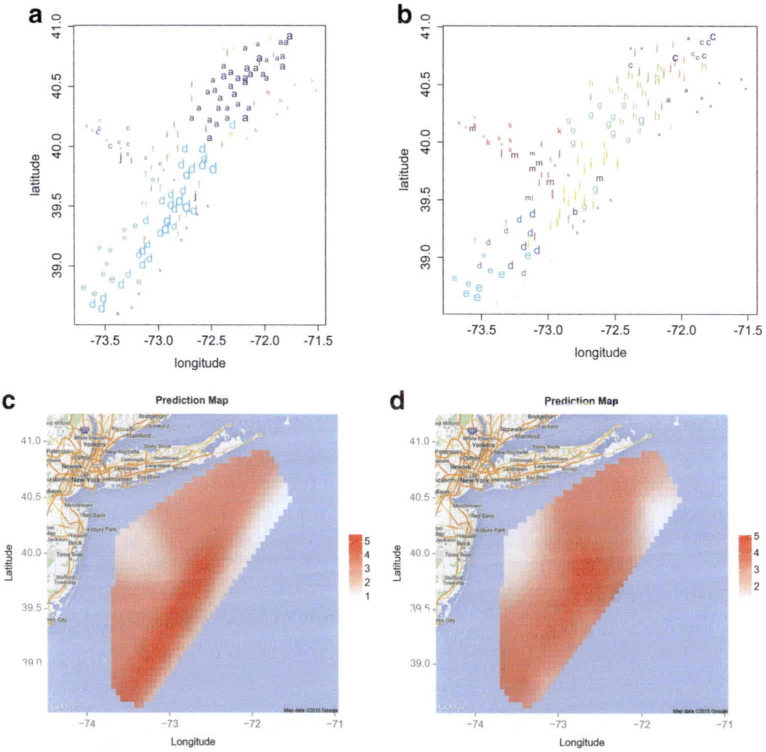


**Fig. 18.5** Estimated $\rho$ and predictive maps produced by the sPPM based on $g^1(\cdot)$ and $g^2(\cdot)$ as applied to the scallops data. (**a**) Estimated partition associated with $g^1$, (**b**) Estimated partition associated with $g^2$, (**c**) Predictive map associated with $g^1$, (**d**) Predictive map associated with $g^2$

## 18.4 Conclusion

Inference for spatial data gives rise to some interesting challenges for nonparametric Bayesian inference. We reviewed some of the recent literature and discussed in some more detail two models based on SSM and on PPMx, respectively.

**Acknowledgements** Peter Müller's research was partially supported by NIH R01 CA132897. Fernando A. Quintana's research was partially funded by grant FONDECYT 1141057. Garritt L. Page's research was partially funded by grant FONDECYT 11121131. Jaeyong Lee was partially supported by Advanced Research Center Program (S/ERC), a National Research Foundation of Korea grant funded by the Korea government (MSIP) (2011-0030811). Seongil Jo's research was supported by Basic Science Research Program through the National Research Foundation of Korea (NRF) funded by the Ministry of Education (NRF-2014R1A6A3A01059555).

## References

Albert, J. H. and Chib, S. (1993). Bayesian analysis of binary and polychotomous response data. *J. Amer. Statist. Assoc.*, **88**(422), 699–679.

Banerjee, S., Carlin, B. P., and Gelfand, A. E. (2015). *Hierarchical modeling and analysis for spatial data*. Chapman & Hall/CRC, Boca Raton, Florida, 2 edition.

Barry, D. and Hartigan, J. A. (1992). Product partition models for change point problems. *The Annals of Statistics*, **20**, 260–279.

Besag, J. (1974). Spatial interaction and the statistical analysis of lattice systems (with discussion). *J. Roy. Statist. Soc. Ser. B*, **36**(2), 192–236.

Christensen, R., Johnson, W., Branscum, A. J., and Hanson, T. (2011). *Bayesian Ideas and Data Analysis: An Introduction for Scientists and Statisticians*. CRC Press.

Clayton, D. and Kaldor, J. (1987). Empirical Bayesian estimates of age-standardized relative risks for use in disease mapping. *Biometrics*, **43**, 671–681.

Cressie, N. A. C. (1993). *Statistics for spatial data*. John Wiley & Sons Inc., New York.

Dahl, D. B. (2006). Model-based clustering for expression data via a dirichlet process mixture model. In M. Vannucci, K. A. Do, and P. Müller, editors, *Bayesian Inference for Gene Expression and Proteomics*, pages 201–218. Cambridge University Press.

Denison, D. G. T. and Holmes, C. C. (2001). Bayesian partitioning for estimating disease risk. *Biometrics*, **57**, 143–149.

Duan, J. A., Guindani, M., and Gelfand, A. E. (2007). Generalized spatial Dirichlet process models. *Biometrika*, **94**, 809–825.

Ferguson, T. S. (1973). A Bayesian analysis of some nonparametric problems. *Ann. Statist.*, **1**(2), 209–230.

Gelfand, A. E., Kottas, A., and MacEachern, S. N. (2005). Bayesian nonparametric spatial modeling with Dirichlet process mixing. *Journal of the American Statistical Association*, **100**, 1021–1035.

Gelman, A., Hwang, J., and Vehtari, A. (2014). Understanding predictive information criteria for Bayesian models. *Statistics and Computing*, **24**(6), 997–1016.

Hartigan, J. A. (1990). Partition models. *Communications in Statistics, Part A – Theory and Methods*, **19**, 2745–2756.

Hegarty, A. and Barry, D. (2008). Bayesian disease mapping using product partition models. *Statistics in Medicine*, **27**(19), 3868–3893.

Ishwaran, H. and James, L. F. (2001a). Gibbs sampling methods for stick-breaking priors. *Journal of the American Statistical Association*, **96**, 161–173.

Ishwaran, H. and James, L. F. (2001b). Gibbs sampling methods for stick-breaking priors. *J. Amer. Statist. Assoc.*, **96**(453), 161–173.

Jo, S., Lee, J., Müller, P., Quintana, F. A., and Trippa, L. (2015). Dependent species sampling models for spatial density estimation. Technical report, Department of Statistics, Seoul National University.

Lee, J., Quintana, F., Müller, P., and Trippa, L. (2013). Defining predictive probability functions for species sampling models. *Stat. Sci.*, **28**(2), 209–222.

Müller, P. and Quintana, F. (2010). Random partition models with regression on covariates. *J. Statist. Plann. Inference*, **140**(10).

Müller, P., Quintana, F., and Rosner, G. L. (2011). A product partition model with regression on covariates. *J. Comput. Graph. Statist.*, **20**(1), 260–278. Supplementary material available online.

Page, G. L. and Quintana, F. A. (2015). Spatial product partition models. Technical report, Pontificia Universidad Católica de Chile.

Pitman, J. and Yor, M. (1997). The two-parameter Poisson-Dirichlet distribution derived from a stable subordinator. *Ann. Probab.*, **25**(2), 855–900.

Quintana, F. A. and Iglesias, P. L. (2003). Bayesian Clustering and Product Partition Models. *Journal of the Royal Statistical Society Series B*, **65**, 557–574.

Reich, B. J. and Fuentes, M. (2007). A multivariate semiparametric Bayesian spatial modeling framework for hurricane surface wind fields. *The Annals of Applied Statistics*, **1**(1), 249–264.

Rue, H. and Held, L. (2005). *Gaussian Markov random fields: Theory and applications*. Monographs on Statistics and Applied Probability, 104. Chapman & Hall, Boca Raton.

Scott, S. L. (2011). Data augmentation, frequentist estimation, and the Bayesian analysis of multinomial logit models. *Statist. Papers*, **52**(1), 639–650.

Sethuraman, J. (1994). A constructive definition of Dirichlet priors. *Statist. Sinica*, **4**(2), 639–650.

Sun, D., Tsutakawa, R. K., and Speckman, P. L. (1999). Posterior distribution of hierarchical models using CAR(1) distributions. *Biometrika*, **86**(2), 341–350.

# Chapter 19
# Spatial Boundary Detection for Areal Counts

Timothy Hanson, Sudipto Banerjee, Pei Li, and Alexander McBean

**Abstract** Two Bayesian nonparametric model-based approaches to areal boundary detection for count outcomes are proposed, compared, and illustrated. The linear predictor in a standard Poisson regression is augmented with random effects stemming from modified stick-breaking representations of the Dirichlet process. The modifications induce spatial correlation among counts from differing counties such that closer counties are more highly correlated. The discrete nature of the Dirichlet process is an advantage in this setting as two counties can have the same random effect, implying no boundary, with positive probability. The methods are compared on counts of patients hospitalized due to pneumonia and influenza in Minnesota.

T. Hanson (✉)
Department of Statistics, University of South Carolina, 216 LeConte College,
1523 Greene St., Columbia, SC 29208, USA
e-mail: hansont@stat.sc.edu

S. Banerjee
Department of Biostatistics, U.C.L.A. School of Public Health, 650 Charles E. Young Dr. South,
Los Angeles, CA 90095, USA
e-mail: sudipto@ucla.edu

P. Li
Medtronic Incorporated, 710 Medtronic Pkwy. N.E., Minneapolis, MN 55432, USA
e-mail: pei.li@medtronic.com

A. McBean
Division of Biostatistics, University of Minnesota, 420 Delaware Street S.E.,
Minneapolis, MN 55455, USA

 
R. Mitra, P. Müller (eds.), *Nonparametric Bayesian Inference in Biostatistics*,
Frontiers in Probability and the Statistical Sciences,
DOI 10.1007/978-3-319-19518-6_19

## 19.1 Introduction

With routine accessibility to geographical information systems (GIS) and large patient record databases, researchers and administrators in public health are increasingly encountering datasets that are aggregated as case counts or rates over *areal* units or regions (e.g., counties, census-tracts or ZIP codes). Aggregated counts are reported to protect patient privacy, as fully georeferenced (e.g., latitude and longitude of an individual's address) outcomes can lead to patient identification, especially in the case of rare diseases.

Statistical models for areal data accommodate sparsely sampled regions by smoothing across and borrowing information from spatial neighbors (see, e.g., Anselin 1988; Le Sage and Pace 2009; Banerjee et al. 2015). Subsequent inferential interest often resides in the formal identification of "barriers" or "boundaries" on the spatial surface or map. The 'boundary' here is a geographical unit border with sharply discrepant outcomes on either side. In particular, interest lies in quantifying statistically significant differences among neighboring regions, identifying the spatial barriers or *difference boundaries* that delineate them. Ultimately, the underlying influences responsible for these boundaries or barriers are typically of scientific and administrative interest. This 'boundary' detection problem is often referred to as "wombling," after a foundational article by Womble (1951). While statistical boundary analysis has been applied extensively to point-referenced and gridded (or lattice) data (e.g. Banerjee and Gelfand 2006), formal statistical inference in areal contexts present unique challenges that we address in this chapter.

Deterministic areal wombling is often carried out using algorithms (e.g. Jacquez and Greiling 2003a,b) that are fast and easy to implement; however these approaches fail to account for sources of uncertainty such as extremes in counts corresponding to thinly populated regions. Such exceptions naturally arise in observed data, and are not due to systematic differences; such variability should be accounted for in modeling. Li et al. (2011) proposed statistical learning for boundaries using the Bayesian information criterion (BIC). In hierarchical model based approaches, Lu and Carlin (2005), Lu et al. (2007), Ma et al. (2010), and Fitzpatrick et al. (2010) investigated estimation of an areal adjacency matrix (defined as $W$ in Sect. 19.2) within a hierarchical framework using priors on the edges. However, inference from these models are usually highly sensitive to prior specifications on certain parameters.

In this chapter we discuss Bayesian hierarchical models that overcome the aforementioned problems, providing inference for areally aggregated health outcome data, including assessment of difference boundaries, using classes of flexible nonparametric Bayesian hierarchical models. Section 19.2 offers a brief review of models for areally referenced count data. Section 19.3 elucidates key issues in areal boundary analysis and reviews the Bayesian nonparametric modeling approaches first presented in Li et al. (2015). Sections 19.4 and 19.5 discuss, respectively, a simulation study and the analysis of a Minnesota Pneumonia & Influenza (P & I) dataset to detect spatial health barriers between neighboring counties in Minnesota. Finally, Sect. 19.6 concludes the chapter.

## 19.2 Hierarchical Models for Areal Data

Let $Y_i$ be the observed random number of patients who experienced a particular clinical outcome in areal unit $i$, $i = 1, \ldots, n$, and let $E_i$ be the fixed expected number of outcomes for that unit. The $E_i$ are typically the expected number of outcomes in the region assuming the outcome is equally likely across space. For rare outcomes, a Poisson approximation to the binomial sampling distribution of outcome counts yields

$$Y_i \overset{ind}{\sim} Poisson(E_i e^{\mu_i}) \ , \ i = 1, \ldots, n, \tag{19.1}$$

where $\mu_i = \mathbf{x}_i'\beta + \phi_i$ represents the log-relative risk, $\mathbf{x}_i$ includes explanatory covariates for region $i$ and $\beta$ are the corresponding regression coefficients.

The $\phi_i$ represent the *spatial random effect* associated with region $i$; they are often modeled using *Markov random fields* (MRF), e.g., Cressie (1993) and Banerjee et al. (2015, Chapter 3) that imply a joint distribution for $\phi = (\phi_1, \phi_2, \ldots, \phi_n)'$:

$$\phi \ \sim N_n\left(0\,,\, \sigma^2 (D - \rho W)^{-1}\right) \ , \tag{19.2}$$

where $N_n$ denotes the $n$-dimensional normal distribution, $D$ is a $n \times n$ diagonal matrix with diagonal elements $m_i$ equal to the number of neighbors of area $i$, and $W = \{w_{ij}\}$ is the adjacency matrix with $w_{ii} = 0$, and $w_{ij} = 1$ if $i$ is adjacent to $j$ and 0 otherwise. In the joint distribution (19.2), $\sigma^2$ is the spatial dispersion parameter, and $\rho$ is a spatial autocorrelation parameter. We term this distribution as conditionally autoregressive with parameters $\rho$ and $\sigma^2$, denoted $CAR(\rho, \sigma^2)$ for short. A sufficient condition for $D - \rho W$ to be positive definite is that $\rho \in (1/\lambda_{(1)}, 1)$, where $\lambda_{(1)}$ is the minimum eigenvalue of $W$ (Banerjee et al. 2015). Note that $\lambda_{(1)} < 0$.

The CAR model has been especially popular in Bayesian inference as its conditional specification is convenient for Gibbs sampling and MCMC schemes. The distribution in (19.2) reduces to the intrinsic conditionally autoregressive (ICAR) prior if $\rho = 1$, or an independence model if $\rho = 0$. The ICAR model induces maximal "local" smoothing by borrowing strength from the neighbors, while the independence model assumes independence of spatial rates and induces "global" smoothing. The smoothing parameter $\rho$ in the CAR prior (19.2) controls the strength of spatial dependence among regions, though it is well-appreciated that a fairly large $\rho$ may be required to deliver significant spatial correlation (Wall 2004).

## 19.3 Bayesian Nonparametric Models for Areal Data

### *19.3.1 Modeling Considerations for Areal Boundary Analysis*

Areal boundary analysis can be approached from different perspectives. For example, Li et al. (2011) treat the problem as one of statistical learning for the edges, where each model represents a different *boundary hypothesis*. Emphasizing speed of execution and ease of use, they consider a leave-one-edge-out mechanism, where each model has exactly one geographical boundary omitted from the adjacency matrix. There are as many models as there are edges and the BIC is used to arrive at a ranking of the boundaries and detect difference boundaries. This fails to account for the joint effects of the edges and what impact deleting one may have on the other. More generally, one can consider models varying in their specification of the neighborhood matrix $W$ that controls spatial smoothing. One extreme case is that the whole map is one big cluster without any difference boundaries. At the other extreme all the geographical edges may in fact be difference boundaries. Any intermediate model that lies between these extremes is completely specified by modifying the original map to delete some edges. Ideally we would like to consider a class of models $\mathscr{M} = \{M_1, \ldots, M_K\}$ representing all possible models or all possible maps derived from $W$ by deleting combinations of geographical edges. In other words, $M_k$ denotes a model with the adjacency matrix $W_k$ that has been derived by changing some of the 1's to 0's in $W$. This amounts to dropping some edges from the original map or, equivalently, combining two regions into one. However, now we encounter an explosion in the number of models: $2^{\mathbf{1}'W\mathbf{1}/2}$ models where $\mathbf{1}' = (1, 1, \ldots, 1)$. This requires sophisticated MCMC model composition, $\text{MC}^3$ algorithms, or other types of stochastic variable selection algorithms for selecting models (see, e.g., Hoeting et al. 1999); these methods are computationally intensive in relatively large maps. Li et al. (2012) reformulate the problem as one of Bayesian hypothesis testing within a class of spatial moving average models adjusting for multiple tests using false discovery rates (FDR). The method, though still computationally intensive, is competitive and provides a benchmark for our simulation studies. Another approach seeks to estimate the adjacency matrix within a hierarchical framework using priors on the adjacency relationships, see Ma et al. (2010). These involve incorporating "edge effects," i.e., random effects corresponding to the edges, in addition to regional effects. These edge effects would be modeled by another CAR model, or some other MRF, leading to rather complex site-edge models (Ma et al. 2010). However, these models often involve weakly identifiable parameters that are difficult to tune causing the MCMC algorithms to be substantially slower in converging to the desired posterior distributions.

Instead of incorporating random "edge effects," we explore an alternative stochastic mechanism that allows us detect difference boundaries by considering probabilities such as $P(\phi_i = \phi_j \,|\, i \sim j)$. Clearly, continuous priors for the $\phi_i$'s do not work as they render $P(\phi_i = \phi_j \,|\, i \sim j) = 0$. Therefore, we seek to model the spatial effects in an almost surely discrete fashion, while at the same time accounting for the spatial dependence. The Dirichlet process (Ferguson 1973), and more generally stick-breaking priors, are a natural Bayesian nonparametric approach to this problem. In Sect. 19.3.3 and 19.3.4 modifications of the Dirichlet process to accommodate areal dependence are explored.

### 19.3.2 Dirichlet Process Mixture Models for Clustered Data

The starting point for both spatial models reviewed below is the Dirichlet process mixture (DPM) model, a very popular choice for analyzing clustered data. This model encourages clustering of observations without borrowing information from geographical neighbors. In the context of (19.1), a DPM prior specifies $\phi_i \stackrel{iid}{\sim} G$, where $G \sim DP(\alpha, G_0)$ is a random distribution modeled as a Dirichlet process (DP) with baseline measure $G_0$ (Ferguson 1973). The data hierarchically follow

$$Y_i|\beta,\phi_i \sim Poisson\left(E_i e^{\mathbf{x}_i'\beta+\phi_i}\right), \;\; \phi_i|G \stackrel{iid}{\sim} G, \;\; G \sim DP(\alpha, G_0), \tag{19.3}$$

where $DP(\alpha, G_0)$ denotes the Dirichlet process prior: a random probability distribution over a measurable space $(\Omega, \mathscr{B})$ such that the random vector $(G(A_1), \ldots, G(A_k))$ is distributed as a finite-dimensional Dirichlet distribution with parameters $(\alpha G_0(A_1), \ldots, \alpha G_0(A_k))$ for any finite measurable partition $(A_1, A_2, \ldots, A_k)$ of $\Omega$. Here $\alpha > 0$ is a real-valued parameter and $G_0$ is probability distribution called the *base measure*. The Dirichlet distribution is a family of continuous multivariate probability distributions parameterized by a vector of positive real numbers. It is the multivariate generalization of the beta distribution and forms a conjugate prior for the multinomial distribution in Bayesian statistics. Further technical details on DPs are available in Ferguson (1973) from a formal probabilistic perspective. More explicitly, we write

$$P(G(A_1), \ldots, G(A_k)\,|\,\alpha, G_0) = \frac{\Gamma(\alpha)}{\prod_{i=1}^k \Gamma(\alpha G_0(A_i))} \times \prod_{i=1}^k \{G(A_i)\}^{\alpha G_0(A_i)-1},$$

For any measurable set $A$, $E\{G(A)\} = G_0(A)$, $var\{G(A)\} = G_0(A)G_0(A^C)/(\alpha+1)$, and so the parameter $\alpha$ represents the degree of concentration of the DP. The posterior distribution for $(G(A_1), \ldots, G(A_k))$ given $\phi = (\phi_1, \ldots, \phi_n)$ is again a Dirichlet distribution:

$$G(A_1), \ldots, G(A_k)\,|\,\alpha, \phi \sim Dir(\alpha^*, \{G_0^*(A_j)\}_{j=1}^k), \tag{19.4}$$

where $n_i = \sum_{j=1}^n 1(\phi_j \in A_i)$ , $\alpha^* = \alpha + n$ and $G_0^*(A_i) = \frac{n_i + \alpha G_0(A_i)}{\alpha+n}$. Making the partitions $(A_1, \ldots, A_k)$ infinitely finer, the infinite measure $G|\phi$ is seen to be a DP in the form $DP\left(\alpha+n, \frac{\sum_{i=1}^n \delta_{\phi_i} + \alpha G_0}{\alpha+n}\right)$, where $\delta_{\phi_i}$ is Dirac point mass at $\phi_i$.

When the $\phi_i$'s are modeled as in (19.3), then marginalizing over the measure $G$ yields the predictive distribution of $\phi_{n+1}$, given $\{\phi_i\}_{i=1}^n$ as

$$\phi_{n+1}\,|\,\{\phi_i\}_{i=1}^n, \alpha, G_0(\cdot) \sim \frac{\alpha G_0(\cdot) + \sum_{i=1}^n \delta_{\phi_i}(\cdot)}{\alpha+n}. \tag{19.5}$$

Thus the number of distinct values of $\phi_j$'s is controlled by the precision parameter $\alpha$. For $i \geq 1$, the observation $\phi_i$ takes on a new value with probability $\alpha/(\alpha+i-1)$, thus the expected number of distinct values of $\phi$ is $\sum_{i=1}^{n} \frac{\alpha}{\alpha+i-1}$. The conditional distributions in (19.5) can be used to determine the joint distribution of $\phi_i$'s conditional upon $G_0$ and $\alpha$, i.e., after integrating out $G$. Blackwell and MacQueen (1973) related the Dirichlet process to a generalized *Polya urn scheme* that leads to effective MCMC sampling strategies (Escobar and West 1995).

The stick-breaking representation of the DP (Sethuraman 1994) says that a draw from the Dirichlet process can be written as $G(\cdot) = \sum_{k=1}^{\infty} p_k \delta_{\theta_k}(\cdot)$ $a.s.$, where $\theta_k \overset{iid}{\sim} G_0$, $p_k = V_k \prod_{j<k}(1-V_j)$, and $V_k \overset{iid}{\sim} Beta(1,\alpha)$. The $p_k$'s are called the "stick-breaking" weights (their infinite sum equals 1) and the $\theta_k$'s are called atoms. By Sethuraman's representation it is immediate that $G$ is a.s. discrete, implying $P(\phi_i = \phi_j) > 0$ for a random sample $\phi_1, \ldots, \phi_n | G \overset{iid}{\sim} G$.

In practice, the infinite sum is often replaced by the sum of the first $K$ $(K \leqslant n)$ terms, since the probability masses $p_1, p_2, \ldots$ decay rapidly. We can simply let $V_K = 1$ to truncate the sum (Ishwaran and Zarepour 2000) yielding a finite approximation. Many authors simply choose $K$ to be a number large enough that there exists some empty components during the MCMC run or by examining the size of the last weight $p_K$ under the prior. Following Reich and Fuentes (2007), we choose $K$ according to the latter. For concerns regarding truncation bias, exact sampling can be executed using slice-sampling (Kalli et al. 2011). Note however, for areal data, $K$ is naturally bounded above by the number of areal units and so the infinite representation is not necessary or even appropriate.

The stick-breaking representation is an extremely rich framework that includes the DP but allows for immediate generalizations; in particular the introduction of dependence is straightforward and intuitive. The dependent Dirichlet process (DDP, MacEachern 2001) introduces dependence through either the stick-breaking weights, atoms, or both. De Iorio et al. (2004) induced dependence across related random distributions by consideration of ANOVA-type models for the atoms; De Iorio et al. (2009) and Hanson and Jara (2013) extended this idea to atoms that are regressions on covariates for the analysis of survival data. Chung and Dunson (2009) consider weights that are regressions on covariates.

In terms of spatially varying Dirichlet processes, Gelfand (2005) developed a DDP for point-referenced spatial data through an underlying Gaussian process base measure $G_0$ yielding a probability distribution defined on a space of surfaces that yields almost surely discrete realizations with countable support; Duan et al. (2007) extended this model by allowing different surface selection at different sites. Griffin and Steel (2006) proposed an order-based DDP which induced dependence in the stick-breaking weights across predictors (including space) via permutations. Reich and Fuentes (2007) developed a spatial stick-breaking prior to analyze hurricane surface wind fields where weights are spatially correlated. Petrone et al. (2009) and Rodríguez et al. (2010) develop spatial DPs where the weights follow a copula representation. Zhou et al. (2015) consider a spatial model where marginal distributions follow the survival DDP of De Iorio et al. (2009), but a copula induces dependence for georeferenced data. The local Dirichlet process (Chung and Dunson 2011), developed to accommodate predictor-dependent weights in a DDP with identical margins, offers an approach to the localized spatial "sharing" of atoms that could conceivably be extended to the areal setting through a suitable definition of what a neighborhood is at each areal location; also see Theorem 4 in Dunson et al. (2007) for a related idea. However, as presented, the aforementioned spatial DP approaches do not directly apply to areal data; we now consider such modifications.

The spatial models for areal data reviewed next employ the DP for two fundamental reasons: (1) the DP naturally allows for clustering among regions in that $P(\phi_i = \phi_j) > 0$, and (2) the DP provides a rich, robust model for the spatial effects $\phi$. The models we propose below correspond to a subclass of stick-breaking process priors that includes the DP as a special case. In particular, we construct an areally referenced stick-breaking process (ARSB) and an areally referenced Dirichlet process (ARDP) for areal data allowing natural formal boundary analysis. These models, described next, are easily adapted to multivariate settings using multivariate SAR or CAR models (e.g. Banerjee et al. 2015).

### 19.3.3 Areally Referenced Spatial Stick-Breaking Prior

We adapt the point-referenced spatial stick-breaking approach of Reich and Fuentes (2007) to areal data by incorporating spatial dependence in the DP by introducing additional weights that borrow strength across the neighbors using CAR priors.

The spatially varying random effects $\phi_i$ are each assigned a stick-breaking prior $G^{(i)}$ whose weights $(p_{i1},\ldots,p_{iK})$ are given by $p_{i1} = w_{i1}V_1, p_{ik} = w_{ik}V_k\prod_{j<k}(1-w_{ij}V_j)$, $i = 1,\ldots,n$, $k = 1,2,\ldots,K$; the weights depend not only on the usual stick-breaking $(V_1,\ldots,V_K)$, but also on "location" weight parameters $(w_{i1},\ldots,w_{iK})$. Since the CAR marginals have support over the entire real line, we introduce a transformation $\text{logit}(w_{ik}) = z_{ik}$ and take the $(z_{1k},\ldots,z_{nk})$ to be distributed as CAR yielding a MRF on the location weights and encouraging smoothing across neighbors. Of course, any other link mapping the unit interval to the real line could be used. For example Cai et al. (2013) generalize the logit link in the ARSB model for Poisson data of Li et al. (2015) to allow for spatially varying regression coefficients, and apply their model to the spatial assessment of low birth-weight across South Carolina counties. Larger values of $\rho$ induce greater smoothing and setting $\rho = 1$ gives the ICAR prior. This prior is improper as $D-W$ is singular, but for a map without islands this issue can be resolved by imposing the additional constraint $\sum_{i=1}^{n} z_{ik} = 0$.

The ARSB model truncated to $K$ terms with Poisson outcomes is

$$
\begin{aligned}
Y_i &\sim Poisson(E_i e^{\mu_i}),\ \mu_i = \mathbf{x}_i'\beta + \phi_i,\ \phi_i \sim G^{(i)}, \\
G^{(i)}(\cdot) &= \sum_{k=1}^{K} p_{ik}\delta_{\theta_k}(\cdot),\ \theta_k \overset{iid}{\sim} N(0,\sigma_s^2) \\
p_{ik} &= w_{ik}V_k\prod_{j<k}(1-w_{ij}V_j), \\
V_k &\overset{iid}{\sim} Beta(1,\alpha),\ \{\log\tfrac{w_{ik}}{1-w_{ik}}\} \sim CAR(\rho,\sigma_k^2),
\end{aligned}
\tag{19.6}
$$

for $i=1,\ldots,n$ and $k=1,\ldots,K$; define $p_{i1}=w_{i1}V_1$. Recall that the $\alpha$ parameter stochastically controls the number of distinct elements of $\phi$ among the $n$ observations. The covariance between $\phi_i$ and $\phi_j$ can is derived as follows.

$$\begin{aligned} Cov(\phi_i,\phi_j) &= E_{\mathbf{p}}[Cov(\phi_i,\phi_j)\,|\,\mathbf{p}] + Cov(E[\phi_i\,|\,\mathbf{p}],E[\phi_j\,|\,\mathbf{p}]) \\ &= \sigma_s^2 E_{\mathbf{p}}\left[\Sigma_{k=1}^{\infty} w_{ik}V_k w_{jk}V_k \prod_{l<k}(1-(w_{il}+w_{jl})V_l + w_{il}w_{jl}V_l^2)\right] \end{aligned}$$

Letting $c_i = E[w_{is}]$ and $c_{ij} = E[w_{is}w_{js}]$, the marginal covariance between $\phi_i$ and $\phi_j$ is

$$\begin{aligned} &\sigma_s^2 E_{\mathbf{p}}\left\{\Sigma_{k=1}^{\infty} w_{ik}V_k w_{jk}V_k^2 \prod_{l<k}(1-(w_{il}+w_{jl})V_l + w_{il}w_{jl}V_l^2)\right\} \\ &= \sigma_s^2 \Sigma_{k=1}^{\infty} c_{ij}EV^2 \prod_{l<k}(1-(c_i+c_j)EV + c_{ij}EV^2) \\ &= \sigma_s^2 \Sigma_{k=1}^{\infty} c_{ij}E[V^2](1-(c_i+c_j)E[V] + c_{ij}E[V^2])^{k-1} \\ &= \sigma_s^2 c_{ij}E[V^2]\frac{1}{(c_i+c_j)E[V] - c_{ij}E[V^2]}. \end{aligned}$$

$EV$ and $EV^2$ are just functions of $\alpha$ according to $V_k \stackrel{iid}{\sim} Beta(1,\alpha)$.

### *19.3.4 Areally Referenced Dirichlet Process*

The ARSB model incorporates dependence between the discrete distributions on different regions but does not yield identical marginal distributions for the $\phi_i$. The ARDP, described next, maintains the marginal distribution of each spatial random effect $\phi_i$ to be a regular univariate DP, incorporating the spatial dependence between these DPs via a copula representation for the weights.

Consider spatial random effects $\phi_i$, $i=1,\ldots,n$, arising marginally from an identical random measure $G$, where $G \sim DP(\alpha, G_0)$. We introduce spatial dependence between these DPs by constructing dependent uniform $(0,1)$ random variables. Suppose $\gamma_1,\ldots,\gamma_n$ are jointly distributed as a $CAR(\rho,\sigma_\gamma)$, and $F^{(1)}(\cdot),\ldots,F^{(n)}(\cdot)$ denote the cumulative distribution functions of the marginal distributions of each component of the CAR random vector. Marginally, each $F^{(i)}(\gamma_i)$ is uniform $(0,1)$ but they are *dependent* through $\gamma_1,\ldots,\gamma_n$.

We formulate our hierarchical ARDP model as follows.

$$
\begin{aligned}
Y_i &\sim Poisson(E_i e^{\mu_i}),\ \mu_i = \mathbf{x}_i'\beta + \phi_i,\\
\phi &= (\phi_1,\ldots,\phi_n)' \sim G_n,\ G_n = \sum_{u_1,\ldots,u_n} \pi_{u_1,\ldots,u_n}\delta_{\theta_{u_1}}\ldots\delta_{\theta_{u_n}},\\
\pi_{u_1,\ldots,u_n} &= P\left(\sum_{k=1}^{u_1-1} p_k < F^{(1)}(\gamma_1) < \sum_{k=1}^{u_1} p_k,\ldots,\sum_{k=1}^{u_n-1} p_k < F^{(n)}(\gamma_n) < \sum_{k=1}^{u_n} p_k\right),\\
\theta_k &\overset{iid}{\sim} N(0,\sigma_s^2),\ \boldsymbol{\gamma} = (\gamma_1,\ldots,\gamma_n)' \sim N_n(\mathbf{0},\sigma_\gamma^2(D-\rho W)^{-1})\\
p_k &= V_k \prod_{j<k}(1-V_j),\ V_k \overset{iid}{\sim} \text{Beta}(1,\alpha).
\end{aligned}
\tag{19.7}
$$

where $p_1 = V_1$, $k = 1,2,\ldots,K$, and $\sigma_\gamma^2(D-\rho W)^{-1}$ is the covariance matrix of a proper CAR distribution ($\rho < 1$). Using the cumulative distribution function of the $\gamma_i$'s to model the weights is an adaptation of the hybrid Dirichlet process (Petrone et al. 2009) and the latent stick-breaking process (Rodríguez et al. 2010), where point-referenced continuous spatial copulas are used to model weight dependency; the latter uses ordered atoms. In contrast, we model dependent areal counts using copula-based MRF's. Bayesian nonparametric copula-based approaches that model dependence on the observables directly (vs. latent surfaces or random effects) are given by Li et al. (2015) for areal data and Zhou et al. (2015) for georeferenced data.

Like the hybrid Dirichlet process and the latent stick-breaking process, the marginal distribution of $G^{(i)}(\phi_i)$, for each $i$, follows an identical DP

$$
G^{(i)}(\phi_i) - \sum_{k=1}^{K} \sum_{u_1,\ldots,u_i=k,\ldots,u_n} \pi_{u_1,\ldots,u_i=k,\ldots,u_n}\delta_{\theta_{u_1}}\ldots\delta_{\theta_{u_i=k}}\ldots\delta_{\theta_{u_n}} = \sum_{k=1}^{K} p_k\delta_{\theta_k}, \tag{19.8}
$$

where $p_k = \Sigma_{k=1}^{K} P\left(\Sigma_{t=1}^{k-1} p_t < F^{(i)}(\gamma_i) < \Sigma_{t=1}^{k} p_t\right) = V_k \prod_{j<k}(1-V_j)$. The covariance between $\phi_i$ and $\phi_j$ is given by

$$
\begin{aligned}
\text{Cov}(\phi_i,\phi_j) &= \sigma_s^2 \sum_{l=1}^{K} P(u_i - u_j = l)\\
&= \sigma_s^2 \sum_{l=1}^{K} P\left(\Sigma_{k=1}^{l-1} p_k < F^{(i)}(\gamma_i) < \Sigma_{k=1}^{l} p_k, \Sigma_{k=1}^{l-1} p_k < F^{(j)}(\gamma_j) < \Sigma_{k=1}^{l} p_k\right)\\
&= \sigma_s^2 \sum_{l=1}^{K} p_l P\left(F^{(i)^{-1}}(\Sigma_{k=1}^{l-1} p_k) < \gamma_i < F^{(i)^{-1}}(\Sigma_{k=1}^{l} p_k) \mid \right.\\
&\qquad \left. F^{(j)^{-1}}(\Sigma_{k=1}^{l-1} p_k) < \gamma_j < F^{(j)^{-1}}(\Sigma_{k=1}^{l} p_k)\right),
\end{aligned}
$$

where $(\gamma_i, \gamma_j)$ follows a bivariate normal distribution with covariance specified by the CAR model. Posterior inference for the ARSB and ARDP models are based upon Markov chain Monte Carlo simulations, presented in the Appendix.

### *19.3.5 A Practical FDR-Based Method to Select Difference Boundaries*

To obtain a threshold for detecting difference boundaries, our approach treats the spatial boundary analysis problem as one of multiple hypothesis testing. For each pair of adjacent regions, say $i$ and $j$, we test $\phi_i = \phi_j$ against $\phi_i \neq \phi_j$. This produces as many hypotheses as there are edges. Recently, several authors have advocated the use of the FDR to adjust for multiplicities in hypothesis testing problems (e.g., Benjamini and Hochberg 1995; Efron et al. 2001; Storey 2002, 2003).

We identify a boundary $(i, j)$ as a difference boundary if the posterior probability that $P(\phi_i = \phi_j \,|\, Y)$ exceeds a certain threshold $t$, where $Y = (Y_1, Y_2, \ldots, Y_n)'$. For each pair of neighboring regions, we construct $A_{(i,j)}(Y;t) = \{Y : P(\phi_i \neq \phi_j \,|\, Y) > t\}$, a *critical region* that indicates evidence in favor of $(i, j)$ being a difference boundary. The choice of $t$ will control the FDR below a level $\delta = 0.05$. If $Z_{(i,j)} = I(\phi_i = \phi_j)$ and $v_{(i,j)} = P(Z_{(i,j)} = 0 \,|\, Y)$, then the FDR is

$$FDR = \frac{\sum_{i\sim j} Z_{(i,j)} I(v_{(i,j)} > t)}{\sum_{i\sim j} I(Z_{(i,j)} > t)} \quad \text{where } i \sim j \text{ if } w_{ij} \neq 0\,. \tag{19.9}$$

Estimation of (19.9) is straightforward. It is obtained as the posterior expectation

$$\widehat{FDR} = \mathrm{E}[FDR \,|\, Y] = \frac{\sum_{i\sim j}(1 - v_{(i,j)}) 1(v_{(i,j)} > t)}{\sum_{i\sim j} 1(v_{(i,j)} > t)}\,, \tag{19.10}$$

where $v_{(i,j)}$ is computed as a Monte Carlo mean of the posterior samples for $Z_{ij}$. Rejection rules can be then constructed to bound the FDR at target level $\delta$: reject if $v_{(i,j)} > t$, where

$$t = \sup\left\{ u : \frac{\sum_{i\sim j} I(v_{(i,j)} > u)(1 - v_{(i,j)})}{\sum_{i\sim j} I(v_{(i,j)} > u)} \leq \delta \right\}.$$

Li et al. (2012) explored FDR-based methods in conjunction with a parametric class of smoothed moving average models using a Bayesian spike-slab prior to adjust for multiple tests; these models require rather awkward constraints on the random effects. Their approach required estimating as many models as there are geographical boundaries making it computationally expensive. For example, for testing county boundaries in the state of Minnesota, they had to estimate 211 models. More importantly, their method does not provide posterior estimates from any single model: obtaining model-averaged estimates is not straightforward. These drawbacks are circumvented with the ARDP and ARSB.

## 19.4 A Simulation Study

To evaluate our methods, we conducted a simulation study using the template of a Minnesota county map in Li et al. (2011). There are $n = 87$ counties in Minnesota, and 211 pairs of neighboring counties. We simulated 50 datasets on a map of Minnesota, where the state was divided into six regions. Each dataset was generated from (19.1), where $\mu_i$ was one of five different means corresponding to the five different shades mapped on Fig. 19.1; the darker shades correspond to higher means. To add some irregularity, we also included one county (Sherburne county shaded white in Fig. 19.1) that has all its boundaries as true difference boundaries. This resulted in "six" different clusters on the map and 47 "true difference boundaries" delineating the clusters with substantially different means.

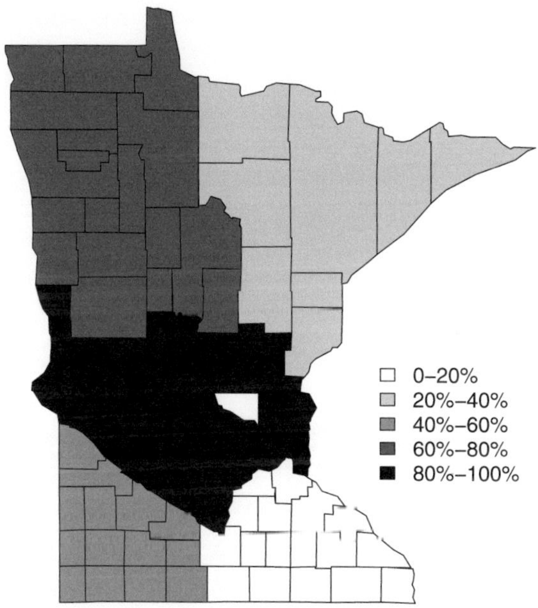


**Fig. 19.1** A map of the simulated data with the *grey-scales* showing the six different clusters, each having its own mean. There are 47 boundary segments that separate regions with different means (*shades*). The percentages reflect the quantiles for the distribution of the outcomes

For every pair of geographical neighbors $(i, j)$, we computed the posterior probability $P(\phi_i \neq \phi_j \mid Y)$ and chose the $T = 35, 40, 45, 50$ and 55 edges with the highest posterior probabilities. As there are 47 true difference boundaries, these choices encompass settings where we could, theoretically have obtained 100% accuracy (when $T = 35, 40, 45$) and also where we are assured of a few false positives (when $T = 50, 55$).

The prior specification and computational details of the ARDP model are in the Appendix. The parameter $\alpha$ can be fixed based upon the expected number of clusters a priori. In this case, six clusters suggest a value of $\alpha$ around 1.25. We experimented with $\alpha$ ranging from 0.25 to 1.75 without much change in posterior inference. The results presented here correspond to $\alpha = 0.5$, leading to an expected number of clusters of about 3, half the number of true clusters. We also fixed $\rho = 0.98$ in the ARDP model (ICAR is inappropriate since the covariance matrix must be proper and nonsingular). Higher values of $\rho$ provide more smoothing; values greater than 0.95 are recommended (Wall 2004; Banerjee et al. 2015). For the ARSB model, we used ICAR ($\rho = 1$) coupled with the sum-to-zero constraint. We assumed a regression structure with only an intercept (i.e., $\mathbf{x}_i \equiv 1$) and placed a flat prior on the corresponding $\beta$. The vague prior $\Gamma(0.01, 0.01)$ was specified for the precision parameters $\tau_s$ and $\tau_\gamma$ . In both models, the stick-breaking truncation was set at $K = 15$ terms.

We compared the performance of DPM, ARSB, and ARDP with three existing methods: the deterministic boundary likelihood value (BLV) algorithm of Jacquez and Greiling (2003a,b) using the `BoundarySEER` software (see http://www.biomedware.com) with default thresholds set from a BLV histogram; the model-based approach of Lu and Carlin (2005), which we call the "LC method"; and the class of discrete spatial moving average (SMA) models outlined in Li et al. (2011). We ran these models within the `R` statistical software environment running three parallel chains for each model and dataset. Convergence was diagnosed after 12,000 iterations of burn-in using Gelman-Rubin diagnostics and autocorrelation plots from the `coda` package in `R`. A subsequent $5{,}000 \times 3 = 15{,}000$ samples were used for posterior inference. On a workstation using a Intel dual core 4 GHz processor, each model took less than 5 h of CPU time to deliver its entire inferential output for all the 50 simulated datasets.

Table 19.1 presents the average detection rates for these different methods applied to the 50 simulated datasets. The DPM and the BLV methods do not explicitly borrow strength across neighbors, while the other four methods in Table 19.1 exploit the adjacency structure of the underlying map. There is little difference between the ARDP and ARSB, but both methods slightly outperform the others in both sensitivity and specificity under all five scenarios. In addition, while the performance of the SMA model is perhaps comparable, it is computationally onerous and less robust to prior assumptions (Li et al. 2011) than ARDP or ARSB.

The LC method is based upon a parametric CAR model that does not render itself to probabilistic boundary analysis (since $P(\phi_i = \phi_j)$ will always be zero for $i \neq j$). However, one could fit parametric CAR models and use the posterior expectation of the absolute differences of the rates $E(\|\eta_i - \eta_j\| \,|Y)$, where $\eta_i = \frac{\mu_i}{E_i}$ acts as a boundary difference score. Higher values indicate spatial barriers between units $i$ and $j$. The DPM, ARSB, and ARDP models not only yield estimates of $\eta_i$, as in the "LC" method, but they also deliver nonzero posterior probabilities $P(\phi_i = \phi_j \,|Y)$. Therefore, we used the posterior expectation metric to compare its performance. The SMA model does not deliver posterior estimates of spatial effects from a single model. Hence, we exclude it from this comparison.

**Table 19.1** Sensitivity and specificity in the simulation study (50 datasets generated on a Minnesota map) for the ARDP, ARSB, DPM, LC, and BLV methods

| T | Method | Sensitivity | Specificity | T | Method | Sensitivity | Specificity |
|---|---|---|---|---|---|---|---|
| 35 | ARDP | 0.768 | 0.998 | 40 | ARDP | 0.822 | 0.990 |
| | ARSB | 0.771 | 0.991 | | ARSB | 0.821 | 0.989 |
| | DPM | 0.737 | 0.989 | | DPM | 0.791 | 0.991 |
| | BLV | 0.711 | 0.990 | | BLV | 0.778 | 0.979 |
| | LC | 0.702 | 0.989 | | LC | 0.767 | 0.976 |
| | SMA | 0.740 | 0.998 | | SMA | 0.818 | 0.991 |
| 45 | ARDP | 0.881 | 0.971 | 50 | ARDP | 0.927 | 0.962 |
| | ARSB | 0.878 | 0.972 | | ARSB | 0.930 | 0.968 |
| | DPM | 0.870 | 0.968 | | DPM | 0.897 | 0.952 |
| | BLV | 0.831 | 0.964 | | BLV | 0.869 | 0.944 |
| | LC | 0.813 | 0.959 | | LC | 0.859 | 0.941 |
| | SMA | 0.872 | 0.975 | | SMA | 0.901 | 0.955 |
| 55 | ARDP | 0.940 | 0.943 | | | | |
| | ARSB | 0.941 | 0.940 | | | | |
| | DPM | 0.895 | 0.915 | | | | |
| | BLV | 0.891 | 0.920 | | | | |
| | LC | 0.881 | 0.917 | | | | |
| | SMA | 0.925 | 0.930 | | | | |

Table 19.2 presents the results for four of the methods. The deterministic BLV method detects 89.6 % of the boundaries. The promise of our stochastic models is evident from the superior performances of the ARDP and the ARSB models. Since we know the true boundaries in Fig. 19.1, we can assess the performances of these approaches in detecting the true boundaries. We find that the DPM, ARDP, and the ARSB models are each able to detect about 90 % of the true boundaries; the ARDP model performs slightly better than the other two.

**Table 19.2** Assessment of the true wombling boundaries with those produced by LC, ARDP, and ARSB based on $P(\phi_i = \phi_j \mid Y)$ and $E(\|\eta_i - \eta_j\| \mid Y)$ in the simulation study

| | Assessment using $P(\phi_i = \phi_j \mid Y)$ | Assessment using $E(\|\eta_i - \eta_j\| \mid Y)$ |
|---|---|---|
| LC | – | 78.7 % |
| DPM | 89.3 % | 82.2 % |
| ARDP | 91.4 % | 88.3 % |
| ARSB | 89.1 % | 83.3 % |

Using the posterior expectation metric, we again find that the proposed ARSB and ARDP models outperforming the LC method; they outperform the DPM model as well in terms of the posterior expectation metric.

Finally, we compare the estimated FDR computed from (19.10) and the true FDR in (19.9). Figure 19.2 plots the estimated FDR from the ARSB model against the number of edges selected from a cutoff value $t$. We also plot the realized FDR (dashed line) based on the boundary detections and the underlying truth. For each

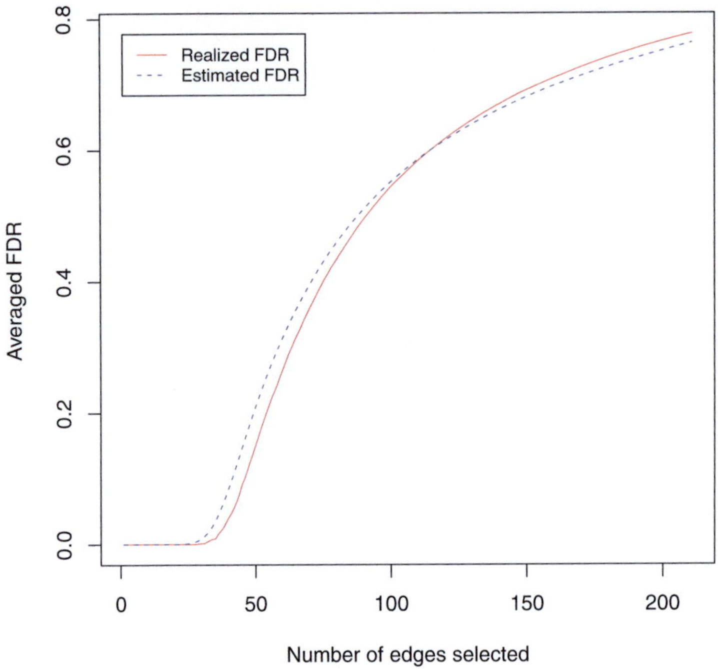


**Fig. 19.2** The realized FDR and the estimated FDR curves in the simulation study. The x-axis is number of edges selected as difference boundaries

number of edges selected, the FDR curves are averages over the 50 simulated datasets. Overall, we find that the FDR is well estimated by the proposed approach. The slight overestimation when number of edges selected is $< 100$ and the underestimation for $> 100$ indicates conservatism since only the lower values are of interest when the FDR is controlled by a certain target, say 10 %. This figure is almost indistinguishable for the ARDP model (not shown) and also the SMA model (reported in Li et al. 2012).

## 19.5 Analysis of Minnesota *P&I* Dataset

We applied our method to the Minnesota Pneumonia and Influenza (*P&I*) diagnosis dataset. P & I rank as the eighth leading cause of death in the United States and the sixth leading cause in people over 65 years of age, with Pneumonia consistently accounting for the overwhelming majority of deaths of the two. Together, they cost the U.S. economy in 2005 an estimated $40.2 billion. Identifying difference boundaries that perform well with regard to sensitivity and specificity can help identify so-called "health barriers" more accurately and buttress an active surveillance program for an influenza-like illness. Reported difference boundaries can provide information to the surveillance system to limit unnecessary entry or egress of people from affected areas, thereby thwarting the spread of infection. Furthermore, health policy analysts can exploit difference boundaries to better coordinate between local administrators and execute plans for hospital needs and antiviral or vaccine interventions. Finally, difference boundaries can lead to better identification of lurking covariates or latent factors that may better explain the discrepant hospitalization rates between neighboring counties.

We analyzed a dataset consisting of Minnesota residents above 65 years of age who were enrolled in the Medicare fee-for-service program as of December 31, 2001. The Medicare Denominator file for 2001 was used to define the cohort. The medicare provider analysis and review (MedPAR) manages patient records based on date of discharge and supplied information regarding hospitalizations resulting from *P&I*. Rates of *P&I* hospitalization are traditional measures of the impact of influenza virus in the elderly population. We identify the 'boundaries' that separate the more affected areas from the less affected areas.

If $Y_i$ and $O_i$ are the observed number of hospitalizations and the population in county $i$ respectively, then $E_i = \frac{\sum_{k=1}^{n} Y_i}{\sum_{k=1}^{n} O_i} O_i$ is the expected number of cases (under the assumption of no spatial variation in rates), where $n$ is the total number of counties. The choropleth map of the raw data is shown in Fig. 19.3. The high-valued SMR (standard mortality ratio) counties are scattered over the map, with a clump in the southwest and some isolated regions surrounded by sparsely inhabited counties that also have lower counts.

We employed the models in Sect. 19.4 to detect boundaries on the *P&I* hospitalization map. The same prior specification and model settings were applied here as the simulation study, except we took $\alpha = 1$, a customary choice when one does not seek a prior distribution on this parameter (Hanson 2006) or has no a priori information about the number of clusters. Three parallel MCMC chains were executed on the same computing environment as described in Sect. 19.4. Convergence was diagnosed after $10{,}000$ iterations of burn-in using Gelman-Rubin diagnostics and autocorrelation plots and a subsequent $5{,}000 \times 3 = 15{,}000$ samples were used for posterior inference. Each model consumed less than 10 min of CPU time to produce its entire inferential output for the Minnesota Pneumonia and Influenza dataset with very little difference between the ARSB, ARDP, and the (non-spatial) DPM model.

Health administrators may prefer to use a "top bracket" of most likely difference boundaries for policy formulation. The top 50 difference boundaries detected by each model are highlighted in Fig. 19.4. Table 19.3 presents a comprehensive "lookup table" containing the names of adjacent counties that have been ranked in decreasing order according to $1 - P(\phi_i = \phi_j \,|\, Data)$ from the ARDP model. Instead of selecting this "bracket" arbitrarily, statisticians may prefer a threshold obtained by controlling the FDR. Setting $\delta = 5\,\%$ yields Numbers 1–33 as difference boundaries, while setting $\delta = 10\,\%$ detects Numbers 1–42 as difference boundaries. This table offers an easy reference for health administrators and officials to identify the more substantial spatial health barriers in the state.

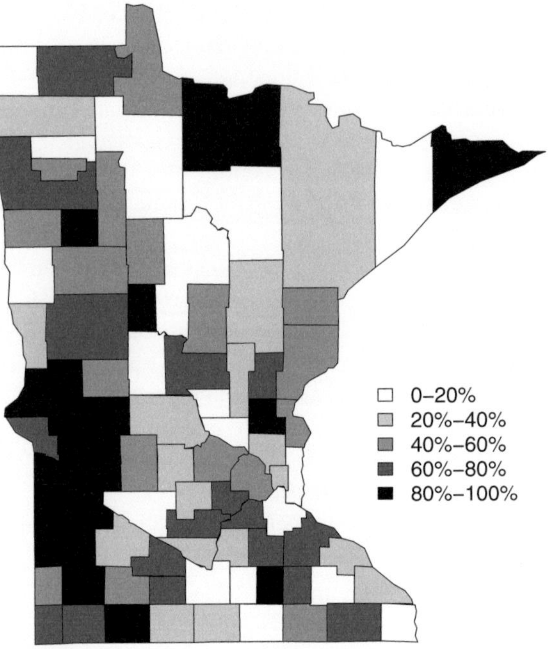


**Fig. 19.3** Choropleth map of the SMR in MN (P&I) dataset. The percentages reflect the quantiles for the distribution of the SMR

About 90 % of the boundaries listed in Table 19.3 are detected by all four models. As a specific example consider Cook and Koochiching county. The outcome variable in the former is substantially higher than its only neighbor, Lake, while Koochiching county is separated from all its neighbors due to its extremely high *P&I* SMR, even after being smoothed by the model. Among the 50 difference boundaries detected by the ARDP model, 47 are also detected by the ARSB model. The three county-pairs that went undetected by ARSB were Goodhue and Olmsted, Freeborn and Steele, and Big Stone and Traverse. The ARSB model detected boundaries between counties Becker and Wadena, Cotton Wood and Jackson, and Cook and Lake.

The map in Fig. 19.3 does not display clustering as pronounced as did the simulation example. It does, however, reflect well on our models that the rankings in Table 19.3 are very consistent with competing methods. The agreement between the ARDP and the SMA in terms of identifying the difference boundaries using FDR-based thresholds is very strong with over 95 % agreement in boundary selection.

Lastly, the minimum predictive loss approach of Gelfand and Ghosh (1998) was implemented to compare the four models. Specifically, for each posterior sample $\beta^{(l)}$ and $\phi^{(l)}$, $l = 1, \ldots, L$ obtained using Markov chain Monte Carlo, we generate replicates for each data point as $y_{rep,i}^{(l)} \sim Poisson(\mu_i^{(l)})$, where $\mu_i^{(l)} = \mathbf{x}_i'\beta + \phi_i^{(l)}$. Preferred models should perform well under a decision-theoretic *balanced loss func-*

**Fig. 19.4** Difference boundaries detected by various models in the Minnesota (P&I) dataset

*tion* that penalizes both departure from fit and departure from smoothness as reflected by variation in replicates. We compute the departure from fit, say G, as $\sum_{i=1}^{n}(y_{obs,i} - \mu_{rep,i})$, where $\mu_{rep,i} = (1/L)\sum_{l=1}^{L} y_{rep,i}^{(l)}$, and the departure from smoothness, say P, as $\sum_{i=1}^{n} \sigma_{rep,i}^{2}$, where $\sigma_{rep,i}^{2} = (1/L)\sum_{l=1}^{L}(y_{rep,i}^{(l)} - \mu_{rep,i})^2$. This yields a model comparison score, D = G + P, with lower values of D suggesting better model performance. The D scores are summarized in Table 19.4 which reveals that all four models perform almost equally well here in terms of the criterion based upon departure from the 'fit' and the departure from the 'smoothness'. The LC model performs slightly better than the rest of the three and the nonspatial DPM model is slightly inferior for this specific dataset.

**Table 19.3** Names of adjacent counties that have significant boundary effects from the ARDP model. The numbers in the first column are the ranks according to $P(\phi_i = \phi_j \,|\, Y)$

| | | | |
|---|---|---|---|
| 1 | Beltrami, Koochiching | 26 | Koochiching, Lake of the Woods |
| 2 | Cass, Wadena | 27 | Isanti, Mille Lacs |
| 3 | Douglas, Pope | 28 | Chippewa, Renville |
| 4 | Freeborn, Steele | 29 | Murray, Pipestone |
| 5 | Goodhue, Olmsted | 30 | Becker, Mahnomen |
| 6 | Itasca, Koochiching | 31 | Rice, Waseca |
| 7 | Kandiyohi, Pope | 32 | Blue Earth, Brown |
| 8 | Koochiching, St. Louis | 33 | Dodge, Olmsted |
| 9 | Pope, Stearns | 34 | Chisago, Isanti |
| 10 | Anoka, Isanti | 35 | Redwood, Yellow Medicine |
| 11 | Dakota, Goodhue | 36 | Pennington, Polk |
| 12 | Lincoln, Pipestone | 37 | Goodhue, Wabasha |
| 13 | Murray, Redwood | 38 | Pope, Swift |
| 14 | Steele, Waseca | 39 | Morrison, Todd |
| 15 | Renville, Yellow Medicine | 40 | Fillmore, Olmsted |
| 16 | Cottonwood, Murray | 41 | Cook, Lake |
| 17 | Jackson, Martin | 42 | Douglas, Grant |
| 18 | Kandiyohi, Swift | 43 | Mahnomen,Norman |
| 19 | Pope, Stevens | 44 | Grant, Wilkin |
| 20 | Todd, Wadena | 45 | Mahnomen, Polk |
| 21 | Lyon, Redwood | 46 | Jackson, Nobles |
| 22 | Murray, Nobles | 47 | Morrison, Todd |
| 23 | Isanti, Sherburne | 48 | Dodge, Olmsted |
| 24 | Otter Tail, Todd | 49 | Big Stone, Traverse |
| 25 | Clay, Otter Tail | 50 | Morrison, Stearns |

**Table 19.4** Predictive loss criterion under all four models for Minnesota (P&I) dataset. G is a goodness-of-fit term while P is a penalty term which penalized departure from "smoothness"

| | G | P | D |
|---|---|---|---|
| LC | 0.47 | 2.81 | 3.29 |
| DPM | 0.82 | 2.76 | 3.58 |
| ARDP | 0.76 | 2.73 | 3.49 |
| ARSB | 0.64 | 2.77 | 3.41 |

## 19.6 Conclusion and Future Work

The paper presented a class of nonparametric Bayesian hierarchical models for detecting difference boundaries on maps. An advantage of the new approach is that it permits the probabilistic estimation of an edge as a difference boundary, and improves the percentage of true detection. A disadvantage is that the model cannot be easily fit into any existing commercial software. We fit these models in `R` (www.r-project.org), and we hope to collect these models in an `R` package in the near future.

The ARDP and ARSB models in conjunction with the FDR controlled threshold selection provide a major improvement over earlier work by Li et al. (2012). However, issues related to optimal selection of boundaries warrants further investigation, especially regarding the sensitivity of the inference to FDR-based cutoffs and to prior specifications. Further extensions can be formulated by incorporating classes of loss functions, as discussed by Müller et al. (2006), for a more comprehensive decision-theoretic framework. Such developments may, in turn, lead to more definitive conclusions regarding the performance of these models in maps that display weaker clustering patterns.

**Acknowledgements** This work was supported by NSF/DMS 1106609, and NIH grants 1-RC1-GM092400-01, 1R03CA165110, and 1R03CA176739-01A1.

## Appendix

Posterior inference for our models are based on MCMC posterior simulations. There are two main strategies used. The first avoids computing parameters characterizing $G$ by marginalizing it out and relying on the Polya urn scheme of Blackwell and MacQueen (1973). A limitation of this approach is that it is only applicable when the prior can be characterized by a generalized Polya urn mechanism. Ishwaran and James (2001) proposed the blocked Gibbs Sampler that directly sampled from the posterior of the random measure, avoiding the marginalization over $G$. We use the blocked Gibbs Sampler with some Metropolis-Hasting steps nested in it to update all random parameters in our model. We truncate the infinite sum in $G$ by the first $m$ terms. We only provide details for the ARDP model. That for the ARSB model is similar (and even simpler), while algorithms for the DPM model may be found in Escobar and West (1995).

We place a flat prior on parameter $\beta$ and reparameterize the variance parameters with its inverse, $\tau_s = \sigma_s^{-2}, \tau_\gamma = \sigma_\gamma^{-2}$, then place a conjugate gamma prior of the precision parameters $\tau_s$ and $\tau_\gamma$. The likelihood of the model is expressed as

$$L = \prod_{i=1}^{n} Poisson(Y_i \,|\, \mathbf{x}_i'\beta + \phi_i) \tag{19.11}$$

The posterior density given the data $\mathbf{Y} = \{Y_i\}$ is proportional to the likelihood multiplied by all the prior distributions: $L(Y_i \,|\, \beta, \{\phi_i\})p(\beta)p(\phi \,|\, \tau_s)p(\gamma | \tau_\gamma)p(\mathbf{V})p(\tau_s)p(\tau_\gamma)$. Note that $\phi_i = \theta_{u_i}$ and we updated $\phi_i$ by updating $\theta$ and $u_i$. The MCMC algorithm proceeds as follows.

1. Update $\beta \,|\, \boldsymbol{\theta}, \boldsymbol{\gamma}, \mathbf{V}, \tau_s, \tau_\gamma$: The full conditional distribution only depends on the likelihood due to the flat prior. Sample candidate $\beta^*$ from $N(\beta, K_\beta I)$ ($K_\beta = 0.05$ worked well), then accept the candidate $^*$ with probability

$$\min\left\{1,\frac{\exp(\Sigma_{i=1}^{n}(-E_i\exp(\mathbf{x}_i'\beta^*+\phi_i)+y_i(\mathbf{x}_i'\beta^*+\phi_i)))}{\exp(\Sigma_{i=1}^{n}(-E_i\exp(\mathbf{x}_i'\beta+\phi_i)+y_i(\mathbf{x}_i'\beta+\phi_i)))}\right\} \tag{19.12}$$

2. Update $\theta_j\,|\,\boldsymbol{\beta},\boldsymbol{\gamma},\mathbf{V},\tau_s,\tau_\gamma$: Sample candidate $\theta_j^*$ from $N(\theta_j,K_\theta)$ ($K_\theta=0.05$ worked well), then accept the candidate $\theta_j^*$ with probability

$$\min\left\{1,\frac{\exp(\Sigma_{i:u_i=j}(-E_i\exp(\mathbf{x}_i'\beta+\theta_j^*)+y_i(\mathbf{x}_i'\beta+\theta_j^*))-\frac{\tau_s}{2}\theta_j^{*2})}{\exp(\Sigma_{i:u_i=j}(-E_i\exp(\mathbf{x}_i'\beta+\theta_j)+y_i(\mathbf{x}_i'\beta+\theta_j))-\frac{\tau_s}{2}\theta_j^{2})}\right\} \tag{19.13}$$

3. Update $\boldsymbol{\gamma_i}\,|\,\boldsymbol{\beta},\boldsymbol{\theta},\mathbf{V},\tau_s,\tau_\gamma$: Sample candidate $\gamma^*$ from $N(\gamma_i,K_\gamma)$ ($K_\gamma=0.01$ worked well), compute the corresponding candidate $\mathbf{u}^*$ through $u_i^*=\Sigma_{j=1}^{n} jI(\Sigma_{k=1}^{j-1}p_k< F^{(i)}(\gamma_i)<\Sigma_{k=1}^{j}p_k)$, then accept the candidate $\gamma^*$ with probability

$$\min\left\{1,\frac{\exp\left(-\frac{1}{2}\gamma^{\star\prime}\Sigma_2^{-1}\gamma^\star\right)\exp(-E_i\exp(\mathbf{x}_i'\beta+\theta_{u_i^*})+y_i(\mathbf{x}_i'\beta+\theta_{u_i^*}))}{\exp\left(-\frac{1}{2}\gamma'\Sigma_2^{-1}\gamma\right)\exp(-E_i\exp(\mathbf{x}_i'\beta+\theta_{u_i})+y_i(\mathbf{x}_i'\beta+\theta_{u_i}))}\right\} \tag{19.14}$$

4. Update $\mathbf{V}\,|\,\boldsymbol{\beta},\boldsymbol{\theta},\boldsymbol{\gamma},\tau_s,\tau_\gamma$: Sample candidate $\mathbf{V}^*$ from $N(\mathbf{V},0.01I_m)$, compute the corresponding $\mathbf{p}^*$ and $\mathbf{u}^*$. Accept the candidate $\mathbf{V}^*$ with probability

$$\min\left\{1,\frac{\Pi_{k=1}^{m}(1-V_k^*)^{\alpha-1}\Pi_{i=1}^{n}\exp(-E_i\exp(\mathbf{x}_i'\beta+\theta_{u_i^*})+y_i(\mathbf{x}_i'\beta+\theta_{u_i^*}))}{\Pi_{k=1}^{m}(1-V_k)^{\alpha-1}\Pi_{i=1}^{n}\exp(-E_i\exp(\mathbf{x}_i'\beta+\theta_{u_i})+y_i(\mathbf{x}_i'\beta+\theta_{u_i}))}\right\} \tag{19.15}$$

5. Update $\tau_s\,|\,\boldsymbol{\beta},\boldsymbol{\theta},\boldsymbol{\gamma},\mathbf{V},\tau_\gamma$ : Sample from $Gamma\left(\frac{n}{2}+a,\frac{\Sigma_{i=1}^{n}\phi_i^2}{2}+b\right)$, where $a=b=0.01$, which is the conjugate gamma full conditional distribution for $\tau_s$.
6. Update $\tau_\gamma\,|\,\boldsymbol{\beta},\boldsymbol{\theta},\boldsymbol{\gamma},\mathbf{V},\tau_s$ : Sample from $Gamma\left(\frac{n}{2}+c,\frac{\boldsymbol{\gamma}'(D-\rho W)\boldsymbol{\gamma}}{2}+d\right)$, where $c=d=0.01$, which is the conjugate gamma full conditional distribution for $\tau_\gamma$.

## References


Anselin, L. (1988). *Spatial Econometrics: Methods and Models*. Kluwer Academic Publishers, Boston, MA.

Banerjee, S., Carlin, B.P., and Gelfand, A.E. (2015). *Hierarchical Modeling and Analysis for Spatial Data, Second Edition*. Chapman and Hall/CRC Press, Boca Raton, FL.

Banerjee, S. and Gelfand, A.E. (2006). Bayesian Wombling: Curvilinear gradients assessment under spatial process models. *J. Amer. Statist. Assoc.* **101**, 1487–1501.

Benjamini, Y. and Hochberg, Y. (1995). Controlling the false discovery rate: A practical and powerful approach to multiple testing. *J. Roy. Statist. Soc. B* **57**, 289–300.

Blackwell, D. and MacQueen, J.B. (1973). Ferguson distributions via Polya urn schemes. *Ann. Statist.* **1**, 353–355.

Cai, B., Lawson, A., Hossain, M., Choi, J., Kirby, R., and Liu, J. (2013). Bayesian semiparametric model with spatially-temporally varying coefficients selection. *Statistics in Medicine*, 32, 3670–3685.

Chung, Y. and Dunson, D. B. (2009). Nonparametric Bayes conditional distribution modeling with variable selection. *J. Amer. Statist. Assoc.* **104**, 1646–1660.

Chung, Y. and Dunson, D.B. (2011). The local Dirichlet process. *Ann. Instit. Statist. Math.* **63**, 59–80.

Cressie, N. (1993). *Statistics for Spatial Data Revised Edition*. Wiley, Hoboken, NJ.

De Iorio, M., Johnson, W.O., Müller, P., and Rosner, G.L. (2009). Bayesian Nonparametric Nonproportional Hazards Survival Modeling, *Biometrics*, **65**, 762–771.

De Iorio, M., Müller, P., Rosner, G.L., and MacEachern, S.N. (2004). An ANOVA model for dependent random measures. *J. Amer. Stat. Assoc.*, **99**, 205-215.

Duan, J., Guindani, M., and Gelfand, A.E. (2007). Generalized spatial Dirichlet process models. *Biometrika* **94**, 809–825.

Dunson, D.B., Pillai, N.S., and Park, J-H. (2007). Bayesian density regression. *J. Roy. Statist. Soc. Ser. B* **69**, 163–183.

Efron, B., Tibshirani, R., Storey, J. D., and Tusher, V. (2001). Empirical Bayes analysis of a microarray experiment. *J. Amer. Statist. Assoc.* **96**, 1151–1160.

Escobar, M.D. and West M. (1995). Bayesian density estimation and inference using mixtures. *J. Amer. Stat. Assoc.* **90**, 577–588.

Ferguson T.S. (1973). A Bayesian analysis of some nonparametric problems. *Ann. Statist.* **1**, 209–230.

Fitzpatrick, M., Preisser, E., Porter, A., Elkinton, J., Waller, L., Carlin, B., and Ellison, A. (2010). Ecological boundary detection using Bayesian area wombling. *Ecology*, 91, 3503–3514.

Gelfand A.E., Kottas A., and Maceachern S. N. (2005). Bayesian nonparametric spatial modelling with Dirichlet process mixing. *J. Amer. Statist. Assoc.* **100**, 1021–1035.

Gelfand, A.E. and Ghosh, S.K. (1998). Model choice: A minimum posterior predictive loss approach *Biometrika* **85**, 1–11.

Griffin, J.E. and Steel, M.F.J. (2006). Order-based dependent Dirichlet processes. *J. Amer. Statist. Assoc.* **101**, 179–194.

Hanson, T. (2006). Modeling censored lifetime data using a mixture of gammas baseline. *Bayesian Analysis*, 1, 575-594.

Hanson, T. and Jara, A. (2013). Surviving fully Bayesian nonparametric regression models. In *Bayesian Theory and Applications*, pp. 593–615. P. Damien, P. Dellaportas, N. Polson, and D. Stephens, eds. Oxford University Press: Oxford.

Hoeting, J. A., Madigan, D., Raftery, A.E., and Volinsky, C.T. (1999). Bayesian model averaging: A tutorial. *Statist. Sci.* **14**, 382–401.

Ishwaran H. and James L. F. (2001). Gibbs sampling methods for stick-breaking priors. *J. Amer. Statist. Assoc.* **96**, 161–173.

Ishwaran, H. and Zarepour, M. (2000). Markov chain Monte Carlo in approximate Dirichlet and beta two-parameter process hierarchical models. *Biometrika*, **87** 371-390.

Jacquez, G.M. and Greiling, D.A. (2003a). Local clustering in breast, lung and colorectal cancer in Long Island, New York. *Int. J. Health. Geogr.* **2**,3.

Jacquez, G.M. and Greiling, D.A. (2003b). Geographic boundaries in breast, lung and colorectal cancers in relation to exposure to air toxics in Long Island, New York. *Int. J. Health. Geogr.* **2**,4.

Kalli, M., Griffin, J.E., and Walker, S.G. (2011). Slice sampling mixture models. *Statist. Comp.* **21**, 93–105.

Le Sage, J. and Pace, K. (2009). *Introduction to Spatial Econometrics*. Chapman and Hall/CRC, Boca Raton, FL.

Li, L., Hanson, T., and Zhang, J. (2015). Spatial extended hazard model with application to prostate cancer survival. *Biometrics*, **71**, 313–322.

Li, P., Banerjee, S., Hanson, T., and McBean, A. (2015). Bayesian hierarchical models for detecting boundaries in areally referenced spatial datasets. *Statistica Sinica*, 25, 385–402.

Li, P., Banerjee S., and McBean A.M. (2011). Mining edge effects in areally referenced spatial data: A Bayesian model choice approach. *Geoinformatica* **15**, 435–454.

Li, P., Banerjee S., McBean A.M. and Carlin, B.P. (2012). Bayesian areal wombling using false discovery rates. *Statistics and its Interface* **5**, 149–158.

Lu, H. and Carlin, B.P. (2005). Bayesian areal wombling for geographical boundary analysis. *Geogr. Anal.* **37**, 265–285.

Lu, H., Reilly, C., Banerjee, S., and Carlin, B.P. (2007). Bayesian areal wombling via adjacency modeling. *Environ. Ecol. Statist.* **14**, 433–452.

Ma, H., Carlin, B.P., and Banerjee, S. (2010). Hierarchical and joint site-edge methods for Medicare hospice service region boundary analysis. *Biometrics* **66**, 355–364.

MacEachern, S.N. (2001). Decision theoretic aspects of dependent nonparametric processes. In *Bayesian Methods with Applications to Science, Policy and Official Statistics* (Edited by E. George), 551–560. Eurostat.

Müller, P., Parmigiani, G., and Rice, K. (2006). FDR and Bayesian multiple comparisons rules. In *Bayesian Statistics 8*. Ed(s) J.M. Bernardo, S. Bayarri, J.O. Berger, A.P. Dawid, D. Heckerman, A.F.M. Smith and M. West. Oxford University Press.

Petrone, S., Guindani, M., and Gelfand, A.E. (2009). Hybrid Dirichlet mixture models for functional data. *J. Roy. Statist. Soc. B.* **71**, 755–782.

Reich, B. and Fuentes, M. (2007). A multivariate semiparametric bayesian spatial modeling framework for hurricane surface wind fields. *The Ann. Appl. Statist.* **1**, 249–264.

Rodríguez, A., Dunson, D., and Gelfand, A. (2010). Latent stick-breaking processes. *Journal of the American Statistical Association*, 105, 647–659.

Sethuraman, J. (1994). A constructive definition of Dirichlet priors. *Statist. Sinica* **4**, 639–650.

Storey, J. (2002). A direct approach to false discovery rates. *J. Roy. Statist. Soc. B* **64**, 479–498.

Storey, J. D. (2003). The positive false discovery rate: A Bayesian interpretation and the q-value. *Ann. Statist.* **31**, 2013–2035.

Wall, M. (2004). A close look at the spatial structure implied by the CAR and SAR models. *Journal of Statistical Planning and Inference*, 121, 311–324.

Womble, W.H. (1951). Differential systematics. *Science* **114**, 315–322.

Zhou, H., Hanson, T., and Knapp, R. (2015). Marginal Bayesian nonparametric model for the time-to-infection of a threatened amphibian. *Biometrics*, in press.

# Part VI
# Causal Inference and Missing Data

# Chapter 20
# A Bayesian Nonparametric Causal Model for Regression Discontinuity Designs

George Karabatsos and Stephen G. Walker

**Abstract** For non-randomized studies, the regression discontinuity design (RDD) can be used to identify and estimate causal effects from a "locally randomized" subgroup of subjects, under relatively mild conditions. However, current models focus causal inferences on the impact of the treatment (versus non-treatment) variable on the mean of the dependent variable, via linear regression. For RDDs, we propose a flexible Bayesian nonparametric regression model that can provide accurate estimates of causal effects, in terms of the predictive mean, variance, quantile, probability density, distribution function, or any other chosen function of the outcome variable. We illustrate the model through the analysis of two real educational data sets, involving (resp.) a sharp RDD and a fuzzy RDD.

## 20.1 Introduction

A basic objective in scientific research is to infer causal effects from data. Randomized studies are the gold standard of causal inference. In an ideal randomized study, the investigator randomly assigns each subject into one of the treatment conditions, with equal probability, and each subject complies with her/his treatment assignment. Then, treatment subjects are the same as non-treatment subjects, in terms of the distribution of all observed and unobserved pretreatment covariates, aside from sampling error (e.g., Rubin 2008); and the outcome variable is independent of

G. Karabatsos (✉)
University of Illinois-Chicago, Chicago, IL, USA
e-mail: gkarabatsos1@gmail.com

S.G. Walker
The University of Texas at Austin, Austin, TX, USA
e-mail: s.g.walker@math.utexas.edu



R. Mitra, P. Müller (eds.), *Nonparametric Bayesian Inference in Biostatistics*,
Frontiers in Probability and the Statistical Sciences,
DOI 10.1007/978-3-319-19518-6_20

the chosen treatment intervention, conditionally on the treatment variable (Dawid 2002). Then the causal effect is given by a comparison of the outcome variable under the treatment intervention, against the outcome variable under the non-treatment intervention.

Often, it is necessary to estimate causal effects from a non-randomized, observational study, because a randomized study can be infeasible due to financial, ethical, or time constraints (Rubin 2008). However, causal inference from a non-randomized study is more challenging. This is because without randomization, treated and non-treated subjects differ almost-surely in terms of the pretreatment covariates.

The regression discontinuity design (RDD) (Thistlewaite and Campbell 1960; Cook 2008) is a type of non-randomized design where a continuous-valued assignment variable (Lee and Lemieux 2010) assigns each subject to the treatment (non-treatment, resp.) condition, whenever her/his observed value of the assignment variable equals or exceeds (resp. is less than) a fixed cutoff value. Under relatively mild conditions, notably when subjects have imperfect control of the assignment variable, the RDD provides a "locally-randomized experiment." Then treatments are "as good as randomly assigned" for the subgroup of subjects with assignment variable values near the cutoff (Lee 2008), making the causal effect identifiable for that subgroup. As proven in Goldberger (2008), the RDD can empirically produce causal effect estimates that are similar to those estimates of a standard randomized study (Aiken et al. 1998; Buddelmeyer and Skoufias 2004; Black et al. 2005; Schochet 2009; Berk et al. 2010; Shadish et al. 2011).

The RDD has existed for over 50 years, with little initial interest (Cook 2008). However, since 1997, more than 74 RDD-based empirical studies have emerged from these fields (Lee and Lemieux 2010; Bloom 2012; Wong et al. 2013), for at least three reasons (Van der Klaauw 2008; Lee and Lemieux 2010). First, many non-randomized studies employ treatment assignment rules that can be easily conceptualized as RDDs. Second, the empirical results of RDDs are intuitive and can be easily conveyed graphically, say, by a plot of the outcomes against the assignment variable. Third, the identification of causal effects in an RDD requires weaker and hence more credible assumptions, compared to the stronger assumptions that are required by other popular causal models, mentioned below. This gives the researcher the flexibility to choose from a range of causal estimation methods.

The other popular causal models for non-randomized studies assume a "potential outcomes" (counterfactual) framework of causal inference (e.g., Rubin 1974, 1978). This is typically done assuming using notation that is simplified by the Stable Unit Treatment Value Assumption (SUTVA), which implies no interference between subjects and no versions of treatments (Rubin 1990). The popular models make further assumptions of unconfoundedness (i.e., treatment and non-treatment outcomes are independent of treatment assignments, conditionally on all pretreatment covariates) and overlap (i.e., there is a chance to receive either the treatment or the non-treatment, conditionally on any value of the pretreatment covariates) (Imbens 2004). These models are defined by a regression of the outcome variable, on variables of treatment receipt and observed pretreatment characteristics, and/or involves matching/weighting subjects on the observed pretreatment variables and/or on propensity scores (e.g., Imbens 2004). The regression may also be on a hypothesized set of unobserved pretreatment covariates, in order to study the sensitivity of causal effect estimates over varying degrees of hidden bias (e.g., Rosenbaum and Rubin 1983), i.e., over changes in the distribution of these covariates. However, it may be argued that for typical non-randomized studies, unconfoundedness and overlap are not very credible assumptions (e.g., Imbens 2004; Lee 2008). Even SUTVA is questionable.

For RDDs, the mainstream causal models are linear, polynomial, or local-linear models that employ a regression of the outcome variable on the assignment variable. Such models aim to provide causal inferences in terms of mean comparisons of treatment outcomes and non-treatment outcomes, and to provide sufficiently flexible modeling of the regression function (Imbens 2004; Lee and Lemieux 2010), in a neighborhood around the cutoff. However, in many settings, it may also be of interest to base causal inferences on comparisons of additional features of the outcome variable, such as the variance, quantiles (percentiles), and/or the entire probability density function.

To address these open issues, we propose a Bayesian nonparametric regression model (Karabatsos and Walker 2012) for causal inference in RDDs. It is an infinite-mixture model that allows the entire probability density of the outcome variable to change flexibly as a function of covariates. Our model can provide inferences of causal effects in terms of how the treatment variable impacts the mean, variance, a quantile, probability density function (p.d.f.), distribution function, and any other chosen function of the outcome variable. Finally, the accurate estimation of causal effects relies on an appropriate model for the data. Karabatsos and Walker (2012) showed that their Bayesian nonparametric regression model tended to have better predictive performance than other parametric and flexible nonparametric regression models of common usage, over many real data sets.

Also, our model can be extended to handle causal inferences from a fuzzy RDD (Trochim 1984). In contrast to a standard "sharp" RDD, a fuzzy RDD involves a study where not all subjects adhere to the treatment assignment rule. This is because, for example, some subjects do not comply with their respective treatment assignments, or because some subjects receive treatments for which they are not eligible.

In Sect. 20.2, we review the data assumptions that are required to identify and estimate causal effects from an RDD. Unlike all previous guides to performing causal inference from RDDs (e.g., Imbens and Lemieux 2008; Lee and Lemieux 2010; Bloom 2012; Wong et al. 2013), we do not rely on the potential outcomes approach. Instead we focus on the extended conditional independence approach, which addresses the problem of causal inference entirely by the concepts of standard probability theory (Dawid 2000, 2002). We should also mention that we will not address SUTVA, as this assumption only makes sense in the potential outcomes framework (Dawid 2000).

In Sect. 20.3, we describe our Bayesian nonparametric model that can estimate causal effects from the various RDDs. In Sect. 20.4, we illustrate our model through the analysis of two educational data sets, involving (resp.) a sharp RDD and a fuzzy RDD. Section 20.5 concludes with a short discussion of the free user-friendly software that can be used to implement our Bayesian nonparametric approach to RDDs, and a discussion of possible extensions of our approach to multivariate RDDs, involving cutoffs in more than one dimension.

Throughout, we denote by $\Pr(X \in A)$ the probability of an event $A$, for a given random variable, $X$. Also, we assume that a continuous (resp., discrete) random variable $X$ admits a cumulative distribution function (c.d.f.), denoted by a capital letter, such as $F(x) = \Pr(X \leq x)$, with corresponding probability density (resp., mass) function or p.d.f., denoted by a lower case letter, with $f(x) = \int \mathrm{d}F(x)$. We use the notation $X \sim F(x)$, $X \sim F$, or $X \sim f(x)$ to refer to $X$ as having the distribution $F$. Accordingly, we denote by $\mathrm{n}(\cdot\,|\,\mu,\sigma^2)$ as the density of the normal $\mathrm{N}(\cdot\,|\,\mu,\sigma^2)$ c.d.f. with mean and variance $(\mu,\sigma^2)$; with $\Phi(\cdot) = \mathrm{N}(\cdot\,|\,0,1)$ the c.d.f. of the normal $\mathrm{n}(\cdot\,|\,0,1)$ p.d.f.; $\mathrm{ga}(\cdot\,|\,a,b)$ and $\mathrm{ig}(\cdot\,|\,a,b)$ (resp.) denotes the p.d.f.s of the gamma $\mathrm{Ga}(\cdot\,|\,a,b)$ distribution and inverse gamma $\mathrm{IG}(\cdot\,|\,a,b)$ distribution (c.d.f.), with shape parameter $a$ and rate parameter $b$; and $\mathrm{U}(\cdot\,|\,a,b)$ is the c.d.f. of a uniform distribution.

## 20.2 Identifying Causal Effects in an RDD

A non-randomized study from an RDD involves three variables that are observable from each individual from a sample of $n$ subjects, indexed by $i = 1,\ldots,n$. They are the *outcome variable*, $Y$; a binary *treatment variable* $T$, where $T = 1$ refers to treatment *receipt* and $T = 0$ refers to non-treatment receipt; and a continuous-valued *assignment variable* $R$. Each subject $i$ is assigned the treatment whenever $R_i \geq r_0$, and is assigned the non-treatment whenever $R_i < r_0$, given a known fixed cutoff $r_0$. As will be further described below, in a fuzzy RDD, being assigned treatment does not imply receiving treatment. The treatment assignment variable is thus $\mathbf{1}_{R\geq r_0}$, with $\mathbf{1}_{(\cdot)}$ the indicator function. An RDD study gives rise to a sample data set, $\mathscr{D}_n = \{(r_i,t_i,y_i)\}_{i=1}^n$, including derived observations $\mathbf{1}_{r_i\geq r_0}$, and possibly observations of $p$ pretreatment covariates, $\mathbf{x}_i = (x_{1i},\ldots,x_{pi})^\top$. Also, we introduce a non-random regime parameter, $\Psi_T \in \{\emptyset,0,1\}$. An "idle" or "observational" regime is indicated by $\Psi_T = \emptyset$, when the joint distribution of $(R,T,Y)$ "arises naturally." An intervention regime is indicated by setting $\Psi_T = 0$ or 1. Here, $\Psi_T = 1$ indicates the treatment intervention, and $\Psi_T = 0$ indicates the nontreatment intervention. The observational regime is none other than the RDD itself, whereas the intervention regimes are hypothetical.

A characterizing assumption of the RDD is that the conditional probability of treatment receipt is discontinuous at $r_0$. That is:

$$\textbf{Assumption RD: } \lim_{r\downarrow r_0} \Pr(T = 1\,|\,r) \neq \lim_{r\uparrow r_0} \Pr(T = 1\,|\,r). \tag{20.1}$$

There are two types of RDDs. In the classical, *sharp RDD* (Thistlewaite and Campbell 1960; Cook 2008), the probability function $\Pr(T = 1\,|\,r)$ has a discontinuous jump of size 1 at $R = r_0$, with point mass probability function:

$$f(t\,|\,r) = \Pr(T = t\,|\,R = r) = \mathbf{1}^t_{r \geq r_0}(1 - \mathbf{1}_{r \geq r_0})^{1-t}. \tag{20.2}$$

Then the treatment receipt is identical to treatment assignment, with $T = \mathbf{1}_{R \geq r_0}$.

In the *fuzzy RDD* (Trochim 1984), $\Pr(T = 1\,|\,R = r)$ has a discontinuous jump that is smaller than 1 at $r_0$. Then $f(t\,|\,r)$ is not a point mass density. The smaller jump results from imperfect treatment adherence (e.g., treatment non-compliance), where some of the subjects of the given study were either assigned $\mathbf{1}_{r \geq r_0} = 0$ but received $T = 1$, or assigned $\mathbf{1}_{r \geq r_0} = 1$ but received $T = 0$.

For either type of RDD, a typical measure of the causal effect is defined by the difference of conditional means (expectations) of $Y$ at $R = r_0$:

$$\tau = \lim_{r \downarrow r_0} \mathbb{E}(Y\,|\,R = r, \Psi_T = 1) - \lim_{r \uparrow r_0} \mathbb{E}(Y\,|\,R = r, \Psi_T = 0) \tag{20.3a}$$

$$= \mathbb{E}(Y\,|\,R = r_0, \Psi_T = 1) - \mathbb{E}(Y\,|\,R = r_0, \Psi_T = 0) \tag{20.3b}$$

where $\mathbb{E}(Y\,|\,r,t) = \int y\,\mathrm{d}F(y\,|\,r,t)$. Also if $r_0^+ = \lim_{r \downarrow r_0} r$, $r_0^- = \lim_{r \uparrow r_0} r$, and $r_0^+ = r_0^- = r_0$, then $R$ is continuous. A motivation for restricting to $R = r_0$ in the definition of the causal effect (20.3b) is that this is the only effect that can be directly estimated from given data ($\mathscr{D}_n$) of an RDD.

In general, for any choice of function $H\{\cdot\}$ of $Y$, the causal effect is given by:

$$\tau_H = \lim_{r \downarrow r_0} \mathbb{E}(H\{Y\}\,|\,r, \Psi_T = 1) - \lim_{r \uparrow r_0} \mathbb{E}(H\{Y\}\,|\,r, \Psi_T = 0) \tag{20.4a}$$

$$= \mathbb{E}(H\{Y\}\,|\,r_0, \Psi_T = 1) - \mathbb{E}(H\{Y\}\,|\,r_0, \Psi_T = 0). \tag{20.4b}$$

Therefore, depending on the choice of function $H\{\cdot\}$, causal effects are not only interpretable in terms of the mean of $Y$ (when $H\{Y\} = Y$), but also in terms of the variance ($H\{Y\} = \{Y - \mathbb{E}(Y\,|\,r,t)\}^2$), cumulative distribution function (c.d.f.) $F(y\,|\,r,t)$ ($H\{Y\} = \mathbf{1}_{Y \leq y}$), probability density function (p.d.f.) ($f(y\,|\,r,t)$), survival function $1 - F(y\,|\,r,t)$, and so on. Inverting the c.d.f. obtains $F^{-1}(u\,|\,r,t)$, for $u \in [0,1]$. Then causal effects can also be interpreted in terms of the $u$th quantile of $Y$.

If $R$ is discrete, then obviously $r^+ \neq r^- \neq r_0$. Then Eq. (20.4) [including (20.3)] still provides a measure of causal effect, which may require additional extrapolation in its estimation.

Next we describe how a causal effect $\tau_H$ is identified from the sharp RDD.

### 20.2.1 Identification in the Sharp RDD

The sharp RDD can be characterized in terms of the extended conditional independence framework of causal inference, extending the ideas from (Dawid 2002, Sections 6.2, 7). In this RDD, it is typically assumed that the joint p.d.f. of $(R,T,Y)$, conditionally on $\Psi_T = \psi_T$, is given by:

$$f(r,t,y\,|\,\psi_T) = f(r) f(t\,|\,r,\psi_T) f(y\,|\,r,t). \tag{20.5}$$

This p.d.f. (20.5) gives rise to the conditional independence properties (a) $R \perp\!\!\!\perp \Psi_T$ and (b) $Y \perp\!\!\!\perp \Psi_T \,|\, R,T$, where $\perp\!\!\!\perp$ denotes conditional independence (Dawid 1979). Property (a) states that the distribution of $R$ is the same in both observational and interventional circumstances. Meanwhile, (b) is a causal property, because it says that the distribution of $Y$ is unaffected by the choice of interventional regime $\Psi_T$, conditionally on $(R,T)$ (Dawid 2002). Properties (a) and (b) together imply that $R$ is a sufficient covariate (Dawid 2002, 2010).

Motivated by the causal property (b), we now focus on the conditional p.d.f.:

$$f(t,y\,|\,r,\psi_T) = f(t\,|\,r,\psi_T)f(y\,|\,r,t). \tag{20.6}$$

If the intervention parameter takes on a null value $\Psi_T = \emptyset$, then the joint distribution of the random variables $(T,Y)$ arises naturally. Then (20.6) reduces to the joint p.d.f. $f(t,y\,|\,r) = f(t\,|\,r)f(y\,|\,r,t)$, where $f(t\,|\,r)$ is the point mass density (20.2).

In contrast, an intervention that sets $\Psi_T = t_0 \in \{0,1\}$, modifies $f(t\,|\,r,\psi_T)$ to $\mathbf{1}(t = t_0)$, in the joint p.d.f. (20.6). Also, recall that the causal effect is estimable only conditionally on $R = r_0$. Then the conditional p.d.f. of $Y$ can be written as:

$$f(y\,|\,r_0,t_0) = f(y\,||\,r_0,t_0), \tag{20.7}$$

where $||$ denotes "conditioning by intervention" (Lauritzen 2000). The equality in (20.7) holds by virtue of the causal property mentioned above. Then for a general choice of function $H\{\cdot\}$, the causal effect is given by a comparison of $\mathbb{E}(H\{Y\}\,|\,r_0,t_0)$, for $t_0 = 0,1$, including the p.d.f.s $f(y\,|\,r_0,t_0)$. For example, the causal effect in terms of the difference, as in the general definition (20.4).

Now we turn to the issue of identifying the causal effect from data. Suppose that there is a reason to believe that in the absence of treatment, subjects close to the threshold $r_0$ are similar. Then for the sharp RDD, the causal effect $\tau_H$ is identified by assumption RD (20.1) and:

$$\textbf{Continuity at } r_0 \text{ for } \Psi_T = 0\text{:}\;\; f(y\,|\,r,\Psi_T = 0) \text{ is continuous in } r \text{ at } r_0 \text{ for all } y. \tag{20.8}$$

Equation (20.8) is a density version of the assumption in Hahn et al. (2001), who only look at mean shifts and hence assume that

$$\mathbb{E}(Y\,|\,r,\Psi_T = 0) \text{ is continuous in } r \text{ at } r_0. \tag{20.9}$$

We believe it is important to model mean shifts not by having a mean shift model but rather by modeling the density of the observations and then picking out the mean from this. This is the correct approach. Our contribution involves replacing identifying assumption (20.9) by (20.8), and hence we need to model the density nonparametrically.

Assuming (20.8) rather than only (20.9) allows the treatment effect to exhibit itself in more ways than a mean shift. For example, a variance shift would also be informative, even in the absence of a mean shift. The model we employ is quite general and allows many aspects of treatment effect to be explored by studying any differences between density estimate either side of the cut-off at $r_0$. However, we can obviously estimate the key mean shift, having modeled the density functions either side of the cut-off point $r_0$, simply by estimating the means of the two density functions.

Authors such as Lee (2008) and Lee and Lemieux (2010) further elaborated on the continuity assumption (20.8). They showed that if subjects have *imprecise control* of $R$ at $r_0$, then this continuity condition holds, and that treatments are "as good as randomly assigned" for the subgroup of subjects having values of the assignment variable $R$ located in a small neighborhood around $r_0$.

### 20.2.2 Identification in the Fuzzy RDD

In the fuzzy RDD, the probability function $\Pr(T = 1 \mid r)$ has a discontinuous jump that is smaller than 1 at $r_0$, meaning that $R$ does not determine $T$. Then $\mathbf{1}_{R \geq r_0}$ and $T$ are distinct variables since the event $\mathbf{1}_{R \geq r_0} \neq T$ is possible; and $T$ and $Y$ may both depend on unobserved confounding variables, collectively labeled as $U$. Given these considerations, we may extend the joint p.d.f. (20.5) to:

$$f(r,t,y,u \mid \psi_T) = f(r \mid u) f(t \mid r,u,\psi_T) f(y \mid r,t,u) f(u). \qquad (20.10)$$

This p.d.f. admits the conditional independence properties (a) $R \perp\!\!\!\perp \Psi_T \mid U$ and (b) $Y \perp\!\!\!\perp \Psi_T \mid R,T,U$ (and condition (a) can be strengthened to $(R,U) \perp\!\!\!\perp \Psi_T \mid U$). Here, (b) is a causal property, and (a) and (b) together imply the assumption that $R$ is a sufficient covariate, conditionally on $U$. But for the purposes of making causal inferences from real data, we cannot condition on $(R,T,U)$ because $U$ is unobserved.

However, for the fuzzy RDD, the assumptions RD (20.1) and continuity (20.8), along with certain additional assumptions also imply that the causal effect (20.4) is identified by the ratio:

$$\tau_H = \frac{\lim_{r \downarrow r_0} \mathbb{E}(H\{Y\} \mid r) - \lim_{r \uparrow r_0} \mathbb{E}(H\{Y\} \mid r)}{\lim_{r \downarrow r_0} \mathbb{E}(T \mid r) - \lim_{r \uparrow r_0} \mathbb{E}(T \mid r)}, \qquad (20.11)$$

for a general function $H\{\cdot\}$ of $Y$. The numerator of (20.11) is the Intention to Treat (ITT) effect. The denominator is a measure of treatment adherence, which decreases as noncompliance increases. In the sharp RDD, where $\mathbf{1}_{R \geq r_0} = T$, the denominator is 1 (i.e., perfect adherence), and then the ITT effect coincides with the causal effect $\tau_H$. See Hahn et al. (2001) for more details.

## 20.3 Estimating Causal Effects in an RDD

Here, we propose our Bayesian nonparametric model for causal inference in RDDs.

### *20.3.1 Bayesian Nonparametric Model*

For the sharp RDD, our Bayesian nonparametric model (Karabatsos and Walker 2012) is defined by:

$$f(y_i \,|\, r_i, t_i; \boldsymbol{\zeta}) = \sum_{j=-\infty}^{\infty} \mathrm{n}(y_i \,|\, \mu_j, \sigma_j^2)\,\omega_j\{\eta(r_i,t_i), \sigma(r_i,t_i)\},$$

$$i = 1, \ldots, n, \tag{20.12a}$$

$$\omega_j(\eta, \sigma) = \Phi\left(\frac{j-\eta}{\sigma}\right) - \Phi\left(\frac{j-1-\eta}{\sigma}\right) \tag{20.12b}$$

$$\eta(r,t) = \beta_0 + \beta_1 r + \beta_2 t \tag{20.12c}$$

$$\sigma^2(r,t) = \exp(\lambda_0 + \lambda_1 r + \lambda_2 t) \tag{20.12d}$$

$$\mu_j, \sigma_j^2 \,|\, \mu_\mu, \sigma_\mu^2, b_\sigma \sim \mathrm{N}(\mu_j \,|\, \mu_\mu, \sigma_\mu^2)\mathrm{IG}(\sigma_j^2 \,|\, 1, b_\sigma),\ j = 0, \pm 1, \pm 2, \ldots \tag{20.12e}$$

$$\mu_\mu, \sigma_\mu^2 \sim \mathrm{N}(\mu_\mu \,|\, \mu_0, \sigma_0^2)\mathrm{U}(\sigma_\mu \,|\, 0, b_{\sigma\mu}) \tag{20.12f}$$

$$b_\sigma, \boldsymbol{\beta}, \boldsymbol{\lambda} \sim \mathrm{Ga}(b_\sigma \,|\, a_0, b_0)\mathrm{N}(\boldsymbol{\beta}, \boldsymbol{\lambda} \,|\, \mathbf{0}, v\mathbf{I}) \tag{20.12g}$$

where the mixture weights $\omega_j\{\eta(r,t), \sigma(r,t)\}$ sum to 1 at each value of $(r,t)$. Also, the terms (20.12a) and (20.12b) may be deconstructed via the generation of a latent indicator variable $Z \sim \mathrm{N}(\eta, \sigma^2)$, and then taking $Y \sim \mathrm{N}(\mu_j, \sigma_j^2)$ if $j-1 < Z \leq j$.

The model (20.12) allows the entire probability density of the outcome variable $Y$ to change flexibly as a function of covariates. The parameter $\sigma(r,t)$ measures the multimodality of $f(y \,|\, r,t)$ (Karabatsos and Walker 2012). Specifically, as $\sigma(r) \to \infty$, the density $f(y \,|\, r,t)$ becomes more multimodal, with weights $\omega_j\{\eta(r,t), \sigma(r,t)\}$ converging to a discrete uniform distribution; and as $\sigma(r,t) \to 0$, the density $f(y \,|\, r,t)$ becomes more unimodal, and "local," with $f(y \,|\, r,t) \approx \mathrm{n}(y_i \,|\, \mu_j, \sigma_j^2)$ and

$$\omega_j\{\eta(r,t), \sigma(r,t)\} \approx 1 \ \text{ if } j-1 < \eta \leq j.$$

Furthermore, the model has a discontinuity at $r_0$ due to the presence of the term $T$ in both (20.12c) and (20.12d). The effect, controlled by the coefficients $(\boldsymbol{\beta}, \boldsymbol{\lambda})$, is to reallocate the weights either side of $r_0$, resulting in different densities either side of this value. Obviously, there is a discontinuity if and only if either of the coefficients $(\lambda_2, \beta_2)$ is nonzero. The normal prior $\mathrm{N}(\boldsymbol{\beta}, \boldsymbol{\lambda} \,|\, \mathbf{0}, v\mathbf{I})$ consists of a prior variance parameter $v$, which controls for both the prior support for the range of the mixture density component indices $j = 0, \pm 1, \pm 2, \ldots$ (via the parameter $\beta$), and for the range of the level of multimodality in $f(y \,|\, r,t)$. As $v \to \infty$, a wider range of component densities and multimodality is supported; and as $v \to 0$, $f(y \,|\, r,t)$ becomes a normal density.

When prior information is limited about the model parameters, we may attempt to specify non-informative priors, for example, by choosing $\mu_0 = 0, \sigma_0^2 \to \infty$, $a_0 \to 0$, $b_0 \to 0$, and $v = 10^5$, and by choosing $b_{\sigma\mu}$ according to prior knowledge about range of the $Y$ variance. For instance, if $Y$ is known to have a variance of 1, then $b_{\sigma\mu} = 5$ provides a vague prior choice. For such choices of prior parameters, the Bayesian model (20.12), over 22 real data sets, demonstrated very good predictive accuracy, and better predictive accuracy compared to many other regression models, and compared to the Bayesian model under different choices of prior (Karabatsos and Walker 2012).

The model (20.12) has infinite-dimensional parameter, $\zeta = ((\mu_j, \sigma_j^2)_{j=-\infty}^{\infty}, \mu_\mu, \sigma_\mu^2, b_\sigma, \boldsymbol{\beta}, \boldsymbol{\lambda})$, with prior density $\pi(\boldsymbol{\zeta})$. A set of data $\mathscr{D}_n = \{(y_i, r_i, t_i)\}_{i=1}^n$ updates the prior $\pi(\zeta)$ to a posterior density, given by

$$\pi(\boldsymbol{\zeta} \,|\, \mathscr{D}_n) = \frac{\prod_{i=1}^n f(y_i \,|\, r_i, t_i; \boldsymbol{\zeta})\pi(\zeta)}{\int \prod_{i=1}^n f(y_i \,|\, r_i, t_i; \zeta)\mathrm{d}\Pi(\boldsymbol{\zeta})},$$

with $\Pi(\boldsymbol{\zeta})$ (and $\Pi(\boldsymbol{\zeta} \,|\, \mathscr{D}_n)$, resp.) the c.d.f. of $\pi(\boldsymbol{\zeta})$ (of $\pi(\boldsymbol{\zeta} \,|\, \mathscr{D}_n)$). Also, let $F(y \,|\, r, t; \boldsymbol{\zeta})$ be the c.d.f. of $f(y \,|\, r, t; \boldsymbol{\zeta})$. Then the posterior predictive density, $f_n(y \,|\, r, t)$, and the conditional posterior predictive expectation ($\mathbb{E}_n$) and variance ($\mathbb{V}_n$) of the outcome $H\{Y\}$ are given (resp.) by:

$$\begin{aligned}
f_n(y \,|\, r, t) &= \int f(y \,|\, r, t;\, \boldsymbol{\zeta})\mathrm{d}\Pi(\boldsymbol{\zeta} \,|\, \mathscr{D}_n), \\
\mathbb{E}_n(H\{Y\} \,|\, r, t) &= \int \{\textstyle\int H\{y\}\mathrm{d}F(y \,|\, r, t; \boldsymbol{\zeta})\}\mathrm{d}\Pi(\boldsymbol{\zeta} \,|\, \mathscr{D}_n), \\
\mathbb{V}_n(H\{Y\} \,|\, r, t) &= \int \left[\textstyle\int \{H\{y\} - \mathbb{E}_n(H\{Y\} \,|\, r, t)\}^2 \mathrm{d}F(y \,|\, r, t;\, \boldsymbol{\zeta})\right] \mathrm{d}\Pi(\boldsymbol{\zeta} \,|\, \mathscr{D}_n).
\end{aligned}$$

Depending on the choice of function $H\{\cdot\}$, the posterior mean $\mathbb{E}_n$ and variance $\mathbb{V}_n$ of the conditional expectation $\mathbb{E}(Y \,|\, r, t)$, variance $\mathbb{V}(Y \,|\, r, t)$, c.d.f. $F(y \,|\, r, t)$ at a point $y$, are given (resp.) by $\mathbb{E}_n\{\mathbb{E}(Y \,|\, r, t)\}$ and $\mathbb{V}_n\{\mathbb{E}(Y \,|\, r, t)\}$; $\mathbb{E}_n\{\mathbb{V}(Y \,|\, r, t)\}$ and $\mathbb{V}_n\{\mathbb{V}(Y \,|\, r, t)\}$; and $\mathbb{E}_n\{F(y \,|\, r, t)\} = F_n(y \,|\, r, t)$ and $\mathbb{V}_n\{F(y \,|\, r, t)\}$. For assessing the fit of the Bayesian model to data, a standardized residual for each observation $y_i$ may be computed by

$$z_i = \{y_i - \mathbb{E}_n(Y \,|\, r_i, t_i)\}/\{\mathbb{V}_n(Y \,|\, r_i, t_i)\}^{1/2}.$$

If $|z_i| > 2$, then $y_i$ can be judged as an outlier.

### 20.3.2 Estimating Causal Effects with the Bayesian Model

For our Bayesian model, the posterior estimates of the causal effect of $T$ on $H\{Y\}$, conditionally on $R = r_0$, are given as follows, under the assumptions RD (20.1) and continuity at $r_0$ (20.8).

For the sharp RDD, the estimate $\widehat{\tau}_H$ of the causal effect is given by

$$\widehat{\tau}_H^{(S)} = \mathbb{E}_n(\tau_H^{(S)}) = \mathbb{E}_n(H\{Y\} \mid r_0, T=1) - \mathbb{E}_n(H\{Y\} \mid r_0, T=0), \tag{20.13}$$

with posterior variance $\mathbb{V}_n(\tau_H^{(S)}) = \mathbb{V}_n(H\{Y\} \mid r_0, 1) + \mathbb{V}_n(H\{Y\} \mid r_0, 0)$. Then $\widehat{\tau}_H^{(S)} \pm 2[\mathbb{V}_n(\tau_H^{(S)})]^{1/2}$ provides an approximate 95 % posterior confidence band around $\widehat{\tau}_H^{(S)}$.

When inferring the causal effect in terms of the $u$th quantile, via $\widehat{\tau}_H^{(S)} = F_n^{-1}(u \mid r, 1) - F_n^{-1}(u \mid r, 0)$, we may judge whether $\widehat{\tau}_H$ is significantly different from zero by using a P-P plot (Wilk and Gnanadesikan 1968) to check for non-overlap of the 95 % posterior credible intervals $(F_{n(0.025)}(y \mid r,t), F_{n(0.975)}(y \mid r,t))$ at $u$, for $T = 0, 1$, and over a wide range of points $y \in \mathscr{Y}$. Here, $F_{n(0.025)}(y \mid r,t)$ ($F_{n(0.975)}(y \mid r,t)$, resp.) denotes the posterior 2.5th percentile (97.5th percentile, resp.) of $F(y \mid r,t)$.

For the fuzzy RDD, the causal effect estimate $\widehat{\tau}_H$, in terms of the ratio estimator (20.11), may be obtained by two independent regressions. The first involves estimating the numerator using our regression model (20.12), after replacing the covariate $T$ with $\mathbf{1}_{R \geq r_0}$. The second involves a regression of $T$ on $(R, \mathbf{1}_{R \geq r_0})$ to estimate the denominator, via the posterior predictive expectations:

$$\mathbb{E}_n(T \mid r, \mathbf{1}_{r \geq r_0} = a) = \int \Pr(T = 1 \mid r, a;\, \boldsymbol{\zeta}_T) \mathrm{d}\Pi(\boldsymbol{\zeta}_T \mid \mathscr{D}_n),\ a = 0, 1.$$

Our model (20.12) can be extended to binary regression, by modeling the response density by:

$$\begin{aligned} &\Pr(T_i = 1 \mid r_i, \mathbf{1}_{r_i \geq r_0};\, \boldsymbol{\zeta}_T) \\ &\quad = \int_0^\infty \left[ \sum_{j=-\infty}^{\infty} \mathrm{n}(t_i^* \mid \mu_j, \sigma_j^2) \omega_j \{\eta(r_i, \mathbf{1}_{r_i \geq r_0}), \sigma(r_i, \mathbf{1}_{r_i \geq r_0})\} \right] \mathrm{d}t_i^*, \ \ i = 1, \ldots, n. \end{aligned} \tag{20.14}$$

analogous to (20.12a). This provides flexible modeling of the inverse link function by a covariate dependent, infinite mixture of normal c.d.f.s (Karabatsos and Walker 2012).

As before, denote $\pi(\boldsymbol{\zeta} \mid \mathscr{D}_n)$ as the posterior density for the model (20.12) for the $Y$ outcome; and denote $\pi(\boldsymbol{\zeta}_T \mid \mathscr{D}_n)$ as the posterior density for the version of the model for the $T$ outcome, using (20.14). Then both posterior densities admit the conditional independence property $\boldsymbol{\zeta} \perp\!\!\!\perp \boldsymbol{\zeta}_T \mid \mathscr{D}_n$, so then we can write $\pi(\boldsymbol{\zeta}, \boldsymbol{\zeta}_T \mid \mathscr{D}_n) = \pi(\boldsymbol{\zeta} \mid \mathscr{D}_n)\pi(\boldsymbol{\zeta}_T \mid \mathscr{D}_n)$. This means that the posterior densities of both models, $\pi(\boldsymbol{\zeta} \mid \mathscr{D}_n)$ and $\pi(\boldsymbol{\zeta}_T \mid \mathscr{D}_n)$, can be estimated either separately or jointly.

For the fuzzy RDD, an estimate of the causal effect, in terms of the ratio (20.11), is given by the posterior average of the ratio:

$$\mathbb{E}_n(\tau_H^{(F)}) = \int \left\{ \frac{\mathbb{E}(h\{Y\} \mid r_0, 1;\, \boldsymbol{\zeta}) - \mathbb{E}(h\{Y\} \mid r_0, 0;\, \boldsymbol{\zeta})}{\mathbb{E}(T \mid r_0, 1;\, \boldsymbol{\zeta}_T) - \mathbb{E}(T \mid r_0, 0;\, \boldsymbol{\zeta}_T)} \right\} \pi(\boldsymbol{\zeta}, \zeta_T \mid \mathscr{D}_n) \mathrm{d}(\boldsymbol{\zeta}, \boldsymbol{\zeta}_T),$$

for a given choice of function $h\{\cdot\}$, where the differences in the ratio above are based on values of $\mathbf{1}_{r\geq r_0} = 0, 1$. A computationally fast (but somewhat ad-hoc) first-order Taylor approximation to $\mathbb{E}_n(\tau_H^{(F)})$ is given by:

$$\widehat{\tau}_H^{(F)} = \widehat{\tau}_H^{(S)} / \{\mathbb{E}_n(T \mid r_0, \mathbf{1}_{r\geq r_0} = 1) - \mathbb{E}_n(T \mid r_0, \mathbf{1}_{r\geq r_0} = 0)\} = \widehat{\tau}_H^{(S)} / \mathbb{E}_n(D_T). \quad (20.15)$$

For example, given the choice of function $H\{Y\} = \mathbf{1}_{Y\leq y}$, we have the causal effect defined by a comparison of c.d.f.s at a point $y$, weighted by $\mathbb{E}_n(D_T)$, with

$$\widehat{\tau}_{\mathbf{1}(Y\leq y)}^{(F)} = \{F_n(y \mid r, 1) - F_n(y \mid r, 0)\} / \mathbb{E}_n(D_T).$$

The second-order approximation is given by

$$\widehat{\tau}_H^{(F[2])} = \{\widehat{\tau}_H^{(S)} / \mathbb{E}_n(D_T)\} + [\{\widehat{\tau}_H^{(S)} \mathbb{V}(D_T)\} / \{\mathbb{E}_n(D_T)\}^3\}],$$

with $\mathbb{V}(D_T) = \mathbb{V}_n(T \mid r_0, 1) + \mathbb{V}_n(T \mid r_0, 0)$. The posterior variance $\mathbb{V}_n(\tau_H^{(F)})$ has first-order approximation:

$$\mathbb{V}_n(\tau_H^{(F)}) \approx \{\widehat{\tau}_H^{(S)} / \mathbb{E}_n(D_T)\}^2 [\mathbb{V}_n(\tau_H^{(S)})(\widehat{\tau}_H^{(S)})^{-2} + \mathbb{V}(D_T)\{\mathbb{E}_n(D_T)\}^{-2}]. \quad (20.16)$$

These approximations are derived from standard results involving the distribution of the ratio of two random variables (e.g., Stuart and Ord 1998, p. 351). Then $\widehat{\tau}_H^{(F)} \pm 2\{\mathbb{V}_n(\tau_H^{(F)})\}^{1/2}$ gives a 95 % posterior interval around $\widehat{\tau}_H^{(F)}$. Also, when inferring the causal effect $\widehat{\tau}_H^{(F)}$ in terms of treatment and non-treatment differences at the $u$th quantile, we may judge whether $\widehat{\tau}_H$ is significantly different from zero by using a P-P plot of the 95 % posterior intervals $F_n(y \mid r, a) \pm 2\{\mathbb{V}_n(\tau_{\mathbf{1}(Y\leq y)}^{(F)})\}$, over points $y \in \mathbb{R}$, and then checking for nonoverlap for these intervals at point $u$.

Alternatively, it may be of interest to investigate the sensitivity of the causal effect estimate $\widehat{\tau}_H^{(F)}$ to variations of treatment adherence (e.g., compliance). This can be achieved by estimating the ratio $\widehat{\tau}_H^{(F)}$ for each of a set of fixed nonzero values (e.g., $1, .9, .8, \ldots, -1$) for the denominator, with each estimate having posterior variance

$$\mathbb{V}_n(\tau_H^{(F)}) \approx (\widehat{\tau}_H^{(S)} / \mathbb{E}_n(D_T))^2 \{(\mathbb{V}_n(\tau_H^{(S)}))(\widehat{\tau}_H^{(S)})^{-2}\}.$$

Using Markov Chain Monte Carlo (MCMC), Gibbs sampling methods, along with a slice sampling step for $\sigma_\mu$, can be used to estimate all of the aforementioned posterior quantities (Karabatsos and Walker 2012). We use Rao-Blackwell (RB) methods to estimate all the posterior linear functionals, such as $\mathbb{E}_n(H\{Y\} \mid r, a)$, $\mathbb{V}_n(H\{Y\} \mid r, a)$, $r_i$, $\mathbb{E}_n(T \mid r, a)$ (for $a = 0, 1$), $\mathbb{V}_n(T \mid r_0, a)$, $\mathbb{E}_n(\tau_H^{(S)})$, $\mathbb{V}_n(\tau_H^{(S)})$, $\mathbb{E}_n(\tau_H^{(F)})$, and $\mathbb{V}_n(\tau_H^{(F)})$ (Gelfand and Mukhopadhyay 1995).

## 20.4 Illustrative Applications

The Bayesian nonparametric model was illustrated through the analysis of two data sets, using menu-driven software that was developed by the first author (Karabatsos 2014a,b). The first data set was collected from four Chicago University schools of education, which established a new curriculum that aims to train teachers to help improve Chicago public schools. This data set involved a sharp RDD. The second data set, obtained from Angrist and Lavy (2008), involves a fuzzy RDD, from a study of the effect of class size on student achievement (Angrist and Lavy 1999). For each data set, it seems reasonable to make the assumptions of RD (20.1) and continuity (20.8) at $r_0$ (i.e., imprecise control at $r_0$), in order to identify the causal effects of treatment on the outcome, conditionally on $r_0$ (see Sect. 20.2).

For both data sets, the Bayesian nonparametric model assumed the same vague priors that were mentioned in Sect. 20.3.2. According to standardized residuals, the model under these priors provided good predictive accuracy for each data set.

All posterior estimates of this model, reported in the next two subsections, are based on 40K MCMC samples. These samples were obtained from every fifth iterate of a run of 200K MCMC sampling iterations, after discarding the first 2K burn-in samples. This provided accurate posterior estimates according to standard convergence assessments (Geyer 2011). Specifically, univariate trace plots displayed good mixing of model parameters and posterior predictive samples, and all posterior predictive estimates obtained 95 % MC confidence intervals with half-width sizes near .01.

### 20.4.1 Learning Math Teaching: Time Series Data

For the first data set, the aim is to estimate the effect of the new teacher education curriculum on math teaching ability, among $n = 347$ undergraduate teacher education students attending one of four Chicago universities. This data set involves a sharp RDD, an interrupted time-series design (Cook and Campbell 1979) using an assignment variable of time, ranging from fall semester 2007 through spring semester 2013. The new curriculum (treatment) was instituted in Fall 2010 (the cutoff, $r_0$), and the old teacher curriculum (non-treatment) was active before then. The outcome variable ($Y$) is the number-correct score on the 25-item Learning Math for Teaching (LMT) test (LMT 2012). Each of the students completed the LMT test (89.9 % female; 135 and 212 students under the old and new curriculum), after finishing a course on teaching algebra. Among them, the average LMT score was 12.9 (s.d. $= 3.44$), with Cronbach's alpha reliability 0.63. The LMT scores were transformed to z-scores, having sample mean 0 and variance 1.

Using our Bayesian model, we analyzed the data to estimate the effect of the new curriculum, versus the old curriculum, on student ability to teach math (LMT score), at the Fall semester 2010 cutoff. The model included the LMT test z-score as the dependent variable ($Y$), and included covariates of the assignment variable

($R$), given by TimeF10 = Year − 2010.6, and of the treatment assignment variable CTPP = $\mathbf{1}_{\text{Year} \geq 2010.6}$. The cutoff 2010.6 is the time midpoint between Spring 2010 (2010.3) and Fall 2010 (2010.9).

For the model, R-squared was 0.99, and nearly all the standardized residuals ranged between −1 and 1 over the 347 observations, with one residual slightly exceeding 2. Figure 20.1 presents the model's posterior predictive density estimate of the LMT outcome, for the new curriculum (treatment) and for the old curriculum (non-treatment), at Fall 2010. The new curriculum, compared to the old, increased the LMT scores, by shifting the density of LMT scores to the right. This shift corresponds to an increase in the mean (from −0.17 to −0.13), the tenth percentile (−2.01 to −1.97), and the 25th percentile (−1.43 to −1.31), but these increases were not statistically significant from zero according to 95 % credible intervals of the predictive mean and of the posterior c.d.f. estimates. Also, each density presents two modes (clusters) of students, indicating the presence of a latent binary covariate.

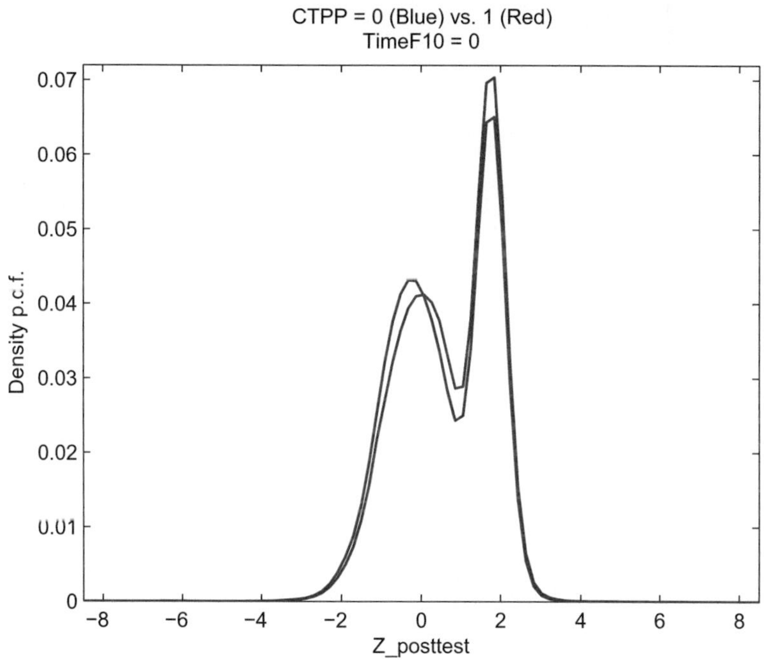


**Fig. 20.1** Posterior predictive density estimates of $Y$, under treatment ($T = 1$, *red*), and under non-treatment ($T = 0$, *blue*)

## *20.4.2 Maimonides' Data: Fuzzy RDD*

The twelfth-century rabbinic scholar Maimonides proposed a rule that specifies a maximum class size of 40, under the belief that smaller class sizes promote higher student achievement (see Hyamson 1937, p. 58b). Specifically, for a given class $c$ in school $s$, the rule assigns average class size ($\text{Psize}_{sc}$) as a function of beginning-of-the-year school enrollment ($e_s$), according to the prediction equation $\text{Psize}_{sc} = e_s/\text{floor}[((e_s - 1)/40) + 1]$. The rule (equation) assigns students of a school into a single classroom when the school's enrollment is less than 41, assigns students into two classrooms of average size 20.5 when school enrollment reaches 41; assigns students into three classrooms of average size 27 when enrollment reaches 81; and so on. The cutoff number 20.5 distinguishes between small and large classes.

Here, we study the effect of class size on average class verbal achievement, through the analysis of data on fourth grade students who each attended one of 2,056 classes in Israeli public schools during 1991. These schools used Maimonides' rule to allocate students into classrooms. Demographic statistics are reported in Angrist and Lavy (1999) (three other classes were not analyzed because they had missing achievement data). For the Bayesian model, the dependent variable ($Y$) is average class verbal score (avgverb), which we transformed to z-scores with sample mean 0 and variance 1. The covariates include the assignment variable ($R$), defined by the rule-predicted class size centered at the cutoff 20.5 (i.e., Psize205 = Psize − 20.5), and include the indicator of large (vs. small) class assignment, Plarge = $\mathbf{1}_{\text{Psize}\geq 20.5}$. Now, while Maimonides' rule may assign a given class to be a large (small, resp.), the class could become small (large, resp.). For example, one school in the data set had an enrollment of 41, leading to some students receiving a large class of 21, and other students receiving a small class of 20. Therefore, the data arise from a fuzzy RDD, and for the data analysis, we also consider a variable defined by the indicator of large class receipt, large = $\mathbf{1}_{\text{classize}\geq 20.5}$. We also fit the Bayesian model, with the treatment ($T$) variable, large, as the dependent variable, and with covariates Psize205 and Plarge.

For the avgverb ($Y$) dependent variable, the Bayesian model obtained an R-squared of 0.88, with standardized fit residuals ranging from −1.1 to 1.3 over the 2,056 observations. Thus the model had no outliers. For the treatment ($T$) dependent variable, large, the Bayesian model had no outliers, and estimated 0.93 as the denominator of the causal effect estimator (20.11). Figure 20.2 presents the model's posterior predictive density estimates of the avgverb outcomes, for the treatment versus the non-treatment, each divided by 0.93. It was found that large class size (versus small) causally increased the verbal score, in terms of the 5th percentile (−2.45 to −2.19), 10th percentile (−1.81 to −1.66), 25th percentile (−0.94 to −0.88), and causally decreased the score in terms of the 75th percentile (0.90 to 0.79), 90th percentile (1.48 to 1.37), and 95th percentile (1.76 to 1.66). Each of these estimates is based on taking the predictive quantile estimates of the Bayesian model for avgverb ($Y$), and dividing them by 0.93. Also, each density presents two modes (clusters) of students, indicating the presence of a latent binary covariate.

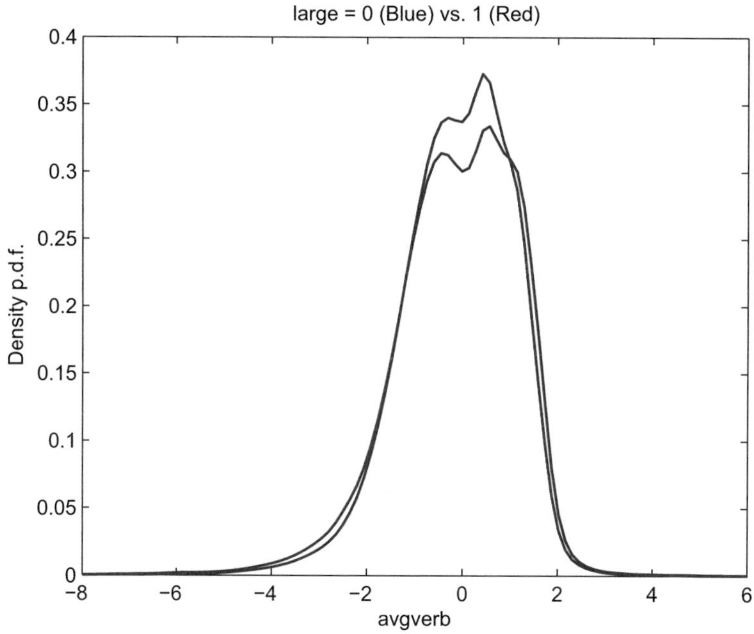


**Fig. 20.2** Posterior predictive density estimates of $Y$, under treatment ($T = 1$, *red*), and under non-treatment ($T = 0$, *blue*)

## 20.5 Conclusions

We proposed and illustrated a flexible Bayesian nonparametric regression model for causal inference in RDDs. Such designs identify causal effects under relatively mild conditions. While the existing linear models for RDDs only focus on mean causal effects, the Bayesian model provides inferences of causal effects in terms of the mean, variance, distribution function, quantile, probability density, or any other functional of the outcome variable.

In future work, the Bayesian nonparametric regression modeling approach will be extended to handle RDDs involving a multivariate assignment variable, $\mathbf{R} \in \mathbb{R}^K$, which assigns to the treatment condition (versus nontreatment) if and only if $\mathbf{R} \in S$, for some set $S$. In this case, the measure of the causal effect ($\tau_H$) no longer depends on a single cutpoint $r_0$, but instead depends on multiple cutpoints, defined by the boundary points of $S$. In principle, it is straightforward to extend the model to handle a multivariate RDD, because then the model would simply include $\mathbf{R}$, along with a 0-1 indicator of the event $\mathbf{R} \in S$, as covariates. A future study will carefully study how causal effects can be summarized over the multiple boundary points of $S$, via the model's posterior predictive distribution.

For the Bayesian nonparametric model discussed in this chapter, a user-friendly and menu-driven software is freely available, entitled: "Bayesian Regression: Nonparametric and Parametric Models" (Karabatsos 2014a,b). This free software package can be downloaded and installed from:

http://tigger.uic.edu/~georgek/HomePage/BayesSoftware.html.

The Bayesian nonparametric model can be easily specified for data analysis, by clicking the menu options "Specify New Model" and "Infinite probits regression model." Afterwards, the item response (dependent) variable, covariates, and prior parameters can be easily selected (clicked) by the user. Then, to run for data analysis, the user clicks the "Run Posterior Analysis" button to run the MCMC sampling algorithm for a chosen number of sampling iterations. Immediately after the completion of the MCMC run, the software automatically opens a text output file containing the results of the data analysis, including summaries of the posterior distribution of the model, obtained from the MCMC samples. The software also allows the user to conveniently check for MCMC convergence, through a menu option that can be clicked to construct trace plots, and through another menu option that can be clicked to run a batch means analyses to construct 95 % Monte Carlo confidence intervals of the posterior estimates of the model parameters. Other menu options of the software allow the user to construct plots and (additional) text output of the (MCMC estimated marginal) posterior distributions of the model parameters (e.g., box plots), and allow the user to output text and residual plots that report the fit of the model in greater detail. Additional menu options allow the user to construct posterior predictions of the model, as a function of the covariates, in terms of the mean, variance, quantile, probability density, distribution function, or other chosen functions of the outcome variable.

Currently, the software provides the user a choice of 59 statistical models, including a large number of Bayesian nonparametric regression models. The software allows the user to specify Dirichlet process (DP) mixture models, and more generally, mixture regression models based on the stick-breaking process (Ishwaran and James 2001). The mixing can be done either on the intercept parameter, or on the entire vector of regression coefficient parameters, depending on the user's choice. The latter mixture model gives rise to a Dependent Dirichlet (DDP) process mixture model (see DeIorio et al. 2004). In principle any one of these DP or DDP mixture models can also be used to perform causal inferences from an RDD design, using the inference methods discussed earlier in this chapter.

**Acknowledgements** This research is supported by NSF grant SES-1156372. Thanks to Phillip Dawid and Peter Müller for helpful comments. The results of this paper were presented in a University of Texas statistics seminar during Fall of 2013; and in sessions on causal analysis for the ISBA conference during Summer of 2014 in Cancun; for the Society of Research on Educational Effectiveness (SREE) Spring 2014 conference in Washington, D.C.; and for JSM 2013 at Montreal.

## References

Aiken, L., West, S., Schwalm, D., Carroll, J., and Hsiung, S. (1998). Comparison of a randomized and two quasi-experimental designs in a single outcome evaluation efficacy of a university-level remedial writing program. *Evaluation Review*, **22**, 207–244.

Angrist, J. and Lavy, V. (1999). Using Maimonides' rule to estimate the effect of class size on scholastic achievement. *The Quarterly Journal of Economics*, **114**, 533–575.

Angrist, J. and Lavy, V. (2008). Replication data for: Using Maimonides' rule to estimate the effect of class size on student achievement. http://thedata.harvard.edu/dvn/dv/JAngrist/. Accessed: 2014-06-05.

Berk, R., Barnes, G., Ahlman, L., and Kurtz, E. (2010). When second best is good enough: A comparison between a true experiment and a regression discontinuity quasi-experiment. *Journal of Experimental Criminology*, **6**, 191–208.

Black, D., Galdo, J., and Smith, J. (2005). Evaluating the regression discontinuity design using experimental data. Unpublished manuscript.

Bloom, H. (2012). Modern regression discontinuity analysis. *Journal of Research on Educational Effectiveness*, **5**, 43–82.

Buddelmeyer, H. and Skoufias, E. (2004). *An evaluation of the performance of regression discontinuity design on PROGRESA*. World Bank Publications.

Cook, T. (2008). Waiting for life to arrive: A history of the regression-discontinuity design in psychology, statistics and economics. *Journal of Econometrics*, **142**, 636–654.

Cook, T. and Campbell, D. (1979). *Quasi-Experimentation: Design and Analysis Issues for Field Settings*. Rand McNally, Chicago.

Dawid, A. (1979). Conditional independence in statistical theory. *Journal of the Royal Statistical Society, Series B*, **41**, 1–31.

Dawid, A. (2000). Causal inference without counterfactuals. *Journal of the American Statistical Association*, **95**, 407–424.

Dawid, A. (2002). Influence diagrams for causal modelling and inference. *International Statistical Review*, **70**, 161–189.

Dawid, A. (2010). Beware of the DAG! *Journal of Machine Learning Research-Proceedings Track*, **6**, 59–86.

DeIorio, M., Müller, P., Rosner, G., and MacEachern, S. (2004). An ANOVA model for dependent random measures. *Journal of the American Statistical Association*, **99**, 205–215.

Van der Klaauw, W. V. (2008). Regression–discontinuity analysis: A survey of recent developments in economics. *Labour*, **22**, 219–245.

Gelfand, A. and Mukhopadhyay, S. (1995). On nonparametric Bayesian inference for the distribution of a random sample. *Canadian Journal of Statistics*, **23**, 411–420.

Geyer, C. (2011). Introduction to MCMC. In S. Brooks, A. Gelman, G. Jones, and X. Meng, editors, *Handbook of Markov Chain Monte Carlo*, pages 3–48, Boca Raton, FL. CRC.

Goldberger, A. (2008/1972). Selection bias in evaluating treatment effects: Some formal illustrations. In D. Millimet, J. Smith, and E. Vytlacil, editors, *Modelling and evaluating treatment effects in economics*, pages 1–31, Amsterdam. JAI Press.

Hahn, J., Todd, P., and der Klaauw, W. V. (2001). Identification and estimation of treatment effects with a regression-discontinuity design. *Econometrica*, **69**, 201–209.

Hyamson, M. (1937). *Annotated English translation of Maimonides Mishneh Torah, Book I (The Book of Knowledge)*. Jewish Theological Seminary, New York.

Imbens, G. (2004). Nonparametric estimation of average treatment effects under exogeneity: A review. *The Review of Economics and Statistics*, **86**, 4–29.

Imbens, G. W. and Lemieux, T. (2008). Regression discontinuity designs: A guide to practice. *Journal of Econometrics*, **142**, 615–635.

Ishwaran, H. and James, L. (2001). Gibbs sampling methods for stick-breaking priors. *Journal of the American Statistical Association*, **96**, 161–173.

Karabatsos, G. (2014a). Bayesian Regression: Nonparametric and parametric models, version 2014x. http://www.uic.edu/~georgek/HomePage/BayesSoftware.html.

Karabatsos, G. (2014b). Bayesian Regression: Nonparametric and parametric models, version 2014b. Software users manual. http://www.uic.edu/~georgek/HomePage/BayesSoftware.html.

Karabatsos, G. and Walker, S. (2012). Adaptive-modal Bayesian nonparametric regression. *Electronic Journal of Statistics*, **6**, 2038–2068.

Lauritzen, S. (2000). Causal inference from graphical models. In O. Barndorff-Nielsen, D. Cox, and C. Kluppelberg, editors, *Complex Stochastic Systems*, pages 63–107, London. CRC Press.

Lee, D. (2008). Randomized experiments from non-random selection in U.S. house elections. *Journal of Econometrics*, **142**, 675–697.

Lee, D. and Lemieux, T. (2010). Regression discontinuity designs in economics. *The Journal of Economic Literature*, **48**, 281–355.

LMT (2012). *Learning Mathematics for Teaching (LMT) Assessment*. University of Michigan, Ann Arbor, MI.

Rosenbaum, P. and Rubin, D. (1983). Assessing sensitivity to an unobserved binary covariate in an observational study with binary outcome. *Journal of the Royal Statistical Society, Series B*, **45**, 212–218.

Rubin, D. (1974). Estimating causal effects of treatments in randomized and nonrandomized studies. *Journal of Educational Psychology*, **66**, 688–701.

Rubin, D. (1978). Bayesian inference for causal effects: The role of randomization. *Annals of Statistics*, **6**, 34–58.

Rubin, D. (1990). Neyman (1923) and causal inference in experiments and observational studies. *Statistical Science*, **5**, 472–480.

Rubin, D. (2008). For objective causal inference, design trumps analysis. *The Annals of Applied Statistics*, **2**, 808–840.

Schochet, P. (2009). Statistical power for regression discontinuity designs in education evaluations. *Journal of Educational and Behavioral Statistics*, **34**, 238–266.

Shadish, W., Galindo, R., Wong, V., Steiner, P., and Cook, T. (2011). A randomized experiment comparing random and cutoff-based assignment. *Psychological Methods*, **16**, 179.

Stuart, A. and Ord, K. (1998). *Kendall's Advanced Theory of Statistics: Volume 1: Distribution Theory*. Wiley, New York.

Thistlewaite, D. and Campbell, D. (1960). Regression-discontinuity analysis: An alternative to the ex-post facto experiment. *Journal of Educational Psychology*, **51**, 309–317.

Trochim, W. (1984). *Research design for program evaluation: The regression-discontinuity approach*. Sage, Newbury Park, CA.

Wilk, M. and Gnanadesikan, R. (1968). Probability plotting methods for the analysis of data. *Biometrika*, **55**, 1–17.

Wong, V., Steiner, P., and Cook, T. (2013). Analyzing regression-discontinuity designs with multiple assignment variables: A comparative study of four estimation methods. *Journal of Educational and Behavioral Statistics*, **38**, 107–141.

# Chapter 21
# Bayesian Nonparametrics for Missing Data in Longitudinal Clinical Trials

Michael J. Daniels and Antonio R. Linero

**Abstract** We discuss the problem of performing inference on a causal effect of interest, such as an intention-to-treat effect, in the context of longitudinal clinical trials with informatively missing data. Addressing this problem requires the modeling of infinite-dimensional nuisance parameters; modeling these nuisance parameters poorly can result in substantial bias in the original estimation problem. Additionally, the presence of informative (nonignorable) missingness results in effects of interest being unidentified in the absence of strong, unverifiable, assumptions. We argue that Bayesian nonparametric methods are natural in this setting because they (1) allow for flexible modeling and (2) allow for uncertainty in untestable assumptions to be taken into account through the use of informative priors elicited from subject matter experts. We further argue that a sensitivity analysis to assess the impact of unverifiable assumptions is essential. Flexible Bayesian approaches which incorporate the longitudinal structure of the data are presented in the context of categorical and continuous outcomes, and strategies for sensitivity analysis are discussed in both cases. The methods are illustrated on data from a clinical trial designed to assess the efficacy of treatments for acute Schizophrenia.

M.J. Daniels (✉)
Department of Statistics & Data Sciences and Department of Integrative Biology, University of Texas at Austin, Austin, TX 78712, USA
e-mail: mjdaniels@austin.utexas.edu

A.R. Linero
Department of Statistics, Florida State University, Tallahassee, FL 32306-4330, USA
e-mail: arlinero@stat.fsu.edu

 
R. Mitra, P. Müller (eds.), *Nonparametric Bayesian Inference in Biostatistics*, Frontiers in Probability and the Statistical Sciences,
DOI 10.1007/978-3-319-19518-6_21

## 21.1 Introduction

In longitudinal clinical trials, typically one only observes a subset of the data intended to be collected. In regulatory settings, one is usually interested in assessing the effect of a treatment on a well-defined causal effect of interest, such as the *intention-to-treat* effect of randomization to different treatment regimes. The presence of missing data results in such causal effects being unidentified in the absence of strong, untestable assumptions.

This chapter provides a Bayesian nonparametric perspective on the problem of estimating these effects. Proper analysis in the presence of missing data inherently depends on subjective assessments made by the analyst. Here, we find the Bayesian approach particularly attractive, as it allows for the principled incorporation of subject-matter expertise directly into the analysis through an informative prior. Quoting Hogan et al. (2014), to address identifiability issues, one might:

(i) Make an assumption—for example, that the missing data are missing at random (MAR)—and assume it holds with no further critique.
(ii) Fit several different models to the joint distribution of the response and missingness and assess how inferences change.
(iii) Fit an under-identified model and focus on obtaining uncertainty regions for the effect of interest (Manski 2009; Vansteelandt et al. 2006).

Unless an assumption like MAR is known to hold due to subject-matter considerations, we view approach (i) to be unwise. We will primarily consider approach (ii). Our focus will be on methods which consider different identifying assumptions, but which—crucially—leave the model for the observed data generating distribution unchanged. We term an assessment of the impact of missingness assumptions on inferences a *sensitivity analysis*. The Bayesian approach is then used to incorporate uncertainty in the identifying assumptions and allow the analyst to reach a final inference. The use of informative priors on unidentified components of the model is also closely related to the frequentist approach of accounting for uncertainty about missingness assumptions via bounds; for example, Hogan et al. (2014) illustrate that basing inference on uncertainty regions often coincides with a particular choice of prior distribution.

Unlike many applications of Bayesian nonparametrics, our interest is in low-dimensional functionals of the underlying distribution, such as the mean response at completion of the study. In the absence of missing data the use of Bayesian nonparametrics here might be criticized as being unnecessarily complex; for example, one might assume that the distribution of the response is multivariate Gaussian with the inferences robust to violations of this assumption. Unfortunately, the presence of missing data requires us to model quantities which would normally be considered nuisance parameters; moreover, the accuracy with which we estimate these nuisances has a direct impact on the quality of inference about the quantities we are interested in. Under ignorability, for example, one needs to estimate either the conditional regression functions of a response given its history, or the probability of completing the study given the response; in the absence of parametric assumptions, both options require the estimation of an infinite-dimensional object. The nonparametric Bayes approach is then a natural approach to take, with the Bayesian paradigm being used (a) to construct flexible models which are appropriately shrunk toward simpler parametric models and (b) to allow for uncertainty in underlying assumptions.

### *21.1.1 Notation and Definitions*

The $J \times 1$ vector $y_i = (y_{i1}, \dots, y_{iJ})$ is a vector of observations intended to be collected on subject $i = 1, \dots, n$, and $r_i$ is the vector of observed data indicators such that $y_{ij}$ is observed if $r_{ij} = 1$. The response vector $y_i$ can be partitioned into $y_{\text{obs},i} = (y_{ij} : r_{ij} = 1)$ and $y_{\text{mis},i} = (y_{ij} : r_{ij} = 0)$. We assume that complete data $c_i = (y_i, r_i)$ is drawn jointly from some joint density $p(y, r)$. The observed data is written $o_i = (y_{\text{obs},i}, r_i)$. We will let $o_{1:n} = (o_1, \dots, o_n)$ denote the collection of observed data on all subjects, and similarly let $c_{1:n}$ denote the collection of complete data.

Let $\psi(f)$ denote a target functional of interest of the *full data response model* $f(y) = \int p(y, r)\ dr$. Our focus will primarily be on mean functionals $\psi_j = \int y_j f(y)\ dy$. The model for the conditional distribution of missingness given $y$ will be written $\pi(r \mid y)$ and is called the *missing data mechanism.*

Missing data is often classified based on the form of the missing data mechanism (Rubin 1976). Missing completely at random (MCAR) corresponds to the missingness not depending on the full data response,

$$\pi(r_i \mid y_i) = \pi(r_i), \qquad i = 1, \dots, n.$$

MCAR may hold, for example, if budget constraints force analysts to follow up on only a random subset of the respondents; it is typically only a reasonable assumption if the missingness was designed by the investigator. Missing at random (MAR) corresponds to the missingness being independent of the missing responses, conditional on the observed responses,

$$\pi(r_i \mid y_i) = \pi(r_i \mid y_{\text{obs},i}), \qquad i = 1, \dots, n.$$

MCAR is implied by MAR. We say the missingness is missing not at random (MNAR) if the missingness is not MAR. MNAR missingness is sometimes called *informative missingness.*

$\Pi$ will denote the prior on $(p, c_{1:n})$. We will abuse notation writing, for example, $p(r \mid y_{\text{obs}})$ for the conditional probability of $r$ given $y_{\text{obs}}$, and $\Pi(df \mid o_{1:n})$ for the marginal posterior of $f$ given the observed data, when it will not create any ambiguity.

For likelihood-based inference, a more informative categorization is *ignorable* versus *nonignorable* missingness. For Bayesian inference, missingness is ignorable (Rubin 1976) when the following conditions hold:

1. Missingness is MAR.

2. $f$ and $\pi$ are a priori independent, $\Pi(df, d\pi) = \Pi(df) \times \Pi(d\pi)$.

When these conditions hold, it can be shown that:

$$\Pi(df, d\pi \mid o_{1:n}) = \Pi(df \mid y_{\text{obs},1}, \ldots, y_{\text{obs},n}) \times \Pi(d\pi \mid o_{1:n}).$$

As such, only the full data response model $f(y)$ needs to be modeled to make inference on functionals $\psi(f)$. Missingness is nonignorable when any of the conditions above do not hold. If MAR holds but $f$ and $\pi$ are a priori dependent then one has nonignorable MAR missingness. For nonignorable missingness, the joint distribution $p(y, r)$ needs to be modeled.

In the setting of a longitudinal study, missingness is said to be *monotone* if missingness is solely due to dropout, i.e., $r_j = 0$ implies $r_k = 0$ for all $k > j$. If this is not true, we call the missingness *non-monotone*; in longitudinal studies with non-monotone missingness, non-dropout missingness is also called *intermittent*. For monotone missingness, the information in the observed data indicators $r$ is completely contained in the number of observed responses $s = \max\{j : r_j = 1\}$. In this setting (and in longitudinal data analysis in general), it is convenient to summarize the response history up to $j$ by $\bar{y}_{ij} = (y_{i1}, \ldots, y_{ij})$.

Monotonicity of missingness simplifies analyses, but does not always hold. Generalizations of the concepts of MAR and ignorability provide a default way to address intermittent missingness. The missing data $y_{\text{mis}}$ is said to be *partially missing at random* (PMAR) (Harel and Schafer 2009) given $h(r)$ if there exists some coarsening $h(r)$ of $r$ such that

$$p(r \mid y, h(r)) = p(r \mid y_{\text{obs}}, h(r)).$$

This expresses the notion that any dependence between $y_{\text{mis}}$ and $r$ can be captured through the dependence between $y_{\text{mis}}$ and $h(r)$. If $h(r)$ is the dropout time $s$, then we are making the assumption that the only aspect of the missingness which is dependent on $y_{\text{mis}}$ is the dropout time. Similarly, when the functions $p(r \mid y, h(r))$ and $p(y, h(r))$ are a priori independent, missingness is said to be *partially ignorable* given $h(r)$, and one is free to model only the coarsening $h(r)$ and the response $y$ when conducting inference.

### 21.1.2 Literature on Bayesian Nonparametrics in Missing Data Models

Bayesian approaches deal very naturally with missing data. Because the Bayesian approach quantifies all uncertainties via probability, one may address missingness by treating the missing data in the same manner one treats any unknown parameter. The literature on this topic is too large to do justice; for comprehensive summaries see Little and Rubin (1986), Daniels and Hogan (2008) and Molenberghs et al. (2014). Under ignorability, likelihood-based inference does not depend on the missing data mechanism, so frequently the analyst will use whatever model they would have used had there been no missing data. The missing data may then be accommodated in a straight-forward manner during computations using data augmentation. This procedure aligns so closely with the Bayesian perspective that missing data can often be handled automatically by black-box Gibbs sampling platforms like BUGS or JAGS. Unfortunately, the simplicity of the computations often leads investigators to think of missingness as a nuisance that can be handled in a default manner.

Similarly, many Bayesian nonparametric methods have been proposed which capture time-varying dependencies. Dunson (2006, 2007) proposed a dynamic Dirichlet process (dDP) to model the change in underlying latent-trait distributions over time in a flexible manner. Ren et al. (2008) proposed a dynamic hierarchical Dirichlet process (dHDP) in the spirit of the hierarchical Dirichlet process (Teh et al. 2006) to give a more flexible sharing of atoms across time. Dunson and Herring (2006) proposed a functional Dirichlet process to flexibly model and cluster the responses of individuals based on their latent trajectories. Beyond the Dirichlet process and similar constructions, time varying analogues of the Indian Buffet Process have also been considered (Gael et al. 2009; Williamson et al. 2010). In principle, these methods can be easily adapted to accommodate ignorable missing data.

Bayesian nonparametrics have also been used to facilitate multiple imputation (MI) (Little and Rubin 1986; Rubin 1987). MI methods impute the missing data from the predictive distribution $\Pi(dc_{1:n} \mid o_{1:n})$ a fixed number of times. The completed datasets are then analyzed separately, with the inferences obtained for each dataset combined in a principled manner. General wisdom holds that the imputation model should be "big enough" to preserve whatever relationships might be studied later. Bayesian nonparametric methods are attractive because they allow for the preservation of complex relationships in the data. In this direction, Si and Reiter (2013) proposed a Dirichlet mixture model for multiple imputation of high-dimensional survey data. However, it is often the case that the model used for imputation is not compatible with the model used for inference; for details see Meng (1994) and Daniels et al. (2014).

#### 21.1.2.1 Likelihood Factorizations

Likelihood-based approaches for nonignorable missingness can largely be categorized by how the joint distribution $p(y,r)$ is factorized when setting up the model. Selection models (Heckman 1979; Diggle and Kenward 1994) are based on the factorization

$$p(y,r) = f(y) \times \pi(r \mid y).$$

Selection models often possess the "benefit" of fully identifying the effects of interest. We feel that full identification of $p(y,r)$ masks inherent identifiability difficulties. This can be overcome by making the model suitably nonparametric; for example, in a cross-sectional setting, Scharfstein et al. (2003) used a Dirichlet process model for the response $f(y)$, with the selection model parametrized by a weakly-identified sensitivity parameter. By contrast, in the context of spatial statistics, Pati et al. (2011) developed a Gaussian process model for a spatial response and addressed informative choice of sampling locations with a selection model, completely identifying the joint. In general, it is difficult to specify a selection model in non-simple settings that allows a true sensitivity analysis.

Pattern-mixture models (Little 1993, 1994; Hogan and Laird 1997) consider the opposite factorization,

$$p(y,r) = g(y \mid r) \times \phi(r).$$

In the absence of further assumptions, the model $\phi(r)$ is fully identified by the data while the mixture model $g(y \mid r)$ is not. A benefit of pattern-mixture models is that they force the analyst to confront identifiability issues directly. $g(y \mid r)$ may be identified by specifying parametrically how information is shared across missingness patterns or by making modeling assumptions in unidentified patterns, or may be left partially identified to facilitate a sensitivity analysis (Daniels and Hogan 2000; Thijs et al. 2002; Daniels and Hogan 2008).

Shared parameter models (Wu and Carroll 1988; Henderson et al. 2000), introduce latent variables such that

$$p(y,r) = \int p(y,r \mid b)\, G(db). \tag{21.1}$$

To simplify the structure, it is typical to assume independence of $y$ and $r$ conditional on the random effect $b$, although Fieuws and Verbeke (2006) note potential pitfalls. If $b$ has support $\{1,2,\ldots,K\}$, then this is referred to as a latent class model (Roy 2003).

An advantage of shared parameter models is their ability to express complex relationships between $y$ and $r$ in a small number of latent variables. This is especially useful in multivariate longitudinal settings (Dunson and Perreault 2001). The form of (21.1) suggests modeling the random effects distribution $G(db)$ nonparametrically. This possibility was remarked on by Dunson (2007). Random measures have been used to model random effects distributions in this spirit (Kleinman and Ibrahim 1998). Shared parameter models have similar full identification problems as selection models. Some relatively recent work (Creemers et al. 2010; Njagi et al. 2014) shows how to introduce sensitivity parameters into shared parameter models, but we feel it is not in the spirit of the shared parameter model specification.

## 21.2 Our Framework

Our approach starts from the following factorization, termed the *extrapolation factorization* by Daniels and Hogan (2008):

$$p(y,r) = p_{\text{mis}}(y_{\text{mis}} \mid y_{\text{obs}}, r) \times p_{\text{obs}}(y_{\text{obs}}, r). \tag{21.2}$$

The second component on the right-hand side is the observed data model and the first component on the right-hand side is the extrapolation model. This factorization is consistent with the pattern mixture model factorization,

$$\begin{aligned} p(y,r) &= g(y \mid r) \times \phi(r) \\ &= \underbrace{g(y_{\text{mis}} \mid y_{\text{obs}}, r)}_{p_{\text{mis}}} \times \underbrace{g(y_{\text{obs}} \mid r) \times \phi(r)}_{p_{\text{obs}}}. \end{aligned}$$

We advocate specifying a Bayesian nonparametric prior $\Pi_{\text{obs}}$ directly on the observed data distribution $p_{\text{obs}}$.

Because the observed data likelihood is $p_{\text{obs}}(y_{\text{obs}}, r)$, the choice of $p_{\text{mis}}$ does not impact the fit of the model to the observed data, i.e., $(y_{\text{obs}}, r)$ only informs directly about $p_{\text{obs}}$. Therefore, the impact of unverifiable assumptions on inferences must be honestly assessed. This can be done by varying parameters which do not impact the fit of $p_{\text{obs}}$ to the observed data and/or by specifying informative priors for these parameters; such parameters are called sensitivity parameters. The absence of $p_{\text{mis}}$ in the observed data likelihood also suggests that when assessing model fit one should only use criteria that are invariant to choice of the extrapolation distribution—see Property I in Daniels et al. (2012).

Sensitivity parameters are directly related to the extrapolation factorization. We give a nonparametric version of the definition given by Daniels and Hogan (2008) and Hogan et al. (2014). Other definitions are given by Robins (1997) and Vansteelandt et al. (2006). Let $\mathscr{P}_{\text{obs}}$ and $\mathscr{P}_{\text{mis}}$ denote appropriate spaces of distributions associated with $p_{\text{obs}}$ and $p_{\text{mis}}$ respectively. Let $\xi \in \Xi$ and define $g : \{\mathscr{P}_{\text{obs}} \times \Xi\} \to \mathscr{P}_{\text{mis}}$ such that $g(p_{\text{obs}}, \xi) = p_{\text{mis}}$. We will call the function $g(p_{\text{obs}}, \xi)$ an *identifying restriction*. The parameter $\xi$ is called a *sensitivity parameter*. Some properties of sensitivity parameters include:

1. Sensitivity parameters are unidentified in the sense that

$$\Pi(d\xi \mid p_{\text{obs}}, o_{1:n}) = \Pi(d\xi \mid p_{\text{obs}}).$$

   Additionally, the likelihood $L = \prod_i p(y_{\text{obs},i}, r_i)$ does not depend on the sensitivity parameter.
2. For fixed and known $\xi$, $p_{\text{mis}}$—and hence the joint $p(y,r)$—is identified, provided that $g(p_{\text{obs}}, \xi)$ is suitably smooth in $p_{\text{obs}}$.

Our approach is to elicit clinically meaningful informative priors on the sensitivity parameters $\xi$; hence, it is essential that the function $g(p_{\text{obs}}, \xi)$ be interpretable to clinicians. Our approach is to anchor our analysis at a well-understood identifying restriction. In our examples, we will choose $g(p_{\text{obs}}, \xi)$ such that $g(p_{\text{obs}}, \mathbf{0})$ corresponds to the MAR assumption, with deviations of $\xi$ from $\mathbf{0}$ interpreted as deviations from MAR. Additionally, $\xi$ itself will be easily interpretable as, for example, a location-shift parameter or an effect on the log-odds ratio of dropout.

An additional benefit of working directly with the extrapolation factorization is that different types of dropout can be accounted for by allowing the extrapolation distribution $p_{\text{mis}}$ to depend on cause of dropout; hence, assumptions about the missing data can be allowed to depend on the reason for dropout. This is illustrated in Sect. 21.6.

Our overall approach can be broken down into two steps.

(i) Choose a prior $\Pi_{\text{obs}}$ for $p_{\text{obs}}$ to obtain a good fit to the data without overfitting, conducting model selection in a manner which is invariant to $p_{\text{mis}}$.
(ii) Identify $p_{\text{mis}}$ by specifying a family of clinically meaningful identifying restrictions $g(p_{\text{obs}}, \xi)$. Then, vary $\xi$ smoothly to assess the impact of assumptions on inferences, and conclude with a final inference by placing an informative prior on $\xi$.

Section 21.3 focuses on (i), while Sect. 21.4 focuses on (ii).

## 21.3 Examples of Models for the Observed Data

As with Bayesian nonparametric models in general, it is important to control the flexibility of our models by using the prior to achieve shrinkage toward simpler parametric models. In this section, we describe priors $\Pi_{\text{obs}}$ for $p_{\text{obs}}$ which induce shrinkage toward simple models and leverage the longitudinal structure of the data. For simplicity, we assume in the examples to follow that any intermittent missingness is partially ignorable given the dropout time $s_i = \max\{j : r_{ij} = 1\}$.

### *21.3.1 Longitudinal Binary Responses*

The joint distribution of binary longitudinal data and the missing data indicators can always be specified nonparametrically in terms of a "big" multinomial distribution. However, due to sparsity, the cell probability estimates will be unstable and highly variable. Informative Dirichlet priors on the cell probabilities can induce stability. A better option is to factor this distribution to exploit the longitudinal structure of the data; the following specification, which shrinks toward a lag-1 Markov model, follows Wang et al. (2010).

We specify saturated models for the distribution $o_i = (\bar{y}_{is_i}, s_i)$ using the following two sets of factors:

$$\{p(y_j = 1 \mid s \geq j, \bar{y}_{j-1}) : j = 1, \ldots, J\}$$
$$\{p(s = j-1 \mid s \geq j-1, \bar{y}_{j-1}) : j = 2, \ldots, J\}.$$

We parameterize these models as follows:

$$\begin{aligned} p(y_1 = 1) &= \alpha_1 \\ p(y_2 = 1 \mid s \geq 1, y_1 = y) &= \alpha_{2,y} \\ p(y_j = 1 \mid s \geq j, y_{j-1} = y, \bar{y}_{j-2}) &= \alpha_{j,\bar{y}_{j-2},y} \\ p(s = 1 \mid y_1 = y) &= \gamma_{1,y} \\ p(s = j-1 \mid s \geq j-1, y_{j-1} = y, \bar{y}_{j-2}) &= \gamma_{j-1,\bar{y}_{j-2},y}, \end{aligned} \tag{21.3}$$

for $j = 2,\ldots,J$ and $y = 0,1$. Note that the parameters $(\alpha,\gamma)$ specify a saturated model for $p_{\text{obs}}$. The model proposed here is rich enough to provide an exact fit to the observed data. Unfortunately, the number of parameters increases exponentially in $J$ and there will be many combinations of $\bar{y}_{j-1}$ (i.e., cells) which will be sparse or empty. This has been called the *curse of dimensionality* (Robins and Ritov 1997). Although these parameters are not themselves of interest, our interest is often in some functional of $p(y_J = 1)$, which is a function of all the parameters in (21.3). To overcome this, we specify informative priors for these parameters. Given that each parameter is a Bernoulli probability, we specify beta priors for these parameters such that the distribution is centered on a first order Markov model for $y$, with an unknown shrinkage parameter. More specifically, we use the following priors:

$$\alpha_{j,\bar{y}_{j-2},y} \sim beta(m^{(\alpha)}_{j,y}/\eta^{\alpha}_{j,y},(1-m^{(\alpha)}_{j,y})/\eta^{\alpha}_{j,y})$$
$$\gamma_{j-1,\bar{y}_{j-2},y} \sim beta(m^{(\gamma)}_{j-1,y}/\eta^{\gamma}_{j-1,y},(1-m^{(\gamma)}_{j-1,y})/\eta^{\gamma}_{j-1,y}),$$

for $j = 2,\ldots,J$ and $y = 0,1$. Note for all values of $\bar{y}_{j-2}$, the prior mean, $m^{(\alpha)}_{j,y}$ is the same. As the $\eta$'s tend to 0 the model for the observed data approaches a first order Markov model. Wang et al. (2010) then place Unif$(0,1)$ priors on all $m$ parameters and specify independent uniform shrinkage priors (Daniels 1999) on the $\eta$ parameters.

To illustrate the effect of shrinkage, Fig. 21.1 presents an analysis on the cell probabilities in the Breast Cancer Prevention Trial given by Wang et al. (2010), giving the probability of depression under an assigned treatment at each time. We see that estimated posterior probabilities of depression given each possible history is stabilized substantially relative to the saturated model.

### *21.3.2 Longitudinal Continuous Responses*

We describe an approach for continuous data with a somewhat different spirit than the approach used in Sect. 21.3.1. The general approach described here can be applied to data types other than continuous data through the introduction of latent continuous variables. For consistency with the analysis in Sect. 21.6 we describe the method assuming monotone dropout with dropout time $s = \max\{j : r_{ij} = 1\}$.

Our goal is to place a prior $\Pi_{\text{obs}}$ on $p_{\text{obs}}$; the previous section effectively constructed this prior directly on $\mathscr{P}_{\text{obs}}$. We instead construct $\Pi_{\text{obs}}$ indirectly by considering a *working prior* $\Pi^{\star}$ on $\mathscr{P}$, the space of models for the complete data distribution $p(y,s)$. Rather than using $\Pi^{\star}$ for inference, however, we use it to *induce* the prior on $p_{\text{obs}}$,

$$\Pi_{\text{obs}}(A) = \Pi^{\star}(p_{\text{obs}} \in A). \tag{21.4}$$

To understand this approach, we can think of $\Pi^{\star}$ as a prior on a *working model* $p^{\star}$ with $p^{\star}$ giving its own interpretation of how the observed data was generated. This is depicted in Fig. 21.2. Because $\Pi^{\star}$ is a prior on $\mathscr{P}$, this prior could be used

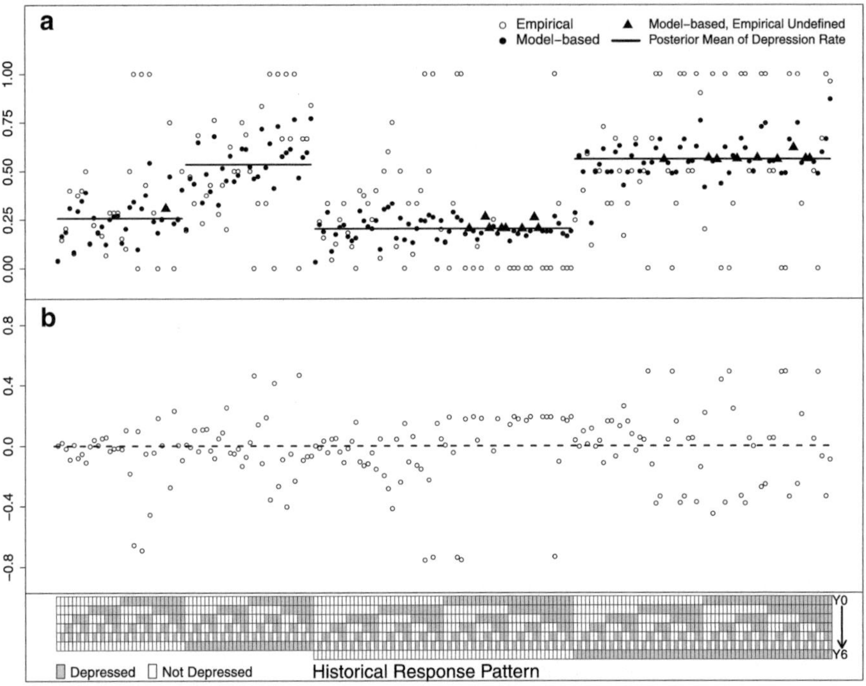


**Fig. 21.1** (**a**) The empirical rate and model-based posterior mean of $p(y_j = 1 \mid \bar{\mathbf{y}}_{j-1}, S \geq j)$ and $j = 6,7$. (**b**) The difference between the empirical and model-based posterior mean of the depression rate. The $x$-axis is the pattern of historical response data $\bar{\mathbf{y}}_{j-1}$. (**a**) Conditional depression rate. (**b**) Shrinkage difference

to identify $p_{\text{mis}}$; however, we will not use $\Pi^{\star}(dp \mid o_{1:n})$ for inference, instead using only the induced posterior $\Pi_{\text{obs}}(dp_{\text{obs}} \mid o_{1:n})$ and handling the extrapolation model $p_{\text{mis}}$ separately, as in Sect. 21.4 below. This discards any information $\Pi^{\star}$ gives about $p_{\text{mis}}$.

There are several reasons for wanting to take the indirect approach described above. First, in light of the curse of dimensionality, borrowing information across timepoints $j$ and dropout times $s$ is required for sample sizes encountered in practice. In constructing models which share information across timepoints, we have found it easier to construct reasonable priors on $\mathscr{P}$ than to construct priors on $\mathscr{P}_{\text{obs}}$. The working-model analogy can provide intuition about how a given prior shares information across time and dropout patterns. Second, strategies for posterior computation can leverage the working prior/working model analogy in order to conduct efficient posterior inference. If inference is being conducted through MCMC, the top path in Fig. 21.2 suggests a data-augmentation algorithm. Inference algorithms can then be based on an appropriate full data model.

A final benefit is that theoretical properties of the induced prior $\Pi_{\text{obs}}$ in the presence of missing data, such as weak and strong consistency, follow readily from standard sufficient conditions on $\Pi^{\star}$ when there is no missing data (Linero 2015a,b);

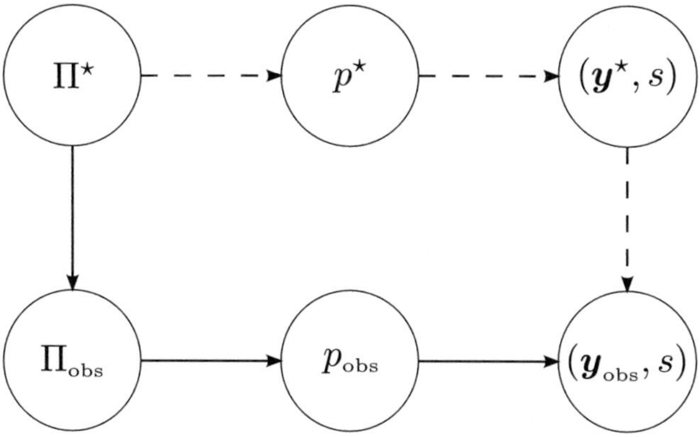


**Fig. 21.2** Graphical depiction of the working-model approach; the working prior $\Pi^{\star}$ induces the prior $\Pi_{\text{obs}}$ on $p_{\text{obs}}$, while having a latent interpretation has having the data come from a working model $p^{\star}$. The *solid path* describes the model for the observed data, while the *dashed path* describes a latent explanation for the observed data

this is particularly convenient if $\Pi^{\star}$ is chosen to be a prior for which these sufficient conditions have already been verified.

#### 21.3.2.1 A Dirichlet Process Mixture Working Prior

We describe a working prior based on Dirichlet process mixtures which was used by Linero and Daniels (2015). The prior $\Pi^{\star}$ is a truncated Dirichlet process mixture model such that if $p^{\star} \sim \Pi^{\star}$ then

$$p^{\star}(y,s) = \sum_{k=1}^{K} w_k f_{\theta^{(k)}}(y)\pi_{\gamma^{(k)}}(s \mid y),$$

where $(\theta^{(k)}, \gamma^{(k)}) \stackrel{\text{iid}}{\sim} H$ for some base distribution $H$ and $w = (w_1, \ldots, w_K)$ is given a truncated stick-breaking prior. Typically $f_{\theta}(\cdot)$ will be a normal kernel, with $\theta = (\mu, \Sigma)$. One possible choice for $\pi_{\gamma}(s \mid y)$ is the sequential hazard model (Diggle and Kenward 1994),

$$\pi_{\gamma}(s \mid y) = \text{expit}(\zeta_s + \lambda_s^T \bar{y}_s) \prod_{j<s} (1 - \text{expit}(\zeta_j + \lambda_j^T \bar{y}_j)),$$

where $\text{expit}(x) = [1 + e^{-x}]^{-1}$ is the logistic function and $\gamma = (\zeta, \lambda)$. $p^{\star}$ is then a mixture of models which satisfy the MAR assumption. While this may seem overly restrictive, any joint $p(y,s)$ can be approximated arbitrarily well by an appropriate mixture of MAR (or even MCAR) models.

A possible choice for the prior on $(\mu^{(k)}, \Sigma^{(k)})$ is the standard normal-inverse-Wishart prior. We instead favor a prior which reparametrizes $\Sigma^{(k)}$ in terms of autoregressive parameters (Daniels and Pourahmadi 2002); given that $y_i$ belongs to mixture component $k$, we may write

$$y_{ij} = \mu_j^{(k)} + \sum_{\ell=1}^{j-1} \phi_{\ell j}^{(k)} (y_{i\ell} - \mu_\ell^{(k)}) + \varepsilon_{ij}, \qquad \varepsilon_{ij} \sim \mathcal{N}(0, \rho_j^{2(k)}).$$

The reparametrization in terms of $(\mu^{(k)}, \phi^{(k)}, \rho^{(k)})$ offers more possibilities for the practitioner to exploit the longitudinal structure of the data; for example, $\phi^{(k)}$ terms associated with higher-order lag terms may be shrunk toward 0 to achieve shrinkage toward a structured covariance matrix.

In Sect. 21.6, we impose the restriction that $\pi_\gamma(s \mid y) = \text{expit}(\zeta_s + \lambda_s y_s) \prod_{j<s}[1 - \text{expit}(\zeta_j + \lambda_j y_j)]$. The $\lambda_j^{(k)}$ and $\zeta_j^{(k)}$ were then given $\mathcal{N}(\mu_\lambda, \sigma_\lambda^2)$ and $\mathcal{N}(\mu_\zeta, \sigma_\lambda^2)$ priors. In the spirit of Gelman et al. (2008), the $y_{ij}$'s were standardized to have mean 0 and standard deviation 0.5, and the parameters $\mu_\zeta$ and $\mu_\lambda$ were both given Cauchy priors with location 0 and scales 5 and 2.5 respectively. Inverse-gamma priors were used on $\sigma_\zeta^2$ and $\sigma_\lambda^2$. In principle, however, we might treat $(\zeta, \lambda)$ in the same manner as $(\mu, \phi)$, shrinking to, rather than assuming, the lag-1 model within class.

## 21.4 Identifying Restrictions and Sensitivity Parameters

We now discuss the choice of identifying restrictions $g(p_{\text{obs}}, \xi)$. Identifying restrictions are a convenient and parsimonious way to identify the extrapolation distribution for monotone missingness.

The available case missing value (ACMV) restriction (Little 1994) equates the unidentified distributions to their identified counterparts,

$$p(y_j \mid \bar{y}_{j-1}, s = k) = p(y_j \mid \bar{y}_{j-1}, s \geq j), \tag{21.5}$$

for $j > k$. Molenberghs et al. (1998) showed that this restriction is equivalent to MAR for monotone missingness,

$$p(s = j \mid y) = p(s = j \mid \bar{y}_j).$$

A parsimonious way to introduce sensitivity parameters is to embed the ACMV restriction into a class of non-future dependent (NFD) restrictions. The non-future dependent restrictions introduced by Kenward et al. (2003) identify all but one conditional distribution for each dropout pattern $(s)$ and assume missingness does not depend on future responses. In particular,

$$p(y_j \mid \bar{y}_{j-1}, s = k) = p(y_j \mid \bar{y}_{j-1}, s \geq j-1), \tag{21.6}$$

for $j-1>k$. For monotone missingness, under the above restriction, the probability of the $j$'th response being missing conditional on having seen the $(j-1)$'st response depends on the past ($\bar{y}_j$), the present ($y_{j+1}$), but not the future ($\{y_k : k > j+1\}$),

$$p(s=j \mid y) = p(s=j \mid \bar{y}_j, y_{j+1}).$$

Table 21.1 gives a schematic representation of the NFD assumption. Under NFD, for each value of $s$ there is one unidentified distribution, $p(y_j \mid \bar{y}_{j-1}, s=j-1)$. Thus, the right-hand side in (21.6) is unidentified, unlike in (21.5). Setting $p(y_j \mid \bar{y}_{j-1}, s= j-1) = p(y_j \mid \bar{y}_{j-1}, s \geq j)$ shows that ACMV is a special case of NFD.

**Table 21.1** Schematic representation of NFD when $J=4$

| | $j=1$ | $j=2$ | $j=3$ | $j=4$ |
|---|---|---|---|---|
| $S=1$ | $p_1(y_1)$ | ? | $p_{\geq 2}(y_3 \mid \bar{y}_2)$ | $p_{\geq 3}(y_4 \mid \bar{y}_3)$ |
| $S=2$ | $p_2(y_1)$ | $p_2(y_2 \mid y_1)$ | ? | $p_{\geq 3}(y_4 \mid \bar{y}_3)$ |
| $S=3$ | $p_3(y_1)$ | $p_3(y_2 \mid y_1)$ | $p_3(y_3 \mid \bar{y}_2)$ | ? |
| $S=4$ | $p_4(y_1)$ | $p_4(y_2 \mid y_1)$ | $p_4(y_3 \mid \bar{y}_2)$ | $p_4(y_4 \mid \bar{y}_3)$ |

Subscripting $p$ with $j$ or $\geq j$ denotes conditioning on the events $s=j$ and $s \geq j$
The distributions above the dividing line correspond to $p_{\text{mis}}$

In what follows, we review several approaches to identify these distributions. First, we consider an exponential tilting approach (Birmingham et al. 2003; Scharfstein et al. 1999). We take

$$p(y_j \mid \bar{y}_{j-1}, s=j-1) \propto p(y_j \mid \bar{y}_{j-1}, s \geq j) \exp[q_j(\bar{y}_j; \xi)] \tag{21.7}$$

This assumption is equivalent to the assumption that

$$\log \left\{ \frac{\text{Odds}(s=j-1 \mid \bar{y}_j, s \geq j)}{\text{Odds}(s=j-1 \mid \bar{y}'_j, s \geq j)} \right\} = q_j(\bar{y}_j; \xi) - q_j(\bar{y}'_j; \xi) \tag{21.8}$$

provided that $\bar{y}'_{j-1} = \bar{y}_{j-1}$. The tilting function $q_j(\bar{y}_j, \zeta)$ represents the effect on the scale of log-odds ratios of $y_j$ on dropout holding $\bar{y}_{j-1}$ fixed. For example, one might take $q_j(\bar{y}_j; \xi) = \xi_j y_j$ with $\xi_j$ a sensitivity parameter which can be interpreted as the effect of changes in $y_j$ on the probability of dropout at time $j-1$ on the log-odds scale. The identifying restriction $g(p_{\text{obs}}, \xi)$ is determined by the combination of NFD and the exponential tilting assumption. This approach was used with the model in Sect. 21.3.1 by Wang et al. (2010).

Next, we consider a transformation based approach. Suppose that $y$ is continuously valued. We assume the existence of a transformation $\mathscr{T}_j(y_j; \bar{y}_{j-1}, \xi)$ such that

$$[y_j \mid \bar{y}_{j-1}, s=j-1] \stackrel{d}{=} [\mathscr{T}_j(y_j; \bar{y}_{j-1}, \xi) \mid \bar{y}_{j-1}, s \geq j], \tag{21.9}$$

where $\stackrel{d}{=}$ denotes equality in distribution. The transformation $\mathscr{T}_j$ can be thought of as a correction for missingness in the following sense. Suppose that two subjects, $A$ and $B$, are identical up-to time $s=j-1$, but that $A$ continues on study ($s \geq j$) but

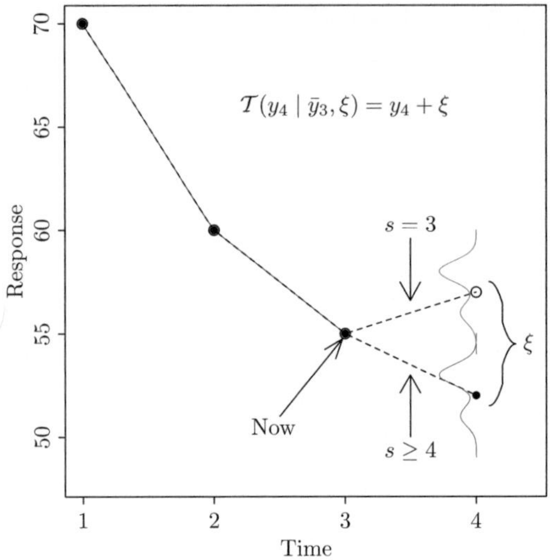


**Fig. 21.3** Depiction of coupling interpretation of transformation approach; two subjects have the same response value at times $j = 1,2,3$, but one continues $(s \geq 4)$ while the other does not $(s = 3)$. The conditional distribution for both observations is the same after applying the correcting transformation

$B$ does not $(s = j-1)$. $\mathscr{T}_j^{-1}$ functions as a correction to the response of $B$ such that the response at time $j$ has the same conditional distribution for both subjects. This is expressed in Fig. 21.3. A commonly used choice of $\mathscr{T}_j$ is a location-scale shift

$$\mathscr{T}_j(y_j; \bar{y}_{j-1}, \xi) = \xi_{1j} y_j + (1 - \xi_{1j}) E_p[y_j \mid \bar{y}_{j-1}, s \geq j] + \xi_{2j},$$

which has the effect of shifting the conditional mean of $y_j$ by $\xi_{2j}$ and scaling the conditional standard deviation of $y_j$ by $\xi_{1j}$.

The transformation method may be less appropriate for non-continuous responses. For binary responses, it may be more appropriate to consider transforming the mean,

$$E_p[y_j \mid \bar{y}_{j-1}, s = j-1] = \mathscr{T}_j\left(E_p[y_j \mid \bar{y}_{j-1}, s \geq j]; \xi\right).$$

For example, one might set

$$E_p[y_j \mid \bar{y}_{j-1}, s = j-1] = \text{expit}\left\{\text{logit}\left(E[y_j \mid \bar{y}_{j-1}, s \geq j]\right) + \xi_j\right\},$$

which gives $\xi_j$ the interpretation of a shift in probability of success on the scale of log odds-ratios. See, for example, (Daniels and Hogan 2008, page 248) or (National Research Council 2010, page 92) for examples of this approach.

### 21.4.1 Incorporation of Information on Reason for Dropout

In longitudinal clinical trials with attrition the analyst often has access to information on cause of dropout for given individuals. Presumably, a subject who was removed from study due to an incidental protocol violation ought to be handled differently than a subject who dropped out specifically because his prescribed treatment was ineffective.

One approach (Hogan and Laird 1997) is to divide causes of dropout into "informative" and "noninformative" causes, and regard the true dropout time of an individual to be censored if the cause of dropout is noninformative. Another approach is to model the cause of dropout itself and incorporate it into the analysis (Linero and Daniels 2015).

### 21.4.2 Intermittent Missingness

The identifying restrictions outlined above are useful for monotone missingness. As mentioned in Sect. 21.1.1, invoking the partial ignorability assumption with $h(r) = \max\{j : r_j = 1\}$ allows the use of techniques for monotone missingness even in the presence of non-monotone missingness.

As with any other assumption about the missing data, partial ignorability must be assessed on subject-matter grounds and may or may not be reasonable in any given situation. Alternative assumptions for intermittent missingness may be used when it fails, such as *sequential explainability* (Vansteelandt et al. 2007) or *selection-bias permutation missingness* (Robins et al. 2000). Briefly, the sequential explainability assumption states that

$$p(r_j = 1 \mid \bar{r}_{j-1}, y) = p(r_j = 1 \mid \bar{o}_{j-1}),$$

where $\bar{r}_{j-1}$ and $\bar{o}_{j-1}$ denote the missingness pattern and observed data history up-to time $j-1$. One may then introduce sensitivity parameters to allow for smooth departures from sequential explainability in order to facilitate a sensitivity analysis. Unlike MAR, sequential explainability is not sufficient to identify $p(y, r)$; it is, however, strong enough to identify the marginal distributions $f(y_j)$ and hence effects of interest such as the marginal means $\psi_j = \int y_j f(y_j)\, dy_j$. An appropriate modification of the inference algorithm given in Sect. 21.5 can be extended to the case of sequential explainability.

## 21.5 General Strategy for Posterior Inference

By constructing models based on the extrapolation factorization, we can outline a general three step strategy for posterior inference. For each iteration of our MCMC scheme we do the following.

1. Generate a sample $p_{\text{obs}}$ from our Markov chain targeting $\Pi_{\text{obs}}(dp_{\text{obs}} \mid o_{1:n})$.
2. Generate $\xi$ from its prior $\Pi(d\xi \mid p_{\text{obs}})$.
3. Combine $p_{\text{obs}}$ and $\xi$ via the chosen identifying restriction to get $p(y,r) = f(y) \times \pi(r \mid y)$ and calculate desired functionals $\psi(f)$.

Steps 1 and 2 are standard, with the approaches in Sects. 21.3.1 and 21.3.2 allowing for sampling of $p_{\text{obs}}$, potentially using packages such as JAGS or BUGS. Step 3 is more problematic, as quantities of interest are complicated functionals of $(p_{\text{obs}}, \xi)$.

To address Step 3 in general, Linero and Daniels (2015) estimate linear functionals $\psi(f) = \int t(y)f(y)\,dy$ via Monte Carlo at each iteration by forward-sampling a large number of observations from $p(y,s)$. This Monte Carlo approach is frequently used in the context of estimating causal effects via the so-called G-computation formula, and has a long history (Robins 1986, 1989; Robins et al. 2000). A very similar strategy was used by Scharfstein et al. (2014). Our use of G-computation is computationally expensive, though we have not found it prohibitive and we believe there is scope for improving it substantially.

The steps needed to forward-sample from $p(y,r)$ under NFD are:

(i) Draw $(\bar{y}_s, s)$ from $p_{\text{obs}}$.
(ii) Draw $y_{s+1}$ from $p(y_{s+1} \mid \bar{y}_s, s)$, making use of our identifying restriction.
(iii) For each $j > s+1$:

   a. Draw $r_j \sim \text{Bern}(\rho)$ where $\rho = p(s \geq j \mid \bar{y}_{j-1}, s \geq j-1)$.
   b. If $r_j = 1$, draw $y_j$ from $p(y_j \mid \bar{y}_{j-1}, S \geq j)$, otherwise draw $y_j$ from $p(y_j \mid \bar{y}_{j-1}, s = j-1)$ making use of our identifying restriction.

This algorithm is tractable for the models described in Sects. 21.3.1 and 21.3.2, though for small $J$, it is not needed for the model in Sect. 21.3.1 as the G-computation formula can be calculated exactly; see Wang et al. (2010) for details.

## 21.6 Example Data Analysis

We illustrate the proposed approach on data from a clinical trial designed to assess the efficacy of a treatment for acute schizophrenia. A more detailed analysis of this data, including a detailed simulation study to assess frequentist properties, is given by Linero and Daniels (2015). The trial was double-blinded with subjects randomized to one of three treatments: a test drug (81 subjects), an active control drug (45 subjects), and a placebo (78 subjects). The Positive and Negative Syndrome Scale (PANSS) was used to assess severity of schizophrenia symptoms in subjects.

Interest was primarily in the causal effects of the baseline randomization at each time point (known as the *intention-to-treat effect*)

$$\psi_j = E_p[y_j - y_1 \mid \text{on drug}] - E_p[y_j - y_1 \mid \text{not on drug}], \qquad j = 2, \ldots, 6.$$

Moderate attrition was observed, with dropout rates of 33 %, 19 %, and 25 % on the test drug, placebo, and active control arms, respectively.

We use the working-model approach for continuous data outlined in Sect. 21.3.2, employing the non-future dependence assumption. Figures 21.4 and 21.5 provide some motivation for the use of the Dirichlet process working prior. Displayed are two latent classes of observations on the Placebo arm of the study, consisting mostly of completers ($s = 6$), along with a symmetric heatmap displaying the posterior pairwise probability of individuals belonging to the same cluster according to the Dirichlet process. The heatmap reveals several non-overlapping groups. These plots suggest that a standard multivariate Gaussian model is likely to be inadequate.

Information was also available on cause-of-dropout, with dropouts on the placebo and test drug arms more likely to be for informative reasons. The transformation method was used, with the additional restriction that the transformation is not applied if the dropout was for a noninformative reason. A location shift transformation was used, as in Fig. 21.3, i.e., $\mathscr{T}_j(y_j; \bar{y}_{j-1}, \xi_{j-1}) = y_j + \xi_{j-1}$. Formally, we assumed

$$[y_j \mid \bar{y}_{j-1}, S = j-1] \stackrel{d}{=} \vartheta[y_j + \xi_{j-1} \mid \bar{y}_{j-1}, s \geq j] + (1-\vartheta)[y_j \mid \bar{y}_{j-1}, s \geq j],$$

where $\vartheta$ is a modeled probability of informative missingness. $\vartheta$ was given a Uniform$(0,1)$ prior and estimated from the information on cause of dropout in the data. Figure 21.6 presents an assessment of how well our model matches various aspects of the observed data, and the extent to which the model-based credible intervals match model-free credible intervals. We find little disagreement between the model and model-free approaches.

To get a sense of the overall impact on inferences of missingness through the sensitivity parameters, $\zeta_j$, we consider a sensitivity analysis wherein we restrict $\xi_j = \xi_{j'}$ but allow $\xi$ to vary across treatments with $\xi_A$ for the active treatment and $\xi_P$ for the placebo. Inferences for fixed sensitivity parameters are given in Fig. 21.7; we see that generally as the role of informative missingness becomes more drastic (i.e., as $\xi_P = \xi_A$ increases) the posterior probability $\Pi(\psi_6 > 0 \mid o_{1:n})$ tends toward 1.

Once we understand the role of the sensitivity parameters by conducting a sensitivity analysis, a final inference can be reached through the use of informative priors on the sensitivity parameters. Each of the $\xi_j$ were given a Uniform$(0,8)$ prior to reflect the prior belief that deviations due to informative missingness were not thought to plausibly exceed one conditional standard deviation; similar calibration was done in Daniels and Hogan 2008, Section 10.2.8. Final results of the analysis are given in Fig. 21.8. In this case, we see that the presence of informative missingness masks the effect of the treatment if MAR is incorrectly assumed. The magnitude of the masking is of course dependent on the prior chosen for $\xi_j$; had our prior placed more mass on larger values of the $\xi_j$ then one would obtain more extreme results.

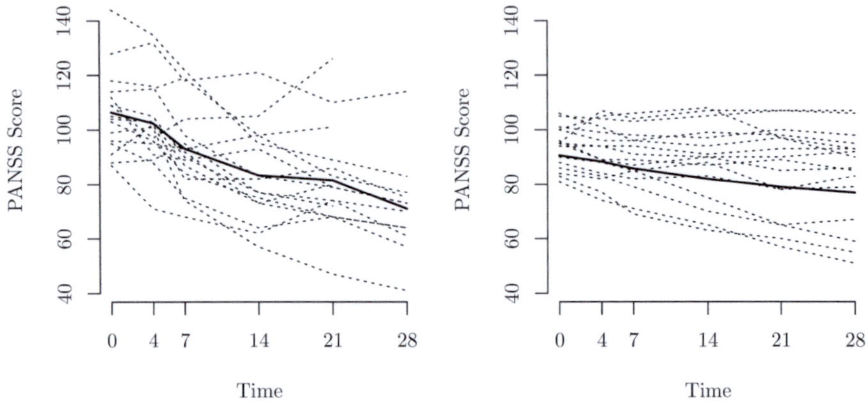


**Fig. 21.4** Trajectories of two latent classes of subjects on the placebo arm of the trial (*dashed*) with mean response over time (*solid*). Each figure contains 16 trajectories for the purpose of comparison

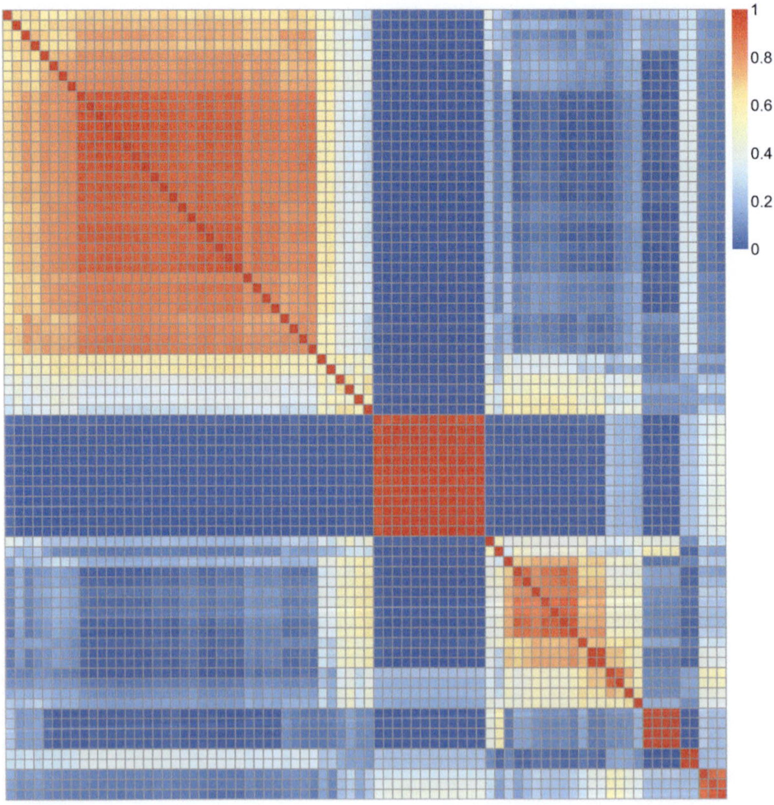


**Fig. 21.5** Clustering of subjects in the placebo arm of the study; groups in Fig. 21.4 were chosen to be completers with uniformly high pairwise probability of clustering together, with low posterior probability of overlap

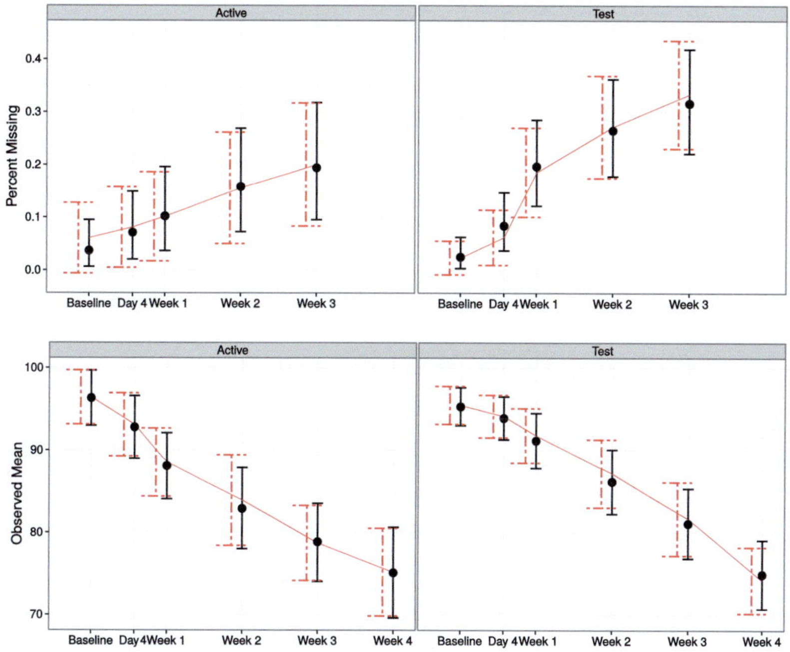


**Fig. 21.6** *Top*: Modeled (*solid*) and model-free (*dashed*) estimates and intervals for the marginal dropout probabilities at each time. *Bottom*: The same plots, but with respect to the marginal mean at each time. *Error bars* represent 95 % confidence/credible intervals

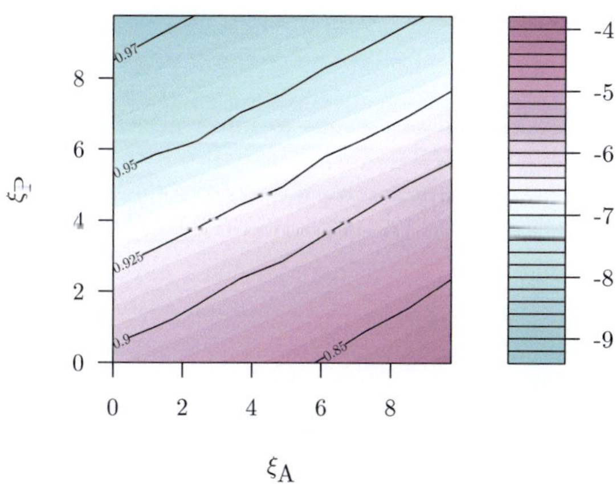


**Fig. 21.7** Contour plot giving inference for the effect $\psi_6$ for different values of the sensitivity parameters $\xi_A$ and $\xi_P$. *Coloring* is associated with the posterior mean $E[\psi_6 \mid \text{Observed Data}, \xi]$ while the *solid lines* give contours of the posterior probability that $\psi_6 > 0$

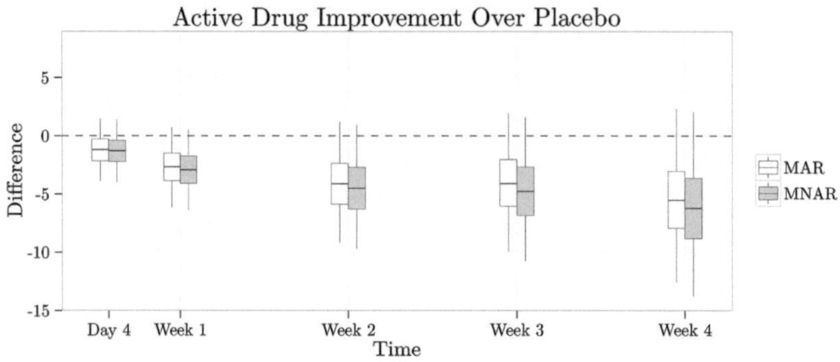


**Fig. 21.8** Posterior means, quartiles, and 95 % credible intervals for the $\psi_j$'s

The final posterior mean of $\psi_J$, the intention-to-treat effect at completion of the study, was $(-6.2)$ with 95 % credible interval $(-12.6, 2.3)$.

## 21.7 Open Issues

In this section, we point out some areas for future work.

The observed data models proposed in Sect. 21.3 could be modified to account for ordinal responses. In particular, for the models in Sect. 21.3.2, it would require the introduction of latent variables (Kottas et al. 2005; Johnson and Albert 1999), potentially with identifying restrictions constructed on the scale of the latent variables. A challenge in this setting is to ensure that the identifying restrictions are clinically meaningful. Related extensions to multivariate longitudinal data would have similar issues with identifying the extrapolation distribution.

We discussed a common strategy for non-monotone missingness earlier, using identifying restrictions (with potential sensitivity parameters) for dropout and partial ignorability for intermittent missingness. Alternative strategies, mentioned in Sect. 21.4.2, which rely on different models for the observed data distribution and different types of identifying restrictions need further exploration.

It is not uncommon to include additional covariates in the analysis that we do not want to include in our causal estimand of interest, but are predictive of missingness. Such covariates, called auxiliary covariates, can make an (auxiliary variable) MAR assumption more realistic and potentially reduce the range of sensitivity parameters. Such covariates are often included by specifying (i) a model to impute the missing responses which includes these covariates and (ii) a separate model which uses the completed datasets for inference (Rubin 1987). These two models are often not compatible (Meng 1994). In recent work Daniels et al. (2014) took a Bayesian approach which ensured compatibility and obtained the desired inference model. Related Bayesian nonparametric extensions related to those developed here have yet to be developed.

In order to conduct a meaningful sensitivity analysis, it is essential to be able to introduce interpretable sensitivity parameters. Accomplishing this while maintaining tractable inference algorithms—in particular, being able to forward-sample $p(y,r)$ as discussed in Sect. 21.5—is often the primary obstacle to our approach; how to do this in complicated models is essentially an open problem.

There are direct connections of our recommended approach for missing data to causal inference. In particular, a Bayesian nonparametric model can be used for the observed data. Untestable assumptions, similar to the specification of the extrapolation distribution here, need to be specified with sensitivity parameters. A recent illustration in the setting of assessing the causal effect of mediators can be found in Kim et al. (2015).

## References

Birmingham, J., Rotnitzky, A., and Fitzmaurice, G. M. (2003). Pattern-mixture and selection models for analysing longitudinal data with monotone missing patterns. *Journal of the Royal Statistical Society, Series B.*, **65**, 275–297.

Creemers, A., Hens, N., Aerts, M., Molenberghs, G., Verbeke, G., and Kenward, M. G. (2010). A sensitivity analysis for shared-parameter models for incomplete longitudinal outcomes. *Biometrical Journal*, **52**(1), 111–125.

Daniels, M., Wang, C., and Marcus, B. (2014). Fully Bayesian inference under ignorable missingness in the presence of auxiliary covariates. *Biometrics*, **70**(1), 62–72.

Daniels, M. J. (1999). A prior for the variance in hierarchical models. *Canadian Journal of Statistics*, **27**(3), 567–578.

Daniels, M. J. and Hogan, J. W. (2000). Reparameterizing the pattern mixture model for sensitivity analyses under informative dropout. *Biometrics*, **56**(4), 1241–1248.

Daniels, M. J. and Hogan, J. W. (2008). *Missing data in longitudinal studies: Strategies for Bayesian modeling and sensitivity analysis*. CRC Press.

Daniels, M. J. and Pourahmadi, M. (2002). Bayesian analysis of covariance matrices and dynamic models for longitudinal data. *Biometrika*, **89**(3), 553–566.

Daniels, M. J., Chatterjee, A. S., and Wang, C. (2012). Bayesian model selection for incomplete data using the posterior predictive distribution. *Biometrics*, **68**(4), 1055–1063.

Diggle, P. and Kenward, M. G. (1994). Informative drop-out in longitudinal data analysis. *Applied statistics*, pages 49–93.

Dunson, D. B. (2006). Bayesian dynamic modeling of latent trait distributions. *Biostatistics*, **7**(4), 551–568.

Dunson, D. B. (2007). Bayesian methods for latent trait modelling of longitudinal data. *Statistical Methods in Medical Research*, **16**, 399–415.

Dunson, D. B. and Herring, A. H. (2006). Semiparametric Bayesian latent trajectory models. Technical report, ISDS Discussion Paper 16, Duke Univ., Durham, NC, USA.

Dunson, D. B. and Perreault, S. D. (2001). Factor analytic models of clustered multivariate data with informative censoring. *Biometrics*, **57**(1), 302–308.

Fieuws, S. and Verbeke, G. (2006). Pairwise fitting of mixed models for the joint modeling of multivariate longitudinal profiles. *Biometrics*, **62**(2), 424–431.

Gael, J. V., Teh, Y. W., and Ghahramani, Z. (2009). The infinite factorial hidden Markov model. In *Advances in Neural Information Processing Systems*, pages 1697–1704.

Gelman, A., Jakulin, A., Pittau, M. G., and Su, Y.-S. (2008). A weakly informative default prior distribution for logistic and other regression models. *The Annals of Applied Statistics*, pages 1360–1383.

Harel, O. and Schafer, J. L. (2009). Partial and latent ignorability in missing-data problems. *Biometrika*, **96**(1), 37–50.

Heckman, J. J. (1979). Sample selection bias as a specification error. *Econometrica: Journal of the Econometric Society*, pages 153–161.

Henderson, R., Diggle, P., and Dobson, A. (2000). Joint modelling of longitudinal measurements and event time data. *Biostatistics*, **1**(4), 465–480.

Hogan, J. W. and Laird, N. M. (1997). Mixture models for the joint distribution of repeated measures and event times. *Statistics in medicine*, **16**(3), 239–257.

Hogan, J. W., Daniels, M. J., and Hu, L. (2014). A bayesian perspective on assessing sensitivity to assumptions about unobserved data. In G. Molenberghs, G. Fitzmaurice, M. G. Kenward, A. Tsiatis, and G. Verbeke, editors, *Handbook of Missing Data Methodology*. CRC Press.

Johnson, V. E. and Albert, J. H. (1999). *Ordinal Data Modeling. Statistics for Social Science and Public Policy*. New York: Springer-Verlag.

Kenward, M., Molenberghs, G., and Thijs, H. (2003). Pattern-mixture models with proper time dependence. *Biometrika*, **90**, 53–71.

Kim, C., Daniels, M. J., and Roy, J. A. (2015). A framework for Bayesian nonparametric inference for causal effects of mediation. Technical Report.

Kleinman, K. P. and Ibrahim, J. G. (1998). A semiparametric Bayesian approach to the random effects model. *Biometrics*, pages 921–938.

Kottas, A., Müller, P., and Quintana, F. (2005). Nonparametric Bayesian modeling for multivariate ordinal data. *Journal of Computational and Graphical Statistics*, **14**(3), 610–625.

Linero, A. R. (2015a). Bayesian nonparametric analysis of longitudinal studies in the presence of informative missingness. Technical Report.

Linero, A. R. (2015b). *Nonparametric Bayes: Inference Under Nonignorable Missingness and Model Selection*. Ph.D. thesis, University of Florida.

Linero, A. R. and Daniels, M. J. (2015). A flexible Bayesian approach to monotone missing data in longitudinal studies with nonignorable missingness with application to an acute schizophrenia clinical trial. *Journal of the American Statistical Association*, **in press**.

Little, R. J. A. (1993). Pattern-mixture models for multivariate incomplete data. *Journal of the American Statistical Association*, **88**(421), 125–134.

Little, R. J. A. (1994). A class of pattern-mixture models for normal incomplete data. *Biometrika*, **81**(3), 471–483.

Little, R. J. A. and Rubin, D. B. (1986). *Statistical analysis with missing data*. John Wiley & Sons.

Manski, C. F. (2009). *Identification for prediction and decision*. Harvard University Press.

Meng, X. L. (1994). Multiple-imputation inferences with uncongenial sources of input. *Statistical Science*, pages 538–558.

Molenberghs, G., Michiels, B., Kenward, M. G., and Diggle, P. J. (1998). Monotone missing data and pattern-mixture models. *Statistica Neerlandica*, **52**, 153–161.

Molenberghs, G., Fitzmaurice, G., Kenward, M. G., Tsiatis, A., and Verbeke, G. (2014). *Handbook of Missing Data Methodology*. CRC Press.

National Research Council (2010). *The Prevention and Treatment of Missing Data in Clinical Trials*. The National Academies Press.

Njagi, E. N., Molenberghs, G., Kenward, M. G., Verbeke, G., and Rizopoulos, D. (2014). A characterization of missingness at random in a generalized shared-parameter joint modeling framework for longitudinal and time-to-event data, and sensitivity analysis. *Biometrical Journal*, **56**(6), 1001–1015.

Pati, D., Reich, B. J., and Dunson, D. B. (2011). Bayesian geostatistical modelling with informative sampling locations. *Biometrika*, **98**(1), 35–48.

Ren, L., Dunson, D. B., and Carin, L. (2008). The dynamic hierarchical Dirichlet process. In *Proceedings of the 25th international conference on Machine learning*, pages 824–831. ACM.

Robins, J. (1989). The control of confounding by intermediate variables. *Statistics in medicine*, **8**(6), 679–701.

Robins, J. M. (1986). A new approach to causal inference in mortality studies with sustained exposure periods – application to control of the healthy worker survivor effect. *Math Modeling*, **7**, 1393–1512.

Robins, J. M. (1997). Non-response models for the analysis of non-monotone non-ignorable missing data. *Statistics in Medicine*, **16**(1), 21–37.

Robins, J. M. and Ritov, Y. (1997). Toward a curse of dimensionality appropriate(CODA) asymptotic theory for semi-parametric models. *Statistics in medicine*, **16**(3), 285–319.

Robins, J. M., Rotnitzky, A., and Scharfstein, D. O. (2000). Sensitivity analysis for selection bias and unmeasured confounding in missing data and causal inference models. In *Statistical models in epidemiology, the environment, and clinical trials*, pages 1–94. Springer.

Roy, J. (2003). Modeling longitudinal data with nonignorable dropouts using a latent dropout class model. *Biometrics*, **59**(4), 829–836.

Rubin, D. (1976). Inference and missing data. *Biometrika*, **63**, 581–592.

Rubin, D. B. (1987). *Multiple Imputation for Nonresponse in Surveys*. Wiley.

Scharfstein, D., McDermott, A., Olson, W., and Wiegand, F. (2014). Global sensitivity analysis for repeated measures studies with informative dropout: A fully parametric approach. *Statistics in Biopharmaceutical Research*, **6**(4), 338–348.

Scharfstein, D. O., Rotnitzky, A., and Robins, J. M. (1999). Adjusting for nonignorable dropout using semiparametric nonresponse models. *Journal of the American Statistical Association*, **94**, 1096–1146.

Scharfstein, D. O., Daniels, M. J., and Robins, J. M. (2003). Incorporating prior beliefs about selection bias into the analysis of randomized trials with missing outcomes. *Biostatistics*, **4**(4), 495–512.

Si, Y. and Reiter, J. P. (2013). Nonparametric Bayesian multiple imputation for incomplete categorical variables in large-scale assessment surveys. *Journal of Educational and Behavioral Statistics*, **38**(5), 499–521.

Teh, Y. W., Jordan, M. I., Beal, M. J., and Blei, D. M. (2006). Hierarchical Dirichlet processes. *Journal of the American Statistical Association*, **101**(476).

Thijs, H., Molenberghs, G., Michiels, B., Verbeke, G., and Curran, D. (2002). Strategies to fit pattern-mixture models. *Biostatistics*, **3**(2), 245–265.

Vansteelandt, S., Goetghebeur, E., Kenward, M., and Molenberghs, G. (2006). Ignorance and uncertainty regions as inferential tools in a sensitivity analysis. *Statistica Sinica*, **16**, 953–979.

Vansteelandt, S., Rotnitzky, A., and Robins, J. (2007). Estimation of regression models for the mean of repeated outcomes under nonignorable nonmonotone nonresponse. *Biometrika*, **94**(4), 841–860.

Wang, C., Danies, M. J., Scharfstein, D. O., and Land, S. (2010). A Bayesian shrinkage model for incomplete longitudinal binary data with application to the breast cancer prevention trial. *Journal of the American Statistical Association*, **105**, 1333–1346.

Williamson, S., Orbanz, P., and Ghahramani, Z. (2010). Dependent Indian buffet processes. In *International conference on artificial intelligence and statistics*, pages 924–931.

Wu, M. C. and Carroll, R. J. (1988). Estimation and comparison of changes in the presence of informative right censoring by modeling the censoring process. *Biometrics*, pages 175–188.

# Index





R. Mitra, P. Müller (eds.), *Nonparametric Bayesian Inference in Biostatistics*, Frontiers in Probability and the Statistical Sciences, DOI 10.1007/978-3-319-19518-6

MIX
Papier aus verantwortungsvollen Quellen
Paper from responsible sources
FSC® C105338

If you have any concerns about our products,
you can contact us on
**ProductSafety@springernature.com**

In case Publisher is established outside the EU,
the EU authorized representative is:
**Springer Nature Customer Service Center GmbH**
**Europaplatz 3, 69115 Heidelberg, Germany**

Printed by Libri Plureos GmbH
in Hamburg, Germany